Animal Behavior

Animal Behavior

Concepts, Methods, and Applications

SECOND EDITION

Shawn E. Nordell
Washington University in St. Louis

Thomas J. Valone
Saint Louis University

New York   Oxford

OXFORD UNIVERSITY PRESS

Oxford University Press is a department of the University of Oxford.
It furthers the University's objective of excellence in research,
scholarship, and education by publishing worldwide. Oxford is a
registered trade mark of Oxford University Press in the UK and
certain other countries.

Published by Oxford University Press
198 Madison Avenue, New York, New York 10016

Library of Congress Cataloging-in-Publication Data

CIP data is on file at the Library of Congress
ISBN 978-0-19-027674-4

9 8 7 6 5 4 3 2

Printed by Webcom Inc., Canada

For

Buck, Ernie, Kirby, Grace, Max, and Louie

Brief Contents

Contents

Chapter 4 Behavioral Genetics   61

Chapter 5   Sensory Systems and Behavior   87

Chapter 6   Communication   113

Chapter 7   Learning, Neuroethology, and Cognition   147

Chapter 10   Dispersal and Migration   241

Chapter 11   Habitat Selection, Territoriality, and Aggression   273

Chapter 12   Mating Behavior   301

Chapter 13   Mating Systems   335

Chapter 15　　Sociality　391

Preface

Our love of science grew out of our own undergraduate experiences, when we were first immersed in the research process. We learned to develop interesting questions and testable hypotheses, read and reference the associated primary literature, collect and analyze data, and present and discuss our conclusions. In writing this book, our goal was to share that excitement and that process in order to inspire and guide new generations of students in animal behavior research.

The book's framework is based on developing mandates that aim to shift science education from a memorization-based pedagogy to a conceptual approach that promotes scientific inquiry. Both the National Academies of Science and the American Association for the Advancement of Science[1] have called for teaching approaches that stimulate curiosity about the natural world, emphasize experimental design, and foster a conceptual understanding of biological systems. In addition, these bodies have underscored the importance of developing students' quantitative reasoning and communication skills. This book responds to all of these important mandates by taking a *conceptual approach that highlights the process of science and the real-world applications of animal behavior research*.

The Approach: Concepts, Methods, and Applications

Our approach involves three major components:

- Each chapter is organized around four to six major concepts.
- Each chapter emphasizes the process of science by deconstructing research studies to show the experimental design and methodology that drives the science of animal behavior.
- Each chapter contains one or more examples of how the major concepts are being applied to societal issues.

Throughout the book, we use an accessible, question-driven style to engage students. Incorporating Tinbergen's four questions to show the diverse nature of animal behavior research, we infuse proximate and ultimate examinations of behavior throughout each chapter.

Concepts
- Each chapter is built around *broad organizing concepts*. We present the concepts up front, at the beginning of each chapter, to provide a framework for understanding and evaluating the empirical research examples presented.
- These organizing concepts represent *higher-order critical thinking* and allow students to see the "big picture" of how research has led to our understanding of behavior. For example, rather than titling a section

---

[1]National Research Council Committee on Undergraduate Biology Education to Prepare Research Scientists for the 21st Century, *BIO2010: Transforming Undergraduate Education for Future Research Biologists* (Washington, D.C.: National Academies Press, 2003); C. A. Brewer & D. Smith, *Vision and Change in Undergraduate Biology Education: A Call to Action* (Washington, D.C.: American Association for the Advancement of Science, 2011).

"Evolution," we offer a complete concept: "Evolution by natural selection favors behavioral adaptations that enhance fitness." This clear presentation of concepts synthesizes and summarizes the information provided on complex topics, allowing students to make connections, see the broader implications of those connections, and apply their learning.

Methods

- We use carefully selected *Featured Research* studies from the primary literature to demonstrate the *experimental designs and methods scientists use to address research questions*. Our selection of studies includes both classic and contemporary work that aims to highlight the global diversity of researchers who are contributing to the study of animal behavior. In our selection process, we paid particular attention to the seminal and growing contribution of women and underrepresented groups to the science of animal behavior.
- We *deconstruct each research example* to allow students to become immersed in the process of animal behavior research as a *rigorous quantitative science*. We use a *hypothesis-testing model* to illustrate the scientific process and incorporate sufficient detail into our presentation of the methodology for instructors and students to comprehend and critically evaluate the rationale, process, and results of the given research.
- We create opportunities for students to *develop their quantitative skills* by interpreting graphs from *actual data*. We present or summarize actual data in the "Results" figures and include either individual data points or error bars so that students can see the variation in behavior.

Applications

- We relate the major concepts presented to real life through the *Applying the Concepts* feature, which shows how animal behavior research is applied to various societal issues, as well as human behavior. For example, we demonstrate how animal behavior findings are used in conservation or outside of academia, as in the manipulation of predation risk to mitigate crop damage.

NEW to This Edition

Each chapter from the first edition has been revised using the extremely constructive suggestions offered by many reviewers and adopters. Throughout the textbook, we have better integrated the major concepts with their associated featured research studies, increased our coverage of proximate questions, and updated our empirical examples.

- **New chapter—Sensory Systems and Behavior (Chapter 5).** This new chapter examines how animals gather information from their environment and how that information influences their behavior. This chapter greatly enhances the coverage of proximate research in behavior studies and allows students to better appreciate the diversity of sensory systems found in different taxa.
- **New "Applying the Concepts" features on human behavior.** These examples expand the applications of animal behavior research to human behavior, reinforcing the relevance to students of the major concepts in animal behavior.

- **New "Evaluate" questions and exercises.** These additions, which appear at the end of each Scientific Process box, are designed to more fully engage students in the research described, allow students to develop their quantitative skills, and promote in-class discussions of the material.
- **New photos throughout the book.** These additions enhance the visual presentation of the research and illustrate the diversity of species discussed.
- **Enhanced supplements program.** For the Second edition, we have expanded and improved upon the already robust ancillary package that accompanies the text. Enhancements include an expanded test item file, refined and expanded PowerPoint Lecture Notes slides for each chapter, and an updated interface for the Video Library and Guide.

Changes by Chapter

Chapter 1—The Science of Animal Behavior: We moved the history of animal behavior to the first chapter, along with a discussion of proximate and ultimate research studies and Tinbergen's four questions. These changes better set the stage for how knowledge in the field has developed.

Chapter 2—Methods for Studying Animal Behavior: This chapter has been moved up to improve the continuity from Chapter 1. We also moved the section on hypothesis testing to this chapter to better introduce experimental design and scientific literacy in our explanation of how research is disseminated and communicated. Finally, we expanded the coverage of mathematical models to better demonstrate how these are used to generate hypotheses.

Chapter 3—Evolution and the Study of Animal Behavior: This chapter was relocated to follow the chapter on methods and precede the chapter on behavioral genetics, providing a more logical flow. In addition, we have expanded our coverage of frequency-dependent selection, game theory modeling, and individual and group selection.

Chapter 4—Behavioral Genetics: We added a new concept on heritability and reorganized the discussion and examples of gene expression and quantitative trait loci (QTL) mapping. We also updated and expanded the discussion of fire ant genetics and social organization and greatly expanded the coverage of animal personalities. New material includes an additional empirical example on spiders and a discussion of a meta-analysis of research that relates to a mathematical model explaining the evolution of personalities.

Chapter 5—Sensory Systems and Behavior: NEW to the second edition, this chapter focuses on proximate research that examines a variety of sensory systems, including chemoreception, mechanoreception, and photoreception. It also covers the detection of electric and magnetic fields and the co-evolution of sensory systems in a predator–prey system.

Chapter 6—Communication: This chapter has been moved up to enhance the continuity from sensory systems to communication. We also added new studies on environmental influences on signals, including research on ant chemical pheromone trails and fish visual signals, and the new concept of extended phenotype signals, with empirical examples, for more complete coverage of communication.

Chapter 7—Learning, Neuroethology, and Cognition: We have added a discussion and empirical examples of animal teaching to bolster understanding of the social learning concept. In addition, we updated and improved our treatment of cognition with new examples that better reflect recent developments in the field. Finally, we expanded our coverage of brain and behavior research with examples of research on brain size and cognition in fish.

Chapter 8—Foraging Behavior: We have added a new concept that deals with individuals obtaining food from other individuals, focusing on kleptoparasitism and producer–scrounger models, and have provided relevant empirical studies.

Chapter 9—Antipredator Behavior: We expanded our proximate coverage of links between stress hormones and behavior and predation risk. This chapter also offers new coverage of evasive actions animals use once attacked by a predator, including an empirical example of a startle display in butterflies.

Chapter 10—Dispersal and Migration: We have expanded our coverage of the evolution of migratory behavior, including the addition of an empirical example of heritability of migratory behavior in birds.

Chapter 11—Habitat Selection, Territoriality, and Aggression: We have greatly expanded our discussion of game theory by adding a new section on self-assessment and mutual assessment models. We have also expanded our coverage of proximate work on hormonal influences on aggression by including associations between juvenile hormone and wasp aggression.

Chapter 12—Mating Behavior: We have expanded the discussion of Fisher's runaway process.

Chapter 13—Mating Systems: We have added an examination of human mating systems.

Chapter 14—Parental Care: We have reorganized this chapter so that parent-offspring conflict theory provides a theoretical framework for understanding variations in parental care exhibited by individuals in a population. We have also added the new concept of co-evolution involving interspecific brood parasitism and egg rejection behavior.

Chapter 15—Sociality: This chapter has undergone a substantial revision. We have expanded our coverage of kin selection with the addition of empirical proximate work on kin discrimination and have revised concepts relating to cooperative reproduction to emphasize kinship and ecological constraints (e.g., habitat saturation) with several new empirical examples. We have also expanded our coverage of the division of labor hypothesis and the evolution of sterile castes.

Teaching and Learning Features

Scientific Process boxes: To further emphasize the process of science, each chapter contains one or more Scientific Process boxes. These features present detailed descriptions of research within a scientific framework, clearly and concisely laying out each step of the process: research question, hypotheses, predictions, methods, results, and conclusions. Students can thus easily follow the research example at a glance, from its conception (the original research question) to the use of the scientific method (the creation of testable hypotheses, the experimental protocol, and the evaluation of data) and ultimately the findings of the work (conclusions of the study). The details contained in these examples illustrate and reinforce the rigorous nature of animal behavior research.

Applying the Concepts: Each chapter contains examples showing how animal behavior research is being applied to real-life situations, including applications to human behavior. This allows students to understand the importance and relevance of "pure" scientific research to larger societal issues. By including these features in every chapter, students have the opportunity to see the broader implications and importance of research.

Toolboxes: These boxes explain essential skills or complex terms in the field of animal behavior research with the aim of building students' intellectual and practical toolkit. They do not appear in every chapter but are included as needed and can provide students who lack essential knowledge typically acquired in prior courses, such as how to construct and interpret phylogenies, with the additional *background information* needed.

Chapter Summary and Beyond: At the end of each chapter, we provide a brief summary of the concepts covered. In doing so, we also point students to recent papers that further develop the ideas presented in the chapter. No textbook can be exhaustive, and so these papers are ideal for students or instructors seeking additional information about a concept.

Chapter Questions: At the end of each chapter, we provide critical thinking questions and assignments designed to foster class discussions. Answers to even-numbered questions are provided in the textbook, while answers to odd-numbered questions can be found online in the Instructor's Support Package (see below). These answers provide explanations in response to chapter questions, as well as background information that can aid in leading class discussions.

Data: We include data summaries as they typically appear in the primary literature (with individual data points plotted on line graphs and the inclusion of means and standard error bars in bar graphs) rather than the typical cartoon summary often found in textbooks. Thus, students can see that variation in behavior is ubiquitous in research.

Images: The book contains research examples from all over the world, covering a diverse range of taxa. We have strived to include ample representation from major taxa (invertebrates, amphibians and reptiles, fish, birds, and mammals) in each chapter. Because most students have a limited knowledge of animal diversity, we have included an image of each featured species in addition to brief natural history descriptions of that species. The textbook's website (www.oup.com/us/nordell) includes over 300 additional images provided by researchers that illustrate research protocols, habitat features, and study organisms.

Support Package

Oxford University Press offers a comprehensive ancillary package for instructors who adopt *Animal Behavior: Concepts, Methods, and Applications*, Second Edition. The Ancillary Resource Center (ARC), located at www.oup-arc.com/nordell, contains the following teaching tools:

- **Video library and guide:** To help students better understand animal behavior research and visualize the science in each chapter, and to provide instructors with an easy-to-use resource for videos tied directly to the concepts covered in the text, we have assembled links to three types of short videos:
  1. Natural history videos that illustrate specific behaviors of the study organisms in the featured studies

2. Process-of-science videos that illustrate the methodology described in featured research

3. Interviews with the scientists whose research is featured in the book

- We have carefully selected these videos because even though animal behavior is a visual endeavour, students often have little experience with the diversity of taxa studied or these species' behavior. Each video link is accompanied by a brief description of its content and information on the video's length. We hope that instructors will incorporate these videos into their classes to help students better understand animal behavior and research on that behavior.

- **Digital image library:** The image library includes electronic files in PowerPoint format of every illustration, graph, photo, figure caption, and table from the text, in both labeled and unlabeled versions. Images have been enhanced for clear projection in large lecture halls. The library also includes more than 300 additional photos from researchers depicting study organisms, research protocol set-ups, data collection, and habitats. We provide these photos and a guide to their content for each chapter so that instructors can easily incorporate them into their courses.

- **Lecture notes for each chapter:** Editable lecture notes in PowerPoint format make preparing lectures faster and easier than ever. Each chapter's presentation includes a succinct outline of key concepts and featured research studies, and incorporates the graphics from the chapter. New to the Second Edition, in the PowerPoint lecture notes, we have written inquiry questions for each chapter that create opportunities for students to assess their critical thinking skills. These graphically based questions allow students to develop and practice their quantitative and cognitive skills. Additionally, each chapter's presentation includes links to relevant videos from the video library and guide to make it easier for faculty to incorporate these materials into their lectures.

- **Test bank:** The authors have created a test-item file that includes over 350 exam questions in multiple-choice and short-answer formats. The test-item file now also includes in-class activity questions and short-answer questions that faculty can use to foster an active learning classroom.

- **Suggested further readings:** To assist instructors in identifying classic, contemporary, and relevant readings from the primary literature, we have assembled a list of over 25 suggested further readings taken directly from the journal *Behavioral Ecology*. These readings, like the video library, are tied directly to the concepts covered in the text. In addition, we provide a pedagogical outline to help students develop their ability to critically evaluate the primary literature.

Contact your local OUP sales representative or visit www.oup-arc.com/nordell to learn more and gain access to these instructor resources.

Acknowledgments

This book is the product of our long collaboration in academics and life, but we have many people to thank for providing us with immense support. We thank our scientific mentors who helped us develop our love and understanding of science. For Shawn, this includes her master's advisor, Donald Thomson, who introduced her to the wonderful world of field research, and her Ph.D. advisor, Astrid

Kodric-Brown, who taught her critical thinking without being critical. For Tom, this includes his undergraduate advisor, Tom Caraco, who introduced him to the wonderful science of animal behavior, and his Ph.D. advisor, Michael Rosenzweig, who taught him how to be a scientist. Shawn would also like to thank her pedagogical mentor, Mary Stephen, who helped her develop her understanding of and research on the scholarship of learning and teaching. We would both like to thank our many, many wonderful students who graciously provided constructive feedback and input regarding our teaching and their learning of our favorite discipline. We also thank our Saint Louis University colleagues for their support throughout the process, particularly the Department of Biology, Don Brennan, and Joe Weixlmann. Finally, we thank Robin Carter, Scott Freeman, Sallie Marston, Manuel Molles, and Bob Ricklefs for sharing their wisdom and advice regarding the world of writing textbooks, and Stan Braude for his extensive, detailed suggestions regarding the second edition.

We also wish to express our deep appreciation to the wonderful team at Oxford University Press USA who guided us through the development of this second edition: Jason Noe, senior editor, for helping us develop and execute the plan for the revisions for this edition and guiding us graciously through the process; John Haber, developmental editor for the first edition, who taught us the fine art of clarity in writing and provided continuous good humor despite missed deadlines; Lauren Mine, developmental editor for both the first and second editions, who always championed our cause to ensure our vision was realized and was vital for enhancing and clarifying the message of the book; Jennifer Weiss, developmental editor for the second edition, who helped to edit several chapters; Micheline Frederick, production editor, for meticulously shepherding the manuscript through copyediting and production; Andrew Heaton, associate editor, for conscientiously following up on so many details; David Jurman, marketing manager; Frank Mortimer, director of marketing; Patrick Lynch, editorial director; and John Challice, vice president and publisher. We also thank those in production for creating such a beautiful book, as well as our eagle-eyed art editor, Ayelet Arbel.

We could not have done this project without the amazing and continued support and encouragement from our dear friends and family. We thank our friends in St. Louis (especially Vera and Joe Brandt, Bob and Caroline Cordia, Jean and Bill Curtis, Diane and Frank Lockhart, Sharon Matlock, Denise Mandle, Lyn Magee, Chris Baumann, and Chris Sebelski) and Tucson (especially Ellen Tuttle, Marcy Tigerman, and Marcy Wood), for whom we express great love and gratitude. Their unfailing enthusiasm was appreciated more than we can ever say and helped us through many deadlines. We also thank our families both past and present who supported our endeavors with love and appreciation. Shawn would like to thank her Academic Ladder Writing Club group members, who supported her writing on a daily basis.

Finally, we thank Buck, Ernie, Kirby, Grace, and Max, whose boundless energy and wagging tails kept us smiling, and who have taught us more about animal behavior than they will ever know.

We are grateful to the following reviewers, whose comments helped shape this edition:

Bryan Arnold	Illinois College
David Bailey	St. Norbert College
Amber Harris Bozer	Tarleton State University
Stan Braude	Washington University
Miranda Brenneman	Coastal Carolina University

Nicole Cameron	Binghamton University
Michelle Cawthorn	Georgia Southern University
Melloni Cook	University of Memphis
Micheal Dent	University at Buffalo
Teresa Dzieweczynski	University of New England
Ryan Earley	University of Alabama
Miles Engell	North Carolina State University
Preston Foerder	University of Tennessee at Chattanooga
Sharon Gilman	Coastal Carolina University
Charles Gunnels	Florida Gulf Coast University
Shala Hankison	Ohio Wesleyan University
Sarah Huber	College of William and Mary
Michael King	Stetson University
William Kroll	Loyola University Chicago
David Luther	George Mason University
Heidi Lyn	University of Southern Mississippi
Karen Mabry	New Mexico State University
Jay Mager	Ohio Northern University
Duane McClearn	Elon University
Lindsey Renee Merkham	Oregon State University
Charles Morgan	University of St. Joseph
Peter Nonacs	University of California, Los Angeles
Brian Palestis	Wagner College
Eric Patterson	Georgetown University
Alexandra Protopopova	Texas Tech University
Corinna Ross	Texas A&M University–San Antonio
Stephen Shuster	Northern Arizona University
Adam Smith	George Washington University
Nancy Solomon	Miami University
Robert Sorge	University of Alabama at Birmingham
Donald Sparling	Southern Illinois University, Carbondale
Heather Spence	Hunter College
Andrew Stoehr	Butler University
Malini Suchak	Canisius College
Thomas Trott	Suffolk University
Daniel Welsh	Fitchburg State University
Gerald Wilkinson	University of Maryland
Alyson Zeamer	University of Texas at San Antonio

We also continue to be grateful to the reviewers of the first edition:

Elizabeth Archie, University of Notre Dame; Suzanne Baker, James Madison University; Peter Bednekoff , Eastern Michigan University; Russell Benford, Northern Arizona University; Andrew R. Blaustein, Oregon State University; Joel Brown, University of Illinois at Chicago; Theodore E. Burk, Creighton University; Prassede Calabi, University of Massachusetts–Boston; Blaine Cole, University of Houston; Francine Dolins, University of Michigan–Dearborn; Richard Duhrkopf, Baylor University; Emily DuVal, Florida State University; Fred C. Dyer, Michigan State University; Janice Edgerly-Rooks, Santa Clara University; Miles Engell, North Carolina State University; Ann Fraser, Kalamazoo College; Sharon Gill, Western Michigan University; Harold Gouzoules, Emory University; Blaine D. Griffen, University of

South Carolina; Sylvia L. Halkin, Central Connecticut State University; Jodee Hunt, Grand Valley State University; Valerie James-Aldridge, University of Texas–Pan American; Clint Kelly, Iowa State University; Astrid Kodric-Brown, University of New Mexico; William Kroll, Loyola University Chicago; David Lahti, Queens College; Tracy Langkilde, Pennsylvania State University; Susan Lewis, Carroll University; Catherine Lohmann, University of North Carolina at Chapel Hill; Jeff Lucas, Purdue University; Karen Mabry, New Mexico State University; John C. Maerz, University of Georgia; Tara Maginnis, University of Portland; Frank F. Mallory, Laurentian University; Mary Beth Manjerovic, University of Central Florida; Lauren Mathews, Worcester Polytechnic Institute; Marion McClary, Fairleigh Dickinson University; Kevin J. McGraw, Arizona State University; Daniela Monk, Washington State University; Cy L. Mott, Kentucky Wesleyan College; James Nieh, University of California, San Francisco; Brian Palestis, Wagner College; Daniel R. Papaj, University of Arizona; Stephen Pruett-Jones, University of Chicago; Rick Relyea, University of Pittsburgh; Lanna Ruddy, SUNY Geneseo; Mike Ryan, University of Texas; Debbie Schlenoff, University of Oregon; Toru Shimizu, University of South Florida; Joseph Sisneros, University of Washington; Donald Sparling, Southern Illinois University; Eric Strauss, Loyola Marymount University; Thomas Terleph, Sacred Heart University; Kevin Theis, Michigan State University; Jeffrey Thomas, Queens University of Charlotte; Robert M. Turnbull, University of Southern Mississippi–Gulf Coast; Al Uy, University of Miami; E. Natasha Vanderhoff, Jacksonville University; Sean Veney, Kent State University; Margaret Voss, Penn State University; David Westneat, University of Kentucky; and Michele Jade Zee, Northeastern University.

Animal Behavior

The Science of Animal Behavior

We can observe animal behavior everywhere. In backyards and parks, squirrels collect and bury nuts to eat later. Each spring, birds, frogs, and crickets sing to attract mates. Under the eaves of houses and other buildings, wasps make nests, where the queen lays eggs and workers care for offspring. Farther from our home, male giraffes in Africa exchange blows by swinging their heads and necks against one another in competition for females (Figure 1.1), while fish in the Great Barrier Reef form large schools to minimize predation (Figure 1.2). And each fall and spring, millions of birds migrate between temperate and tropical regions.

This book will introduce you to the wonders of animal behavior. In this chapter, we begin by exploring the ways that animals and their behaviors are an important part of the world. We then introduce the science of animal behavior and the scientific method. We'll see how scientists test hypotheses to learn about the natural world and examine a fundamental characteristic of animal behavior research: it looks at both the proximate mechanisms that generate that behavior as well as the ultimate reasons for why it evolved. We'll examine the varied approaches and perspectives used to study animal behavior. Finally, we discuss that although most scientific hypotheses about behavior are developed in the context of evolution, many nonscientists try to explain animal behavior based on the idea that animals possess human emotions.

Figure 1.1. Giraffe behavior. An aggressive interaction between two males.

Figure 1.2. Schooling fish. Snapper form schools to minimize predation.

1.1 Animals and their behavior are an integral part of human society

Understanding the behavior of animals has always been important to people. Beginning in prehistoric times and continuing for tens of thousands of years, humans painted images of animals on cave walls all over the world. These drawings are detailed enough to allow us to identify different species (both extinct and extant), and many images depict animals exhibiting behaviors such as eating, sleeping, and engaging in acts of aggression. Recent research indicates that these paintings were likely intended to present realistic depictions of animals and their behavior rather than to evoke symbolic connotations (Pruvost et al. 2011). Because humans in prehistoric times relied on animals for food, knowledge about animal behavior was important for survival (Shipman 2010).

Animals and animal behavior are still an integral part of society. Millions of people live and interact with animals daily. The American Veterinary Medical Association reports that in 2011, in the United States alone, over 80 million households contained a pet (American Veterinary Medical Association 2012). This number includes approximately 139 million freshwater fish, 90 million cats, 73 million dogs, 18 million other small mammals, 16 million birds, and 11 million reptiles. Most owners value their pets for companionship: we, for example, enjoy both watching our pets behave and interacting with them.

Many people work with animals. Cattle, chickens and turkeys, hogs, and sheep are important agricultural products; horses are used by ranchers, law enforcement officials, and the horse-racing industry; and dogs have long been used in both police work and the military (e.g., Ensminger 2012) (Figure 1.3). In all of these cases, people manage animal behavior to accomplish a task. In a different way, the behavior of animals is also integral to

Figure 1.3. Police working dogs. Dogs are used as trackers and substance detectors.

medicine. It helps researchers assess and learn about sensory, motor, and cognitive functions. For instance, behavioral changes often reflect the effects of neurochemical agents, neurotoxins, or hormonal changes, which can be more easily studied in animals than in humans. Research on memory, cognitive function, and learning often involves measuring and recording animal behavior.

In addition, animals provide entertainment for millions of people. According to the World Association of Zoos and Aquariums, more than 700 million people visit zoos and aquariums worldwide each year (Gusset & Dick 2011). Many movies and popular television shows feature animal behavior, and entire television networks, such as Animal Planet and National Geographic Wild, are dedicated to animals.

Yet the study of animal behavior involves much more than interacting with pets, working with animals, or watching animals in movies or on television. Rather, animal behavior researchers use the scientific method to understand the behaviors we observe. So what exactly do we mean by *behavior*?

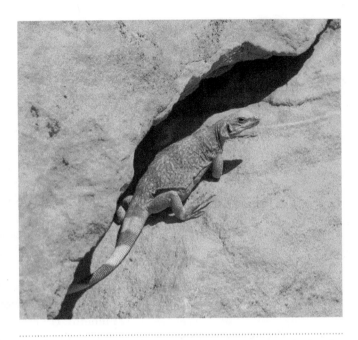

Figure 1.4. Lizard thermoregulation. Ectotherms, like this chuckwalla (*Sauromalus varius*), move to cool locations as their body temperature rises.

Recognizing and defining behavior

We define **animal behavior** as any internally coordinated, externally visible pattern of activity that responds to changing external or internal conditions (Beck, Liem, & Simpson 1991; Levitis, Lidicker, & Freund 2009). Let's examine each part of this definition in turn.

Internally coordinated refers to internal information processing such as endocrine signaling, sensory information processing, or the action of neurotransmitters. When two male giraffes meet during the breeding season, such processes coordinate their aggressive behavior. However, when a lion's mane moves away from its body because of a gust of wind the animal is *not* exhibiting a behavior, because it has no control over this movement.

Externally visible activity refers to patterns that we can observe and measure. For example, we can observe a squirrel eating an acorn and can quantify this behavior. We cannot externally observe the variation in a lizard's heart rate.

We can, however, observe an animal's *behavioral response to changing conditions*. For example, male crickets, frogs, and birds vocalize in response to changes in day length, temperature, or moisture at specific times of the year. Similarly, during a summer day, a desert lizard moves from the top of a hot rock to the underside of a cool ledge to reduce its body temperature (Figure 1.4). Lizards are ectotherms; as such, they cannot regulate their body temperature internally but can do so behaviorally.

animal behavior Any internally coordinated, externally visible pattern of activity that responds to changing external or internal conditions.

Measuring behavior: elephant ethograms

Behaviors must be measurable—that is, we must be able to quantify our observations with numbers according to specific rules (Martin & Bateson 1993)—and different people must be able to recognize a behavior independently. Often, such characterization begins when a trained observer completes an **ethogram**, which is a formal description or inventory of an animal's behaviors. Ethograms typically list or catalogue defined, discrete behaviors that a particular species exhibits.

Featured Research

ethogram A formal description of an animal's behaviors.

Figure 1.5. Captive Asian elephant. Ethograms are often constructed to describe and quantify the behavior of captive animals.

time budget A summary of the total time and relative frequency of different behaviors of an individual.

Researchers can use an ethogram and measure how many times a behavior occurs (its frequency), the length of time of a behavior (its duration), the frequency of the behavior per unit time (its rate), or the vigor or forcefulness of the behavior (its intensity). Sometimes only certain behaviors, such as social behaviors (e.g., grooming self, grooming others, chasing others), are quantified. Typically, a more complete list of behaviors that occur over a specific time period is recorded (e.g., sleeping, grooming self, grooming others, walking, climbing tree, eating food, drinking). From these observations, a researcher can then determine both the total and the relative time that an animal engaged in each behavior—in other words, the measure of the behavior divided by the overall time spent observing the animal. The resulting **time budget** indicates the total time and relative frequency of each behavior.

Ethograms are commonly constructed for animals in captivity and can be used as a baseline for healthy behavior. Paul Rees created an ethogram for captive Asian elephants (*Elephas maximus*) (Figure 1.5) at the Chester Zoo in the United Kingdom (Rees 2009). When elephants were in their outdoor enclosure, he recorded their behavior every five minutes for an entire day once a week for 11 months. His ethogram contains 15 behaviors, including feeding, standing, and digging (Table 1.1). In addition, he measured stereotypic behavior, or captivity-induced behavioral anomalies that are used as an index of the welfare of captive animals (Mason 2010). These behaviors often include repetitive behaviors that lack an apparent purpose, such as head bobbing or pacing. Rees found that captive elephants spent about a quarter of their time feeding, and that stereotypic

TABLE 1.1 Elephant ethogram. This ethogram describes the behavior of captive elephants (*Source:* Rees 2009).

BEHAVIOR	DESCRIPTION
High-frequency behaviors	
Dusting	Collecting soil and throwing it over the body/rubbing it onto the skin (while standing still or walking), including digging into the soil for this purpose
Feeder ball	Feeding or attempting to feed at a metal feeder ball containing small quantities of food
Feeding	Collecting food with the trunk and placing it in the mouth while standing still or walking (does not include suckling or activity at the feeder ball)
Locomotion	Walking (except feeding or stereotyping)
Playing	Chasing another elephant/mock fighting with another elephant (but not as a result of an antagonistic encounter or as part of courtship)
Standing	Standing motionless (not while stereotyping or dusting)
Stereotypy	Repetitive behavior with no obvious purpose: weaving, head bobbing, pacing backward and forward or in an arc, walking in circles
Suckling	Calf suckling from mother or other female (measured separately from feeding)
Low-frequency behaviors	
Aggression	Hitting/pushing as a result of an antagonistic encounter (but not as a part of play)
Bathing	Standing/lying in pool/squirting water from pool over body with trunk
Digging	Digging in soil using the foot (but not as a part of a dusting behavior)
Drinking	Collecting water in the trunk and squirting it into the mouth
Lying down	Lying down on the ground (on side or prone)
Rolling	Rolling in soil or mud (but not as a part of playing with another individual)
Sex	Courting or being courted/mounting another elephant or being mounted by another elephant of either sex

behavior was negatively correlated with feeding behavior; in other words, the more time spent feeding, the less time spent pacing or head bobbing. Rees suggests that using widely spaced feeders to supply food slowly and at random times could reduce the frequency of stereotypic behavior.

We turn next to a discussion of the scientific method: the process scientists use to understand the behaviors they observe.

1.2 The scientific method is a formalized way of knowing about the natural world

When you think of science, you may think first of an impressive body of knowledge—the facts that you are asked to learn in class. Have you ever considered where that knowledge came from? It was obtained by human beings seeking to understand the natural world. Scientists engage in the **process of science**, which involves observing events, organizing knowledge, and providing testable explanations (Mayr 1982). The process of science is fundamental to our understanding of the natural world.

process of science Observing events, organizing knowledge, and providing explanations through the formulation and testing of hypotheses.

Scientific discoveries continue to occur today at a tremendous pace. Every week, new discoveries appear on television, in magazines, and on the Internet. As instructors, we look back at our courses from just five years ago and marvel at how much has changed in so short a time. You and your instructors may be involved in new discoveries yourselves.

Why does scientific progress occur so rapidly? What is different about scientific knowledge compared with knowledge in the humanities? As we see next, the process of science means constructing hypotheses that make testable predictions.

The importance of hypotheses

Some disciplines—like English, history, and philosophy—are part of the humanities. Others, like anthropology and sociology, are social sciences. Still others, like biology and chemistry, are natural sciences. In all these disciplines, scholars seek to gain a better understanding of the world. In the humanities, researchers are interested in understanding the human experience. The social sciences involve knowledge of human behavior and societies. And in the natural sciences, researchers strive to understand the natural world.

You may have noticed differences between your courses not only in the information presented but also in how you are expected to think about that information. For example, the humanities discuss art, literature, and how humans think and act. You might be expected to present your own perspective while critically evaluating the perspectives of others. Often, two people may arrive at very different perspectives using similar evidence.

Historians, for example, strive to understand why certain events happened by interpreting personal journals, letters, government publications, and newspaper articles. A historian forms a thesis, or narrative, to explain events and then interprets relevant documents to support that thesis (Hult 2002). For instance, a historian might be interested in what motivated Thomas Jefferson, John Adams, James Madison, and other founding fathers of the United States to declare independence from England and form their own system of government. Were economic considerations, such as preservation of individual wealth, the primary motive (Beard 1913), or did convictions about individual liberty and rights play a larger role (McDonald 1958)? By examining the Declaration of Independence,

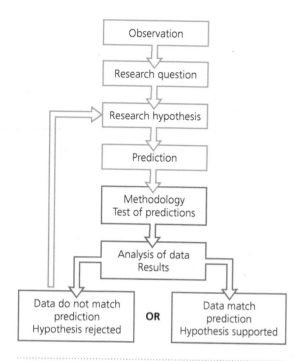

Figure 1.6. The scientific method. This flowchart summarizes the scientific method.

▪ **research question** A formal statement of an unknown that begins the scientific method.

▪ **hypothesis** An explanation based on assumptions that makes a testable prediction.

▪ **alternate hypothesis** The statistical hypothesis that the proposed explanation for an observation does significantly affect the behavior of the organism.

▪ **null hypothesis** The statistical hypothesis that an observation results from chance. Also called the hypothesis of no effect.

the U.S. Constitution, the *Federalist Papers*, and letters written by key figures, historians have developed these and other interpretations. In each case, research involves discovering, reevaluating, and interpreting the evidence.

Scientists, like other scholars, begin with questions, but they formulate these questions into hypotheses. Hypotheses are explanations that make predictions that can be tested. Because these tests can be repeated and confirmed by other scientists, their results are much less subject to debate. This is why science sometimes seems like just an accumulation of facts. Yet these facts are the results of scientific studies that have been confirmed repeatedly by the scientific community over many years.

The scientific method

Scientists use the scientific method to learn about the natural world. The scientific method is a formalized process that involves the testing of hypotheses (Figure 1.6). This process often begins with an observation of a single event or pattern that requires explanation. This observation forms the basis of a **research question**—a brief statement of something that we would like to understand. For example, suppose you were walking through your neighborhood and observed that some yards had many American robins (*Turdus migratorius*) feeding in them, while other yards had very few. Robins, common songbirds found on lawns throughout North America, feed on invertebrates such as earthworms and beetle grubs. This observation might lead to the following research question (Scientific Process 1.1):

> **Research question:** Why is there variation in the number of robins feeding in different yards?

The identification of patterns like this can be accomplished with careful observation and mere human curiosity. Throughout this book, you will see how researchers have identified different behavioral patterns that have led to a variety of research questions.

The next step in the scientific process is the formulation of a hypothesis, or, more formally, a research hypothesis. You may think of a hypothesis as an educated guess, but this definition is far too simplistic. Rather, a research **hypothesis** is an explanation that is based on assumptions and produces a testable prediction. Research hypotheses are evaluated using two statistical hypotheses that reflect the two possible outcomes. One outcome is that the proposed explanation for the observation *does* have a significant effect; this is the **alternate hypothesis**, or H_a. The other possible outcome is that the proposed explanation does *not* have a significant effect; this is the **null hypothesis**, or H_0, the hypothesis of no effect. These terms were coined by Sir Ronald Fisher (1966), a British geneticist, evolutionary biologist, and statistician. Together, the null and alternate hypotheses are known as statistical hypotheses.

For example, to explain why more robins feed in some yards than in others, you might hypothesize that the quantity of food varies between yards and that this variation affects robin abundance. This leads to two statistical hypotheses:

> **Alternate hypothesis:** The amount of food in a yard determines the number of robins feeding there.
> **Null hypothesis:** The amount of food in a yard does not determine the number of robins feeding there.

SCIENTIFIC PROCESS 1.1

Robin abundance and food availability

RESEARCH QUESTION: *Why is there variation in the number of robins feeding in yards?*

Hypothesis: The amount of food in a yard determines the number of robins feeding there.

Prediction: Yards with more food will have more robins.

Methods: The researchers:

- Counted the number of robins at the same time of day in each of 30 yards
- Quantified the amount of food (earthworms) available to robins in each yard by examining several 900 cm² sampling areas
- Mixed 40 g of yellow ground mustard seed into 4 L of water and then poured the mixture into each sampling area so that it was absorbed into the soil
- Counted the number of earthworms that emerged in each sampling area over a period of 10 minutes

Results:

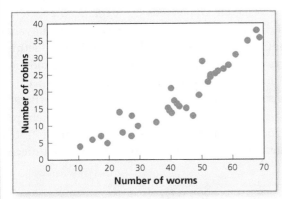

Figure 1. The greatest numbers of robins were found in yards with the most worms.

OR

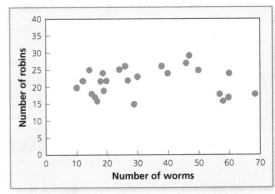

Figure 2. There is no relationship between the number of robins and the number of worms in yards.

Conclusion 1: The hypothesis is supported. Yards with more earthworms have more robins.

Conclusion 2: The hypothesis is not supported. Yards with more earthworms do not have more robins.

The alternate hypothesis assumes that the amount of food in a yard is the only factor that determines the number of robins feeding there. Both the alternate and null hypotheses make predictions, although they are different:

Alternate hypothesis prediction: Yards with more food will have more robins.

Null hypothesis prediction: There is no relationship between the amount of food in a yard and the number of robins in the yard.

The last step in the scientific process is to evaluate the research hypothesis by testing the prediction of the null hypothesis. One way to do this is to make new observations by collecting data from many different yards on the number of

EVALUATE

Design a sampling protocol for counting the number of robins in yards. Draw a graph that shows a negative relationship between the number of robins and the number of worms in yards.

robins and the number of earthworms (a common food of robins) in each yard. With these data, you can now test the prediction of the null hypothesis. There are only two possible outcomes. If there is no relationship between the amount of food in a yard and the number of robins, then you *fail to reject the null hypothesis*. When this occurs, you also *reject the alternate hypothesis*, because this hypothesis does not provide an adequate explanation for the pattern you observed. Alternatively, if you find that yards with more food have more robins, then you *reject the null hypothesis* and *fail to reject the alternate hypothesis*. You have thus *found support* for the alternate hypothesis.

When scientists find support for a research hypothesis, we do not conclude that the hypothesis has been "proven" to be correct. Scientists can never prove that a hypothesis is correct (Popper 1959) because random chance or an untested variable (i.e., a different alternate hypothesis) could also have led to the observed phenomenon. So all a scientist can do is test the hypothesis repeatedly and always fail to reject it. As a result, all hypotheses are tentative explanations of observed phenomena. They may eventually be rejected and replaced by hypotheses that make more accurate predictions. Scientists can, and do, test multiple hypotheses simultaneously to formulate explanations of observations of the natural world that can facilitate rapid scientific advancements (Platt 1964).

What do you do when you fail to reject the null hypothesis? When this occurs, you need to develop a new research hypothesis to explain the original observation—in this case, differences in the number of robins feeding in yards. This might involve a slight modification to the rejected hypothesis or something radically different. For instance, you might now postulate that robins avoid areas with predators like cats. Your new hypothesis would predict that yards with no predators will have high numbers of robins, while yards with predators will have few robins feeding there. This new hypothesis makes a completely different prediction than your original hypothesis—and this prediction, too, can be tested. This time, you would gather data on the numbers of cats and robins in different yards (Scientific Process 1.2).

Negative results and directional hypotheses

As we just saw, sometimes scientists do not reject the null hypothesis, a situation known as **negative results**. Negative results indicate that the alternate hypothesis does not explain the behavior being examined. This is an important part of the scientific process. It does not indicate a failure by the scientist, because we have learned that one possible explanation is incorrect. Instead, scientists need to develop a new alternate hypothesis to explain the observation.

Hypotheses can also be directional or nondirectional, depending on their specificity. A directional alternate hypothesis predicts specifically how the variable under examination will affect a particular behavior—positively or negatively—whereas a nondirectional alternate hypothesis usually offers no specific prediction of how the variable will affect behavior. Researchers use a nondirectional hypothesis when they do not have a specific prediction about an animal's response. For example, when a moving individual determines that a predator is nearby, it could quickly flee to a safer location or it could become motionless to avoid detection. In one scenario, the speed of the individual's movement increases as a result of detecting a predator, while in the other scenario, the speed of the individual's movement decreases. In both cases, the speed of movement changes once the individual has observed a predator, but the direction of the change is different. Therefore, the appropriate hypothesis is nondirectional, since

negative results An outcome in which the null hypothesis is not rejected, and thus the alternate hypothesis is rejected.

SCIENTIFIC PROCESS 1.2

Robin abundance and predators

RESEARCH QUESTION: *Why is there variation in the number of robins feeding in yards?*

New Hypothesis: The presence of predators affects the number of robins feeding in yards.

New Prediction: Yards with no predators will have a high number of robins, whereas yards with predators like cats will have few robins.

Methods: The researchers:

- Placed a cat (predator) in a cage in the center of half the yards (treatment)
- Placed an empty cage in the center of the other yards (control)
- Counted the number of robins in each yard at the same time of day

Results:

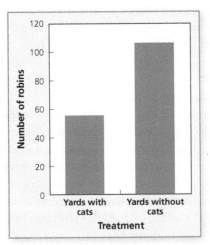

OR

Figure 1. Yards without cats had more robins.

Figure 2. Yards with and without cats had similar numbers of robins.

Conclusion 1: The hypothesis is supported. Yards without cats have more robins.

Conclusion 2: The hypothesis is not supported. Yards without predators do not have more robins.

a specific prediction about movement (increase or decrease) may not be possible. The null hypothesis would state that the presence of a predator would not affect the speed of movement of the individual, while the alternate hypothesis would state that the presence of a predator would affect the speed of movement of the individual.

EVALUATE
Develop one or more additional hypotheses to explain variation in the number of robins in yards. Draw a graph of positive results.

Correlation and causality

Let's say that the data do allow you to reject the null hypothesis. Suppose you found a positive relationship between the number of earthworms and the number of robins feeding in yards. Clearly, these data support your alternate hypothesis that the amount of food in a yard determines the number of robins feeding there. The data you collected represent a **correlation** between the two variables measured: they co-vary, or vary together, predictably. Correlations can be positive, as in this case, when both variables display either low or high values simultaneously, or negative, when one variable displays high values while the other displays low

correlation Two variables that vary together predictably.

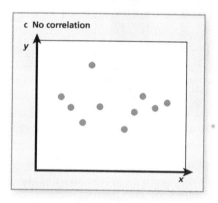

Figure 1.7. Correlations. Correlations can be (a) positive, when both variables (x, y) increase, or (b) negative, when one variable increases as the other decreases. In addition, there may be (c) no correlation between two variables when they vary independently.

scientific theories Hypotheses that have been tested many times, by many different scientists, and have not been rejected. Scientific theories provide a conceptual framework that explains many phenomena and are well supported by observations and experimental tests.

values. In addition, correlations may be zero, which occurs when two variables vary independently of each other (Figure 1.7).

However, we still have work to do to test the hypothesis more fully. The reason is that correlation does not demonstrate causality. That is, even if two variables are correlated, one does not necessarily cause the other. There may be other reasons why the number of robins and the number of worms in yards co-vary positively. Perhaps some yards have shrubs that are excellent nesting sites for robins while others do not, and the yards with shrubs also have more earthworms. It would be unclear whether the robins preferred certain yards because of their earthworms, their shrubs, both, or some other factor entirely. Because correlation does not demonstrate causation, researchers often try to rule out plausible alternative explanations. They also try to find a mechanism that might underlie the correlation. Scientists place a premium on mechanistic explanations for patterns because mechanisms help to explain why the pattern occurs and thus strengthen their conclusions. Because it is very difficult to rule out all other possibilities, it can be very difficult to establish causality through observation alone. Many scientists therefore test hypotheses using controlled experiments, as we will see in Chapter 2, but even experiments cannot prove a hypothesis is correct.

Hypotheses and theories

The more times a hypothesis has been tested and not rejected, the more confidence the scientific community has in the explanation. Hypotheses that make many predictions, have been tested hundreds or thousands of times by many different scientists, and have not been rejected come to be known as **scientific theories**. Scientific theories provide a conceptual framework that explains many phenomena and are well supported by observations and experimental tests. Examples include the cell theory of living organisms, the germ theory of disease, and Charles Darwin's theory of evolution by natural selection. In sum, scientific theories are well-substantiated explanations that form the basis of our understanding of the natural world (National Academy of Sciences 1998).

Note that the word *theory* has a very special meaning in science. For the general public, a theory often means a mere guess. You might have a "theory" about who the murderer is on a television show or who ate the chocolate chip cookies in your dorm. In contrast, scientific theories are hypotheses that have been tested repeatedly. They are more general than specific hypotheses and tend to unify and simplify our knowledge by synthesizing information into a larger framework. For example, cell theory states that cells are the fundamental unit of all organisms.

Social sciences and the natural sciences

So far, we have discussed the humanities and the natural sciences. What about the social sciences, such as psychology? Social scientists create and test hypotheses using the scientific method to study human behavior and societies. Their methodology often includes collecting data from surveys, questionnaires, and public opinion polls, as well as conducting experiments to obtain information about human values, perceptions, and behavior. For example, social scientists might be interested in understanding why there is a higher smoking rate among adults in Missouri than among adults in Arizona. They might hypothesize that people in states with more antismoking education programs have a better understanding of the negative health effects of tobacco and so are less likely to smoke. To evaluate this hypothesis, the researchers would first need to collect data on the antismoking educational programs used in these states and the effect of these programs on a smoker's intention to quit, quitting success rate, and so forth. Alternatively, social scientists could conduct an experiment. Smokers might be randomly assigned to receive either antismoking educational material or a fictional article (as a control). Again, the researchers would collect data on each smoker's intention to quit, quitting success rate, and so forth. Many other factors might also affect tobacco use, such as income or social factors, and these too would need to be considered.

Where does the study of animal behavior fall in this framework? Animal behavior courses are typically offered by biology or psychology departments. Animals are a part of the natural world, and so it makes sense that animal behavior is a part of the natural sciences. Why, then, is animal behavior also studied in psychology departments? The answer is that psychologists have two ways to study human behavior: (1) directly or (2) indirectly, through the study of animals, in the same way that medicine uses animal models to understand human health. Psychological research on the evolution of human behavior often involves the study of our closest relatives. Animal behavior research, whether conducted by biologists or psychologists, uses the scientific method to formulate and test hypotheses.

1.3 Scientists study both the proximate mechanisms that generate behavior and the ultimate reasons why the behavior evolved

What kinds of research questions can we ask about animal behavior? Consider the migration behavior of birds. Many birds travel long distances between summer breeding and winter nonreproductive grounds. The seasonal movement of individuals between these locations is known as migration behavior. We can ask several questions about this behavior—for example, what causes migration? The behavior often is initiated by seasonal changes in day length. These changes trigger hormonal responses that lead to increased feeding, fat deposition, and the onset of long-distance movement (Ramenofsky & Wingfield 2007). In some species, migration behavior is genetically determined and requires no learning, while in others, migration routes must be learned by following experienced individuals (e.g., Lishman et al. 1997) (Figure 1.8).

We can also ask, "Why do birds migrate?" Migration allows individuals to track resources and thus avoid places where resource availability is greatly limited in some seasons; in short, it promotes survival (Cox 1985). To understand the evolution of migration behavior for a particular population or species, researchers often examine how migration patterns differ among different populations of a species or between closely related species (Outlaw et al. 2003; Milá, Smith, & Wayne 2006).

Niko Tinbergen (1963) summarized the different types of research by outlining four basic questions that can be asked about animal behavior:

1. What is the mechanism that causes the behavior?
2. How does the behavior develop?
3. What is the function of the behavior?
4. How did the behavior evolve?

Figure 1.8. Migration. Sandhill cranes (*Grus canadensis*) learn migration routes by following experienced individuals.

APPLYING THE CONCEPTS 1.1

Human infant crying

We can adopt Tinbergen's (1963) approach to more fully understand not only animal but human behaviors. For example, Debra Zeifman summarized over 100 studies on human infant crying from the perspective of Tinbergen's four questions (Zeifman 2001):

1. **What is the mechanism that causes the behavior?**
   Research has revealed three common triggers of infant crying: hunger, discomfort (or pain), and separation from an adult.

2. **How does the behavior develop?**
   Infants cry at birth but also exhibit distinct changes in crying over time. The amount of crying increases and peaks at six weeks of age and then declines until about four months of age and remains relatively constant until age one. In addition, in the first six months, crying typically occurs when the infant is separated from a caretaker. In the second half of the first year, crying more often occurs when the caregiver is nearby and also when the infant is in the presence of a stranger or unfamiliar surroundings, suggesting a growing awareness of its local environment.

3. **What is the function of the behavior?**
   Crying is metabolically costly and, in historical and some modern indigenous societies, may attract predators. What are the selective benefits of this behavior that exceed its costs? Crying is a form of communication that elicits a beneficial response from a caregiver, including the provision of nutrition and enhanced safety. The level of communication is rather sophisticated because several studies have shown that parents can successfully infer an infant's identity (their child versus another child) and health (pain versus hunger) by hearing it cry.

4. **How did the behavior evolve?**
   Many mammalian young produce vocalizations that elicit response from parents: parent–offspring communication is widespread. Nonhuman primate mothers carry their offspring constantly and feed them continuously; offspring crying is rare except when the offspring is physically separated from its mother. Examinations of infant care among aboriginal societies show similar patterns: high levels of contact (over 90% of the time), frequent feeding, and little infant crying. In contrast, in modern Western societies, infants are carried less often (about 33% of the time), are fed intermittently (every 2–4 hours), and cry significantly more often. Interestingly, parental response to crying also differs: crying elicits rapid responses in nonindustrialized societies but is more often ignored in Western societies. These patterns suggest that typical infant care in Western societies might contribute to the increased infant crying observed there. ■

---

proximate explanation An explanation that focuses on understanding the immediate causes of a behavior.

ultimate explanation An explanation of behavior that requires evolutionary reasoning and analysis.

Tinbergen's four questions provide a framework for the study of behavior. Answers to Questions 1 and 2 are often referred to as **proximate explanations**, because they focus on understanding the immediate causes of a behavior (Mayr 1961). These explanations often incorporate studies of genetics, sensory systems, neurons, hormones, and learning. Studies of the genetic basis and hormonal triggers of migration behavior of birds are examples of proximate explanations. Answers to Questions 3 and 4 are known as **ultimate explanations**, because they require evolutionary reasoning and analysis (Mayr 1961). Understanding the effects of migration on survivorship and why migration evolved are examples of ultimate explanations. Researchers have argued that a complete study of behavior requires answers to all four questions (Applying the Concepts 1.1).

1.4 Researchers have examined animal behavior from a variety of perspectives over time

From Darwin forward, animal behavior research has been based on evolutionary thinking. However, during this time a large number of rich and diverse research foci have placed different emphases on Tinbergen's four questions and have developed varied approaches and methodologies. In this section, we briefly review how animal behavior research has changed over time.

Darwin and adaptation

Darwin pioneered the scientific study of behavioral adaptation in three books: *On the Origin of Species* (1859), *The Descent of Man and Selection in Relation to Sex* (1871), and *The Expression of the Emotions in Man and Animals* (1872). In the first book, Darwin discusses instinct, or behaviors neither learned nor requiring experience. One intriguing example he describes is cuckoo reproductive behavior. Many species of cuckoo lay their eggs in the nests of other species, which then care for the young (Figure 1.9). How could such behavior evolve by natural selection? Darwin notes that while most female birds lay an egg every day, female cuckoos do not: they lay one egg every several days. He reasons that if a female cuckoo were to raise her own clutch of many eggs, she would be faced with a prolonged period of incubation and a clutch of chicks whose ages would differ greatly. Such a scenario would be costly for both the female and the chicks. Darwin hypothesizes that selection would favor an individual that laid her eggs in the nests of others, circumventing this problem.

Darwin's work on animal behavior is wide ranging, both taxonomically (e.g., ants, birds, and bees) and topically (e.g., reproduction, aggression, and instinct). Although some of Darwin's descriptions are based on anecdotes and others are anthropomorphic, he always explains behavior using evolutionary reasoning and in terms of adaptation. His work represents the beginning of the evolutionary basis of animal behavior research, a tradition that has continued through to today.

Figure 1.9. Cuckoo behavioral adaptation. A common redstart (*Phoenicurus phoenicurus*) feeds a larger cuckoo chick. Many species of cuckoo lay their eggs in the nests of other species.

Early comparative psychology

George Romanes, a contemporary of Darwin, approached the study of animal behavior from the perspective of a comparative psychologist. Psychologists study the mental processes and behavior of humans. Comparative psychologists, like Romanes, study animal behavior in order to understand human behavior. In *Animal Intelligence*, Romanes systematically examines the reasoning ability of a wide variety of animal taxa, ranging from protozoans to mammals (Romanes 1882). For him, an animal displays a "mind," or conscious action, rather than mere reflex reaction, when it uses previous experiences to learn and modify its behavior in an adaptive manner. For example, if a protozoan encounters a barrier and moves around it, Romanes interprets this behavior as evidence of a conscious or intelligent choice, as opposed to a reflex movement.

Romanes's work stimulated other researchers to focus on examining animal consciousness. He eventually published a "mental tree" to describe a hierarchy of reasoning among taxa (Romanes 1888). While Romanes's construction is simplistic, the patterns it summarizes are surprisingly congruent with current research regarding which animals display higher stages of cognitive development. These include tool use in monkeys and communication in Hymenoptera, one of the largest orders of insects (Burghardt 1985).

Many of the examples in Romanes's book are anecdotal or anthropomorphic, and they were often provided by nonscientists. So although it was one of the first attempts to systematically examine behavior throughout the animal kingdom, it was heavily criticized. For instance, C. Lloyd Morgan emphasized the need for "accurate observation" along with the methods and principles of modern psychology (Morgan 1894). Morgan, too, used animals as a way to understand human consciousness. However, he urged a more cautious approach, proposing that the simplest psychological process possible should be used to interpret an animal's behavior—an idea that came to be known as Morgan's canon (Morgan 1903;

Burghardt 1985; Costall 1993). Although Morgan's canon became a cornerstone of comparative psychology, Morgan may not have meant it to be literally interpreted as such. Morgan was emphasizing the need for a more scientific approach to the study of the mental evolution of animals, but he acknowledged that one should not necessarily exclude more complex explanations of behavior (Costall 1993).

Comparative psychology in North America

In the 1930s, animal behavior research began to flourish in North America. Two comparative psychologists, Edward L. Thorndike and Margaret Floy Washburn, were particularly noteworthy.

Thorndike, like Lloyd Morgan, stressed the importance of avoiding anecdotal evidence. He urged the use of repeated observations, under standardized conditions, to obtain appropriate sample sizes (Galef 1998; Stam & Kalmanovitch 1998). He thus pioneered the use of standardized methodology and the experimental method in the study of animal learning. He is best known for the use of "puzzle boxes," from which an animal could see food that lay outside its reach. These boxes had levers, pulleys, and treadles that the animal had to manipulate in order to escape the box and reach the food. Thorndike observed and quantified how animals learned by trial and error to escape the apparatus to obtain food. Cats, for instance, quickly learned how to escape and escaped more quickly with repeated testing (Thorndike 1911). Thorndike called this increased ability a time curve, today known as a learning curve.

Margaret Floy Washburn was the first woman to earn a Ph.D. in psychology, and the second woman to be inducted as a fellow into the National Academy of Sciences. In addition to her extensive research on the motor theory of consciousness, she wrote a groundbreaking textbook that emphasized the experimental method for comparative psychology. *The Animal Mind* (Washburn 1908) systematically summarizes research on a wide variety of taxa and helped to train a generation of comparative psychologists (Russo & Denmark 1987). For example, it summarizes all the known research on sensory discrimination and spatial perception in organisms ranging from amoebas to vertebrates.

Behaviorism

In the early part of the twentieth century, John Watson spearheaded the development of behaviorism, a field of comparative psychology that studies behavior independent of animal mental states or consciousness. Behaviorists focus on what they can actually observe (Watson 1913; Watson 1924), including the specific stimuli that provoke a response. In this stimulus-response paradigm, both stimuli and responses can be external (behavior) or internal (physiology). Because Watson emphasized the benefits of using variables that could be controlled, he promoted a strong emphasis on laboratory work.

Two of the most influential behaviorists were Ivan Pavlov, a Russian digestive physiologist, and B. F. Skinner. Pavlov studied salivary gland secretions and the role they play in digesting food (Pavlov 1927). He noticed that dogs would often salivate profusely just prior to being fed and discovered that this behavior could also be produced in the absence of food if he first paired the presentation of food with a novel stimulus, like the ringing of a bell. After several such pairings, dogs associated the novel stimulus with food and would salivate after hearing it, a behavior he called a conditional reflex. Classical or Pavlovian conditioning occurs whenever a novel stimulus (like the ringing of a bell) is paired with an existing stimulus (here, the food) and elicits a particular response (salivation). Eventually the novel stimulus alone elicits the same response as the existing stimulus.

Skinner also worked on laboratory animals in highly controlled conditions. He developed novel tools and procedures to study the development of behavior. Most famously, he is known for the creation of **operant chambers**, or **Skinner boxes**, which house a single animal (Figure 1.10). Skinner examined how the behavior of animals could be modified by pairing a particular behavior with either a reward or a punishment (Skinner 1938). Behaviors could be enhanced if animals were given food rewards (a positive reinforcement) for performing the behavior. Alternatively, these behaviors could be reduced if they resulted in a painful shock (a positive punishment). In this process, called operant conditioning, an animal learns to associate a behavior with a particular consequence (either positive or negative).

Figure 1.10. Skinner box. A Skinner box used to study rat behavior.

Classical ethology

In direct contrast to behaviorists, classical ethologists study the behavior of wild animals in nature by observation and experimentation. In general, classical ethologists are interested not, like psychologists, in understanding human minds and behavior, but rather in studying animals and their behavior for their own sake. Karl von Frisch, Konrad Lorenz, and Niko Tinbergen shared the Nobel Prize in Physiology and Medicine in 1973 "for their discoveries concerning organization and elicitation of individual and social behaviour patterns" (Karolinska Institutet 1973).

Von Frisch studied honeybee sensory perception and communication (von Frisch 1956). He discovered that European honeybees (*Apis mellifera*) use vision, olfaction, and taste, to find food. Even more astonishingly, he found that bees communicate the distance and direction of distant food sources to other individuals in the hive by performing a "waggle dance."

Lorenz studied instinctive behavior in birds, particularly imprinting, in which a young animal learns the characteristics of its parent. Many birds, such as goslings, "imprint" on the sight and sound of their mother or other objects they see at hatching (Lorenz 1950). Lorenz found that there is a restricted period of time during which imprinting can occur.

Tinbergen studied instinct—behaviors that are under strong genetic control. Instinctive behaviors termed fixed action patterns are invariant and unlearned; once initiated, they are brought to completion. Much of his work attempted to understand the mechanisms, or releaser stimuli, that initiate a fixed action pattern. One classic example is egg-rolling behavior in graylag geese (*Anser anser*) (Tinbergen 1951). A bird incubating eggs will attempt to retrieve any egg that has been displaced from the nest. Tinbergen demonstrated that the sight of any egg-shaped object near a nest initiates egg retrieval.

Interdisciplinary approaches

Psychologists and ethologists have conducted animal behavior research increasingly as an interdisciplinary endeavor over the last few decades. For instance, we've seen the development of evolutionary psychology and cognitive ethology, disciplines that both emphasize the study of animal thought processes and cognition using an evolutionary framework. Evolutionary psychologists seek to understand human thinking and behavior, while cognitive ethologists focus on understanding the behavior of animals and often integrate information from neuroscience. Both fields assume that natural selection has shaped brain architecture

operant chamber　An enclosure used to study behavioral conditioning.

skinner box　An enclosure used to study behavioral conditioning. Also known as an **operant chamber**.

and thought processes in an adaptive manner. Evolutionary psychologists, such as John Tooby and Leda Cosmides, suggest that we can best explain human behavior by understanding how natural selection shaped neural mechanisms in the past (Griffiths 2008).

Cognitive ethologists also seek to understand how natural selection has acted on mental processes and cognition in order to better understand the behavior of animals. In doing so, they often conduct comparative studies using broad taxonomic comparisons, frequently carrying out research in natural environments. Donald Griffin is credited with the development of cognitive ethology thanks to his book *The Question of Animal Awareness* (Griffin 1976). Today, cognitive ethologists study topics such as learning and memory, counting ability, symbol recognition, and vocal communication in organisms as diverse as bees, crows, rodents, and primates (Bekoff, Allen, & Burghardt 2002).

Behavioral ecology, which developed in the 1960s and 1970s and continues today, focuses on the evolution of behavior by studying its function. As such, the focus of behavioral ecology is on understanding the adaptive value of behavior. Behavioral ecology research often begins with theory (either mathematical or computer models) to examine the fitness benefits and costs of different behaviors. These models are used to predict the behavior that maximizes fitness. These predictions are then tested empirically. What behavioral ecologists learn from this process helps us understand how natural selection has molded behavior (Davies, Krebs, & West 2012).

You can already see that various perspectives place different emphases on Tinbergen's four questions. For example, classical ethologists tend to focus on proximate questions, while behavioral ecologists often emphasize ultimate explanations of behavior. Throughout this book you will read examples of both proximate and ultimate research studies.

1.5 Anthropomorphic explanations of behavior assign human emotions to animals and can be difficult to test

anthropomorphism The attribution of human motivations, characteristics, or emotions to animals.

Anthropomorphism is the tendency to attribute human motivations, characteristics, or emotions to animals. For example, a television documentary might describe a female coyote (*Canis latrans*) as "worried" when her mate is late returning to the den and as "happy" when her mate returns with food. Another show might describe a female meerkat (*Suricata suricatta*) as "sad" when her grown offspring leave the den to establish their own territories. Of course, many cartoons depict animals as essentially humans in animal form—talking, singing, showing feelings, and scheming to achieve different goals.

Perhaps because anthropomorphism is so prevalent in the media, it is also very common among the public (Applying the Concepts 1.2). People often attempt to explain the behavior of animals by attributing their actions to anger, jealousy, or happiness. For example, imagine that a bee stung your leg as you tried to swat it away. Using anthropomorphic thinking, you might believe that the bee stung you because you made it angry. This hypothesis may or may not be true, but how could you possibly test it? What prediction does this hypothesis make? It is difficult, if not impossible, to characterize the emotional state of a bee.

One hypothesis to explain the behavior of the bee does make a prediction: the bee perceived your attempt to swat it as an attack, leading to antipredator

APPLYING THE CONCEPTS 1.2

What is behind the "guilty look" in dogs?

Have you ever come home to find food missing off the counter or the garbage strewn all over the floor? Many dog owners have experienced this and have immediately scolded their dog for disobedience. They often report that their dogs display a "guilty look" because they "know" they have been disobedient (Figure 1). These owners describe their dog's guilty look as avoiding eye contact, lying down and rolling on its side or back, wagging its tail low and quickly, holding its head down, licking or pawing, and moving away from the owner. But do dogs experience "guilt"? This is an anthropomorphic explanation of a behavior. Alexandra Horowitz is a cognitive scientist who studies dog behavior. She designed an experiment to examine the "guilty look" in dogs (Horowitz 2009).

Horowitz recruited 14 dog owners for her experiment. Each dog was exposed to four experimental treatments. At the beginning of each trial, the owner showed the dog a food treat and gave the dog a command not to eat it. The owner then left the room. The experiment had two conditions: either (1) the researchers gave the dog the treat (the dog ate the treat and disobeyed the owner's command), or (2) the researchers removed the treat (the dog could not eat the treat or disobey the owner's command). Once the treat was gone, the owners were called back into the room. They were told either that (1) the dog ate the treat or (2) the dog did not eat the

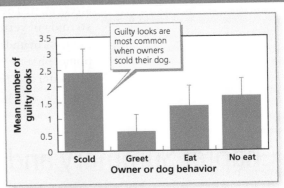

Figure 2. Dog responses. The mean (+ SE) number of guilty looks was highest when dogs were scolded by their owners. (*Source:* Horowitz 2009)

treat (it had been removed), although this information may or may not have been true. Each dog experienced all pairwise combinations of the conditions. Owners were instructed to scold their dogs when told the treat had been eaten and to greet their dogs in a friendly manner when told the treat had not been eaten. All trials were videotaped, and the researchers quantified the number of times each dog displayed behaviors associated with the "guilty look" (avoiding eye contact; lowering head, tail, or ears; moving away from the owner; and so forth).

Horowitz found that the dog's obedience had no effect on its display of "guilty look" behaviors: the mean number of guilty look behaviors was similar whether or not the dog actually ate the treat. Owner behavior did have an effect on dog behavior, however, because "guilty look" behaviors were most common after owners scolded their dog (although in half of these trials, the dog had not disobeyed the command) and least common when they greeted their dog (Figure 2). Horowitz concluded that the dogs did not display more behaviors associated with the "guilty look" after disobeying a command. Horowitz speculates that the "guilty look" likely represents submissive behavior in response to owner scolding. However, she notes that her experiment does not rule out the possibility that dogs could experience guilt. Anthropomorphic explanations of animal behavior remain intriguing but difficult to test. ■

Figure 1. Guilty look.

behavior. In other words, bees sting potential predators to deter attacks. You could test this hypothesis by observing interactions between bees and their predators and then quantifying how often they use their stinger. As a control group, you could compare the interactions between bees and a nonpredator and quantify how often bees sting that species. The hypothesis would predict that bees will more often sting potential predators than they will nonpredators.

While anthropomorphic thinking rarely produces testable predictions, we do not want to dismiss the possibility that animals experience emotions. Indeed, one area of active research, known as cognitive ethology, attempts to understand animal cognitive, or thought, processes. Understanding animal thinking is challenging, but researchers are making progress in its study, as we will see in Chapter 7. In addition, for many animals, there is often a relationship between stress-related behaviors and physiology, as with the production of glucocorticoid steroids in vertebrates (Blas et al. 2007). In these cases, we truly are making progress in understanding animal mental states and emotions. However, anthropomorphic hypotheses remain difficult to test and should be avoided until we are better able to understand animal emotions and cognition.

Chapter Summary and Beyond

Animal behavior is important because it intersects with the lives of many people. We define behavior as the internally coordinated activity patterns of animals that respond to changing external or internal conditions. Levitis, Lidicker, and Freund (2009) provide a nice review of the many definitions of behavior that scientists use. Modern animal behavior research involves more than simply observing animals; it also involves testing hypotheses to explain behavior.

Science is a way of knowing about the natural world that involves use of the scientific method. Scientists use the scientific method to construct hypotheses that make testable predictions. Researchers analyze data to determine whether they support or refute their hypothesis. Science never proves a hypothesis is correct. Rather, it is an ongoing process of hypothesis testing, and that process is crucial to scientific progress. A hypothesis that makes broad predictions and stands up to repeated testing may eventually become a theory. Scientists must consider the limitations of studies that report correlations, because correlation does not demonstrate causation.

Explanations about animal behavior address either proximate or ultimate research questions, as summarized by Tinbergen (1963). While researchers' emphasis on proximate and ultimate explanations has varied, together these explanations provide complementary information and a more complete understanding of behavior. Laland et al. (2011) examine contemporary biological debates in light of the proximate–ultimate dichotomy.

Beginning with Darwin, researchers have studied behavior from many perspectives. Houck and Drickamer (1996) offer a detailed description of the history of the study of animal behavior, including seminal papers.

Dewsbury (1985) provides autobiographies of many important figures in the history of animal behavior research.

Finally, anthropomorphic hypotheses assign human emotions to animals and are difficult to test. For a review of the history and problems associated with anthropomorphism in the scientific study of animal behavior, see Wynne (2004b; 2007).

CHAPTER QUESTIONS

1. Spend 20 minutes quietly observing an active animal. This animal could be a squirrel or bird on campus, a captive animal at the zoo, or your pet. Create an ethogram that describes all of the behaviors you observed.

2. For each of the following, determine whether the example describes a behavior and use the term's definition to explain why. (a) A spider spins a web to catch food. (b) Male fish construct nests to attract females. (c) A fiddler crab waves its claw to fight a rival. (d) A dog salivates just before it is fed dinner. (e) A butterfly moves from a shady to a sunny location.

3. Many disciplines use the word "science" to describe their work. Your college or university may have departments of computer science, political science, food science, or even family and consumer science. Describe one of these disciplines and then determine whether or not you could consider it to be a "science." In this chapter, we discussed how science differs from other disciplines, and you can use that information to develop criteria for this assignment.

4. Why is a hypothesis not simply an educated guess?

5. A caterpillar drops into a creek and is carried downstream while struggling to stay afloat. In what aspects of this

scenario is the caterpillar exhibiting behavior, and in what ways is it not?

6. Many people feed birds because they believe it makes the birds happy. How could you test this hypothesis?

7. Suppose you took your dog to the veterinarian because it was having intermittent diarrhea. One possible explanation for this symptom is that the dog has parasites. Many parasites go through life cycles in which they are only shed from the intestinal tract of the host every few days. The veterinarian has a fecal sample analyzed for parasites. The results come back negative—that is, the laboratory did not find any parasites in the fecal sample. Using the scientific method, can you now rule out parasites as the cause of the diarrhea? What is the rationale for your answer?

Chapter 2

Methods for Studying Animal Behavior

When we were graduate students, we quickly found out that research does not always go as planned. Once, during a study of the mating behavior of round stingrays (*Urolophus halleri*) in the Gulf of California, observations ended abruptly after a large storm created adverse conditions and all the stingrays left the study area. Luckily, when the weather stabilized, the stingrays returned, and we were able to continue (Nordell 1994). More recently, we set up an experiment to study squirrel foraging behavior. We were ready, but the squirrels apparently were not. Although we put out an abundance of a preferred food (hazelnuts) for several weeks, squirrels never came to our study site location. So we had to find a new location.

Many research projects begin with detailed planning. Yet sometimes scientific research begins almost as an "Aha!" moment, as it did for ecologist Stephen Yanoviak (2006). It all started during a project to study deforestation and the spread of mosquito-borne tropical diseases in Peru. Yanoviak was collecting mosquitoes in the rainforest canopy 30 m (over 100 ft) off the ground when he was attacked by ants. He noticed that after he brushed some of the ants off his body, rather than falling to the ground, they appeared to glide back to the tree trunk as if they

Figure 2.1. *Cephalotes* ant. This species of ant can direct its gliding behavior, as ecologist Stephen Yanoviak discovered.

could somehow direct their fall. Later, he made additional observations of this phenomenon by painting the ants' rear legs with white nail polish so that he could better observe them as they fell. He again saw that they seemed not only to glide and direct their movements toward the tree trunk, but even to make 180° turns in the air.

When he headed home, he asked an ant ecologist, Michael Kaspari, and an expert on flying and gliding animals, Robert Dudley, to collaborate with him to examine this behavior more thoroughly. Several years passed before all three could get back to Peru and conduct more "drop" studies. When they did, they found that about 85% of the individuals in the ant species *Cephalotes atratus* seemed to have the ability to direct their gliding flight behavior (Yanoviak, Dudley, & Kaspari 2005) (Figure 2.1). This was an exciting finding, as no one had known that ants could control their descent or direct their gliding flight. So in this case, scientific discovery began quickly and accidentally.

In this chapter we begin by exploring how scientists generate and test hypotheses to answer research questions about animal behavior. We then discuss how animal behavior researchers collect data using observational, experimental, and comparative research methods. Essentially, we examine the nuts and bolts of how animal behavior research is conducted. Next, we discuss important ethical issues in the study of animal behavior. We close the chapter by looking at how scientific knowledge is generated and communicated to the scientific community through peer-reviewed research.

2.1 Animal behavior scientists test hypotheses to answer research questions about behavior

To study the behavior of animals, scientists must form and test hypotheses to answer research questions. These questions can be based on earlier observations, previous knowledge, prior research, a theory, or some combination of these. Let's examine how this process works using the case of wolf spiders.

Hypothesis testing in wolf spiders

Featured Research

The brush-legged wolf spider (*Schizocosa ocreata*) is common in the leaf litter habitat of eastern North American forests. During the brief breeding season each spring, males search for females (Figure 2.2). Females typically mate only once, whereas males mate multiple times, and so unmated females become increasingly rare as the breeding season progresses. How do males find unmated females? Females are cryptic, meaning that their coloration lets them blend in well with the leaf litter. As a result, males search for females slowly, using chemical cues that females deposit on silk threads in the leaf litter. When males find receptive females, they court them with a multimodal display by walking rapidly in a jerky fashion while waving and tapping their forelegs (providing visual cues). They also transmit vibrations through the substrate by rubbing specialized organs together (providing seismic cues). These signals can be sensed by the female and may also

Figure 2.2. Wolf spider. Male brush-legged wolf spiders wave and tap their forelegs to court females.

be used by other nearby males as a cue to the presence of a potential mate. George Uetz, his student Andy Roberts, and collaborator Dave Clark were interested in this mating behavior (Roberts et al. 2006). The research team investigated the following research hypothesis:

Research hypothesis: Male wolf spiders use the behavior of nearby males to find receptive females.

The null hypothesis is that males do not use the presence and behavior of other males as a cue to locate receptive females, while the alternate hypothesis is that they do. The alternate hypothesis predicts that the presence and behavior of a rival male will cause the searching male to move in the rival's direction, search for a female, and begin courtship tapping.

The research team tested these predictions by conducting an experiment. They collected juvenile spiders from a nearby forest and raised them in the lab until they were sexually mature. Focal males (the subjects of study) were tested for their reactions to (1) visual cues only (the sight of a demonstrator male tapping), (2) seismic cues only (the vibrations of another male tapping), and (3) both cues simultaneously. The researchers prompted the demonstrator males to tap by providing them with the silk of females that contained their chemical cues. The researchers recorded the focal male's behavior for three minutes prior to the stimulus, three minutes during the stimulus, and three minutes after the stimulus. This allowed researchers to characterize baseline behavior and examine changes in behavior after exposure to test stimuli. In each phase of a trial, the research team recorded two behaviors of the focal male: its chemoexplore behavior, or behavior to detect chemical signals (slow walking while rubbing the substrate with the pedipalp); and its courtship behavior, or behavior to attract a mate (rapid walking in jerky movements while tapping the substrate). The researchers used 15 focal males in each treatment. In any study, it is important to collect data on many individuals to ensure that data are representative of the population or species. Individuals always exhibit variation, but some individuals may exhibit particularly unusual behavior, perhaps as a result of hunger or disease. The collection of data from many individuals helps prevent researchers from drawing conclusions that are based only on a few nonrepresentative individuals.

The research team found that the treatments had no effect on chemoexplore behavior (Figure 2.3). From these data, the team accepted the null hypothesis: male wolf spiders do not appear to use the visual or vibratory behavior of a courting rival as a cue to the location of a receptive female. Apparently, males just have to find females on their own.

Then a funny thing happened. A year or so later, Roberts, Uetz, and Clark were examining the mating behavior of these spiders in nature and observed males heading in the direction of other males that were tapping in courtship. This was exactly the behavior that they did *not* observe in their lab experiments. What was going on? They remembered that in their previous experiment, they had used a population of males raised in isolation in captivity that had had no experience with rival males. Could prior social interactions be an important prerequisite for this behavior?

To test this new hypothesis, the researchers conducted similar experiments using adult males caught in the wild. This time, they

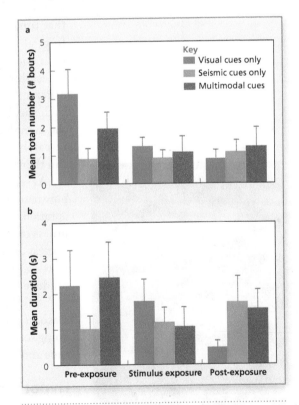

Figure 2.3. Spider response to male cues. There was no effect of experimental treatments on male behavior, as measured by (a) number of chemoexplore bouts or (b) duration of chemoexplore bouts before, during, or after stimulus exposure. The height of each solid bar represents the mean behavior, and the thin lines represent a measure of variation in the data, the standard error of the mean. (*Source:* Roberts et al. 2006)

found that the focal males did display a greater frequency and duration of interactive behaviors (i.e., approach, courtship signaling) during and after exposure to the sight of another male tapping. The findings indicated that male wolf spiders can indeed use the tapping of rival males as a cue for finding females (Clark, Roberts, & Uetz 2012).

This story illustrates an important point. Research is an ongoing process of discovery. The behavior of males in the wild did not seem to match observations of laboratory-reared subjects, which led to a new hypothesis and experiments and a new understanding of spider behavior. This is an excellent example of how science progresses.

Generating hypotheses

At this point you might be wondering, "Where do hypotheses come from?" This is a great question. Many hypotheses are generated by observations in the field that spark the question, "Why did I observe that behavior?" For example, Mike Ryan has long studied túngara frogs (*Engystomops pustulosus*), a species common throughout Central America. Males vocalize to attract females using two calls: a simple "whine" call and a more complex call that is a "whine" plus several "chucks." Ryan had observed that females are more attracted to the complex call (Ryan 1980), and so he wondered why males ever use the simple call (Ryan 2011).

Ryan hypothesized that the complex call might be more costly to produce, and so males only do so rarely. Shortly thereafter, while in the field, he observed that túngara frogs are eaten by frog-eating bats (Figure 2.4)! How do bats find frogs at night? Ryan hypothesized that perhaps these predators locate frogs by their vocalizations, particularly their complex calls. He designed an experiment to test this hypothesis: he played recordings of simple and complex frog calls from different speakers and recorded the number of frog-eating bats that approached each speaker. More than two-thirds of the bats approached the speaker playing the complex call, providing support for the hypothesis (Ryan et al. 1982). This research illustrates how observations of behavior can generate hypotheses that can then be tested.

As seen in this example, researchers often construct hypotheses by considering how natural selection might have acted on the behavior to produce a behavioral adaptation. In this approach, they ask, "How does the observed behavior affect the fitness of the animal?" Fitness here is defined as the survivorship and reproductive success of an individual. (We will explore fitness further when we study evolution, starting in Chapter 3.) Such thinking can help generate alternate hypotheses that produce testable predictions. In the wolf spider example described earlier, the hypothesis that males use rivals as a cue to locate females is based on evolutionary reasoning. It is beneficial for a male to be able to locate receptive females, because doing so allows him the opportunity to court and mate. Any cue that reduces search time and increases encounters with receptive females should be advantageous.

Hypotheses from mathematical models

Another source of hypotheses and predictions are analytical models. Scientists often use mathematical expressions to clarify arguments and assumptions. The resulting models can generate predictions about which behaviors maximize an individual's fitness. All models about behavior are based on assumptions about the ecology and evolution of an organism. The advantage of analytical models is that they allow scientists to easily manipulate their assumptions to produce new predictions about behavior.

Figure 2.4. Frog-eating bat. A bat with its frog prey.

Let's look at one example of a mathematical model that was developed to understand the behavior of gray squirrels (*Sciurus carolinensis*) (Figure 2.5). You may have noticed that squirrels sometimes carry a nut from an open, grassy area to a tree before consuming the nut. Eating in a safe location such as a tree can decrease the likelihood of being killed by a predator—but at what cost? Steve Lima had observed this behavior but noted that carrying items over long distances could also result in increased time and energy costs for the squirrel. He posed the question: When should animals stay out in the open and eat, and when should they carry their food to safety to eat?

Lima and his colleagues (Lima, Valone, & Caraco 1985) hypothesized that food carrying might represent a trade-off between feeding in safety (in a tree) and obtaining high energy intake rates. To develop this hypothesis, they created a mathematical model with two variables: the energy intake rate, or energy consumed per second, and the probability of being killed by a predator, such as a hawk or a dog. The model assumes that a squirrel's fitness increases as energy intake rate increases and as the probability of being killed decreases. It also assumes that a squirrel is at risk of being killed while on the ground but safe while in a tree.

Suppose a squirrel on the ground and away from a tree finds many identical large food items. In Lima's model, for each food item, the squirrel has two options: it can remain at the patch and eat it there, or it can carry items, one at a time, back to the nearest tree and eat there. The first option is riskier, because a predator might attack, but this strategy should lead to high energy intake rates, because the squirrel doesn't spend time and energy traveling back and forth between the food patch and the tree. The second option, carrying food back to a tree, results in a lower energy intake rate but should also be safer. Clearly, a squirrel cannot simultaneously do both. How should the squirrel behave to maximize its fitness? What proportion of items should it carry back to the tree? Lima's model makes two predictions (Figure 2.6). The first prediction is that the larger the food item, the more often it should be carried to a tree. In other words, for a given distance between the food patch and the nearest tree, individuals should carry a greater proportion of food items to the tree as item size increases. This prediction relies on the assumption that larger food items take more **handling time**, or food-processing time. As handling time increases, so does the antipredator benefit in carrying a food item to safety.

The second prediction states that as the distance between the food patch and the tree increases, individuals should reduce their carrying behavior. In other words, for food items of a given size, squirrels should carry a smaller proportion to the nearest tree as the tree's distance from the food patch increases. Although the first prediction is intuitive, the second may appear a bit counterintuitive. Isn't running to the tree supposed to make the squirrel safer? However, more distance means more time spent running back and forth—and thus more time exposed to attack. And that means less antipredator benefit in carrying food to safety.

The two predictions reflect the trade-off between maximizing energy intake rates and minimizing time exposed to predators. Lima and colleagues (one of whom—one of the authors of this book—was then an undergraduate) tested these predictions in a large city park in Rochester, New York. They offered squirrels access to food patches that contained pieces of cookies, a favorite food item of urban squirrels. Each day, the researchers varied the size of the pieces (1, 2, or 3 g)

Figure 2.5. Gray squirrel. Squirrels eat food on the ground but will sometimes carry food to consume it in the safety of a tree.

handling time The amount of time required to manipulate a food item so that it is ready to eat.

Figure 2.6. Food-carrying predictions. The model predicts that the proportion of items carried to safety should increase with food size (blue indicates large food items, orange indicates small food items). For each food size, carrying should decline with increasing distance to safety.

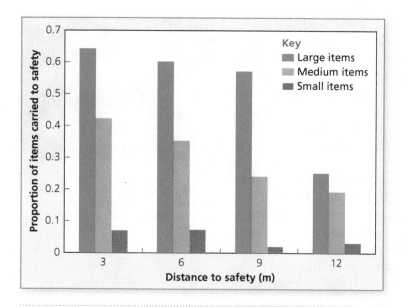

Figure 2.7. Food-carrying behavior. The proportion of large (blue), medium (orange), and small (gray) food items carried to safety for consumption. (*Source:* Lima, Valone, & Caraco 1985)

and their distance from the nearest tree (3, 6, 9, or 12 m). Over the course of two months, all food size and distance combinations were offered several times.

The researchers found that at each distance, squirrels carried the largest item most frequently but rarely carried the smallest item to cover. Additionally, for a given food item size, squirrels carried fewer items to a tree as the distance between the food patch and a tree increased (Figure 2.7). Because the data matched both predictions of the model, Lima and colleagues concluded that the food-carrying behavior they observed in gray squirrels indeed represents a behavioral trade-off between maximizing energy intake rate and minimizing predation risk.

In fact, many species carry food to a safe location before eating it, and other experiments have demonstrated a similar behavioral trade-off in many species of birds (Lima 1985; Valone & Lima 1987). For example, black-capped chickadees (*Poecile atricapillus*), blue jays (*Cyanocitt a cristata*), and white-throated sparrows (*Zonotrichia albicollis*) behaved just like squirrels in various tests. Other species, such as cactus wrens (*Campylorhynchus brunneicapillus*) and eastern towhees (*Pipilo erythrophthalmus*), carried more of the large than the small items but did not carry more items when close to safety. Finally, the carrying behavior of house finches (*Carpodacus mexicanus*) was completely inconsistent with both predictions of the model. Why doesn't the model work for all species? We don't know, but one possibility is that species perceive predation risk in different ways—maybe this could serve as a research project for you.

It is clear from the examples here that hypotheses make testable predictions about behavior. Next, we turn our attention to the different methods used to study behavior and test predictions.

2.2 Researchers use observational, experimental, and comparative methods to study behavior

When conducting research, a scientist must determine the best method to address their research questions and test their hypotheses. Three common methods used in behavioral research are the observational, experimental, and comparative methods (Altmann 1974; Lehner 1998; Bateson & Martin 2007). It is important to note that numerous protocols or data collection techniques can be used with each method and that researchers often employ more than one method in their studies (Applying the Concepts 2.1). In addition, data must be analyzed to see if the patterns observed support the prediction.

The observational method

In the **observational method**, scientists observe and record the behavior of an organism without manipulating the environment or the animal. This method is commonly used both to test hypotheses and to describe behavioral patterns. For

observational method An approach in which scientists observe and record the behavior of an organism without manipulating the environment or the animal.

APPLYING THE CONCEPTS 2.1

Project Seahorse

Amanda Vincent began studying seahorses because she wanted to conduct research outdoors and in the sea. Later, she became fascinated with the animal, particularly its unusual reproductive behavior. For example, only male seahorses get pregnant (Figure 1). Once she started her research, she became engrossed with them. Now Vincent works not only to research seahorses' behavior but also to conserve them.

When she began studying seahorses, Vincent's work involved spending many hours conducting observations. These

Figure 1. Seahorse. This pregnant male big-belly seahorse (*Hippòcampus abdominalis*) is one of many seahorse species studied by Dr. Vincent. It is found only along the coasts of Australia and New Zealand.

animals had rarely been observed in the wild, and very little was known about their behavior. This meant many months of work and many hours each day sitting very still while scuba diving in cold water. Her research began with basic questions about the mating behavior of seahorses. Are they monogamous? What are the behavioral differences between the two sexes of these species? Do males defend territories? To answer these questions, she had to find the seahorses during their reproductive periods. These animals can be very cryptic, or camouflaged, and so blend into their environment. She also had to develop techniques for tagging individuals so she could identify them later and design appropriate measures of behavior.

Because of the research conducted by Vincent, her students, and colleagues, we now know much more about the unique behaviors of seahorses. It turns out that some species are truly monogamous, which is very rare in the animal world and even rarer among fish. Experiments Vincent conducted have demonstrated that mated pairs greet each other in a daily ritual that appears to be important in maintaining that monogamous bond (Vincent 1995a). In addition, Vincent's research indicates that many populations of seahorses and pipefish are decreasing because they are heavily harvested for Chinese medicine, curios, and the hobby aquarium market (Vincent 1995b). In response, Drs. Amanda Vincent and Heather Koldewey founded Project Seahorse to protect seahorse populations while ensuring sustainable livelihoods for those dependent on seahorse fishing. Vincent continues to both study seahorses' behavior (e.g., Vincent, Evans, & Marsden 2005) and work for their conservation (e.g., Vincent 2008). For more information, see www.pbs.org/wgbh/nova/seahorse/vincent.html. ■

example, the observational method is used to construct ethograms, as we saw in Chapter 1. Researchers studying animals in zoological parks frequently use the observational method, because behavior is often an important indicator of well-being or level of stress (Marriner & Drickamer 1994). For instance, changes in the behavior of an animal may indicate changes in its reproductive condition or health (Lindburg & Fitch-Snyder 1994). Researchers studying animals in the wild also use the observational method to test hypotheses, as we see next.

The observational method and reproductive energetics of chimpanzees

Featured Research

A long-term study of chimpanzees (*Pan troglodytes*) conducted by Carson Murray, Elizabeth Lonsdorf, Lynn Eberly, and Anne Pusey illustrates use of the observational method (Murray et al. 2009). Murray and her colleagues investigated the reproductive energetics of free-living female chimpanzees in Gombe National Park in Tanzania. Chimpanzees, medium-sized primates that live in equatorial Africa, prefer to feed on calorie-rich, ripe fruit but will eat a variety of foods. All female mammals face increased energetic demands both during gestation

Figure 2.8. Female chimpanzee. Female chimpanzees face increased energy demands when pregnant and lactating.

(pregnancy) and while feeding their young via lactation (Figure 2.8). In response, many pregnant and lactating females alter their diet to increase their caloric intake. For instance, females of many small primate species (e.g., capuchins and lemurs) will alter their diet to include more nutrient-rich foods. However, prior to Murray et al.'s study, little was known about this kind of behavioral modification in great apes such as chimpanzees. Therefore, Murray's team asked the following question:

Research question: Do pregnant and lactating chimpanzees alter their feeding behavior to meet the increased energy requirements of reproduction?

In order to address this research question, they observed the behavior of three types of adult females: pregnant, lactating, and nonpregnant/nonlactating. They tested the following hypotheses:

Alternate hypothesis: Pregnant and lactating females will consume a diet that provides higher caloric intake than that consumed by nonpregnant, nonlactating females.

Null hypothesis: Pregnant and lactating females will *not* consume a diet that provides higher caloric intake than that consumed by nonpregnant, nonlactating females.

Prediction: Pregnant and lactating females will have a higher percentage of fruit (a high-calorie food) in their diet than will nonpregnant, nonlactating females.

The research team used data collected from one group of chimpanzees in the park to test their prediction. Over a period of 30 years, the group ranged in size from 38 to 64 individuals. Starting in 1973, researchers conducted full-day follows during which they tracked and recorded all the activities of one focal individual for an entire day. They recorded all of its feeding bouts (amount of time spent eating, type of food items eaten) and social interactions during that day. Each day, different individuals were followed, resulting in a large database of chimpanzee behavior. The researchers analyzed over 19,000 hours of data from over 2,300 "follows." From these data, they calculated the percentage of fruit in each individual's diet by dividing the time spent feeding on fruits by the total time spent feeding. Each adult female was classified as either pregnant (back-calculated from birthdates of offspring), lactating, or nonpregnant and nonlactating (Toolbox 2.1).

Murray's team found that the three types of adult females tended to differ in their feeding behavior. The diets of both pregnant and lactating females included more fruit (allowing more energy gain) than did the diet of nonpregnant, nonlactating females (Figure 2.9). The research team concluded that pregnant and lactating females appear to alter their diet to meet the increased energy demands of reproduction. This conclusion was reached through research that relied solely on observations of animals, without any manipulations. An advantage of this method is that researchers can observe naturally occurring behaviors. However, many factors might influence the behavior observed, and so it is difficult to draw firm conclusions about why particular

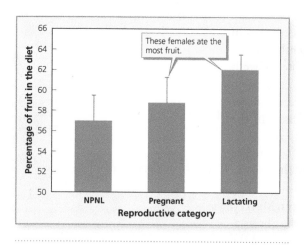

Figure 2.9. Chimpanzee diets. Mean (+ SE) percentage of fruit in the diets of females with different reproductive conditions. Pregnant and lactating females ate more fruit than nonpregnant, nonlactating (NPNL) females. (*Source:* Murray et al. 2009)

behavior patterns occur from observations alone. To do that we need a different method that provides greater control, as we see next.

The experimental method

In the **experimental method**, scientists manipulate or change a variable to examine how it affects the behavior of an animal. The variable that is changed is called the independent variable, and it can be anything that we measure, control, or manipulate. It can be abiotic (e.g., temperature, humidity, wind) or biotic (e.g., habitat, food availability, social interactions). The researcher then measures changes in another variable, the dependent variable, that occur in response to changes in the independent variable. For example, suppose you were examining how food availability affects clutch size (the number of eggs laid by a female during one reproductive attempt). Your experimental treatment might consist of manipulating the amount of food (the independent variable) and then measuring the number of eggs produced (the dependent variable).

In order to determine whether or not your manipulation is actually influencing the dependent variable, it is essential to include a control group. The **control group** does not experience the manipulation but is treated similarly in all other aspects. In the simplest experimental design, only one factor differs between the experimental and control groups. In essence, the control group represents the null hypothesis, and the experimental group represents the alternate hypothesis.

experimental method An approach in which scientists manipulate or change a variable to examine how it affects the behavior of an animal.

control group In an experiment, the group that does not experience a manipulation. It provides a comparison to experimentally manipulated groups.

The experimental method and jumping tadpoles

Featured Research

Verônica de Sousa and her colleagues used the experimental method to study an unusual behavior of an undescribed species of tropical tadpole (*Pseudopaludicola* sp.) (De Sousa, Teresa, & de Cerqueira Rossa-Feres 2011). *Pseudopaludicola* are very small frogs (less than 20 mm snout-vent length) found throughout South America (Laufer & Barreneche 2008). During the rainy season, females lay eggs in the small puddles formed from the footprints of large mammals at the edge of ponds and swamps. The eggs develop into tadpoles that go through metamorphosis in these small puddles. Also in these puddles are aquatic juvenile dragonflies (naiads) that are voracious predators of tadpoles, fish, and invertebrates.

The researchers observed that tadpoles sometimes jump out of their puddles and wondered why they exhibit this behavior. One possibility is that they are attempting to escape from predators. The puddles tend to be closely spaced and numerous, so a tadpole might be able to jump to a safer one. The researchers examined two questions: (1) Is tadpole jumping behavior a response to the presence of a predator? and (2) Does the jumping behavior increase survival? (Scientific Process 2.1)

The researchers collected tadpoles and dragonfly naiads and created small experimental arenas to simulate puddles. To examine the effect of a predator on jumping behavior, they conducted an experiment with three treatments that differed in the presence or absence of the naiad predator and then measured the frequency of tadpole jumping. They found that the frequency of tadpole jumping was significantly higher in the predator-present treatment than in either control treatment. Most jumps occurred in response to naiad movement.

To examine how jumping behavior affects tadpole survival, they conducted another experiment with tadpoles and naiads using two treatments that differed in whether tadpoles could get away from the predator. For this experiment, they measured the number of tadpoles that survived and found that the survival rate was much higher in the treatment in which tadpoles could escape the predator.

SCIENTIFIC PROCESS 2.1

Jumping tadpoles

RESEARCH QUESTION: *Why do tadpoles jump out of puddles?*

Hypothesis: Jumping is an antipredator behavior.

Prediction: Tadpoles will jump most often from puddles that contain a predator.

Methods: The researchers:

- Collected tadpoles and dragonfly naiad predators from the field
- Acclimated tadpoles in individual aquaria in the lab for four to five days
- Created experimental arenas with sandy bottoms to simulate puddles (9 cm diameter, 4.5 cm depth)

Experiment 1: 41 tadpoles each experienced three treatments:

1. Predator treatment (PRE): one naiad in the arena.
2. Predator control treatment (CPRE): an inanimate object (similar in size to a naiad) in arena to control for the presence of a novel object.
3. Control (CON): no object or predator in the arena.

Each trial lasted 45 minutes and was videotaped. Researchers recorded jumping behavior (number of jumps).

Results: Jumping behavior was most common in the predator-present (PRE) treatment.

Experiment 2: 21 tadpoles were randomly assigned to one of two treatments:

1. Screen treatment: a screen was placed over the arena to prevent the tadpole from jumping out.
2. Nonscreened treatment: no screen was placed over the arena, allowing the tadpole to jump out.

Each trial lasted three hours or ended when the tadpole was consumed.

Results: Survival was highest in the nonscreened arena treatment.

Figure 1. Predators and prey. Juvenile dragonflies (naiads) feed on tadpoles. Adult dragonflies are much larger than naiads.

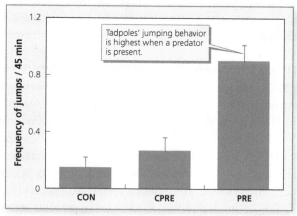

Tadpoles' jumping behavior is highest when a predator is present.

Figure 2. Jumping behavior. The mean (+ SE) frequency of jumping was higher for the predator treatment than for both controls. (*Source:* de Sousa, Teresa, & de Cerqueira Rossa-Feres 2011)

Conclusion: Tadpole jumping behavior is an effective antipredator behavior that enhances survivorship.

EVALUATE
Explain the advantage of using two different controls in this experiment.

These experimental results indicate that naiad predators kill tadpoles and that tadpole jumping is an effective antipredator behavior that may have significant fitness benefits. Since tadpoles are aquatic, wouldn't jumping out of the puddle cause them to desiccate? Observations in the field confirm that tadpoles do spend quite a bit of time out of puddles, and yet they survive. There is often a thin film of water on the substrate that likely reduces the risk of desiccation.

This research study illustrates a strength of the experimental method: specific factors are manipulated to examine their effect on behavior. However, such control often requires research to be conducted in an artificial setting (i.e., the laboratory), which may affect behavior. Another aspect of experimental studies is that they often require greater financial and logistical support than observational studies.

Finally, we examine a third method used to study behavior, particularly how behaviors evolve.

The comparative method

In the **comparative method**, scientists examine differences and similarities between species to understand the evolution of behaviors. Often, sets of closely related species share similar behavioral adaptations because their shared ancestor possessed the behavioral trait. By comparing the behavior of closely related species, researchers can try to understand whether a behavior is ancestral or derived. **Ancestral traits** (also called **plesiomorphic traits**) are found in a common ancestor of two or more species, while a **derived trait** (or **apomorphic trait**) is found in a more recently evolved species and was not present in the common ancestor.

To understand the evolution of behavioral traits, researchers often require knowledge of the evolutionary relationships among taxa (taxonomic groups) such as species. Evolutionary relationships are provided by a **phylogeny**, a hypothesized evolutionary history of a group of species. For any set of species, a phylogeny illustrates our best understanding of the patterns of descent, showing ancestor–descendant relationships that reflect their evolutionary history. For example, two species that are more closely related to one another than to any other species share a recent common ancestor and are known as **sister species**. In a phylogenetic diagram these species are represented by two lines that share a node (the common ancestor).

The comparative method and the evolution of burrowing behavior in mice

How do researchers use the comparative method and phylogenies to examine the evolution of behavior? Consider work conducted by Jesse Weber and Hopi Hoekstra (Weber & Hoekstra 2009). These researchers were interested in understanding the evolution of burrowing behavior of mice species in the genus *Peromyscus*. Species of these small mice exhibit a wide range of burrowing behaviors. Weber and Hoekstra first constructed a phylogeny of seven species using genetic data (Figure 2.10). Next, they conducted behavioral studies of each species to measure different aspects of burrow construction (e.g., entrance length, burrow depth, and total length). They placed a mouse in a large chamber (1.22 m × 1.52 m × 1.07 m) that contained a sandy loam soil and allowed it to construct a burrow for 48 hours. Afterward, they removed the mouse and filled the burrow with polyurethane foam that expanded to create a cast of the burrow, which they then measured.

They found much variation in the burrowing behavior of the seven species studied: two species did not construct burrows, three constructed simple burrows, and two made very complex burrows. Finally, the researchers mapped the burrowing behavior of the species onto their phylogeny.

The patterns in the behavioral studies and the phylogeny provided information about the evolution of burrowing behavior in these species. The analysis indicated that the ancestral trait is the absence of burrowing and that the construction of large, complex burrows is a derived trait. They found that sister species often exhibited significant differences in burrowing behavior (e.g., the plateau mouse and Aztec mouse exhibited different behaviors, as did the deer mouse and the oldfield mouse). Because burrowing behavior was so varied among these

comparative method An approach that examine differences and similarities between species to understand the evolution of behavioral traits.

ancestral trait A trait found in the common ancestor of two or more species. Also called a **plesiomorphic trait**.

derived trait A trait found in an organism that was not present in the last common ancestor of a group of two or more species. Also known as an **apomorphic trait**.

phylogeny A branching diagram showing hypothesized evolutionary relationships among species. Sometimes called a phylogenetic tree.

sister species Two species that are more closely related to one another than to any other species; species that share a recent common ancestor.

Featured Research

Figure 2.10. Mouse phylogeny and burrowing behavior. (a) The molecular phylogeny of seven species of mice identifies sister species. (b) The mean (+ SE) frequency and length of burrow construction for each species demonstrates variation among species. *N* represents the number of burrows sampled. (c) Species range from no burrowing behavior to constructing complex burrows. (*Source:* Weber & Hoekstra 2009)

closely related species, the researchers concluded that variation in this behavior most likely evolved independently in the different species. This conclusion is supported by recent work that indicates a genetic basis for burrowing behavior in these mice (Weber, Peterson, & Hoekstra 2013).

As we've just seen, animal behavior can be examined using a variety of methods. Next, we examine how the study of animal behavior requires careful attention to ethical animal use.

2.3 Animal behavior research requires ethical animal use

In the previous chapter, we examined the process of science as being fundamental to understanding the natural world. That said, scientists are not infallible. We will now look at two ethical issues related to the study of animals: scientific misconduct and animal care.

Scientific misconduct includes the falsification or fabrication of data, purposefully inappropriate analysis of data, and plagiarism. Misconduct can occur in all branches of science and can be identified through peer review. If a problematic paper is published, misconduct is typically identified when other scientists cannot successfully replicate the findings.

Researchers studying animals must also consider how their work affects the individuals they study. The better we understand the behavior of animals, the better we can manage and care for them. However, research can also have negative

scientific misconduct The violation of ethical behavior standards in science. It includes the falsification or fabrication of data, purposefully inappropriate analysis of data, and plagiarism.

impacts on animals. These can arise from simple observations of wild animals, inappropriate housing of animals in laboratories, and the various procedures performed. Scientists and governments have therefore developed ethical standards for animal research.

How research can affect animals

Observing animals in their natural environment often involves the lowest level of invasiveness and stress for individuals. However, even observational field research can negatively affect the animals studied. For example, repeatedly observing individuals can habituate them to humans. If individuals lose their fear of humans, they can become subject to increased predation from poachers or hunters. Manipulations of the animal's environment, such as providing supplemental food, can lead to still other unintended consequences. A large addition of food to the environment could result in a substantial increase in the population, which might result in overcrowding and increased stress (Robb et al. 2008).

Other kinds of behavioral research require more direct contact with animals. For example, field researchers may need to collect blood or tissue samples, quantify parasite loads, measure reproductive condition or morphology, or uniquely mark individuals. These procedures can increase animals' stress levels because they require researchers to trap, restrain, or anaesthetize individuals. Animals in captivity require additional considerations (Figure 2.11), including appropriate housing and husbandry (feeding and care), as well as how the animals will be procured. Researchers also need to determine the fate of the animals when the research is completed. Some animals may need to be sacrificed for examination of their body organs. For example, researchers studying how mating behavior affects neurobiology may need to examine the brains of their study organisms (e.g., Huang & Hessler 2008).

Sources of ethical standards

Most nations have set guidelines for the use of animals in research and teaching. In the United States, a variety of federal agencies and organizations have formulated specific guidelines that scientists must follow to maintain their funding. These include the National Institutes of Health Office of Animal Care and Use (NIH-OACU), the Department of Agriculture Animal and Plant Health Inspection Service (USDA-APHIS), the Association for the Assessment and Accreditation of Laboratory Animal Care (AAALAC), and the American Association for Laboratory Animal Science (AALAS). In Canada, the Canadian Council on Animal Care sets these standards, while in Europe they come from the Council of Europe Conventions on the Protection of Animals and the European Biomedical Research Association.

Scientific societies also have animal care guidelines. These include the guidelines established by the Animal Behavior Society (2006), the American Society of Mammalogists, the American Fisheries Society, the Ornithological Council, and the American Society of Ichthyologists and Herpetologists. In addition, the

Figure 2.11. Animal care. A zoo veterinarian examines a sea lion.

National Research Council's (NRC) Institute of Laboratory Animal Research Commission on Life Sciences publishes a free online book on animal care (Institute for Laboratory Animal Research 1996). In Appendix 3, you'll find further information about the network of agencies that provide guidelines for researchers.

The three Rs

Although there are differences among guidelines in different countries, most follow the 3 Rs: replacement, reduction, and refinement. **Replacement** means using computer modeling, videotapes, or other approaches in place of actual animals. **Reduction** refers to limiting the number of animals subject to disturbance in research or teaching. Such an effort requires much thought and perhaps pilot studies in order to design a study that uses the fewest possible individuals. Finally, **refinement** involves improving procedures and techniques to minimize pain and stress for animals. Almost all animal care guidelines apply only to vertebrates—with one notable exception. In the United Kingdom, animal care guidelines cover the common octopus, *Octopus vulgaris*.

Governmental, research, and educational facilities that receive U.S. federal funding are required to have their research monitored by an institutional animal care and use committee (IACUC). Before conducting research on animals, each scientist must submit an animal care protocol to the committee, which is composed of institutional scientists, a veterinarian, and members of the outside community. The protocol must describe why a species is appropriate for the research and whether nonanimal alternatives are available. For example, is it possible to use videotaped footage from previous work? Could computer modeling replace live animals? The protocol must also justify the number of individuals required, how they will be procured, and specific conditions for animal care (e.g., housing enclosure, food provided, temperature, and light/dark regime). In addition, the scientist must describe procedures that will be used to minimize stress to the animal. Finally, the scientist must address what will be done with the animals after the study is completed. Can they be used for future research? Will animals be released back into the wild at their capture sites, or will they be euthanized?

The IACUC reviews the research protocol to make sure that all animals will receive appropriate care and that the research is well designed, has not been conducted before, and complies with all animal welfare regulations. The committee also assesses the importance of the research question and the quality of the research before allowing it to be conducted. As you can see, ethical considerations are an important factor in animal behavior research. Next, we examine how scientists communicate their research findings to the scientific community.

replacement An ethical guideline that encourages the use of computer modeling, videotapes, or other approaches in place of actual animals.

reduction An ethical guideline that promotes limiting the number of animals subject to disturbance in research or teaching.

refinement An ethical guideline that involves improving procedures and techniques to minimize pain and stress for animals.

2.4 Scientific knowledge is generated and communicated to the scientific community via peer-reviewed research

Every day we hear about important scientific discoveries in areas ranging from human health to the environment. How do we evaluate all this information? How do we determine what is based on valid research and what is just personal opinion? The ability to evaluate scientific information critically and ascertain its validity is called scientific literacy. However, scientific literacy would not be possible if scientists could not evaluate and communicate their research in the first place. Even the most far-reaching discoveries remain unknown until scientists communicate their findings to the world.

The primary literature

Animal behavior research is conducted by investigators and their graduate and undergraduate students at universities, zoological parks, veterinary schools, government and private research institutions, and conservation organizations. After conducting research, scientists write papers and submit them to scientific journals to report their findings. These journals, known as the **primary literature**, are the primary or original source of scientific information (Figure 2.12). The editors of scientific journals use experts to help decide whether to accept or reject a paper for publication, a process known as **peer review**. These reviewers evaluate the importance of the research question and the validity of the hypotheses, methodology, analyses, and conclusions. They also comment on how clearly and concisely the paper is written; if a paper is poorly written, it will be difficult for the reader to evaluate the science. Why are a great many papers rejected? Reviewers scrutinize manuscripts for accuracy, and papers are rejected for such reasons as unsound methodology, incorrect analyses or statistics, poor experimental design, and faulty or inappropriate logic. As a result, the knowledge communicated in the primary literature has passed a rigorous, objective review.

Each discipline has a variety of specialized journals (see Appendix 1). In addition, if the research presents results of broad scientific appeal, it may appear in one of the prestigious and less specialized scientific journals, such as *Science*, *Nature*, *Proceedings of the National Academy of Sciences*, *Proceedings of the Royal Society of London*, or the *Public Library of Science*.

Scientific papers follow a systematic format, which is summarized in Appendix 2 of this book. For instance, the "Methods" section of a paper contains detailed information about how the research was conducted. Such an explanation is important because other scientists must be able to replicate the work in order to test its validity. The paper may also contain figures and tables that summarize important findings. Most natural scientists follow the style manual of the Council of Science Editors (CSE), which describes how to cite references and handle other matters of style. Psychologists and other social scientists follow the American Psychological Association style format (APA), while medical journals tend to use the American Medical Association style format (AMA). Here we follow a modified version of the author-year system of the CSE style (e.g., Nordell & Valone 1998).

The secondary literature

If the results of scientific research are of interest to the public, they may also be reported in newspapers, magazines, and books, as well as on television and over the Internet. Reports like these are called the **secondary literature**, because the information has been presented before and often does not have the same validity as the primary literature (Toolbox 2.2).

Peer review is a hallmark of most research. It allows experts in a field to critically evaluate potential new knowledge. However, hypothesis testing remains the domain of the sciences and provides a means for the rapid advancement of our knowledge about the natural world.

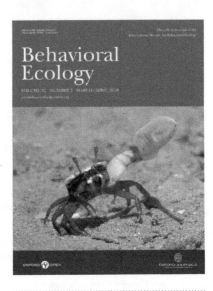

Figure 2.12. Scientific journal. *Behavioral Ecology* is a peer-reviewed journal devoted to animal behavior research.

primary literature The original source of scientific information, typically peer-reviewed scientific journals.

peer review A process in which editors of scientific journals ask experts to review submitted papers to determine whether to accept or reject a paper for publication.

secondary literature A report, which often appears in newspapers, magazines, and books, that summarizes and interprets the primary literature.

Animal sampling techniques

Murray et al. (2009) followed chimpanzees for many full days to collect data on their behavior. More typically, however, researchers collect a partial record of behavioral data using standardized sampling techniques. One common technique is **focal animal sampling**, which provides unbiased data from different individuals (Altmann 1974). In this method, a focal individual is randomly selected and observed for a specified time period, and all pertinent behaviors performed by the individual are recorded. After the time period has expired, a different individual is randomly selected and observed. Focal animal sampling has several advantages:

1. The behavior of different individuals is sampled. If individuals have unique markings or morphological aspects, the researcher can record the behavior of known individuals and identify the total number of unique individuals sampled. If it is difficult to identify individuals, the researcher can move to a different location before selecting a new individual. The researcher can also simply select an individual that is some distance away from the last focal animal observed. This ensures that a number of different individuals are sampled.

2. Because random individuals are selected, instead of individuals that exhibit either unusually high or low levels of a particular behavior, sampling is unbiased.

3. Because data are collected over a recorded amount of time, researchers can easily calculate the frequency of a behavior and the rate at which the behavior was displayed.

4. This technique requires little more than being able to keep track of time and observe and record behavior. Depending on the situation, these tasks might best be accomplished by a pair of researchers: a recorder, who keeps track of time and records the behavior, and an observer, who watches the animal and describes the behaviors displayed.

Another common sampling technique is known as **instantaneous** or **scan sampling**. Here, data are collected from individuals at regular time intervals. For instance, you might be interested in the frequency with which individuals in a group scan their environment (i.e., raise their head to observe their surroundings). You could use scan sampling at 30-second intervals to record the number of group members whose head is raised at a given moment. Over a sufficiently long observation period, these data would provide information on the percentage of time individuals spend scanning the environment. Altmann (1974) provides details of these and other sampling methods.

Scientific literacy

When we hear or read about a scientific discovery, we tend to assume that the information is an accurate and unbiased representation of the original finding. In order to evaluate the accuracy of the report, it is important to be able to refer to the primary literature. How can you determine whether the information you are reading is peer reviewed (from the primary literature), an interpretation of the research (from the secondary literature), or simply someone's personal opinion? Answering this question is an important part of scientific literacy and is critical to being an informed citizen. This information can affect our health, the environment, and how we vote. Here are some of the key questions to ask.

1. **What is the name of the publication, and is it endorsed by a scientific society?**
   Most disciplines have scientific societies whose members are experts in that field. Many scientific societies publish a primary literature publication of research in their discipline.

2. **Who are the authors?**
   What are the researchers' institutional affiliations? Most primary research investigators will be affiliated with a college or university, medical/veterinary/research center, or zoological park or conservation organization. If the scientists are affiliated with an institute, it is important to identify who might be funding their research to determine if there are any conflicts of interest.

3. **Is the article written in the standard scientific format, with an introduction, a methods section, a results section, a discussion or conclusion, and literature cited or references list?**
   Most primary research articles are written using this standard format. Not all journals require the term "introduction" as an actual heading, but they do usually require it as a portion of the paper. Secondary sources will not use this format.

4. **Does the article cite the primary literature in the references?**
   Many secondary literature publications do not cite references, whereas almost all primary literature articles do. Usually the publication will use the author-date system (author's last name, year of publication) within the text.

5. **Does the publication indicate that all articles are peer reviewed?**
   You can go to the website of the publication and see if there are instructions to authors about how the submitted paper will be evaluated. If there is no information about peer review, the publication is probably secondary literature. Does the information on the journal website indicate that it publishes primary research in the discipline?

Chapter Summary and Beyond

Animal behavior researchers generate and test hypotheses. Hypotheses can be developed through observations or through the use of analytical models. To test hypotheses, researchers commonly use observational, experimental, and comparative methods. The observational method is least invasive to the animal and provides data both to describe behavior and to test hypotheses. In the experimental method, researchers manipulate variables to determine their effect on behavior, providing insight into how specific factors affect behavior. The comparative method examines similarities and differences among closely related species to infer patterns of evolutionary change in a trait. The Association of Zoos and Aquariums, in collaboration with the Wildlife Conservation Society, has produced a DVD (*Methods for Animal Behavior Research*) that illustrates research methods for studying animal behavior.

In all research that involves animals, it is imperative to minimize individuals' stress and suffering. Research on animals is monitored and must be approved by institutional (and sometimes federal) animal care and use committees. The three Rs—replacement, reduction, and refinement—are often used as guidelines. Researchers must consider the possibility of replacing animals with computer models, reducing the number of animals required, and refining their techniques to minimize stress and disturbance to the animals used. For a more complete treatment of this subject, see Dawkins (2008) and Allen and Bekoff (2007).

Scientific literacy is the ability to evaluate scientific information critically. Scientific literacy includes understanding how hypothesis testing drives the scientific process, as well as peer review, in which scientists determine whether research is clear, valid, and important enough to be published in the primary literature. Laine and Mulrow (2003) describe the value of peer review and provide insights into how the process works. Papers in scientific journals follow a consistent, systematic format. The public often reads about these research studies in the secondary literature.

CHAPTER QUESTIONS

1. A researcher is interested in studying food color preferences of cardinals, *Cardinalis cardinalis* (a common songbird). She offers a population of cardinals an equal number of small black seeds and large yellow seeds and finds that the birds eat significantly more of the large yellow seeds. From this experiment, the researcher concludes that cardinals prefer to eat yellow seeds rather than black seeds. Critique this experiment. Can you improve on its design?

2. Select three papers about animal behavior from the primary literature. Determine whether the researchers used observational, comparative, or experimental methods in the study. Does the research address proximate or ultimate questions about behavior? Briefly justify your answers.

3. The United Kingdom has added an invertebrate, the common octopus, to its list of animals included in the animal care guidelines. Why are invertebrates in general not included in animal care guidelines? Why do you think the British government included this particular invertebrate in its animal care guidelines? The following references might be useful in your analysis: Boal et al. (2000), Fiorito and Scotto (1992), Kuba et al. (2006), and Mather (2008).

4. In the phylogeny of felids, identify all sister species.

5. Imagine you are conducting an observational study examining the feeding times of individual mallard ducks in small flocks. You want to use focal animal sampling to determine if there are differences in male and female feeding times. How would you randomly select individuals from flocks?

6. Find an article on animal behavior from the primary literature and another from the secondary literature. Describe how you determined that one is an example of the primary literature and the other an example of the secondary literature.

Chapter 3

Evolution and the Study of Animal Behavior

We enjoy watching birds at the feeders in our backyard. Finches, sparrows, chickadees, cardinals, and jays are regular visitors (Figure 3.1). Every so often, so too are Cooper's hawks (*Accipiter cooperii*), predators of small birds. One day, we were watching a group of finches at the feeder when all of a sudden a Cooper's hawk appeared from over the house and attacked. The finches scattered, fleeing quickly to dense shrubs. Three escaped, but one ended up in the talons of the hawk. Such scenarios play out every day in nature—some individuals survive, while others do not. The same variation occurs during mating season: some individuals produce many offspring, while others produce few or none. Variation in survivorship and reproduction is the raw material for adaptive evolution.

Evolutionary biologist Theodosius Dobzhansky once said, "Nothing in biology makes sense except in the light of evolution" (1973). This is because evolution is the foundation of our understanding of the natural world, and animal behavior is one important aspect of that world. In this chapter, we examine the process of evolution by natural selection. We start by describing natural selection in some detail and examine the conditions required for evolution to occur. Although natural selection

Figure 3.1. Birds at a feeder. Birds feeding are at risk of predation.

natural selection The process of differential reproduction and survivorship among individuals within a population.

heritable The characteristic of a trait that can be passed from parents to their offspring because it is genetically based.

evolution Changes in allele frequency or trait values in a population over time.

fitness The relative survivorship and reproductive success (ability to produce viable offspring) of an individual.

Figure 3.2. Dog breeds. There is tremendous variation in body size and morphology in these breeds. (a) Boston terrier; (b) Labrador retriever; (c) greyhounds.

results in changes in populations, it acts on individuals that collectively make up those populations, rather than on the populations themselves. Next we consider the evolution of cooperation and examine how selection can promote behaviors that benefit others. Finally we introduce the concept of sexual selection, a form of natural selection that acts on reproduction. Sexual selection results in many of the morphological and behavioral differences that often exist between the sexes within a species.

3.1 Evolution by natural selection favors behavioral adaptations that enhance fitness

If you have ever visited a dog park, you've likely seen a variety of breeds, such as Boston terriers, Labrador retrievers, and greyhounds (Figure 3.2). Each of these breeds belongs to the same species, *Canis lupus familiaris*, but each breed possesses unique trait features with respect to body size and shape, tail length, head shape, coat texture, and color. The American Kennel Club recognizes over 200 breeds of dogs today. Why are there so many, and where did they come from?

All dog breeds have descended from gray wolf (*Canis lupus*) ancestors as a result of human manipulation of the breeding of individuals over the past 14,000 years (Akey et al. 2010). Over many generations, human breeders have selected individuals that possess certain traits for a particular breed, called the breed standard. Many breeds were created for specific tasks: terriers to hunt vermin, retrievers to find and retrieve hunters' birds, and greyhounds to hunt swift prey. Breeders allowed reproduction only among individuals that possessed certain traits. Differences in breed standards over time eventually led to ever-greater differences among the dog breeds we see today. This process is known as artificial selection because it is done "artificially"—that is, by humans.

Charles Darwin's great insight was that a similar process also occurs naturally, a process that he called natural selection. **Natural selection** is a process in which some individuals in a population survive and reproduce, while others do not. It is the mechanism that results in adaptive evolution. Darwin was not the first to suggest that species evolve, but he (along with Alfred Russell Wallace) was the first to describe the plausible mechanism, natural selection, by which evolution can occur.

Natural selection occurs because there is variation in traits among individuals in a population, and some traits provide individuals with greater reproductive success. When these traits are **heritable**, they are passed genetically from parents to offspring. Natural selection can result in changes in allele frequencies in a population over time—a process that we recognize as **evolution**.

In his book *On the Origin of Species* (1859), Darwin articulated three conditions required for evolution by natural selection:

1. Variation exists among individuals in a population in the traits they possess.
2. Individuals' different traits are, at least in part, heritable. Traits can be passed from parents to their offspring so that offspring resemble their parents in the traits they possess.
3. Traits confer differences in survivorship and reproduction, a measure we call **fitness**: individuals with certain traits will have higher fitness, while

those with other traits will have lower fitness relative to one another. Therefore, the fitness of individuals is not random; it is based on the traits they possess.

Today, we know that gene alleles are the basis of most phenotypic traits (Toolbox 3.1). Random mutations introduce new alleles into populations. Natural selection acts on heritable variation among individuals and can result in changes in allele frequencies and associated trait values in a population. Traits that confer high fitness increase over time, while those that confer low fitness decline, resulting in evolutionary changes.

When traits are heritable, parents will tend to pass on the traits they possess to their offspring via their shared alleles. Many behavioral traits that have been studied are in fact heritable, including mating behavior, feeding behavior, overall activity level, and aggression (Stirling, Réale, & Roff 2002). In the next section, we examine how researchers study the heritability of behavioral traits.

Measures of heritability

Traits must be heritable for natural selection to act on them. How do we quantify the heritability of a trait? Two methods are commonly used:

1. **Parent-offspring regression** analysis examines the similarity between parents and their offspring in terms of the traits they possess. If a trait is heritable, then the trait values of offspring should be similar to the trait values of their parents: that is, there should be a positive relationship between offspring and parent trait values. In this method, offspring trait values are plotted against parent trait values, and the slope of the resulting regression indicates the heritability of the trait.

2. In the **selection experiment** method, different groups of individuals are subjected to differential selection on the trait in question. If artificial selection acting on a trait results in changes in the average trait value in subsequent generations, then the trait is heritable.

Let's examine one case study in which both methods were used to determine the heritability of a behavioral trait—exploratory behavior.

Great tit exploratory behavior

Featured Research

Niels Dingemanse and his colleagues examined the heritability of exploratory behavior in free-living great tits (*Parus major*) (Dingemanse et al. 2002). The great tit is a small, colorful passerine bird found throughout much of Europe and Asia. It is a common inhabitant of woodlands, parks, and gardens, where it feeds on insects and seeds. It nests in tree cavities as well as nest boxes. Because of its widespread distribution, abundance, and use of artificial nesting sites, the great tit has become a favorite study subject for behaviorists.

Previous work indicated that individuals exhibit differences in their exploratory behavior when placed in novel environments: some actively explore their new environment quickly (bold individuals), while others are more reticent and slower to explore (shy individuals) (Verbeek, Drent, & Wiepkema 1994; Drent & Marchetti 1999). Are these behavioral traits heritable? Dingemanse and colleagues used birds from two populations in the Netherlands as part of a long-term research project. The birds bred in artificial nest boxes, which made it easier to capture them and determine relatedness (e.g., mother–offspring relationships). Birds were captured at the nest, uniquely color-banded, and, as adults, placed in

parent-offspring regression A statistical technique used to examine the similarity between parents and their offspring in terms of the traits they possess.

selection experiment An experiment in which different groups of individuals are subjected to differential selection on a trait.

SCIENTIFIC PROCESS 3.1

Heritability of great tit exploratory behavior

RESEARCH QUESTION: *Is exploratory behavior in great tits a heritable trait?*

Hypothesis: Exploratory behavior has a genetic component.

Prediction: There will be a positive correlation between a parent's exploratory behavior and that of its offspring.

Methods: The researchers:

- Placed individuals in an aviary (4.0 m × 2.4 m × 2.3 m) that contained five artificial wooden trees
- Recorded the number of flights and hops in the aviary in the first two minutes as an index of exploratory behavior

Figure 1. Test arena. Crosses indicate position of artificial trees. (*Source:* Dingemanse et al. 2002)

Results:

- The parent-offspring regression on exploratory behavior was positive.

Figure 2. Parent-offspring regression. There was a positive correlation between parent and offspring (mean ± SE) behavior. Size of circles indicates number of individuals. (*Source:* Dingemanse et al. 2002)

Conclusions: Exploratory behavior has a genetic component and is a heritable trait.

EVALUATE

Can exploratory behavior evolve in a population of great tits? Draw a graph of the positive results of a selection experiment for bold and shy behavior.

an aviary that contained novel objects—five artificial wooden "trees." The research team recorded the number of flights and hops that birds took in the aviary in the first two minutes as an index of exploratory behavior. Individuals were then released back into the wild (Scientific Process 3.1).

There was a significant positive correlation between a mother's exploratory score and that of her offspring, demonstrating heritable differences among individuals in this behavior.

Next, the research team conducted a selection experiment for fast and slow exploratory behavior by separating adults into two groups. One group contained those birds with the lowest exploratory behavior scores, while the other contained those birds with the highest scores. Subsequent juveniles from these groups were chosen as the initial breeding generation for two selection lines. In each selection line, nine pairs of birds were used for breeding, and this process was repeated for four generations. Researchers tested individuals from each generation to obtain their exploratory scores. The nine males and females with the highest (fast line) and lowest (slow line) scores of each generation were then selected for breeding.

The research team found strong changes in the exploratory behavior of the two lines in the selection experiment over four generations. By the fourth generation, the average exploratory score for individuals in the fast line was over four times higher than that for individuals in the slow line (Figure 3.3) (Drent, van Oers, & van Noordwijk 2003). This result indicates that exploratory behavior is heritable.

Together, the results of these two experiments conclusively demonstrate that exploratory behavior in great tits is heritable. This conclusion was based on two pieces of evidence: (1) offspring resemble their parents in this behavior, and (2) artificial selection on exploratory behavior produced significant differences in the two artificially selected lines.

Variation within a population

Through the process of natural selection, populations evolve. Such evolution requires variation among individuals in the traits they possess. Why do individuals in populations vary in behavior?

First, as we just saw, individuals differ in their genetic composition. Each generation introduces new variation into populations through gene recombination, the immigration of new alleles into a population, and mutations (Hartl 2000). Because individuals in any population differ genetically, they will also tend to vary in their behavior. Furthermore, changes in environmental conditions can change the fitness of different traits and so maintain much variation in the frequencies of different alleles.

Second, many behaviors develop as a consequence of both genetic and environmental effects. Thus, even close relatives (with similar genes) often exhibit very different behavior as adults when they are exposed to different environmental conditions as juveniles. We will examine this issue in more detail in Chapter 4.

Third, many complex behaviors require learning and so are modified with experience. Because individuals will differ in experience over the course of a lifetime, we will observe differences in their behavior as well. For example, bees need to learn how best to extract nectar and pollen from flowers and then transport them efficiently back to their colony. Dukas (2008) showed that the feeding performance of honeybees (*Apis mellifera*)—that is, their rate of food delivery to the colony—increases dramatically over the first four days of foraging activity as a result of trial-and-error learning. An individual bee delivers food to the colony on its fourth day of activity at a rate almost four times higher than its rate on its first day (Figure 3.4).

Fourth, there might be little or no variation in fitness over a wide range of behaviors. One example is dispersal behavior. Dispersal is the process of moving away from the natal area, or place of birth, to find an adult breeding area or territory. For many species in many

Figure 3.3. Selection experiment. The mean (+ SE) exploratory score of individuals in the artificially selected lines for fast (orange) and slow (blue) exploration scores. (*Source:* Drent, van Oers, & van Noordwijk 2003)

Figure 3.4. Honeybee food delivery. The mean (± SE) rate of food delivery for bees with different experience. Each day, honeybees' rate of food delivery increased, indicating that honeybees gain experience through trial-and-error learning. Inset: honeybee. (*Source:* Dukas 2008)

a

b

Figure 3.5. Owl dispersal. (a) Screech owl. (b) Each line represents the dispersal direction and movement distance of one individual. Individuals (letters) dispersed in all directions. (*Source:* Belthoff & Ritchison 1989)

Featured Research

frequency-dependent selection An evolutionary process in which the fitness of a trait is related to its frequency in a population.

phenotype (P) The observable traits of an individual.

habitats, individuals will experience the same fitness whether they disperse north, south, east, or west. When this is true, we expect to see much variation in dispersal direction within a population. For example, James Belthoff and Gary Ritchison examined the dispersal direction of Eastern screech owls (*Otus asio*) (Figure 3.5a) (Belthoff & Ritchison 1989). They fitted nestlings with radio transmitters and located individuals several times per week as the birds dispersed from their natal territory. The researchers found that there was no particular pattern with respect to dispersal direction, which we would expect if characteristics that determine territory quality and fitness are not associated with compass direction relative to the natal location (Figure 3.5b).

Fifth, individuals in all populations typically differ in size, nutritional status, health, and other traits. These differences can lead to significant variation in behaviors. For instance, many male birds sing complex songs. While song production has a strong genetic component, it also requires learning and can be physically demanding. Thus, in several species, vocal performance increases with age: older males that are in better condition produce different songs from younger males (e.g., Ballantine 2009).

Finally, the fitness of a trait (behavior) may be related to its frequency. In **frequency-dependent selection**, the fitness of a trait depends on its frequency in a population relative to other **phenotypes** (characteristics of an organism). In positive frequency-dependent selection, the fitness of a phenotype increases as it becomes more common (Figure 3.6a); in negative frequency-dependent selection, fitness declines as a phenotype increases in frequency (Figure 3.6b). Negative frequency-dependent selection can maintain different behaviors in a population. For example, some male fish defend territories to attract and mate with females, while other males hide at the edge of these territories and attempt to fertilize the eggs of females attracted to the territory holder. Such "sneaker" male behavior is frequency dependent: it results in high fitness when most males in a population defend territories (a scenario that produces many opportunities to fertilize eggs) but much lower fitness when few males defend territories (producing fewer opportunities for reproduction). As another example, let's examine a research study that investigated frequency-dependent selection in relation to bird predation.

Frequency-dependent selection

Benjamin Fitzpatrick and two high school students, Kim Shook and Reuben Izally, examined frequency-dependent selection in birds foraging on salamanders (Fitzpatrick, Shook, & Izally 2009). The researchers noted that numerous species of small, slender, cryptic salamanders that live on the forest floor throughout North America all have a similar polymorphism. Within each species, some individuals have a dorsal stripe, while others do not. Why might all these different species have a similar polymorphism?

The research team investigated whether frequency-dependent selection as a result of bird predation maintains this polymorphism. They used a mold to create model salamanders out of clay. The models either did or did not possess an ocher-colored dorsal stripe (Figure 3.7). The team then attached a food reward (half a peanut) to the underside of each model using edible library paste and put the models on trays filled with leaf litter. The trays and models were placed in an open area next to a woodlot.

For the first six days, the ratio of striped to unstriped salamander models was 5:45, followed by one day of equal numbers of both morphs. For the next six days, the ratio was reversed (45:5), followed by one final day of equal abundances. Each day, the researchers quantified survival by counting the number of model salamanders that still had their food reward at the end of the day.

The team observed several blue jays (*Cyanocitta cristata*) foraging on the models, so they could confirm that the models were preyed upon. Fitzpatrick and colleagues found that when a morph occurred at low relative frequency, it had higher relative survivorship (Figure 3.8), and vice versa. The rare form always had a survival advantage, whether striped or unstriped. Therefore, frequency did affect survivorship and was negatively frequency dependent. This may explain how this morphological polymorphism is maintained within a species.

As we can see from these examples, a variety of complex factors contribute to variation in behavior within a population. Indeed, evolution by natural selection requires such variation.

Fitness and adaptation

As we have just seen, there is variation among individuals in a population in the traits they possess. Those traits might be the particular morphology, physiology, or behavior of an individual. In general, some traits confer higher fitness (survivorship and reproduction) than others. Traits that result from natural selection and have been selected for their current function are called **adaptations**. Individuals with these traits will tend to survive better and leave more offspring than individuals whose traits do not confer a selective advantage (i.e., yield lower fitness), and so adaptations are identified relative to existing traits in a population (Reeve & Sherman 1993).

The most direct measure of the fitness of an individual is the number of progeny that go on to reproduce that it produces over its lifetime. However, few studies, particularly those that focus on behavior, measure fitness in this way. For one thing, this approach can be both logistically difficult and time consuming. Instead, most studies use indirect measures of fitness. Behavioral researchers often estimate fitness by quantifying parameters such as survivorship, number of mates, body size, growth rate, and feeding efficiency. Such indirect measures are typically positively correlated with more direct fitness measurements (e.g., Ritchie 1990; Taylor et al. 2008).

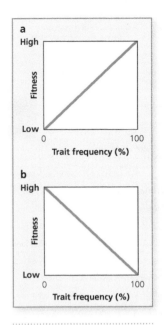

Figure 3.6. Frequency-dependent selection. (a) In positive frequency-dependent selection, the fitness of a trait increases as it becomes more common in a population. (b) In negative frequency-dependent selection, the fitness of a trait declines as it becomes more common in a population.

adaptation A trait that enhances fitness (survivorship and reproduction). Also, an evolutionary process that results in a population of individuals with traits best suited to the current environment.

Figure 3.7. Salamander models. A striped and an unstriped model used in the experiment.

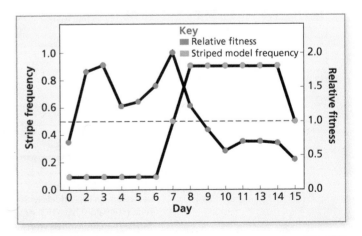

Figure 3.8. Frequency and fitness. The daily frequency of striped models (orange) and their relative fitness (blue). (*Source:* Fitzpatrick, Shook, & Izally 2009)

In this section, we've seen how evolution by natural selection favors behavioral adaptations that enhance fitness. Next, we examine in more detail how traits in populations change over time.

3.2 Modes of natural selection describe population changes

directional selection A situation in which individuals with one extreme (highest or lowest) trait values in a population possess the highest fitness.

disruptive selection A situation in which individuals with either of two extreme (highest and lowest) trait values in a population possess the highest fitness.

stabilizing selection A situation in which those individuals in a population with intermediate trait values have the highest fitness in a particular environment.

Natural selection can cause a change in the frequency of traits in a population over time and so provides an explanation for evolutionary changes. Evolutionary biologists have identified three modes of natural selection, which describe how populations change as a result of the relative fitness values of different trait values. To illustrate this process, consider a population of snails in which individuals exhibit continuous variation in color, ranging from white to pink to red. **Directional selection** occurs when individuals with one extreme trait value possess the highest fitness (Figure 3.9). In this case, let's say red individuals have the highest fitness in a particular environment; as a result, the frequency of red individuals in the population will increase. **Disruptive selection** occurs when individuals with either of two extreme trait values have the highest fitness (Figure 3.10). In this case, red and white individuals both have higher fitness than pink individuals in the same environment. Hence, the frequency of red and white individuals will increase, while the frequency of pink individuals will decline. Last, **stabilizing selection** occurs when individuals with intermediate trait values have the highest fitness in a particular environment. In this case, pink individuals have higher fitness than red or white individuals (Figure 3.11). When

Figure 3.9. Directional selection. (a) Individuals with high trait values have higher fitness than individuals with low trait values. (b) The dashed line is the population before selection. The solid line is the population after selection.

Figure 3.10. Disruptive selection. (a) Individuals with low and high trait values have the highest fitness. (b) The dashed line is the population before selection. The solid line is the population after selection.

Figure 3.11. Stabilizing selection. (a) Individuals with intermediate trait values have the highest fitness. (b) The dashed line is the population before selection. The solid line is the population after selection.

stabilizing selection is at work, the frequency of pink individuals will increase over time. Let's examine one example of each mode of selection.

Directional selection in juvenile ornate tree lizards

For many animals, the ability to move quickly affects survival: faster individuals can better evade predators and can also capture mobile prey more effectively. Therefore, we might expect directional selection to act on the speed of movement. Donald Miles examined this idea by studying selection on locomotor performance in juvenile ornate tree lizards (*Urosaurus ornatus*) (Miles 2004). Ornate tree lizards are found in a variety of habitats in southwestern North America (Figure 3.12). These small lizards (less than 6 cm in snout-vent length) feed on a diverse array of insects and are preyed upon by larger lizards, snakes, and birds.

Miles collected 45 juveniles from one population near Tucson, Arizona, and quantified their locomotor performance in the laboratory. All individuals were tested for two days using a 2 m long raceway with a sand floor. Using gentle prodding, individuals were induced to run down the length of the track. Eight infrared photocells, spaced at 25 cm intervals, recorded their movement and allowed the calculation of mean velocity (average speed over 2 m) and burst velocity (fastest speed over a 25 cm interval). Miles repeated this test eight times for each individual, giving each lizard one hour to recover between trials. He also recorded the body size and mass of each individual, as well as its stride length, which was measured by the distance between successive footfalls in the sand. All individuals were uniquely marked and later released back into the wild at their capture site. Miles then recorded the survival of these individuals over the next six months by visiting the site repeatedly and recapturing surviving individuals. He used survivorship over this time period as a measure of fitness.

Figure 3.12. Ornate tree lizard. These lizards are common in southwestern North America.

Miles found that larger, heavier individuals had high survivorship, as did those with higher mean velocity, higher burst velocity, and longer stride length (Table 3.1). Body size and speed are often correlated, with larger individuals tending to move faster, so which trait was more important in this case? Miles used a regression analysis to determine that locomotor performance was more important than body length in affecting survival. Locomotor performance is a function of stride length (Bonine & Garland 1999). He concluded that in this population, there was strong directional selection on limb length in juveniles: longer limbs produce longer strides that allow individuals to achieve greater velocities, which in turn enhances their survivorship.

TABLE 3.1 Lizard traits. Mean ($\pm$ SE) trait differences between survivors and nonsurvivors (*Source:* Miles 2004).

TRAIT	SURVIVORS	NONSURVIVORS
Snout-to-vent length (mm)	36.6 $\pm$ 3.5	33.6 $\pm$ 4.2
Mass (g)	1.36 $\pm$ 0.43	1.13 $\pm$ 0.46
Initial velocity (m/s)	0.36 $\pm$ 0.16	0.35 $\pm$ 0.15
Burst velocity (m/s)	0.62 $\pm$ 0.18	0.48 $\pm$ 0.14
Mean velocity (m/s)	0.39 $\pm$ 0.14	0.29 $\pm$ 0.11
Stride length (mm)	89.4 $\pm$ 14.6	71.8 $\pm$ 13.1

Disruptive selection in spadefoot toad tadpoles

In disruptive selection, individuals with intermediate phenotypes have lower fitness than individuals with either extreme. This situation can occur when individuals in a population specialize on different resource types and there is competition among individuals for those resources, a situation known as intraspecific competition. For example, imagine a population of individuals that feed on seeds that range in size from small to large. Assume that individuals in a population differ in the type of seeds they consume: one-third eat only small seeds, one-third specialize on intermediate seeds, and one-third consume only large seeds. If intermediate seeds are rare in the environment, then individuals that specialize on intermediate seeds will have the lowest fitness as a result of increased competition for food. Individuals that eat the extremes (either small or large seeds) will have higher fitness and should increase in the population. David Pfennig and colleagues, who have studied disruptive selection in New Mexican spadefoot toad tadpoles (*Spea multiplicata*), examined a similar scenario in several experiments (Pfennig & Pfennig 2005; Pfennig, Rice, & Martin 2007; Martin & Pfennig 2009).

Spadefoot toads are small amphibians (less than 6 cm in length) that live in a variety of habitats in North America. In arid regions of southwestern North America, they breed in temporary rain-filled ponds created by intense summer rainstorms. Once a pond fills with water, eggs hatch and the emerging tadpoles feed in the ponds while they grow and develop over a span of three weeks. Tadpoles consume both detritus (decomposing organic material) and small invertebrates that are typically found in such ponds. Within a single pond population, individuals exhibit a wide range of variation in feeding morphology and behavior. Some individuals (omnivore morphs) feed mainly on detritus on the pond bottom; they have round bodies, smooth mouthparts, and small jaw muscles (Figure 3.13). Other individuals (carnivore morphs) specialize on invertebrates; they have narrower elongated bodies, notched mouthparts, and large jaw muscles. A third group (intermediate morphs) consume both detritus and invertebrates; they possess morphology that is intermediate between that of the two specialists.

Pfennig and colleagues hypothesized that disruptive selection may be operating on tadpole morphology and feeding behavior. Previous work had demonstrated that tadpoles compete for food and that each morph has the highest fitness when it is rare, because rare individuals experience reduced competition (Pfennig 1992). Specialized morphs should experience lower competition, because they compete for food with only a subset of the population—their own morphs and intermediate morphs, but not other specialist morph individuals. In contrast, intermediate morph individuals compete with all individuals in the population for food.

To evaluate their hypothesis, Pfennig's research team tested two predictions:

Figure 3.13. Spadefoot toad tadpoles. Omnivore morph (top) and carnivore morph (bottom).

Prediction 1: Omnivore morphs would feed more efficiently than intermediate morphs on detritus; carnivore morphs would feed more efficiently than intermediate morphs on invertebrates.

Prediction 2: Omnivore and carnivore morphs would have higher fitness than intermediate morphs.

To test the first prediction, the researchers examined the feeding efficiency of tadpoles forced to consume only one food type. In one experiment, they placed lone individuals in small wading pools within cages and allowed them to feed on detritus for eight days. The researchers measured the size of each individual (snout-vent length) at the beginning and end of the experiment to determine its growth rate, which was used as a measure of fitness. The morphology of each individual was quantified (smooth or notched mouth and the width of jaw muscles) to provide a continuous morphology index that ranged from "omnivore morph" through "intermediate" to "carnivore morph." In a second experiment, the research team characterized the feeding performance of individuals on invertebrates by placing lone tadpoles in small containers with ten fairy shrimp (*Thamnocephalus* sp.), a common invertebrate food item for tadpoles. The researchers measured the amount of time that individuals required to capture and eat shrimp, an important aspect of feeding efficiency. They measured the morphology of each individual in this experiment to create a morphology index.

To test the second prediction, the research team marked over 500 individuals from one pond 11 days after it had been filled with water and spadefoot toad eggs had hatched. They measured each individual to quantify a morphology index and then categorized individuals into one of three types: omnivore morph, carnivore morph, or intermediate morph. Individuals of each morph type were marked with a different color elastomer tag, which was injected under the skin, and were then returned to the pond to continue their development. Eight days later, the researchers collected 1,500 individuals from the pond (including many marked individuals) and counted the number of each color of morph to obtain an estimate of each morph's survivorship (one measure of fitness). To obtain a second measure of fitness, the research team also examined 301 unmarked tadpoles from the 1,500 collected. For each, they quantified its morphology index and recorded its size (snout-vent length). Body size in developing tadpoles is correlated with fitness: larger individuals have higher survivorship and fitness (Pfennig & Pfennig 2005).

The researchers found that the feeding performance of each specialized morph was higher than that of intermediate morphs, as they had predicted. For individuals feeding on detritus, omnivore morph individuals exhibited higher growth rates than intermediate morphs, and carnivore morphs more efficiently captured invertebrate prey than did intermediate morphs (Figure 3.14). In addition, both measures of fitness indicated that intermediate morphs had lower fitness than either specialized morph. Intermediate morphs also had a lower probability of survival (indicated by a significantly

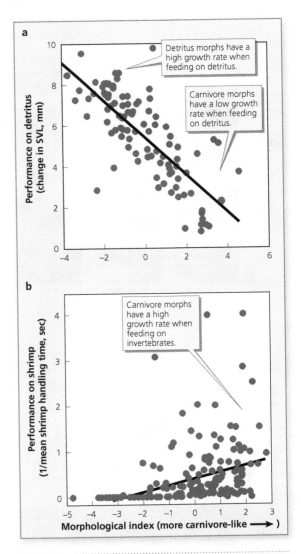

Figure 3.14. Tadpole feeding performance. (a) Growth rate of individuals feeding on detritus as a function of their morphology. Individuals with more carnivore-like morphology showed slower growth. (b) Capture efficiency of individuals feeding on shrimp as a function of their morphology. Individuals with more carnivore-like morphology showed higher capture efficiency. (*Source*: Martin & Pfennig 2009)

lower recapture rate) than both omnivore and carnivore morphs. Among the unmarked individuals, intermediate morphs were smaller than both specialized morphs (Figure 3.15).

The research team concluded that their results supported their hypothesis. Intermediate morph individuals had lower fitness (lower survival and smaller body size) than either omnivore morph or carnivore morph individuals, likely because of competition. Individuals with specialized morphology had fewer competitors and were more efficient feeders than individuals with intermediate morphology.

Stabilizing selection in juvenile convict cichlids

Under stabilizing selection, individuals with intermediate trait values have the highest fitness. For example, consider territorial defense behavior. Many animals defend an area called a territory to protect the food it contains or to attract mates. However, defense of a territory requires time and energy to fight off potential intruders, and these costs increase as territory size increases. Thus, one might anticipate that there would be a territory size of intermediate value that would lead to the highest fitness. Jason Praw and James Grant examined this hypothesis in juvenile convict cichlids (*Amatitlania nigrofasciata*) (Praw & Grant 1999) (Scientific Process 3.2).

Convict cichlids are native to Central America and are a common aquarium fish. Individuals frequently defend small territories and feed on a variety of foods, including invertebrates, small fish, plants, and algae. In order to examine territory defense, Praw and Grant used aquaria that contained a single food patch that fish could defend. Food patches consisted of different numbers of ice cube tray cells. Six patch sizes were used, ranging from one cell up to 121 cells. Commercial fish food pellets were evenly distributed across each patch, which meant that larger patches contained more food than smaller patches.

The researchers allowed a single large, dominant fish to defend a territory around the food patch. These individuals always attempted to defend the entire food patch, which varied in size. Next, the researchers introduced four small, subordinate intruder fish into each aquarium. Intruders attempted to feed from the food patch while the dominant fish defended its territory. For each dominant fish, Praw and Grant recorded the amount of food it ate, the number of intruder chases, and its growth rate over a ten-day period. Because all fish were juveniles (and did not exhibit mating behavior), the researchers focused on growth rate as a measure of fitness. In general, juvenile fish with high growth rates have higher fitness.

The researchers found that fish that defended very small or very large territories had lower growth rates than fish that defended intermediate-sized territories. This occurred because fish that defended very small territories had less food to eat, while those that defended very large territories spent much more time and energy chasing intruders. Individuals that defended intermediate-sized territories had more to eat than fish that

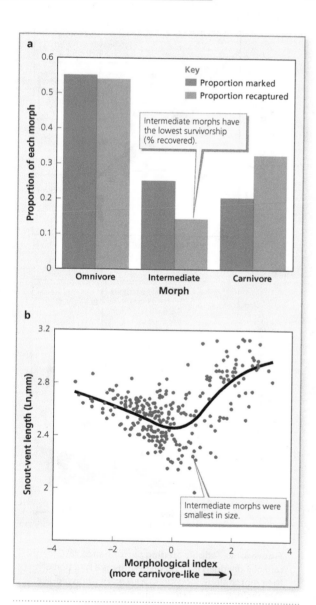

Figure 3.15. Tadpole survivorship and size. (a) The proportion of each morph type recovered in the mark-recapture study. Intermediate morphs had the lowest survival. (b) Unmarked individuals of intermediate morphology were the smallest size, a measure of fitness. (*Source:* Martin & Pfennig 2009)

SCIENTIFIC PROCESS 3.2

Stabilizing selection on territory size in cichlids
RESEARCH QUESTION: *How does territory size in cichlids affect fitness?*

Hypothesis: An intermediate territory size will optimize fitness.

Prediction: Individuals that defend an intermediate-sized territory will have the highest growth rate.

Methods: The researchers:

- Created different-sized food patches
- Placed one patch in an aquarium with a single dominant (focal) fish
- Added four intruders to each aquarium and recorded the territorial defense behavior of the dominant fish and its growth rate

Figure 1. Experimental territories. Each rectangle represents a cell; food is indicated by dots. Patches contain 13–54 mg of food. (*Source:* Praw & Grant 1999)

Results:

- The focal fish defended the entire food patch from intruders.
- Defense behavior increased as territory size increased.
- Fish defending intermediate-sized territories had the highest growth rate.

Figure 2. Patch defense. The chase rate increased with territory size. (*Source:* Praw & Grant 1999)

Figure 3. Growth rate. Fish in intermediate-sized territories had the highest growth rate. (*Source:* Praw & Grant 1999)

Conclusion: Selection will favor individuals that defend intermediate-sized territories, because they will have the highest fitness.

EVALUATE
Explain why the growth rate of fish defending large territories (9 or 11 cells) is lower than the growth rate of fish defending intermediate-sized territories (7 cells).

defended small territories (and so grew faster) but spent less time (and thus energy) chasing intruders than fish with large territories. From this experiment, Praw and Grant concluded that selection would favor individuals that defend territories of intermediate size.

Studying adaptation: the cost-benefit approach

optimal trait value The trait value that confers the highest fitness in a population in a particular environment.

Under all modes of selection, the trait value that confers the highest fitness in a particular environment is called the **optimal trait value**. If the trait under selection is heritable, and if the selection regime remains constant, then populations will evolve toward the optimal trait value (or values) over successive generations. Thus, we see that natural selection is an optimizing process that allows us to understand adaptations by describing the fitness benefits and costs of different phenotypes. This **cost-benefit approach** illustrates a common ultimate explanation method used to study behavioral adaptations: identify fitness costs and benefits of different traits to determine which trait has the highest net benefit.

cost-benefit approach A method used to study behavioral adaptations in which the fitness benefits and costs of different traits are examined to determine which has the highest net benefit (calculated as benefit minus cost).

The simplest cost-benefit approach predicts that a behavior that maximizes fitness, irrespective of how other animals behave, will continue to be passed down from parent to offspring. However, in many situations, the fitness of a behavior will depend on how others behave. Determining the behavior that maximizes fitness in such cases requires the use of game theory, a more sophisticated cost-benefit approach first developed in economics. **Game theory** models find the best solution to an interaction or game between two players (Parker & Hammerstein 1985). They take into account the players, the strategies they adopt, and the payoffs for each one. In economics, the players are people, the strategies are different behaviors, and the payoffs are typically monetary. John Maynard Smith and George Price were the first scientists to apply these economic models to animal behavior (Maynard Smith & Price 1973). In these cases, the players are animals and the strategies are again different behaviors, but the payoffs are described in terms of fitness.

game theory A cost-benefit modeling approach in which an individual's fitness is affected by the behavior of others.

evolutionary stable strategy (ESS) A strategy that, if adopted by a population, cannot be invaded by another strategy because it yields the highest fitness.

The solution to biological games is the **evolutionary stable strategy** (**ESS**); this strategy (a single behavior or multiple behaviors observed in different proportions in the population) cannot be beaten by any alternative strategy, and no player will adopt a different strategy because doing so will result in a lower payoff. Oftentimes, when the fitness of a behavior depends on the behavior of others, fitness will be frequency dependent. Game theory is used to predict the proportion of each behavior that one should observe in a population (Maynard Smith 1982; Riechert & Hammerstein 1983).

We will see the cost-benefit approach, as well as several examples of game theory, used throughout this book. In fact, we have already seen the cost-benefit approach in Chapter 2, when Lima and his colleagues used it as an ultimate explanation for the food-carrying behavior of squirrels that maximizes their fitness.

3.3 Individual and group selection have been used to explain cooperation

individual selection Natural selection acting on individuals.

In the lizard, spadefoot toad, and fish examples described previously, natural selection was assumed to act on individuals. As noted, those individuals that have the highest fitness are more likely to pass their genes and behavior on to the next generation. However, some behavior is not so readily explained by **individual selection**—natural selection at the level of individuals. Many social animals display cooperative

APPLYING THE CONCEPTS 3.1

Do lemmings commit suicide?

Biologists may look skeptically at group selection, but you might get a very different impression from documentaries and television. The idea has even become something of a legend. One common myth is that lemmings commit suicide to prevent overpopulation. Lemmings (Figure 1) are small, herbivorous rodents common in Arctic regions and are one of the subjects of an early Disney nature movie, *White Wilderness*. As the movie describes, approximately every four years, the lemming population increases dramatically. In response, lemmings disperse into new habitats, a behavior that can result in mortality. They are seen swimming across rivers and lakes and even traveling on sea ice floes. In one memorable scene in the movie, individuals appear to jump off a cliff into the Arctic Ocean and swim off to their death. Could such behavior benefit the species as a whole by preventing overpopulation and the exhaustion of resources? If so, is it an example of group selection?

Research indicates that lemming populations do exhibit dramatic population fluctuations in the Arctic and that individuals do disperse during periods of high abundance (Chitty 1996). However, there is no evidence that individuals deliberately commit suicide. Such an explanation, in fact, runs up against problems with the logic of group selection. Individuals that did not cooperate in committing suicide would have

Figure 1. Lemming.

higher fitness than those that did, and "cheating" individuals would quickly come to dominate the population.

In reality, lemmings are simply dispersing to reduce intraspecific competition. Those that disperse onto sea ice floes are not intentionally committing suicide, but rather are moving to find a better habitat. The dramatic "suicide" in the movie was actually staged for the camera (Chitty 1996). ■

behavior, helping others to survive and reproduce. For example, many social insect colonies (e.g., those of bees, termites, and wasps) contain many sterile workers and a single queen who reproduces. The workers defend the colony and deliver food to the queen and her offspring. Why should these individuals forgo their own reproduction and help another reproduce? This question troubled even Darwin. He speculated that selection may act on the entire colony rather than on individuals alone.

Vero C. Wynne-Edwards proposed that cooperation may result from **group selection**, or selection acting on groups (Wynne-Edwards 1962). Wynne-Edwards asked why animals do not overexploit resources. He suggested that they can regulate breeding and population size by defending widely spaced territories or establishing dominance hierarchies in which only the most dominant individuals reproduce—those that successfully establish a territory. If all individuals attempted to maximize their fitness by reproducing, he reasoned, resources would be overexploited, leading to extinction (Applying the Concepts 3.1).

Group selection quickly gained popularity as a way to explain cooperative behavior and is still commonly invoked by nonscientists, but does it explain why individuals cooperate? George Williams said no (Williams 1966). Territoriality does not need to be the result of a group's efforts to limit the use of resources. Individuals may simply compete for available resources. In fact, Williams argued, group selection is based on faulty logic. An individual that did not limit its reproduction (a "cheater") would have a fitness advantage. The offspring of "cheaters" would quickly become a larger and larger portion of the population until none of the individuals exhibited cooperative behavior.

group selection A form of selection that favors particular groups of individuals over other such groups.

direct fitness Genes contributed to the next generation by an individual due to its own reproduction.

indirect fitness The genes contributed to the next generation by an individual as a result of helping non-offspring kin produce additional offspring.

Inclusive fitness The sum of individual's direct and indirect fitness.

kin selection A form of natural selection in which individuals can increase their fitness by helping close relatives, because close relatives share the helper's genes.

multilevel selection A form of natural selection that may result in circumstances when selection on groups is stronger than selection on individuals.

At about the same time, William Hamilton proposed a new explanation for why animals engage in cooperative behavior. Hamilton argued that the total fitness of an individual (its genetic contribution to the next generation) comprised both direct and indirect components. **Direct fitness** is determined by the number of offspring produced by an individual in the absence of help by others, while **indirect fitness** is determined by the offspring of relatives that exist because of the cooperative behavior of the individual. Together, direct and indirect fitness make up **inclusive fitness**. Hamilton's idea has come to be known as **kin selection**, a process in which an individual can increase its total fitness by helping kin, because they share genes from a common ancestor (Hamilton 1964; Maynard Smith 1964). In other words, an individual can increase its genetic contribution to the next generation by helping close relatives that share many alleles through common descent, a situation found in many social insect colonies. Hamilton's insight has had a profound influence on our understanding of behavior, and we examine cooperation and kin selection in much more detail in Chapter 15.

Not everyone was ready to dismiss group selection, however. For one thing, not all cooperating insects display a close genetic relationship among workers, often because two or more queens produce offspring (e.g., Kümmerli & Keller 2007). In response, David Sloan Wilson and others have developed a new model of group selection. They argue that **multilevel selection**, selection on groups, may be stronger than selection on individuals in certain restrictive circumstances: groups must be small, and there must be minimal movement of individuals between groups (e.g., Wilson & Wilson 2007). Empirical support for this model is sparse, however, and so the debate about the importance of multilevel selection for understanding the evolution of cooperation continues (Leigh 2010; Kokko & Heubel 2011).

3.4   Sexual selection is a form of natural selection that focuses on the reproductive fitness of individuals

Natural selection acts on heritable traits that affect survivorship *and* reproduction. **Sexual selection** is a form of natural selection that acts on heritable traits that affect reproduction. Individuals vary in their ability to compete for mates or attract individuals of the opposite sex, which leads to differential reproduction within a sex.

Darwin himself proposed that many traits observed in males and females of a species are due to sexual selection and often result in **sexual dimorphism**, or morphological differences between the sexes. Which traits are involved in sexual selection? Let's consider the case of widowbirds.

sexual selection A form of natural selection that acts on heritable traits that affect reproduction.

sexual dimorphism Morphological differences between the sexes.

Featured Research

Sexual selection in widowbirds

Widowbirds are medium-sized birds found in Africa that often display sexual dimorphism in both color and tail length. Red-collared widowbird (*Euplectes ardens*) females have brown plumage and short tails, while males have black plumage with a bright red collar around their head and extremely long tails (Figure 3.16). In this species, males establish and defend large territories that females visit at the start of the breeding season. During a female visit, the male performs an elaborate courtship flight in which the tail is prominently displayed. Females visit the territories of

Figure 3.16. Red-collared widowbirds. (a) Males have extremely long tails and brighter plumage. (b) Females lack these traits.

several males before selecting one with which to mate, build a nest, and raise offspring. Multiple females can breed in a single male territory.

Sarah Pryke, Staffan Andersson, and Michael Lawes investigated whether sexual selection through female choice could explain the evolution of dramatic coloration and long tails in males (Pryke, Andersson, & Lawes 2001). They examined the traits of the males that females preferred to mate with: Were they those males that had the longest tails or those that had the largest color patches? Prior to the start of the breeding season, the research team captured 43 males from a population in South Africa, placed a unique color band on their leg for identification, measured their tail length and color patch size, and released them. Males were then observed daily both to estimate their territory size and to determine the number of females that nested within each male's territory (his mating success).

The team found that there was variation in male tail length (12–28 cm), color patch size, territory size, and the number of females breeding within a single male's territory (0–9). Pryke and her colleagues determined that male mating success was most strongly and positively correlated with the male's tail length (Figure 3.17) but not with his territory size or color patch.

In this population, females appear to preferentially mate with males possessing the longest tails. Such mate choice by females can help to explain the evolution of sexually dimorphic tail length in this species. Long tails likely are energetically costly (both to produce and in terms of flight costs), and so only those males in the best condition—the highest-quality males in the population—can survive, defend a territory, and display actively to females. If male phenotype (e.g., tail length and overall genetic quality) is heritable, females can enhance the fitness of their offspring by mating with such males. Pryke and colleagues' data, however, cannot explain the evolution of bright color in males by female choice. The researchers suggest that male coloration might play a role in male–male competition for territories, but this hypothesis requires further evaluation.

Sexual selection is an important force that can produce wide variation in reproductive success and fitness among individuals, and affect the evolution of traits. As we will see in Chapter 12, sexual selection is an area of great activity in behavior research.

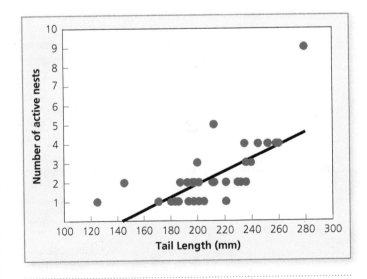

Figure 3.17. Male widowbird mating success. Males with the longest tails had the highest mating success.

TOOLBOX 3.1

Genetics primer

Deoxyribonucleic acid (DNA) condenses to form chromosomes during mitosis and meiosis (Figure 1). Specific stretches of DNA contain genes, which code for polypeptides (e.g., proteins) that have a variety of functions.

Alleles are different versions of a gene (Figure 2), while the locus is the physical location of an allele on a chromosome. Diploid organisms have two copies of each chromosome, and each may have a different allele. If the alleles are the same, then the organisms are homozygous for these alleles; if the alleles are different, then they are heterozygous. During sexual reproduction, each parent contributes only one of its alleles. Therefore, offspring have different allele combinations than their parents.

During meiosis, homologous chromosomes line up in pairs (Figure 3). During this time, they can exchange segments of DNA, a process known as crossing over. This process results in novel combinations of DNA that increase genetic diversity.

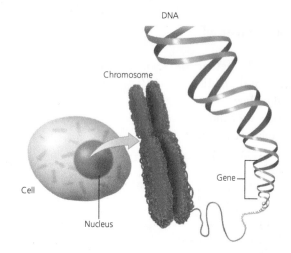

Figure 1. DNA. DNA consists of two helical chains that condense to form chromosomes during mitosis and meiosis. Genes are found in specific stretches of DNA and code for polypeptides.

Genotype:	**PP**	**aa**	**Bb**
	Homozygous for the dominant allele	Homozygous for the recessive allele	Heterozygous

Figure 2. Alleles on chromosomes. The locus (pl., loci) is the physical location of a gene on a chromosome. Alleles at a gene locus can be different versions of the gene.

Figure 3. Recombination. Homologous chromosomes can exchange segments of DNA.

Chapter Summary and Beyond

The behaviors we observe in animals result from evolution by natural selection, which favors adaptations that confer high fitness to individuals. Evolution by natural selection requires that (1) individual differences in traits exist within populations, (2) traits have some heritability, and (3) traits confer differences in survivorship and reproduction in individuals. Populations evolve according to three modes of selection, which describe how trait values differ in terms of fitness. Directional selection results in an increase in one extreme trait value. Disruptive selection leads to an increase in two extreme trait values and a decrease in intermediate values. Stabilizing selection leads to an increase in intermediate trait values over time. However, for a trait to evolve, it must be heritable.

Researchers commonly determine the heritability of traits in two ways. Either they compare trait values of offspring with those of their parents, or they conduct artificial selection experiments to select for a particular trait value. If a trait is heritable, parent–offspring trait value correlations will be positive, and selection experiments will document changes in the trait value over successive generations. Stirling, Réale, and Roff (2002) review studies that indicate that many behaviors are heritable.

Natural selection acts on individuals and causes populations to evolve. Wynne-Edwards (1962) proposed that group selection might be an important evolutionary force to explain the dominance hierarchies and territorial behavior of many species. He envisioned that group selection would prevent the overexploitation of resources. However, his idea requires cooperation by all individuals to limit their reproduction and has been largely discredited. Recent theoretical and empirical work has identified some rare circumstances in which multilevel selection, which includes selection that acts on groups, may be stronger than individual selection. Wilson and Wilson (2007) and Sober and Wilson (2011) provide details about multilevel selection and its relationship to earlier discussions of group selection.

Sexual selection is a form of natural selection that affects differential reproduction among individuals. This important mechanism has led to the evolution of many forms of sexual dimorphism in animals as a result of mating preferences. Clutton-Brock (2007) and Johnstone (2008) provide detailed reviews of sexual selection and mate choice in animals.

CHAPTER QUESTIONS

1. Individuals in a species of moth vary in wing color from white to black, but all individuals prefer to land on the white bark of birch trees. White wing color and the behavioral preference for landing on white birch bark are adaptations, because the wing color of the moth matches the white color of the birch bark and predators have trouble finding white-winged moths on these trees. As a result, white-winged individuals experience lower predation than dark-winged individuals when landing on birch bark. Now consider a population of these moths that live near a power plant that burns coal and emits black soot that coats the bark of birch trees, turning them black. These moths still prefer to land on birch trees, but now white-winged individuals suffer very high predation because the color of their wings contrasts with that of the soot-covered black bark. Dark-winged moths now experience very low predation because they are difficult to see on the dark bark. Is the behavior of landing on birch trees still a behavioral adaptation in this population of moths? How would you test this?

2. A study found tremendous variation in the aggressive behavior of a lobster population off the coast of Maine, with the most aggressive lobsters tending to have the highest reproductive success. The researcher concluded that selection increases the aggressive behavior of individuals in this population. Is this conclusion valid? How might the researcher test this?

3. Imagine you are studying a population of lizards that differ in body mass. Individuals that weigh less than 20 g or more than 30 g have lower fitness than those that weigh 25 g. What mode of selection is operating on the population and what evolutionary change would you predict based on your data? What assumptions did you make to generate your prediction?

4. Consider a population of birds in which males compete with each other to establish territories. Only 50% of adult males successfully establish a territory, and females will only mate with males that possess a territory. In this population, half the adult males do not breed each year. Wynne-Edwards (1962) argued that such territorial behavior and the existence of many nonbreeders could best be explained by group selection. Explain these observations based on individual selection. What predictions follow from the individual selection hypothesis, and what predictions follow from the group selection hypothesis?

5. Describe the similarities and differences between natural selection and sexual selection.

Chapter 4

Behavioral Genetics

Our neighbor's Australian shepherd, Ryker, occasionally escapes from his yard to herd the sheep that live down the road. No one trained Ryker to do this, but he displays herding behavior (Figure 4.1). Our own Labrador retrievers tirelessly retrieve balls, sticks, and toys and are adept in the water. Max and Grace race out each morning to combine retrieving balls with swimming in the pool (Figure 4.2). Why do dog breeds exhibit these differences in their behavior? These animals have a genetic predisposition to exhibit different behaviors—that is, their genes influence their behavior (Spady & Ostrander 2008). Over the past few centuries, Australian shepherds were bred for their herding behavior, while Labradors were bred for retrieving. These differences in behavior among dog breeds came about as a result of the natural variation in early dogs. Breeders could use this natural variation to select dogs with the appropriate traits to breed (artificial selection). Individuals that performed well on specific tasks were selected for future breeding, and because these behaviors have a genetic component, they increased in frequency over time in each breed.

However, genes do not tell the whole story. The environment, too, can strongly affect the behavior of animals. For example, the early social environment of puppies can be a critical factor in their behavioral development. In fact, the American Veterinary Society of Animal Behavior (AVSAB 2008) recommends that owners enroll puppies as early as

Figure 4.1. Herding behavior. Several breeds, such as these sheepdogs, have been bred to herd livestock.

Figure 4.2. Retrieving. Retrievers were bred to retrieve game animals for hunters but, as this one demonstrates, will retrieve almost anything.

eight weeks of age in socialization classes to minimize the risk of behavioral problems such as fear and aggression emerging in adulthood. In addition, although retrievers and sheepdogs naturally exhibit herding and retrieving behaviors, they need to be trained to fully develop these skills.

How, then, do genes and the environment affect animal behavior? Can we ascertain their relative importance in determining variation in behavior among individuals? In this chapter, we see how behavioral geneticists study the genetic and environmental influences on behavior that help to answer these questions. We begin by examining the heritability of behavioral traits.

4.1   Behaviors vary in their heritability

As we discussed in Chapter 3, individuals within and across populations display natural variation in their **phenotype**, or their observable traits, such as behaviors and morphology. This variation is due to differences in their **genotypes** (genetic makeup) and their environment. Behavioral genetics examines how genes and the environment contribute to these differences in behavior by partitioning these effects.

An individual's behavior, or its behavioral phenotypic value (P), is the result of three factors:

1. its genotype (G) at all loci that affect the behavior;
2. the environment (E) it has experienced; and
3. any interactions between them—more formally, gene-environment interactions (G × E).

Given that both genetic and environmental factors affect behavior, how can we assess their relative contributions? One way to do this is by first breaking down the phenotypic variation of a population into its two components:

$$V_P = V_G + V_E$$

where V_P is the variation in the phenotype, V_G is the variation due to genetic effects, and V_E is the variation due to environmental effects. Environmental effects include all of the environmental conditions that individuals experience, as well as their responses to those conditions.

We can now define the **heritability** (H^2) of a trait as the proportion of phenotypic variation in a population that is due to genetic influences:

$$heritability = \frac{V_G}{V_P} = \frac{V_G}{\left(V_G + V_E\right)}$$

This proportion is known as *broad-sense heritability* because it includes all the genetic effects on the phenotype. Multiple genetic factors affect phenotypic variation, and they can be grouped into three types:

phenotype (P) The observable traits of an individual.

genotype (G) All alleles and loci that affect the behavioral phenotype.

heritability The proportion of phenotypic variation in a population that is due to genetic variation.

1. **additive effects (A)**, or the average effect of individual alleles on the phenotype;
2. **dominance effects (D)**, or the interaction between alleles at one locus, in which one allele can mask the expression (phenotype) of the other; and
3. **epistastic effects (epistasis, I)**, or the interaction between genes at different loci, such that one gene modifies the effect of another or acts with it to produce a new phenotype (Conner & Hartl 2004).

We can now define genetic variation as:

$$V_G = V_A + V_D + V_I$$

Both dominance and epistatic effects involve interactions between alleles or genes and are therefore nonadditive. When chromosomes assort independently in meiosis, the effect of both dominance and epistasis on phenotypic variation is influenced by alleles present in individuals in each generation. Additive genetic variation effects, on the other hand, are independent of other genes or the environment. Thus, additive genetic variation is of primary importance for determining how a trait will respond to selection. The proportion of phenotypic variance that is due solely to additive genetic values is narrow-sense heritability (h^2), which is the measure of heritability most commonly reported:

$$h^2 = \frac{V_A}{V_P}$$

Because each parent contributes a single allele per locus to offspring, the phenotypic similarity between offspring and parents depends on the additive effects of alleles. The greater the narrow-sense heritability of a trait, the more likely we are to observe a close resemblance between offspring and their parents in that trait. Therefore, narrow-sense heritability can be determined by examining the similarity of behavior between parents and their offspring.

As we saw in Chapter 3, heritability is commonly determined using parent-offspring regressions, in which the mean trait values of parents are regressed against the mean trait values of their offspring. The slope of the regression can range from zero to one: the higher the slope value, the more offspring resemble their parents. Higher slope values indicate that a greater proportion of the phenotypic variance is additive variance, or variation passed from parents to offspring. Lower slope values indicate that less of the phenotypic variance is due to additive genetic variance and so is not transmitted to offspring. This means that more nonadditive and environmental variance is expressed among offspring.

Not surprisingly, a variety of narrow-sense heritability values have been reported for different behaviors. Examples of low values include those for oviposition behavior in weevils ($h^2 = 0.05$) (Tanaka 2000) and vigilance in marmots ($h^2 = 0.08$) (Blumstein et al. 2010). Examples of higher values include those for defense behavior in snakes ($h^2 = 0.41$) (Garland 1994; Figure 4.3) and courtship behavior in crickets ($h^2 = 0.72$) (Hedrick 1994) (Applying the Concepts 4.1).

In the following sections, we examine the genetic and environmental influences on behavior, including their interaction. We begin with the genetic basis of behavior.

Figure 4.3. Garter snake defensive behavior. In garter snakes, several defensive behaviors, like this open-mouth attack, have high heritabilities.

Dog behavior heritability

In captive breeding programs, individuals are selected as parents for the next generation in order to pass on desired traits. It is crucial, however, to know whether the trait has a heritable component, because if behaviors are affected only by the environment, artificial selection will not be effective. Historically, breeders simply assumed that traits were heritable.

A variety of behavioral tests have been developed to characterize the disposition or temperament of dogs (e.g., Wilsson & Sundgren 1997). For example, behavioral tests are used to assess breeds that will be used as service dogs or in hunting. Individuals must demonstrate the appropriate disposition for the situations they will encounter: service dogs must exhibit high levels of self-confidence (i.e., being nonreactive to novel situations) in a wide variety of environments, while hunting dogs must not react negatively (i.e., fearfully) to gunshots. Silvia Ruefenacht and colleagues used data from 25 years of standardized behavioral field tests on over 3,000 dogs with known pedigrees to determine the heritability of seven different behaviors in German shepherds (Ruefenacht et al. 2002), including self-confidence, reactions to different stimuli, reaction to gunfire, and play behavior. Heritability scores ranged from 0.09 to 0.24, with the highest score being reaction to gunshots. This finding confirms that although selection for specific behavioral traits can be fruitful, the environment also exerts a great deal of influence on behavior. ■

4.2 Behavioral variation is associated with genetic variation

Some of the earliest evidence for the genetic basis of behavior came from captive animals. Early researchers noted that many laboratory strains of mice and rats exhibited consistent differences in behavior (e.g., Yerkes 1913). Because individuals in captivity are usually reared in similar conditions, differences in behavior most likely result from differences in their genotypes (e.g., Plusquellec & Bouissou 2001; Augustsson & Meyerson 2004). Other evidence came from studies of instinct, or innate behaviors—behaviors that are performed the same way each time, are fully expressed the first time they are exhibited, and are present even in individuals raised in isolation. Because all individuals in a species exhibit innate, nonlearned behaviors, these behaviors must have a genetic basis—which also means that they are heritable. Innate behavior includes reflexes, involuntary and often immediate behavioral responses to an external stimulus. One example is the blink reflex: the eyelid of many species, including humans, closes when an object moves quickly toward the eye. This reflex is common in vertebrates, as we learned firsthand when our dog Buck suffered facial nerve paralysis on his right side. The veterinarian confirmed the diagnosis by showing us that Buck's blink response was absent for his right eyelid but fully functioning for the left eyelid.

Examining innate behaviors, Konrad Lorenz and Niko Tinbergen noted that adult graylag geese (*Anser anser*), which lay their eggs in nests on the ground, will extend their neck and use their bill to gently roll displaced eggs back into the nest in a very fixed manner (Lorenz & Tinbergen 1957) (Figure 4.4). Tinbergen, a pioneer of modern animal behavior (see Chapter 1), also examined the innate escape response of newly hatched goslings (Tinbergen 1951). He noted that when a small shadow passed overhead, such as a predator silhouette, the chicks always responded by assuming a characteristic antipredator crouching position and looking up at the sky. In this case, the animals responded with a **fixed action pattern**—a behavior that displays almost no variation and, once started, is not

fixed action pattern A behavior that is invariant, unlearned, and, once initiated, always completed.

Figure 4.4. Graylag geese egg retrieval. Egg retrieval behavior is a fixed action pattern behavior initiated by the sight of an egg outside the nest. (*Source:* Lorenz & Tinbergen 1957)

stopped until completed. These observations led researchers to hypothesize that such behavior must be genetically based.

Behavioral differences between wild-type and mutant-type fruit flies

Featured Research

More direct evidence of a genetic influence on behavior came from the work of John Paul Scott and Margaret Bastock, two researchers who each studied fruit flies (*Drosophila melanogaster*) (Scott 1943; Bastock 1956). In fact, Bastock, a graduate student of Tinbergen, helped open the door to the new field of behavioral genetics.

In her dissertation, Bastock examined behavioral differences between a **wild type**, or the typical form in nature, and a mutant form of *D. melanogaster* called "yellow" because its body is yellow instead of the normal gray (Bastock 1956). This mutant form was rare in nature but arose frequently and reproduced successfully in laboratory stocks. Why aren't yellow forms more common in nature? Bastock wondered whether the gene mutation in yellow flies might induce a behavioral change that resulted in very low reproductive success in the wild. In other words, was genetic variation associated with behavioral variation?

Fruit fly courtship begins with the male orienting toward and then following the female. The male taps her with his foreleg. The foreleg has receptors—molecules on the surface of cells (or the nucleus) that receive chemical information—that identify sex and species. Next, the male begins a wing vibration, a form of courtship song, which is followed by the male licking the female. If the female is receptive, they copulate (Figure 4.5).

To investigate possible links between genetic and behavioral variation, Bastock created different lines of inbred flies whose genomes were very similar except for one thing: the "yellow" gene. She then conducted mating trials using both types of males and wild-type females, and recorded their mating success

wild type The typical form of an organism or gene in nature.

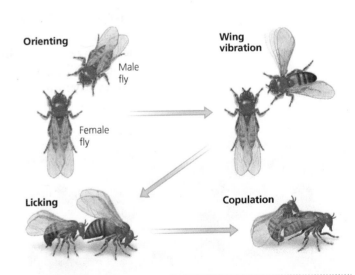

Figure 4.5. Fly courtship. Male fruit fly courtship behavior follows a sequence of events that includes orienting toward the female, vibrating the wings (song production), and licking the female genitalia, which can lead to copulation. (*Source:* Sokolowski 2001)

and courtship behaviors. She found that the wild-type males had overall higher mating success and mated much sooner than the yellow males. In addition, yellow males exhibited significantly less courtship behavior—in particular, wing vibration and licking—than wild-type individuals. Finally, Bastock conducted additional experiments to reject alternate hypotheses that females were responding to differences in the color or differences in the pheromones of the yellow males.

She concluded that important differences in courtship behavior resulted in the observed differences in the mating success of yellow individuals: females mated less with males that produced weak courtship displays. This experiment was one of the first to link variation in genotype with variation in behavioral phenotype. While her work was underappreciated at the time (Cobb 2007), it set the stage for behavioral genetic research of the future. Bastock used the phenotype in order to study genotypes. It took decades, and many technological advances, before researchers could more directly study behavioral genotypes, as we will see next.

Major and minor genes

Understanding the genetic basis of behavior requires unraveling the relationship between specific genes and behavior. Individual genes that are responsible for the majority of phenotypic variation are known as *major genes*. Those that contribute to small amounts of variation are known as *minor genes* (although the combination of many minor genes can have a large influence on a phenotype).

epistasis (l) Interactions between genes at different loci.

pleiotropy A situation in which a single gene affects more than one phenotypic trait.

One explanation for major genes' large influence is **epistasis**: they can affect the expression of many other genes. Another factor is **pleiotropy**: major genes can affect multiple traits, just as Bastock's yellow gene affected both body color and behavior. Because major genes explain much variation in observed behavior, they have been the focus of much work in behavioral genetics. Scientists have adopted several approaches to understanding these relationships: one involves identifying specific gene alleles that directly affect behavior; another involves directly disabling a specific gene to understand how it influences behavior. Let's look at research examples of how each approach has enhanced our understanding of behavioral genetics.

Featured Research

Fire ant genotype and social organization

Solenopsis invicta, the red imported fire ant (Figure 4.6), is native to South America but was accidentally introduced in the 1930s to North America in the cargo of ships sailing from South America. Not only do these ants bite, but they also inflict a painful sting. This exotic species has colonized the southern United States and continues to expand its range (Ascunce et al. 2011). The United States now spends about $6 billion annually on control, medical treatment, and damage to property related to this species.

Fire ants are social (i.e., group-living) insects that live in large colonies and exhibit two distinct kinds of social organization: some colonies possess a single reproductive queen (monogyne colonies), while others contain multiple queens (polygyne colonies). In addition to the queen or queens, each colony has many sterile workers and soldiers, which find food and protect the colony. Polygyne colonies are known to reach extremely high population densities and so have a more severe ecological impact than the monogyne colonies (Resasco et al. 2014). What

Figure 4.6. Fire ant colony. Workers at the colony entrance.

is the proximate explanation for whether a colony is monogyne or polygyne? Is genetic variation associated with variation in social behavior? Kenneth Ross's (1997) work suggests the answer is yes.

Ross collected ants from monogyne and polygyne colonies located in both their introduced range in the United States and their native range in Argentina. He used starch gel electrophoresis analysis to determine that the genotype and allelic frequencies at one locus, *Gp-9*, differed between the two types of colonies. He found that the *Gp-9* genotype had only two alleles, $Gp-9^B$ and $Gp-9^b$. The monogyne colonies all had the $Gp-9^{BB}$ genotype and thus only the $Gp-9^B$ allele, indicating that this allele is fixed in monogyne colonies. The polygyne colonies, however, displayed more variation in their genotype and allelic frequencies. In particular, the reproductive queens were all heterozygotes, $Gp-9^{Bb}$. Thus, in this case, the alleles at one locus appear to regulate the social organization (number of reproductive queens) of a colony.

What does the *Gp-9* gene code for? Genes code for proteins, biochemical compounds that have many functions. Proteins are called the "building blocks" of life because they can be structural components of cells, tissues, and organs. They also function as enzymes by catalyzing chemical reactions, and they are an important component in cell signaling, functioning as neurotransmitters, hormones, or receptors. In this way, proteins play a key role in initiating physiological changes in cells and organ systems that affect the brain and resulting behaviors.

Michael Krieger and Kenneth Ross sequenced the *Gp-9* gene and found that it codes for a pheromone-binding protein (Krieger & Ross 2002). Pheromones are chemicals produced by one individual that affect the behavior of another individual. Chemical recognition is an important cue that allows ants to determine whether individuals are members of their colony or potential intruders. Further study determined that individual workers regulate the number of queens in their hive by accepting queens that produce the appropriate chemical signals and rejecting those that do not (Ross & Keller 2002). The *Gp-9* protein likely plays a role in that recognition.

Subsequent research indicates that these two types of colonies actually differ in several other traits, such as aggressive behavior, colony initiation methods, and sperm production, all of which are not dependent on odor recognition. Are more genes involved? Recent work has revealed that *Gp-9* is part of a supergene—a set of tightly linked genes that affect many traits, including social behavior. The linkage of several hundred genes, including *Gp-9*, is due to a chromosomal inversion that limits recombination within the supergene (Wang et al. 2013). This allows these genes to accumulate on a single chromosome. The multi-queen version of this chromosome, Sb, and the single-queen version of this chromosome, SB, may be analogous to the human sex chromosome Y, as both have limited recombination potential with other chromosomes. This complex system continues to be examined to better understand how genetic variation is associated with phenotypic variation.

Experimental manipulation of gene function: knockout studies

Another approach used to understand how genetic variation is associated with behavioral variation is to disable a gene and examine the effect on behavior, a procedure known as a **knockout technique**. This technique requires a researcher to know the location of a gene and its DNA sequence in order to alter the gene so it cannot function. Individuals with an experimentally inactivated copy of a gene are known as knocked out (or KO) individuals (see Toolbox 4.1), and their behavior is compared with that of wild types. Mario Capecchi, Martin Evans, and

knockout technique A procedure in which a single gene is rendered nonfunctional.

Oliver Smithies received the 2007 Nobel Prize in Physiology and Medicine for their development of this technique. Let's look at a recent study that used the knockout technique in mice to examine how one gene affects behavior.

Featured Research

Anxiety-related behavior and knockout of a hormone receptor in mice

Knockouts are often created by disabling a gene that codes for a receptor for a specific gene product. Arginine vasopressin (AVP) is a peptide hormone, a short polymer of amino acids linked by peptide bonds. This hormone is stored in and released from the posterior pituitary in the brain into the bloodstream, where it plays a role in fluid homeostasis and blood pressure control. AVP is also released into the brain, where it affects social behavior (e.g., social recognition, pair bonding, and paternal behavior) and behavior under stressful conditions. AVP affects only those cells that have a functional AVP receptor. Isadora Bielsky and her colleagues used the knockout technique to study the function of one particular AVP receptor, *AVPR1A* (Bielsky et al. 2004). The *V1aR* gene in mice codes for *AVPR1A*, and so the researchers created a knockout for this gene.

The research team examined responses to stress—collectively known as anxiety-related behaviors. Mice are nocturnal and are therefore averse to bright light. Exposure to bright light conditions produces stress and stimulates movement to dark locations if they are available. In addition, mice display thigmotaxis, or a preference for physical contact, and therefore an avoidance of open areas, probably because such environments carry a high predation risk in the wild. The team used three standard tests to examine the mice's behavior under stress. The tests included placing a subject in (a) an elevated chamber with both open and walled areas, (b) an open arena or field, and (c) an apparatus that had both well-lit and dark areas (Figure 4.7). Each test situation

a Elevated plus maze

b Open field test

c Light/dark box

Figure 4.7. **Stress tests.** Three measures were used to examine the behavioral response to stressful conditions: (a) time spent in open versus walled areas, (b) time spent in the center of an arena versus near the wall, and (c) time spent in a light versus a dark area.

contained a region where individuals were highly exposed and areas where individuals were less exposed to potential predators. Regions like the former are more stressful to and normally avoided by control subjects. Researchers typically measure the amount of time spent in high- versus low-stress areas during tests to quantify how animals behave under different experimental treatments.

In Bielsky and colleagues' study, the researchers compared the behavior of wild-type and knockout mice on the three stress tests. In all three tests, the knockout mice spent more time than the wild-type mice in the light, in the center of the arena, and on the open arms of the elevated-plus maze (Figure 4.8). From these data, the researchers concluded that the *V1aR* gene plays an important role in affecting the behavior of animals placed in stressful situations. Lack of proper gene function modifies behavior so that individuals spend more time in risky locations: these mice do not flee from dangerous conditions. Thus, this gene can strongly affect fitness by affecting habitat choice and movement behavior.

These examples illustrate how behavioral variation is associated with genetic variation. While much of the above work focused on single major genes, most behavior is influenced by many genes. Next, we examine how researchers narrow their focus to identify the different genes that influence a particular behavior.

QTL mapping to identify genes associated with behavior

One approach to identifying genes of interest is to examine **quantitative trait loci** (**QTL**), which are stretches of DNA that either contain or are linked to genes influencing a trait such as behavior. **QTL mapping** is a statistical technique that combines genetic information with trait information to determine which regions of the genome contain the genes that influence the trait. This procedure can provide information on not only the number of genes involved but also their location on chromosomes. Such information can lead to the identification of candidate genes, genes suspected of contributing to a large amount of the phenotypic variation in a specific trait. Let's look at one example.

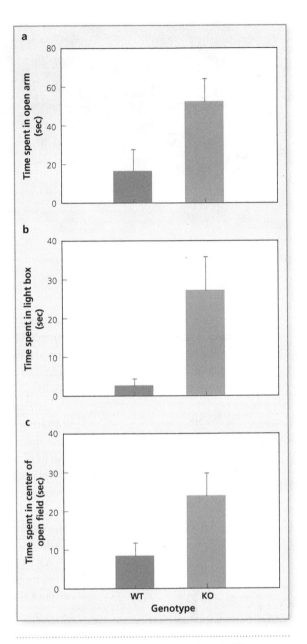

Figure 4.8. Knockout mice behavior. Mean (+ SE) behaviors of wild-type (WT) and knockout (KO) mice. Knockout mice whose AVP receptor was nonfunctional spent more time in the (a) open arm, (b) well-lit areas, and (c) center of the open field compared to wild-type mice. (*Source:* Bielsky et al. 2004)

QTL mapping for aphid feeding behavior

Many insects feed on specific plant species and use a complex suite of feeding behaviors to assess a plant. Marina Caillaud and Sara Via used QTL analysis to examine the number of genes involved with plant choice in pea aphids (*Acyrthosiphon pisum*) (Caillaud & Via 2012). Pea aphids (Figure 4.9) have sucking mouthparts, called stylets, which they use to pierce the soft tissues of plants, such as leaves or stems, to feed on the sap in the phloem. In a population in New York, Caillaud and Via identified two genetically distinct races of pea aphids that feed on different plant species: one specializes on alfalfa, while the other specializes on clover. How do individuals select the correct plant species for feeding? Aphids

quantitative trait loci (QTL) Stretches of DNA that either contain or are linked to genes influencing a trait such as behavior.

Figure 4.9. Pea aphid feeding. Pea aphids feed on plants by extracting sap.

> **QTL mapping** A statistical technique that combines genetic information with trait information to determine which regions of the genome contain the genes that influence the trait QTLs.

explore the plant surface and subepidermic tissues, probe deeply to find phloem sap, and evaluate it prior to ingestion (Caillaud & Via 2000). The duration of these behaviors will differ greatly depending on whether an aphid is on a preferred or nonpreferred plant species.

For the QTL analysis, the researchers created F_1 and F_2 generations that resulted from crossing individuals of the two races. Next, they characterized the behavioral phenotype of plant choice behavior for individuals from the F_1 and F_2 generations, as well as that for each parental race. They then measured total time spent searching for a feeding site (exploration), the time spent on a plant before penetrating it, the latency to inject saliva into the plant to begin the digestive process, and the amount of time spent digesting sap from each plant. A strong preference for a plant was indicated by a rapid time to feed. They assessed the genotype of each individual using 116 amplified fragment-length polymorphism (AFLP) markers (Vos et al. 1995). In this technique, genomic DNA is cut into pieces, or fragments. Known genetic markers (pieces of DNA) are then attached to each piece to create a linkage map showing the position of markers relative to each other, but not their specific location on the chromosome.

Caillaud and Via found from one to three QTLs for the total time spent searching for a feeding site, the time spent on a plant before penetrating it, and the time spent digesting sap from each plant. The proportion of the behavioral variation in plant acceptance behavior explained by each QTL ranged from 7% to over 50%. Given the small number of QTLs and the large proportion of variance for which they accounted, the researchers proposed that perhaps only a few genes are involved with plant selection behavior. They did not find any QTL for latency to inject saliva into the plant, which probably indicates that this particular behavior is influenced by numerous minor genes.

All of these examples illustrate how researchers have identified genetic influences on behavior. We've seen that researchers use both indirect (e.g., the studies on fruit flies and fire ants) and more direct methods (e.g., knockout studies) to examine the relationship between genotypic and behavioral variation. But recall that genes are just one factor that influences behavior: so too does variation in the environment, our next focus.

4.3 The environment influences gene expression and behavior

Recall that variation in behavior among individuals results from both variation in gene alleles and variation in environments. One important way the environment affects behavior is through its effects on gene expression. We've already discussed how gene expression can be affected by genetic factors, but it can also vary due to nongenetic (environmental) factors, such as environmental stress and social conditions, as we see next.

Featured Research

Social environment and gene expression in fruit flies

Recall that genes do not produce specific behaviors but rather code for a diverse array of molecules that affect brain function, through which behaviors are expressed. Animals acquire sensory inputs about both the abiotic and biotic aspects of the environment, including the presence of others, such as conspecifics or predators. These inputs are integrated in the brain and can result in differential **gene expression**. In other words, genes and their products are expressed at different times and in different environmental conditions, depending on sensory inputs.

Research examining environmental influences on gene expression often manipulates the social environment of individuals. Social environment can refer to the presence, type, and number of nearby individuals, as well as interactions with such individuals. How might differences in the social environment affect behavior? Consider that behavioral interactions between two conspecifics often depend on the sex of the participants: male–male interactions may involve aggressive behaviors, while male–female interactions often involve courtship behaviors. Lisa Ellis and Ginger Carney used microarray analysis (Toolbox 4.1) to examine gene expression in male fruit flies exposed to different social environments (Ellis & Carney 2011). In particular, they wanted to determine whether flies possess specific genes for male–female interactions and other genes for male–male interactions, or whether they have instead a set of "socially responsive" genes that are expressed in all social interactions.

Ellis and Carney raised male flies and exposed them to one of two treatment groups or controls. One group was allowed to court a single female, while the other interacted with a rival male for 20 minutes. Control males did not interact with any other fly. Microarray analysis was performed using DNA from the heads of flies in these two groups to examine differences in gene expression in the brain and important sensory organs such as the antennae.

The researchers identified hundreds of genes that were socially responsive. Of these genes, 505 responded to male–male interactions and 281 responded to male–female interactions. However, many of these genes were expressed in both treatments and so are presumed to be socially responsive genes involved in all conspecific interactions. The analysis also identified 240 genes that were responsive only to the male–male interaction treatment but only 16 genes that were uniquely responsive to male–female interaction. Thus, the researchers identified 16 candidate genes associated only with male courtship behavior, while hundreds of other genes were determined to be involved in male–male interactions. Why there is such a difference in the number of genes involved in these different social interactions is unclear, but clearly, environmental conditions strongly affect gene expression, and its resulting behavior, in male flies.

We've just seen how social conditions can affect gene expression, providing insight into the number of genes that affect social behavior. A second focus of environmental effects on gene expression examines birdsong, which we will examine next.

Social environment and birdsong development

All birds produce simple vocalizations, known as calls, which are innate. Many birds, particularly songbirds in the suborder Passeri (of the order Passeriformes), also produce complex vocalizations known as songs, which are used to defend a territory and attract a mate. Birds use their syrinx to produce vocal sounds, and in many songbird species, only males sing (although in tropical species, often both sexes sing).

Environmental exposure to conspecific song is a critical factor for proper song development in birds (e.g., Beecher & Brenowitz 2005). Song learning occurs in

gene expression The process in which gene products are produced.

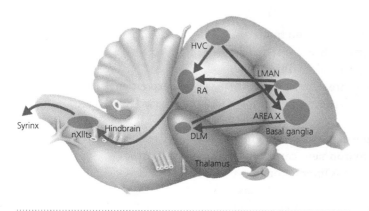

Figure 4.10. Avian song circuit. There are two neural circuits in the songbird brain. The red arrows indicate neural connections within the posterior circuit, which is involved in song production. The blue arrows indicate connections in the anterior circuit, which is involved in song learning. HVC = neostriatal nucleus; RA = robust nucleus of the arcopallium; DLM = medial nucleus of the dorsolateral thalamus; LMAN = anterior neostriatum; nXllts = tracheosyringeal neuron. (*Source:* Clayton, Balakrishnan, & London 2009)

several stages, and species are classified as closed- or open-ended learners. Closed-ended learners must hear a tutor sing its conspecific song shortly after hatching. During this sensitive period, which may last for several weeks, the bird apparently learns the song but does not attempt to reproduce it. After a few weeks, singing begins, and birds produce vocalizations known as a sub-song, which initially differs greatly from, but then converges to the song learned from the tutor. Finally, a bird's song becomes crystallized and does not change. In contrast, a fixed sensitive period does not exist for open-ended learners, which can acquire new song elements throughout life.

Early work on song learning in closed-ended learners relied on socially isolated birds hearing a taped tutor. This approach maximized control of the learning environment and showed that birds deprived of a tutor failed to develop proper song (e.g., Marler & Peters 1977). Additional work revealed that song learning is affected by the type of tutor. For white-crowned sparrows (*Zonotrichia leucophyrs*), the sensitive period lasts longer when a subject is paired with a live tutor rather than a taped tutor (Baptista & Petrinovich 1984). In addition, while the songs of live-tutored birds largely matched those of wild birds, tape-tutored birds produced much simpler songs, with a repertoire only about half the size of that of wild birds (Chaiken, Böhner, & Marler 1993). These results fostered better understanding of the role of environmental influences on song learning (e.g., Beecher et al. 2007). The next step was to understand the neural physiology and genetics of song learning, as we see next.

Social environment and gene expression in birds

The brains of songbirds differ greatly from those of non-songbirds, with songbird brains having enlarged and interconnected areas involved in song memory and production. These areas of the brain, called the song system, consist of a posterior nucleus, which controls sound production (air flow and the syrinx), and an anterior nucleus involved in song learning. An important localized region in the anterior nucleus is Area X (Figure 4.10). Exposure to birdsong, particularly conspecific song, causes numerous genes to be expressed in different regions of the song system. The most important appear to be *FoxP2* and *ZENK* (an acronym for the genes *zif-268*, *Erg-1*, *NGFI-A*, and *Krox-24*) (Mello, Vicario, & Clayton 1992; Bolhuis et al. 2000; Clayton, Balakrishnan, & London 2009).

One way these genes were identified was by impairing their function and observing the effects on song development. Sebastian Haesler and colleagues used this approach to determine the role of *FoxP2*, which encodes the forkhead box protein P2, in zebra finches (*Taeniopygia guttata*) (Figure 4.11) (Haesler et al. 2004). They found that *FoxP2* undergoes increased gene expression in Area X of a bird's brain both when young birds learn to sing and when open-ended learning adults change their song. This suggests that *FoxP2* plays a role in song learning.

Haesler's research team then reduced *FoxP2* levels in Area X before young zebra finches began to learn their song (Haesler et al. 2007). They used a knockdown technique, which involves using a virus to insert short sections of RNA into

Figure 4.11. Zebra finch. Zebra finches learn their songs from tutors.

the *FoxP2* gene at two different locations to reduce its expression, because a knockout technique had not yet been developed for songbirds. Control birds had a short section of RNA inserted into a noncoding region of DNA that had no effect on the gene's expression. Each subject was kept in a sound-isolation chamber with an adult male tutor during the critical period when young finches learn the songs of a tutor. At this time, the songs of both the tutor and the subject were recorded. In order to quantify the similarity of acoustic elements between the vocalizations of two individuals such as subject and tutor (e.g., Nordby, Campbell, & Beecher 2001), the researchers used spectrograms, or sonograms, two-dimensional representations of sound that allow researchers to characterize the acoustic structure of vocalizations (Figure 4.12). Spectrograms can reveal the number of song elements and their characteristic features (e.g., maximum frequency, maximum duration, and intervals between elements).

Figure 4.12. Spectrograms. Three sample spectrograms of tutors (left) and their subjects (right). Different song elements (buzzes and trills) are identified in the tutor song through variation in note frequency and duration. The top subject (a) displays a high degree of song matching to its tutor. The bottom two subjects (b, c) have lower degrees of song matching to their tutors. (*Source:* Nordby, Campbell, & Beecher 2001)

In knockdown birds, *FoxP2* expression was reduced by approximately 70% (Figure 4.13), which affected their song development: acoustic similarity between the tutor and the knockdown subject was much lower than it was in the controls. Knockdown birds tended to omit specific syllables and failed to copy accurately the duration of some song syllables (Figure 4.14). Because birds with experimental down-regulation of the *FoxP2* gene in Area X inaccurately learned their tutor song, the researchers concluded that *FoxP2* is required for normal song development.

Studies of gene expression like the ones we've just examined illustrate one approach used to identify environmental effects on behavior. Another method is to expose closely related individuals to different environments. If such individuals (with similar genes) exhibit different behaviors, then it is likely that their behavioral differences are the result of environmental factors. Next, we examine a research study that illustrates the use of this approach on fish.

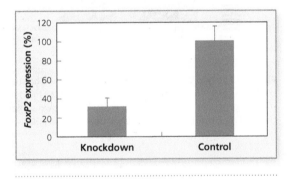

Figure 4.13. *FoxP2* expression. Mean (+ SE) *FoxP2* expression in knockdown and control birds. The expression of the *FoxP2* gene in Area X of the brain was significantly lower in knockdown birds compared to controls. (*Source:* Haesler et al. 2007)

Environmental effects on zebrafish aggression

Featured Research

Christopher Marks and colleagues examined how water conditions during development affect aggressive behavior in adult zebrafish (*Danio rerio*) (Marks et al. 2005) (Scientific Process 4.1). Zebrafish live in a variety of habitats ranging from oxygen-rich fast-flowing streams to oxygen-poor stagnant pools throughout the Himalayas. Oxygen depletion, or hypoxia, is a common stress for aquatic organisms that can affect their development and behavior. Marks and colleagues raised closely related zebrafish in water that had either low or high oxygen levels (hypoxic and normoxic environments, respectively) and then examined aggressive behavior in adult zebrafish when placed in each type

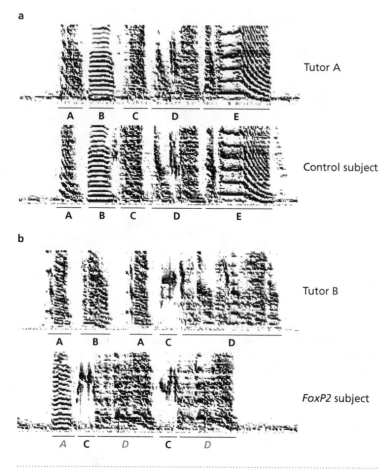

Figure 4.14. *FoxP2* **knockdown spectrograms.** (a) The spectrograms of a control tutor song (top) and subject song display a high degree of song matching. (b) The spectrograms of *FoxP2* knockdown subjects do not match those of their tutor (top) well. (*Source:* Haesler et al. 2007)

of environment. This design allowed the researchers to determine the effect of two different environmental conditions on zebrafish behavior: the developmental environment (DE) and the behavioral test environment (BE).

After the fish had matured, they were tested for aggressive behavior in one of the two developmental environments. Aggression assays were conducted by placing a small mirror on one wall of the test chamber. Fish respond aggressively to their mirror image because they treat the image as they would an intruder. The researchers videotaped the behavior of each fish for two minutes after the mirror was put in place and recorded the amount of time the fish spent butting or nipping the mirror.

Hypoxic-reared fish displayed higher levels of aggression in the hypoxic test chamber, and normoxic-reared fish exhibited higher levels of aggression in the normoxic test chamber. Thus, fish were most aggressive in their rearing conditions. Because the experimental design used full siblings reared in different environments, the results demonstrate how the environment during development can affect behavioral phenotype. Marks and colleagues suggested that aggressive behavior is energetically costly, and that this cost may become particularly high when fish experience a novel oxygen environment. Fish reared in normoxic environments probably experience an oxygen deficiency when in hypoxic conditions and so are incapable of high levels of activity. It is less clear why fish reared in hypoxic environments exhibited slightly lower levels of aggression in normoxic conditions, and so further research on this phenomenon is needed.

The above example illustrates how researchers investigate environmental influences on behavior. It also demonstrates how variation in environments can enhance behavioral variation in a population. In the next section, we examine how genetic and environmental conditions can interact to produce even greater phenotypic variation in populations.

Gene-environment interactions

As we have seen, both genes and the environment can have significant effects on the variation we see in behavioral phenotypes. So how might genes and the environment interact to affect behavioral phenotypes? Suppose you raised two populations of genetically identical aphids, a single genotype, in different environments (say, on different plants) and observed some aspect of their feeding behavior. The range of different behavioral phenotypes observed from the single genotype raised across different environmental conditions is known as the reaction norm. This difference in phenotypes, phenotypic plasticity, indicates that this trait is not fixed but rather is plastic and can respond to variation in the environment, just as we saw for zebrafish aggression.

SCIENTIFIC PROCESS 4.1

Environmental effects on zebrafish aggression

RESEARCH QUESTION: *How does environmental variation affect aggression in zebrafish?*

Hypothesis: Both developmental environment (DE) and behavioral (test) environment (BE) affect aggression.

Prediction: (a) If only DE affects behavior, aggression will be unaffected by BE and will be highest in normoxic (high-oxygen) water. (b) If DE and BE affect behavior but do not interact, aggression will be higher for fish in a normoxic BE and will be highest for fish whose DE was normoxic. (c) If DE and BE interact, aggression will be highest for fish whose BE and DE match.

Figure 1. Predictions. The blue line represents fish raised in a normoxic (high-oxygen) developmental environment, and the orange line represents those raised in a hypoxic (low-oxygen) developmental environment. (*Source:* Marks et al. 2005)

Methods: The researchers:

- Collected eggs from 12 full-sibling clutches (all individuals shared 50% of gene alleles because members of a clutch have the same parents)
- Divided the eggs into two developmental environments (normoxic and hypoxic). The embryos were raised until adulthood (about 75 days of age) with unlimited food. Normoxic development treatment tanks had continuously oxygenated water (high dissolved oxygen content = 6.8 mg/L); hypoxic development treatment tanks had nitrogen gas bubbled into the water (low dissolved oxygen content = 0.8 mg/L)
- Tested fish in either a normoxic or a hypoxic test chamber after a 16-hour acclimation period
- Measured aggression as time spent biting or nipping a mirror image

Results: Hypoxia-raised fish had higher levels of aggression in the hypoxic test chamber, and normoxia-raised fish had higher levels of aggression in the normoxic test chamber.

Figure 2. Results. The blue line represents the normoxic and the orange line the hypoxic development environment. (*Source:* Marks et al. 2005)

Conclusion: Zebrafish display more aggressive behavior in the environment in which they were raised, indicating an interaction between developmental and behavioral environments.

EVALUATE

Explain why the researchers used full siblings in their experiments.

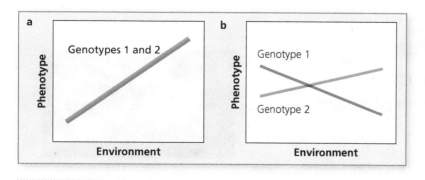

Figure 4.15. Environmentally influenced phenotypes. Genotypes often exhibit variation in phenotype when exposed to different environments, illustrating a reaction norm. (a) Each genotype exhibits the same reaction norm. (b) The two genotypes exhibit different reaction norms because they respond differently to variation in the environment, evidence of a G × E interaction.

rover One type of genetic and behavioral variant in fruit flies. Larval rovers have longer foraging trails than sitters in the presence of food and are more likely to leave a food patch.

Now consider a comparison of the reaction norms of different genotypes. One might expect that each genotype would exhibit a similar response to the different environments (Figure 4.15a). However, genotypes may respond very differently to the different environments, and when this occurs we say there is a significant gene-environment interaction (G×E) (Figure 4.15b). When gene-environment interactions exist, we observe variation among genotypes in their responses to environmental variation. When this is true, phenotypic variation in a population is enhanced and now results from three factors: variation in genotypes (V_G), variation in environments (V_E), and variation that results from their interaction ($V_{G \times E}$). That is:

$$V_P = V_G + V_E + V_{G \times E}$$

Let's examine one of the first studies to document such an effect in fruit flies.

Rover and sitter foraging behavior in fruit flies

One now-classic example of a behavior known to be influenced by gene-environment interaction comes from Marla Sokolowski's work on the foraging behavior of larval fruit flies (*Drosophila melanogaster*). Fruit fly eggs develop during a series of larval stages (Figure 4.16). For several days, the larvae crawl to food sources, eat, and then molt as they increase in size.

sitter One type of genetic and behavioral variant in fruit flies. Larval sitters have shorter foraging trails than rovers in the presence of food and are less likely to leave a food patch.

The larvae of *D. melanogaster* exhibit two behavioral polymorphisms, or variants, known as **rover** and **sitter**. Rovers have longer foraging trails than sitters in the presence of food and are more likely to leave a food patch (Figure 4.17). Sokolowski found that this difference in feeding behavior is due to different alleles at a single gene, the *for* gene, which determines foraging behavior (Sokolowski, Pereira, & Hughes 1997). About 70% of individuals in natural populations have the dominant rover allele (*for^R*) and exhibit the rover phenotype, while 30% are homozygous for the sitter allele (*for^s*) and exhibit the sitter phenotype. Both of these genetic variants are considered wild type because they occur in natural populations. The polymorphism is maintained by frequency-dependent selection: the rover allele has higher fitness in crowded conditions, where it can benefit by traveling farther to find food, while the sitter allele has higher fitness in less-crowded environments. However, in the absence of food, rovers behave like sitters, probably because there is no benefit to increased movement (Graf & Sokolowski 1989). This pattern suggests a significant gene-environment interaction because rover behavior changes strongly with differences in their environment (food abundance), while sitter behavior does not.

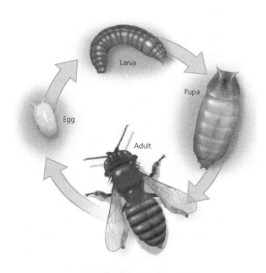

Figure 4.16. *Drosophila* life cycle. Adults lay eggs that hatch within 24 hours. Larvae grow for several days and molt twice. Larvae then encapsulate into a puparium and undergo a four-day metamorphosis prior to emerging as adults.

More recent work examined gene-environment interactions in adult flies. Sokolowski's research team examined the behavior of groups of 25 to 30 adult rovers and sitters exposed to different levels of food availability (Kent et al. 2009). Half were placed in holding vials that contained food (fed environments) for 16 to 18 hours prior to behavioral testing. The other half were placed in a food-deprived environment for the same period of time prior to behavioral testing. The researchers then examined the movement behavior of the flies using a horizontal Plexiglas maze (Figure 4.18). They placed the flies in a vial with food and allowed them to feed for 15 minutes before allowing them to leave the patch and move through the maze. Empty glass vials were placed at all maze exits. The team recorded the proportion of flies in the collecting vials after three minutes, which they called the food-leaving score.

The behavior of fed versus food-deprived rovers and sitters was quite different (Figure 4.19). Fed rovers had a much higher food-leaving score than food-deprived rovers. However, there was little difference between the behavior of fed sitters and that of food-deprived sitters. Here again we see that different environmental conditions have less effect on the behavior of the sitter genotypes compared to rovers.

Both larvae and adult rovers showed marked changes in behavior across different environmental conditions, while sitters did not. What might be causing this difference? The researchers removed the heads from the flies and used mass spectroscopy to determine the compounds stored there. They found that rovers and sitters store energy differently (Figure 4.20). Fed rovers and fed sitters stored the same levels of lipids, but in the food-deprived environment, sitters stored more lipids. The researchers also found a large difference between the carbohydrate storage of rovers and that of sitters in both environments, with sitters always storing more carbohydrates than rovers. Carbohydrate levels for both sitters and rovers were lower in the food-deprived environments, but they were lowest for rovers. This finding indicates that gene-environment interactions affect both behavioral and metabolic traits. Sokolowski's work demonstrates a

Figure 4.17. Rover and sitter foraging trail lengths. This histogram depicts trail lengths in the presence of food for hundreds of larvae. Sitters (blue bars) tend to have much shorter foraging trails than rovers (orange bars). (*Source:* Sokolowski 2001)

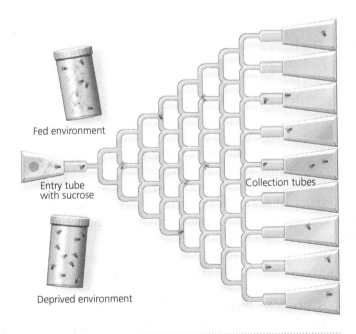

Figure 4.18. Fly maze. Apparatus used to characterize patch-leaving behavior. (*Source:* Kent et al. 2009)

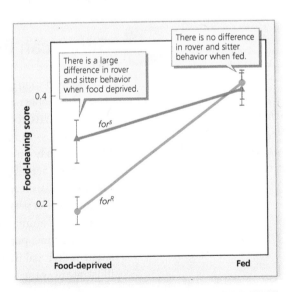

Figure 4.19. Adult rover and sitter food-leaving behavior. Mean (± SE) food-leaving scores for fed and food-deprived flies. Note the difference in the change in sitter (blue) and rover (orange) food-leaving scores between fed and food-deprived flies. (*Source:* Kent et al. 2009)

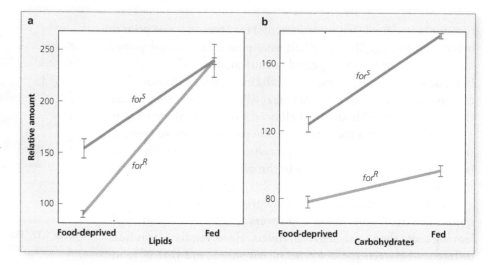

Figure 4.20. Energy storage in rovers and sitters. Change in the heads of rovers (orange) and sitters (blue) in the mean (± SE) relative amount of (a) lipids and (b) carbohydrates between fed and food-deprived individuals. (*Source:* Kent et al. 2009)

gene-environment interaction in both larval and adult fly behavior, showing that rover genotypes exhibit greater changes in behavior across different environments.

In this section we have seen how environmental variation is associated with behavioral variation. One approach to understanding these relationships involves examining how variation in social environments leads to differential gene expression. This type of study has revealed potential candidate genes involved in social behaviors of flies and in song learning among birds. A second approach uses closely related individuals reared in different environments to examine environmental differences in behavior. Finally, we have seen how gene-environment interactions are revealed in studies of flies; articulating these interactions allows researchers to better understand genetic and environmental influences on behavior. Studies of genetic and environmental variations, and their effects on behavior, will continue to inform our understanding of behavioral variation in animals. But the behavioral phenotype of an individual has limits. We examine these limits next in a discussion of animal personalities.

 4.4 ## Genes can limit behavioral flexibility

So far we have seen how genetic factors, environmental factors, and their interaction produce variation in behavior. Genes also appear to limit behavioral variation. We all have met people who are consistently shy, as well as others who are consistently outgoing. Animals, too, display **personalities**—consistent differences in behavior over time or across different environmental contexts. Animal personality traits include variations in boldness (the willingness to take risks and explore novel objects), overall level of activity, and aggressiveness (Sih, Bell, & Johnson 2004; Stamps & Groothuis 2010). As we saw in Chapter 3, exploratory behavior in great tits is heritable, as are other personality traits (Réale et al. 2007). Not only does this mean that the existence of animal personalities can help to explain variation in behaviors among individuals in a population, but it also means that the behavioral variation within a single individual is constrained by its gene alleles. Let's examine two examples that illustrate how the behavioral variation of individuals is limited by their personality.

personality Consistent relative differences in behavior among individuals over time or across different environmental contexts.

Featured Research ▶

Bold and shy personalities in streamside salamanders

Streamside salamanders (*Ambystoma barbouri*) live primarily in the deep pools of shallow streams in eastern North America. The pools are separated by riffles, shallow areas of faster-flowing water, in which salamanders deposit eggs. Developing larvae face two challenges before they undergo metamorphosis and leave the stream. First, they must maximize feeding time—to shorten their development time—before the streams and pools dry up. Second, they must avoid predators such as the green sunfish (*Lepomis cyanellus*). When predators are present, larvae can seek shelter under rocks, but this behavior reduces time spent feeding.

How might salamanders respond to these two challenges? One might expect that the best solution would be to modify behavior given the situation. An individual might, say, maximize feeding time when predators are absent and minimize feeding time when predators are present. That flexibility would produce a negative correlation between behaviors because the two are mutually exclusive: the more time spent hiding, the less time can be spent feeding. But if an individual's personality is heritable, it might constrain its behavior. Are some individuals always more active than others, even when predators are nearby (Figure 4.21)?

To examine this possibility, Andy Sih, Lee Kats, and Erik Maurer examined salamander larvae behavior in the laboratory (Sih, Kats, & Maurer 2003) (Scientific Process 4.2). Each larva was exposed to two experimental treatments that manipulated predation risk. The "fish cue" treatment consisted of adding water from an aquarium that contained four sunfish to the subject's aquarium. Because salamanders can perceive chemical cues from predators, this treatment should increase perceived predation risk for the larvae. In the control "no-fish" treatment, tap water was added. The treatments were separated by two days, and the order was randomized across individuals. The researchers recorded the proportion of time salamander larvae spent away from a refuge.

On average, larvae spent about 75% of the time out of the refuge eating in the no-fish control treatment but only 40% of their time out eating when fish cues were present. Individuals clearly responded to increased predation risk by spending more time hiding, as one might expect. However, examination of their individual behavior revealed another story as well. When the amount of time each larva spent out of the refuge is plotted for both treatments, a significant positive correlation appears, indicating that an individual that spent much time out of the refuge when no fish cues were present also spent much time out of the refuge when fish cues were present. These larvae were acting "bold" in both situations, while "shy" individuals spent little time away from the refuge in both treatments. Few of the salamanders exhibited what might be considered "optimal" behavior: spending much time away from the refuge when no fish cues were present and little time away from the refuge when fish cues were present.

But is there a genetic component to these personalities? The data from each sibling group (sibship) were combined into one value for each treatment. All sibships demonstrated reduced time out of the refuge when fish cues were present. Still, those that spent the most time out of the refuge when no fish cues were present also spent the most time away from the refuge when fish cues were present. Other sibships spent much less time away from the refuge in both treatments. The consistency within but strong differences across sibships suggests that genetic differences underlie the observed behavioral differences. More important, we see how genotype can limit behavioral plasticity.

Animal personalities are not restricted to vertebrates. Let's examine one study of aggressive behavior in spiders.

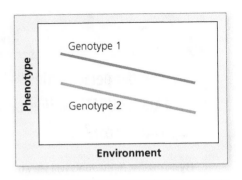

Figure 4.21. Animal personalities. Individuals of different genotype respond to variation in environmental conditions in a consistent manner relative to one another. Here, genotype 1 always exhibits a higher phenotypic value (e.g., activity level) for any environmental context (e.g., variation in predation risk).

SCIENTIFIC PROCESS 4.2

Salamander personalities

RESEARCH QUESTION: *How do salamander larvae respond to predation risk?*

Hypothesis: Individuals will vary their activity level in response to the level of threat.

Prediction: All individuals will exhibit high activity level when predation risk is low and low activity level when predation risk is high.

Methods: The researchers:

- Collected ten egg masses (sibships) to obtain four individual siblings from each
- Placed each larva in an aquarium that contained a refuge: an elevated, opaque disk that covered 50% of the surface area of the container
- Exposed each larva to two experimental treatments:
  - High predation risk, "fish cue" treatment: water from an aquarium containing four sunfish was added to the subject's aquarium.
  - Low predation risk, "no fish" (control) treatment: tap water was added to the subject's aquarium.
- Recorded salamander larvae behavior every 30 minutes for ten hours, noting whether the individual was in or out of its refuge

Results:

- Larvae in the control treatment spent more time out of the refuge eating than did larvae in the high-risk treatment.
- There was a significant positive correlation in the amount of time each larva spent out of the refuge across treatments. Individuals and sibships that spent more time in the open in the control also spent more time in the open in the predator treatment.

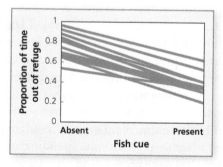

Figure 1. Refuge use. The mean (+ SE) proportion of time spent outside the refuge for control (blue) and high-risk (orange) treatments. (*Source:* Sih, Kats, & Maurer 2003)

Figure 2. Predator effect. There was a positive relationship for time spent out of the refuge with and without fish predator cues. Each circle is one individual. (*Source:* Sih, Kats, & Maurer 2003)

Figure 3. Sibships. Proportion of time out of the refuge for each sibship, without (left) and with (right) fish predator cues. Points connected by a line represent one sibship. (*Source:* Sih, Kats, & Maurer 2003)

Conclusions: Genetic differences across sibships explain the consistent relative differences in activity level across treatments, indicating that salamanders exhibit behavioral personalities.

EVALUATE

How would the data in Figure 3 appear if individual salamanders did not show personalities with respect to activity level?

Aggressive personalities in fishing spiders

In the previous example, we saw that individuals that exhibited high activity levels when predation risk was low also behaved similarly when predation risk was high. In this second example, we see evidence of consistent differences in behavior (personalities) for individuals across two contexts, feeding and mating.

Many female insects and spiders will attack and consume a courting male before copulation, a behavior known as precopulatory sexual cannibalism (Elgar & Schneider 2004). Females that display this behavior incur a cost because they do not mate. How can such a puzzling behavior evolve? The adaptive foraging hypothesis states that females in poor condition (i.e., with low body mass) can benefit from this behavior by obtaining nutrients from the meal, and that this benefit exceeds the costs of not mating as long as the female encounters another male in the future (Newman & Elgar 1991). This hypothesis predicts that female fecundity (fitness) should be positively correlated with her feeding history (including the consumption of conspecific males) and that females in poor condition should be more likely to exhibit sexual cannibalism.

An alternative hypothesis suggests that genetic and behavioral constraints, manifested as an aggressive personality, limit behavioral flexibility such that sexual cannibalism represents a "spillover" of aggression from a feeding to a mating context (Arnqvist & Henriksson 1997). This aggressive spillover hypothesis states that highly aggressive, voracious feeders will have higher feeding rates as juveniles and will subsequently attain large body size as adults, while less aggressive juveniles will attain smaller adult body size due to lower feeding rates. Large adults that are voracious feeders will have higher fecundity than less aggressive feeders, but due to their aggressive, voracious feeding behavior, they will also be more likely to engage in precopulatory sexual cannibalism. This hypothesis predicts positive correlations between the juvenile and adult feeding behavior of individuals (measured as latency to attack heterospecific prey) and the propensity to exhibit sexual cannibalism.

J. Chadwick Johnson and Andy Sih evaluated both hypotheses in the six-spotted fishing spider (*Dolomedes triton*) (Figure 4.22) (Johnson & Sih 2005). Most fishing spiders are semiaquatic and hunt for invertebrate prey by placing their forelimbs on the surface of the water at the edge of a pond or stream. Once a potential victim is detected via ripples on the water, the spider runs across the surface of the water to attack and subdue its prey.

Johnson and Sih examined the foraging behavior of 60 captive female spiders for several months as they matured from juveniles to adults. The researchers determined body size by measuring the cephalothorax of each female and weighed each female prior to all behavioral trials. The condition of each individual was estimated by dividing mass by body size. Twice a week, the research team conducted foraging trials in a large tub filled with water that contained a Styrofoam float from which the females could hunt. At the start of each trial, a cricket (*Acheta domesticus*) was introduced to the water surface 30 cm in front of the female, and the researchers recorded her latency to attack and whether the attack was successful. The latency to attack was used as an index of voracity.

Johnson and Sih found that for both juveniles and adults, aggressive females had higher prey capture success;

Figure 4.22. Fishing spider. Individual with forelimbs on the surface of a pond.

furthermore, juvenile feeding rate was positively correlated with final adult size. In addition, juvenile voracity was positively correlated with adult voracity: individuals' feeding voracity was relatively consistent as juveniles and adults and so exhibited different personalities.

Once the females had matured to adults, they were exposed to four 90-minute courtship trials per week, which continued until the female mated and produced an egg sac. For each trial, a randomly chosen male was introduced 30 cm in front of the female. This presentation never elicited an immediate attack by the female. Males initiated courtship behavior—leg waving and tapping—and the research team recorded the female's latency to attack or whether the pair copulated. During the first five courtship trials, the researchers interrupted any attempted copulations so that they could obtain repeated measures of each virgin female's propensity to attack males. After the fifth trial, females were allowed to copulate and produce an egg sac. Egg sac mass was weighed, and the number of spiderlings that emerged was used as a measure of fecundity.

In courtship trials, precopulatory attacks occurred in 37% of trials, and males were killed in almost half of the attacks. Across females, juvenile and adult voracity were positively correlated with the propensity to engage in precopulatory attacks on males, as predicted by the aggressive spillover hypothesis.

Females consumed up to four males before mating, but the propensity to attack males did not affect the likelihood of egg hatch success. In addition, the number of males eaten was not correlated with the number of spiderlings produced, and females in poor condition were not more likely to attack a male—results not in accord with predictions of the adaptive foraging hypothesis.

In this experiment, females varied considerably in their aggressive feeding voracity on heterospecific prey. Individuals that most rapidly attacked prey had the highest feeding success, obtained the largest adult size, and had the highest fecundity. Such individuals were also the most likely to attack and kill a courting male before copulation. Johnson and Sih suggest that these spiders exhibit different personalities that appear to limit their behavioral flexibility and explain variation in precopulatory sexual cannibalism across females. In contrast, the researchers found no support for the adaptive foraging hypothesis. Future work is required to understand why females vary in feeding voracity and to understand the costs of such behavior.

The existence of animal personalities suggests that behavior is not completely flexible and that genes play a role. Why do personalities exist? And how are they maintained in a population? Recent modeling has made some headway on these questions, as we see next.

Featured Research

Animal personalities: a model with fitness trade-offs

Max Wolf and colleagues argue that most personality traits reflect aspects of risk-taking behavior (Wolf et al. 2007). Boldness, high levels of exploratory behavior, and high overall activity time should increase risk of mortality from predation but also lead to increased resource acquisition. In contrast, shyness, low levels of exploratory behavior, and low overall activity should, in general, be less risky and less rewarding. In this way, animal personalities are tied to a fitness trade-off between current and future reproduction. Effort allocated to current reproduction comes at the cost of future reproduction (we will discuss this trade-off in depth in Chapter 14).

Wolf and colleagues developed a model of risky and shy personalities. They considered a model organism that lives for two years and can reproduce each year. High- and low-quality resources exist in this organism's environment, but high-quality resources are difficult to find. Individuals that do find high-quality resources have higher reproductive success than those that do not. The researchers assumed a trade-off in reproduction between Years 1 and 2 that is mediated by exploratory behavior. If individuals engage in a high level of exploration in Year 1, they will have a low probability of reproducing in Year 1 but a high probability of finding high-quality resources (and thus having high reproductive success) in Year 2. Individuals that spend little time exploring will have a high probability of reproducing in Year 1 but will likely only find low-quality resources.

Wolf and colleagues evaluated the lifetime fitness for individuals that exhibit different levels of exploratory behavior through a series of simulations. Their model shows that an individual with a high level of exploratory behavior can have the same fitness as one with a low level of exploratory behavior because they represent different allocations to current and future reproduction (Wolf et al. 2007). Individuals that explore thoroughly invest more heavily in future reproduction, while those that explore only superficially invest primarily in current reproduction.

Next, the researchers ran a series of simulations to examine the effects of competition for resources and predation risk on the lifetime fitness of individuals. These simulations revealed a general principle with two parts: (1) individuals that invest most heavily in reproduction in Year 2 should be more cautious (shy) to enhance their probability of surviving to reproduce, and (2) individuals that invest more in reproduction in Year 1 should be more bold (aggressive) to enhance their ability to compete for resources. The model suggests that bold, aggressive behavior allows greater short-term resource acquisition, which can be allocated to current reproduction, while shy, reserved behavior allows greater allocation to future reproduction. Because these behaviors produce equal fitness, they will both be maintained in a population. Thus, if boldness and shyness are heritable traits, as appears to be the case for many species, then an individual's alleles will limit its behavioral flexibility to some degree.

The model developed by Wolf and colleagues has not been tested directly. However, in a meta-analysis, Brian Smith and Daniel Blumstein examined 31 publications that studied correlations between personality types and fitness of individuals (measured as reproductive success or survival) (Smith & Blumstein 2008). These studies examined both wild and captive animals and included vertebrates, invertebrates, and terrestrial and aquatic species. Smith and Blumstein's analysis revealed that, on average, bold individuals, especially males, tended to have higher reproductive success than shy individuals but also tended to have a shorter lifespan, giving shyer individuals more opportunities to reproduce: a pattern suggesting equal lifetime fitness for bold and shy individuals that is consistent with the model's predictions. Future tests of this model that focus on a single species will help us better understand the evolution of animal personalities.

Behavioral genetics holds great promise for understanding variation in behavior among individuals, populations, and species. Researchers have made a great deal of progress in identifying major genes and gene-environment interactions that affect behavior. Recent work has also revealed how genes can limit behavioral flexibility. In the coming years, we can expect progress in behavioral genetics to accelerate as we obtain entire genome sequences of additional species.

Molecular techniques

Molecular techniques for studying genes and behavior have become increasingly available. Three of the most prevalent are QTL mapping, creating knockout organisms, and microarray analysis.

QTL MAPPING

Quantitative trait loci (QTL) are portions of the genome that influence certain phenotypic traits. A QTL analysis correlates regions of the genome that have genetic markers with associated behavioral variation among individuals in a family. One way to create a family with behavioral variation is to cross (or mate) individuals that have different behavioral phenotypes and then make quantitative measurements of the behavior of the offspring (Figure 1). This approach also requires information on the individuals' genotypes. We can examine the genotype by cutting the DNA into fragments and attaching genetic markers (short sequences of known DNA) to identify each fragment (Figure 2). Several standard genetic markers are used, such as restriction fragment-length polymorphisms (RFLPs), in which fragments differ in length; simple sequence repeats (SSRs), which are repeating sequences of only a few base pairs (also known as microsatellites); and single nucleotide polymorphisms (SNPs), sequences that differ in only a single base pair nucleotide. These genetic markers must be scattered evenly throughout the genome.

Statistical analyses are used to determine if differences in behavior between individuals correlate with differences in marker genotypes. If there is a correlation, then the QTL that is associated with a behavior is assumed to be linked to the genetic marker.

CREATING KNOCKOUT ORGANISMS

One way to study the effect of a gene is to examine the behavior of animals with and without an active copy of the gene. This technique is often used in mice when the sequence but not the function of a gene is known. By examining behavioral and physiological differences in animals, researchers can gain understanding of the function of the gene.

Knockout mice are transgenic: they have had additional DNA added to their genome that stops a particular gene from functioning. This DNA is created by harvesting embryonic stem (ES) cells from early-stage mouse embryos. These cells are totipotent and so can grow into any type of cell. DNA that will inactivate the gene is inserted into the nuclei of the ES cells, and the cells are then grown in vitro in the lab for several days. Typically, another marker gene is also inserted that allows rapid identification of individuals with the new mutation. For example, this marker gene might produce a unique coat color in mice. The engineered cells are then injected into early-stage mouse embryos, which are implanted in the uterus of a female mouse.

The pups that are born are chimeras, meaning that they develop from two kinds of stem cells—one with the new mutation and those without it. Chimeras are useful only if the new mutation was incorporated into the germ cells, which allows it to be passed to offspring. These mice are then crossed with

Figure 1. QTL analysis. Protocol for QTL analysis. (a) Individuals with different behavioral phenotypes are (b) crossed to create family groups with much behavioral variation. (*Source:* Mackey 2001)

Figure 2. Genetic markers. Genetic markers are attached to DNA fragments for each individual using recombinant DNA techniques.

wild-type individuals to produce offspring with one copy of the mutation. Subsequent inbreeding produces offspring that are homozygous for the mutation. These are knockout individuals.

continued...

Microarray analysis, or gene expression profiling, is another method used to identify genes that influence behavior. Not all genes are activated, or expressed, at all times, so they are not always undergoing transcription (DNA to mRNA) or translation (mRNA to proteins). Microarray analysis measures the amount of mRNA produced as an indicator of gene activation. This technique allows us to examine hundreds if not thousands of genes at once by quantifying which genes are activated in different individuals or tissues.

DNA from different tissues (e.g., different parts of the brain) or from different individuals (e.g., rover or sitter flies) is extracted, and then the mRNA from those cells is collected.

Reverse transcriptase, an enzyme, is used to create complementary cDNA for each mRNA. The cDNA is labeled by color with a fluorescent tag to indicate where it came from. Equal amounts of the labeled cDNA are then placed onto a DNA chip, which is a modified microscope slide that contains hundreds or thousands of known genes of the species, placed in specific locations. The cDNA will hybridize only with its complementary DNA. The chip is then scanned with a laser to quantify all the genes that contain any colored fluorescent dyes. (If more than one type of cDNA is present, the location will appear as an additional color.) The intensity of the dye indicates the intensity of gene expression, with more dye indicating greater expression of that gene.

Chapter Summary and Beyond

Behavioral genetics is the study of the effects of genetic and environmental influences and their interactions on behavior. Anholt and Mackay (2010) provide a comprehensive overview of this field. An important aspect of behavioral genetics is the identification of genes that contribute to large amounts of variation in behavior (Fitzpatrick et al. 2005; Pennisi 2005). The knockout technique (in which a single gene is rendered nonfunctional) has greatly expanded our understanding of how single genes affect behavior. This technique is often used in animal models to gain insight into human pathologies (van der Staay 2006). Both QTL mapping and microarray analysis have been used to identify many genes that influence behavior (Flint 2003). The next step underway is investigation of the expressed proteins associated with specific behaviors (Valcu & Kempenaers 2015).

While genes affect behavior, so too does the environment. The social environment is often manipulated to examine gene expression profiling (Robinson, Fernald, & Clayton 2008), most notably in birdsong. Recent work has begun to reveal how genotypic differences affect selective learning of the song of different tutors (e.g., Mundinger 2010). The recent sequencing of the zebra finch genome should make it possible to use avian knockouts and microarray analysis to understand the effect of specific genes on the singing behavior of birds (Warren et al. 2010). Multidisciplinary research has found behavioral, neural, and genetic parallels between human language development and birdsong development (Bolhuis, Okanoya, & Scharff 2010). Studying gene-environment interactions should help us better understand variation in behavior (Bendesky & Bargmann 2011).

Work on animal personalities has revealed how genotypes can limit behavioral variation among individuals (Stamps & Groothuis 2010). Sih, Bell, & Johnson (2004)

have expanded this finding to a discussion of behavioral syndromes, or suites of correlated behavioral traits across a population. Recent work focuses on the ecological and evolutionary consequences of different personality traits (Wolf & Weissing 2012) and the search for the genetic aspects of personality in humans (Arbelle et al. 2003).

CHAPTER QUESTIONS

1. A researcher examined the behavior of adult minnows from two isolated populations. One population lived in a warm stream with an average temperature of 20°C, while the other lived in a cold stream with an average temperature of 14°C. Individuals from the warm stream population were significantly more aggressive than individuals from the cold stream. The researcher concluded that the difference in aggression was mainly due to environmental conditions. Evaluate this conclusion.

2. Propose a study to determine the heritability of personality in the salamanders studied by Sih, Kats, and Maurer (2003).

3. Briefly describe the information that QTL mapping, microarray analysis, and knockout techniques provide to behavioral geneticists.

4. Identify one research study from the chapter from which you might reasonably conclude that phenotypic variation is more strongly associated with genetic variation than environmental variation.

5. In the salamander system studied by Sih, Kats, and Maurer (2003), explain how the existence of animal personalities indicates that genes limit behavioral plasticity.

6. Heritability values for behavioral traits are typically lower than those for morphological traits. Propose an explanation for this difference.

Chapter 5

Sensory Systems and Behavior

On a warm night in Tucson, Arizona, as we headed out for a walk, the front porch light was surrounded with hundreds of flying and crawling insects (Figure 5.1). Unexpectedly, our dog Grace moved toward a large animal sitting under the porch light: a Sonoran desert toad (*Incilius alvarius*). These toads have skin glands that produce a potent toxin strong enough to kill a dog, so we kept Grace far away from it. Toads prefer moist environments, but the concentration of insects around the light had attracted it to our dry concrete patio.

Many nocturnal insects are strongly attracted to artificial lights with a high ultraviolet (short-wavelength) component. Indeed, when scientists collect nocturnal insects, they often shine a bright ultraviolet light on a white sheet to attract them. Less is known about why nocturnal insects are attracted to such lights. Whatever the explanation, inventors have profited from the knowledge that moths' eyes are less sensitive to longer wavelengths of light. Specially designed "bug lights" that do not emit ultraviolet wavelengths attract fewer moths (van Langevelde et al. 2011). Replacing our standard light bulb with one of these bug lights has eliminated the problem of attracting insects (that attract poisonous toads) to our patio.

In this chapter, we examine how animals perceive their environment. We begin with a basic overview of the nervous system as a proximate

Figure 5.1. Moths at a light. Many nocturnal insects like these moths are attracted to artificial lights.

link between the environment and behavior. We then examine specific sensory systems that allow animals to detect various types of environmental information, focusing on chemicals, light, and sounds. Next, we illustrate differences between human sensory perception and that of several other taxa by examining how animals detect electric and magnetic fields. Finally, we examine co-evolution in sensory systems between species that interact, and we see that sensory systems of species evolve in response to the behavior of other species. Two important themes run through this chapter. The first is that researchers need to understand how animals perceive their environment in order to understand their behavior. The second is that animals living in the same environment may perceive it very differently based on differences in their sensory systems.

5.1 Animals acquire environmental information from their sensory systems

Animals live in a heterogeneous world that varies continuously. Abiotic environmental factors such as light intensity, moisture, and temperature vary widely across space and over time. So do biotic parameters such as the availability of food and mates and the presence of competitors and predators. Such environmental heterogeneity can negatively affect the fitness of individuals that cannot respond appropriately to different environmental conditions.

Environmental information comes in many forms and includes chemical compounds, mechanical sound waves and electromagnetic (light) waves, magnetic fields, and even electrical fields. Animals assess current conditions by gathering information about the environment via their sensory systems. Sensory systems include sensory organs and receptors that have evolved to acquire information from the environment. Sensory receptors are nerve endings that respond to environmental stimuli and transmit information via neurons, nerve cells that receive and transfer electrical and chemical signals. Neurons are composed of many dendrites that detect a stimulus; a cell body (or soma), where the information is integrated; and an axon, which conducts the electrical signal to the axon terminal, where it can be transmitted to other neurons, organs, or muscles and affect behavior (Figure 5.2). Neurons usually have a polarity, so information travels in only one direction.

How does this work? Receptors respond to the energy in a specific environmental stimulus by opening ion channels in the cell membrane that allow an influx of ions. The result is a change in ion concentration inside and outside the cell that leads to a difference in electrical voltage across the membrane. When the difference in electrical charge reaches a specific threshold, an action potential is generated that travels along the axon. When the action potential reaches the end of the axon, it can cause the release of neurotransmitters, such as serotonin, into the synapse, the gap between two neurons. The neurotransmitters travel across this

Figure 5.2. Neuron. The different parts of a neuron. The neural signal travels from the presynaptic to the postsynaptic neuron.

gap and interact with the postsynaptic nerve cell membrane, which can cause its ion channels to open and initiate another action potential in the neuron (Figure 5.3). Subsequently, the neurotransmitters are reabsorbed by the presynaptic neuron. This process continues as action potentials travel from neuron to neuron and to different cells. Action potentials that travel to muscles can cause them to contract, resulting in behavior.

Because species have very different types of biotic interactions, unique evolutionary histories, and may live in different types of environments, the information that is most important to one species may differ substantially from what is most important to another. As a result, we see tremendous variation in the sensory capabilities of different species. A consequence is that each species will have its own specific perceptual environment, or "umwelt" (Von Uexküll 1909).

Throughout this chapter we illustrate important aspects of how animals detect environmental information from their sensory systems. We start by examining the most widespread sensory system: chemoreception.

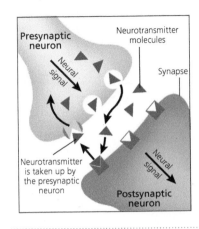

Figure 5.3. Synapse. The movement of neurotransmitters across the synapse. Electrical signals are transformed into chemical signals that move across the synapse.

5.2 Chemosensory systems detect chemicals that are perceived as tastes and odors

The most primitive and universal sensory system is **chemoreception**, the detection of chemical stimuli. Even single-celled organisms such as bacteria can detect and respond to these stimuli (e.g., MacNab & Koshland 1972). In multicellular organisms, chemoreception includes **gustation** (taste), the detection of dissolved chemicals, known as tastants; and **olfaction** (sense of smell), which is the detection of water- or airborne chemicals known as **odorants** and **pheromones**. General odorants are not species-specific and allow animals to locate food and avoid environmental dangers such as predators and fire. Pheromones, by contrast, are species-specific—that is, these organic compounds, produced by an individual, affect the behavior of another individual of the same species. Pheromones typically elicit a specific reaction when detected. They can provide information about the presence of a conspecific food trail, alarm cues indicating predators, or the sex of an individual as a potential mate. We will examine pheromones in more detail in Chapter 6. Sometimes, however, it can be difficult to distinguish whether a particular chemical stimulus is species-specific; in such cases, the stimuli are called odorants until more is known about their specificity.

At its most basic, the chemosensory system involves the binding of a discrete chemical molecule to a receptor on the surface of a sensory neuron. As we discussed above, that binding generates an action potential that is propagated to the central nervous system. Animals detect chemical compounds via odor-binding proteins in sensory structures such as antennae (for invertebrates) and oral and nasal cavities (for vertebrates) (Firestein 2001). Many species facilitate chemical perception by increasing the movement of air or water over their sensory receptors once a chemical has been discerned. This behavior takes the form of sniffing (mammals), tongue flicking (reptiles; Figure 5.4), and enhanced ventilation of water (aquatic species) or air (terrestrial invertebrates). As we

chemoreception The process by which an animal detects chemical stimuli.

gustation The detection of dissolved chemicals, often within the mouth.

olfaction The detection of airborne chemical stimuli.

odorant A gaseous compound that is perceived as odorous.

pheromones Volatile, organic, species-specific compounds that affect the behavior of conspecifics.

Figure 5.4. Snake tongue flick. Reptiles, like this snake, flick their tongues to detect chemical odorants in the air.

will see, researchers often use such behaviors to characterize when a particular chemical has been detected.

How can we assess the chemosensory capabilities of different species? We can begin to answer this question by simply characterizing the diversity of chemoreceptors present in a species. One avenue of research has examined the number of odorant receptor genes across taxa, which varies widely. For instance, there are about 130 odorant receptor genes in the fruit fly (*Drosophila melanogaster*), 400 in humans, 1,100 in mice, over 1,200 in the roundworm (*Caenorhabditis elegans*), and almost 2,000 in the African elephant (*Loxodonta africana*) (Bargmann 2006; Niimura, Matsui, & Touhara 2014). Presumably, variation in receptor types is correlated with chemosensory capability, but this relationship is still being examined.

A more direct way of studying chemosensory capabilities is to record behavioral or physiological responses to the experimental presentation of stimuli. A response is interpreted as an ability to detect the chemical. Examples from rodents and cuttlefish illustrate how researchers use behavioral and physiological responses to understand the chemosensory capabilities of animals.

Featured Research

Sweet and umami taste perception in rodents

Many animals use olfaction to locate food and gustation to determine its quality (e.g., its nutritional value and potential toxicity). There are five basic tastes: sweet, umami (a form of savory taste), salty, sour, and bitter (Chandrashekar et al. 2006). Each of these tastes gives animals different types of information about food (Temussi 2009). For example, sweet tastes provide information about energy content by indicating the presence of carbohydrates, including sugars. Umami tastes provide information about amino acids, the compounds that comprise proteins. Salty tastes can help animals identify salts that allow them to maintain electrolyte balance. Sour and bitter tastes can help animals avoid potentially toxic substances.

Grace Zhao and her colleagues investigated how mammals perceive sweet and umami tastes, the two main attractive taste modalities for many animals, including humans (Zhao et al. 2003). They focused on three candidate taste receptors known as *T1R1*, *T1R2*, and *T1R3*. These receptors span the neuron cell membrane, interact with molecules outside the cell, and activate signaling pathways inside the cell. To determine whether these receptors are associated with sweet and umami tastes, the researchers conducted both behavioral and physiological assays. They used knockout (KO) mice, genetically modified individuals that lacked the DNA sequence coding for one type of receptor. This procedure created KO-1, KO-2, and KO-3 mice that had one nonfunctional receptor (*T1R1*, *T1R2*, and *T1R3*, respectively), while the other two receptors were fully functional.

In the behavioral assay, mice were presented with dissolved sweet (natural and artificial sweetener) or umami (amino acid) tastants or with water as a control for a test period of 30 minutes. When mice licked a stimulus spout, a tastant or water was presented for a five-second trial, and the researchers recorded the rate of spout licking relative to a control with no tastant (i.e., water). The stimuli were presented in a random order during the test period.

A physiological assay measured taste-induced action potentials in anaesthetized mice. The researchers used electrodes to record neural responses when a small amount of each tastant was presented on the subject's tongue. All tests were conducted with each type of KO mouse as well as control mice that were not genetically modified.

Compared with KO-2 and control mice, the KO-1 and KO-3 mice had very low lick rates and showed no action potentials when amino acids were presented (Figure 5.5). This finding indicates that the KO-1 and KO-3 mice could not detect

Figure 5.5. Taste response of mice to amino acids. Compared with controls and KO-2 individuals, KO-1 and KO-3 mice exhibited (a) low mean (+SE) relative lick rates and (b) no action potentials in response to amino acids. (*Source:* Zhao et al. 2003)

Figure 5.6. Response to sucrose. Compared with controls and KO-1 individuals, KO-2 and KO-3 mice exhibited weak neural responses (action potentials) to sucrose. (*Source:* Zhao et al. 2003)

umami taste and that both the *T1R1* and *T1R3* receptors are required for the perception and detection of this taste.

KO-2 and KO-3 mice showed no lick preference and weak neural response to the artificial sweeteners and natural sugars in moderate concentrations (Figure 5.6). In contrast, KO-1 and control mice displayed a strong lick and action potential response to all sweeteners. These data indicate that both *T1R2* and *T1R3* are necessary for the perception of sweet tastes in mice.

This work suggests that both sweet and umami taste perception require multiple functional receptors, and it provides new insights into how animals can gain information about food quality.

Cuttlefish physiological response to odors

Featured Research

Water is an excellent medium for chemical dispersal, and many aquatic organisms use chemoreception to find food, habitats, and conspecifics, as well as to detect predators (Brönmark & Hansson 2000). Cuttlefish are a solitary marine species (Figure 5.7), but they aggregate to spawn, and females lay eggs in communal masses. Jean Boal and her colleagues investigated whether the common cuttlefish (*Sepia officinalis*) could detect odors from conspecific eggs to find a reproductive aggregation (Boal et al. 2010). The research team used the ventilation rate (the rate at which water is respired across the gills) to measure detection of a chemical by an individual.

The study consisted of two experiments using odor samples created from purified extracts of either fresh cuttlefish eggs (less than one hour old) or older eggs (36–48 hours old). Artificial seawater was used as a control odor. Eggs that are less than one hour old should provide a reliable cue for the location of a reproductively active female, whereas older eggs should not.

In the first experiment, Boal and colleagues placed individuals in separate aquaria for 20 minutes to acclimate. After the acclimation period, the researchers counted

Figure 5.7. Cuttlefish. These cephalopods aggregate to spawn.

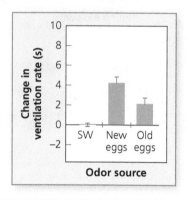

Figure 5.8. Cuttlefish ventilation rate in response to odors. Mean (+SE) ventilation rate increased significantly in response to odors from new eggs compared with odors from old eggs and seawater (SW). (*Source*: Boal et al. 2010)

individuals' number of ventilations per 30 seconds for five minutes to determine the baseline ventilation rate. Odor samples were then added to the aquaria, and subsequent ventilation rates were measured. They found that ventilation rate increased significantly after the addition of odor from new eggs but not after the addition of odor from old eggs. No change in ventilation rate was observed in response to the seawater control (Figure 5.8).

In the second experiment, the research team examined a behavioral response to odor stimuli in a Y-shaped maze apparatus. One arm of the maze contained purified extract of fresh cuttlefish eggs, and the other contained seawater as a control. The researchers measured which arm of the maze the cuttlefish entered first and found that 16 of 22 chose the arm containing the odor of fresh conspecific eggs.

Together these data indicate that cuttlefish can detect and respond to the odor of fresh conspecific eggs. This finding provides a mechanistic understanding of cuttlefish reproductive behavior.

Continual technological advances have allowed the development of even more sophisticated ways to study the chemosensory capabilities of animals. For example, researchers are now able to make electrophysiological recordings of the firing rates of specific sensory neurons in small invertebrates such as mosquitoes (Applying the Concepts 5.1).

APPLYING THE CONCEPTS 5.1

How do mosquitoes find their victims?

Many arthropods, including mosquitoes, feed on the blood of vertebrates, including humans. Early work indicated that many such species were strongly attracted to the odor of their host, and that they could detect the odor of a single host from a distance of 90 m (Vale 1980). Host odors, however, are quite complex; for instance, over 300 compounds have been identified in human sweat (Cork & Park 1996). What odors do mosquitoes use to home in on a potential victim, and can we identify specific sensory neurons they use to detect these odors?

Laboratory and field studies have revealed that many blood-feeding insects are strongly attracted to carbon dioxide and octenol—two compounds exhaled by vertebrates—as well as to several phenols found in sweat (Gibson & Torr 1999). In mosquitoes, the maxillary palp, an appendage on the mosquito's head, is the primary CO_2-sensitive organ. All along this appendage are olfactory receptor neurons within structures known as olfactory sensilla (Figure 1). Tan Lu and his colleagues recorded the firing rate of specific sensory neurons within sensilla in response to odors that were experimentally ventilated (Lu et al. 2007). Within each sensillum, three sensory neurons were identified: one was most sensitive to carbon dioxide, another was most sensitive to octenol, and a third was stimulated by a variety of phenols. By manipulating odor concentrations, the researchers determined that mosquito receptors are highly sensitive to carbon dioxide, responding to a concentration as low as 600 ppm. The

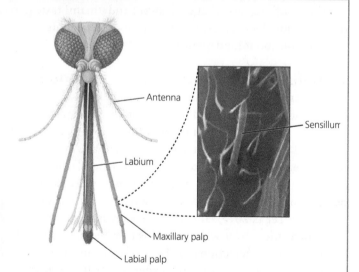

Figure 1. Head of a mosquito showing the maxillary palp and individual sensillum. (*Source:* Lu et al. 2007)

CO_2 in human exhaled air is about 38,000 ppm, but this is rapidly diluted when mixed with surrounding outdoor air, which typically contains 350–450 ppm of CO_2. Such sensitivity helps explain how mosquitoes can detect vertebrate hosts over long distances. ■

Chemosensory systems provide one important source of environmental information. The ability to detect light provides another, as we see next.

Figure 5.9. Light wavelengths visible to humans versus those visible to bees. Species differ in their ability to detect certain wavelengths of light. Note that bees can detect ultraviolet wavelengths that humans cannot.

5.3 Photoreception allows animals to detect light and perceive objects as images

Wavelengths of light—electromagnetic radiation in various frequencies—can be absorbed by, reflected by, or refracted off objects in the environment. Photoreception is the ability to perceive light and physical objects, and it results from visual sensory systems detecting electromagnetic radiation. Vision receptors, or **photoreceptors**, contain photosensitive opsins, proteins that change shape when struck by light and change the permeability of photoreceptor neurons, generating action potentials. Variation in opsin structure confers sensitivity to different wavelengths of light (Stevens 2013). Photoreceptors also vary in their sensitivity to light intensity. For example, in vertebrate eyes, there are two types of photoreceptors: **rods**, which are highly sensitive to a wide spectrum of light at low levels of illumination, and **cones**, which are sensitive to different wavelengths of light and thus color perception.

Across taxa, we see great variation in the number and types of photoreceptor cells. There is a correspondingly wide variation in the wavelengths of light that animals can detect. Most species appear to have either three or four types of photoreceptor cells, but these can have very different sensitivities to light frequencies. Human eyes can detect wavelengths between 380 and 690 nm, a range known as visible light. Many species, however, can detect wavelengths outside of this range. For example, many insects, birds, reptiles, and amphibians can detect ultraviolet wavelengths less than 380 nm, which humans cannot (Figure 5.9).

Let's consider two examples that illustrate how an appreciation of differences in visual sensory systems can aid our understanding of behavior.

photoreceptor Specialized neurons that are sensitive to light and function as visual receptors.

rods A type of photoreceptor that is sensitive to low light levels.

cones A type of photoreceptor for color vision.

Color vision in monarch butterflies

Many animals make extensive use of color information about their environment, especially when foraging or selecting a mate. One group that has a very wide range of light perception capabilities is butterflies. Butterflies have compound eyes containing three opsins that detect ultraviolet, blue, and red-orange wavelengths of light (Stalleicken, Labhart, & Mouritsen 2006). However, the presence of these opsins does not necessarily indicate the ability to perceive and discriminate among these wavelengths (i.e., true color vision). That requires the ability to discriminate among different wavelengths, independent of the intensity of light (Kelber, Vorobyev, & Osorio 2003).

Douglas Blackiston, Adriana Briscoe, and Martha Weiss used behavioral assays to determine whether monarch butterflies (*Danaus plexippus*) (Figure 5.10) have true color vision (Blackiston, Briscoe, & Weiss 2011). They trained individuals to associate a specific color with a food reward by allowing them to feed from an artificial flower of a specific color (red, purple, blue, or yellow). After eight days of training, each butterfly was placed in a cage with an array of 20 artificial flowers: 19 were different intensities of gray (in even incremental shades from nearly black to almost white), and one flower was the color that the butterfly was trained on. The colored flower was closely matched for reflectance (light intensity) with one

Featured Research

Figure 5.10. Monarch butterfly. An insect with true color vision.

Figure 5.11. Male pied flycatcher. Males vary in plumage color, ranging from brown to black, and nest in small cavities or bird boxes.

or two of the gray flowers. This step ensured that the light intensity of the colored flower and the corresponding flowers was similar but that the wavelengths reflected off of them differed. Blackiston and his colleagues measured the time individuals spent probing each flower in the array.

Monarch butterflies probed the colored flower they were trained on significantly more than the gray flower with similar reflectance. In fact, the butterflies spent more than 95% of their time at the flower with the color they were trained on. The performance of individuals trained on four different colors demonstrates that butterflies have the ability to detect a wide range of colors. This adaptation could allow them to track changes in nectar availability in the different flowers found in their summer and winter habitats.

In the next examples, we first see how a study of the visual perception of birds has changed our understanding of their plumage and mating behavior and then how snakes can use their sense of infrared radiation to find prey.

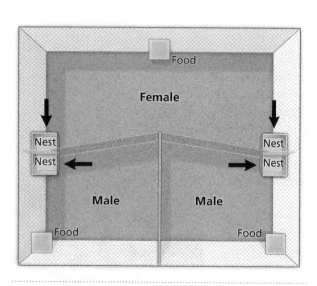

Figure 5.12. Mate choice arena. Outdoor aviary with three compartments. Females could access each nest box, while each male could only access one. Arrows indicate openings to each nest box, while circles indicate the location of food. (*Source:* Siitari et al. 2002)

Featured Research ▶

Ultraviolet plumage reflectance in birds

In many birds, females select a mate based on the male's plumage characteristics. For instance, female house finches (*Carpodacus mexicanus*) prefer to mate with males that have the greatest amount and intensity of red coloration in their plumage (Hill 1990). But in other species, variation in male plumage does not appear to affect female mate choice. Consider the pied flycatcher (*Ficedula hypoleuca*), a small cavity-nesting bird common throughout Europe (Figure 5.11). Male breeding plumage varies in color among individuals, ranging from dull brown to dark black, but previous studies have shown no consistent preference by females for either black or brown males (Alatalo, Lundberg, & Sundberg 1990; Sætre, Dale, & Slagsvold 1994; Sætre et al. 1997).

Recent work, however, indicates that the plumage of many birds reflects ultraviolet (UV) wavelengths (Eaton & Lanyon 2003). Because birds can perceive these wavelengths, they see colors and patterns in avian plumage that humans cannot. One implication of this finding is that many species that traditionally have been categorized as sexually monomorphic are in fact sexually dimorphic because the sexes differ in UV color pattern (Eaton 2005). Another is that variation in male UV plumage traits may also affect female mate choice behavior.

Heli Siitari and her colleagues tested the hypothesis that variation in male UV color affects female mate choice in pied flycatchers (Siitari et al. 2002). They captured adults at the beginning of the breeding season and housed them in a large outdoor aviary that had three compartments: one for a female and one for each of two males. The male compartments were separated by plywood, preventing males from interacting. Each male had access to one of two dual-chamber nest boxes, and the female had access to both (Figure 5.12).

In each of 13 trials, males were matched for percent black coloration, body size, and UV reflectance of the crown and back. One male was then treated with a mixture of UV-absorbing sunscreen and a waterproofing substance that reduced the UV reflectance of his plumage. The other male was the control and was treated with a waterproofing substance that did not reduce the UV reflectance of its plumage (Figure 5.13). Males were allowed to habituate to the aviary before the female was introduced. The experimental and control males exhibited similar levels of courtship display behavior after the female was introduced. Siitari and her colleagues then recorded which nest box the female eventually used to construct a nest, an indication of her mate choice.

The experimental manipulation strongly affected female behavior: 11 of 13 females preferred the control male to the UV-reduced male. Because other plumage color and behavioral traits did not differ significantly between the males, these data strongly suggest that females assess male quality based on the UV coloration of a male's plumage. As such, the study provides new insight into how females use ultraviolet cues in mate choice in this species.

Figure 5.13. Bird plumage in UV. Blue tits are another species with a UV-reflectant crown patch used in courtship. The top row is a blue tit photographed in normal light, and the bottom row is a blue tit photographed using UV light. The left column shows the control, while the right column shows the UV-reduced treatment, which reduced the UV reflectance of the crown patch. (*Source*: Withgott 2000)

Infrared detection in snakes

Featured Research

Earlier, we saw that many species can detect electromagnetic wavelengths shorter than those of visible light. In contrast, other species have evolved the ability to detect longer wavelengths, such as those of infrared radiation (700–1000 nm), which is emitted by warm objects and animals.

The ability to detect infrared wavelengths is relatively rare among species but is found in a diverse set of taxa. For example, beetles in the genus *Melanophila* possess pit organs on their thorax that are sensitive to infrared radiation. These beetles lay their eggs on trees that have recently been killed by fire, and they are known to approach forest fires from a distance of several kilometers. Fires emit smoke, ash, visible light, and infrared radiation, but these beetles seem to rely exclusively on infrared radiation to detect burning trees (Evans 1964; Schmitz & Bleckmann 1998). More recently, researchers have discovered that vampire bats also possess heat-sensitive pits on their nose. Vampire bats feed on the blood of vertebrates, and the ability to detect infrared wavelengths is thought to enhance their ability to locate a vein (Gracheva et al. 2011). But the best-known animals that have the ability to perceive infrared wavelengths are pit viper snakes in the subfamily Crotalinae. These predators ambush warm-bodied endothermic prey via detection of infrared wavelengths.

All crotaline snakes possess facial pit organs containing neuron terminals that detect infrared wavelengths (Figure 5.14). Infrared information is relayed to and processed by the central nervous system, providing a complementary visual system to the one that detects visible light. But how widespread is this system? Can other viper species (non-crotaline vipers) detect infrared radiation?

Adam Safer and Michael Grace compared the infrared detection behavior of two species of crotaline snakes (copperhead, *Agkistrodon contortrix*; and Western diamondback, *Crotalus atrox*) with that of three species of non-crotaline snakes of the viperine genus *Bitis* (puff adder, *Bitis arietans*; Gaboon viper, *B. gabonica*;

Figure 5.14. Pit organ. Western diamondback pit viper with an infrared pit organ under the eye.

Figure 5.15. Behavioral responses to warm and cool targets. Mean (+SE) tongue flick rate. Crotaline snakes exhibited more tongue flicks toward the warm (blue bar) target than the cool (orange bar) target. Viperine snakes exhibited weak responses to both targets. (*Source:* Safer & Grace 2004)

and rhinoceros viper, *B. nasicornis*) (Safer & Grace 2004). All of these snakes hunt mammals whose body temperature typically exceeds ambient temperature. The researchers presented snakes with two white water-filled balloons heated to different temperatures, which were suspended above the floor of the test arena. One balloon was room temperature (20–25°C), while the other was heated to 32–35°C, a temperature similar to that of a live mammal. The balloon targets were similar in size (each 5 × 9 cm) and were placed at one end of a 95 cm diameter test arena with flat black walls.

Individual snakes were placed in the arena and allowed five minutes to acclimate. The pair of targets (placed 6.5 cm apart) was then introduced 15–20 cm in front of the snake. The researchers recorded tongue flicks, head turns, approaches, and strikes at each target.

Safer and Grace found that the crotaline snakes exhibited a greater number of responses to the warm target than to the cool target (Figure 5.15). These snakes struck the warm target in 10 of 31 trials but never struck the cool target (Safer & Grace 2004). In contrast, the behavioral responses of the non-crotaline snakes did not differ with respect to the two targets; these snakes did not strike either target in any of the trials.

These data confirm that crotaline snakes detect and respond to warm objects, and potentially animals, differently than they do to cool objects: they can perceive differences in infrared radiation. However, this infrared imaging system is not universal among viper species, as demonstrated by the lack of response in the *Bitis* species. The ability to perceive infrared wavelengths provides individuals with additional information about their environment that can facilitate the detection of both prey and potential threats.

These studies show how researchers have investigated the ability of species to detect and respond to different wavelengths of electromagnetic radiation in the environment. In addition to chemoreception and photoreception, a third source of environmental information is derived from the detection of mechanical waves, as we see next.

5.4 Mechanoreceptors detect vibrations that travel through air, water, or substrates

mechanoreceptors Sensory receptors sensitive to changes in pressure.

Sounds, like a bird's song or a lion's roar, are mechanical waves that can be heard. The vibrations of an insect walking on a leaf or the waves made by a fish moving through water are mechanical waves that can be felt. **Mechanoreceptors** allow animals to detect mechanical waves or vibrations that travel through air, water, or physical substrates such as the ground. The sensory structures to detect mechanical waves differ greatly across species. For many terrestrial species, auditory receptors—specialized structures that efficiently gather mechanical waves—are found within ears. These receptors are not always located on the head, however; invertebrate species have them on different parts of their body, including their thorax, wings, or legs. As we saw for photoreceptors, taxa vary greatly in their ability to detect different frequencies (sound waves per second) of mechanical waves (Figure 5.16). Frequencies below the lower limit of human hearing (20 Hz) are known as infrasounds, while those above the upper limit (20 KHz) are called ultrasounds.

Let's begin our examination of mechanoreception with two examples of sounds that provide information about conspecifics. The first example involves

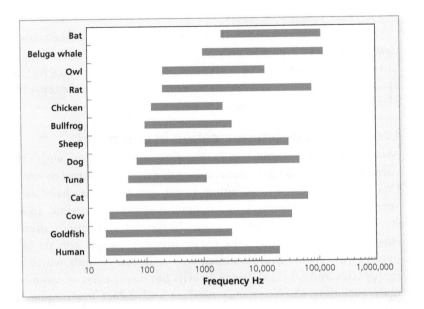

Figure 5.16. **Auditory abilities of species.** Species differ in the range of frequencies their auditory system can detect. Each bar represents the range of frequencies detected.

the detection of high-frequency sounds in insects, while the second involves the detection of low-frequency sounds in elephants.

Ultrasonic song detection in moths

Featured Research

Insects have one set of mechanoreceptors to detect nearby sounds and others to detect distant sounds. Near-field detectors, such as the Johnston's organ in mosquitoes, are usually sensitive to low frequencies, those that are less than 100 Hz. Far-field receptors, such as the tympanal hearing organs found in many insects, are sensitive to a fairly wide range of high frequencies (2–100 KHz), including ultrasounds (Stevens 2013). Thus, insects with far-field receptors can detect the presence of predatory bats that hunt by emitting ultrasounds, as we will see later in the chapter.

In addition, some moth species also produce their own ultrasonic pulses. Why? One possibility is that they are used in courtship. Ryo Nakano and his colleagues (Nakano et al. 2006) examined the possible use of such sounds in courtship behavior in Asian corn borer moths (*Ostrinia furnacalis*). The researchers began by recording the courtship songs and behaviors of male Asian corn borer moths by releasing a single male and eight virgin females into a small cylindrical mesh cage (5.5 × 5 cm). Males landed near females and quickly began producing courtship behaviors and song (Figure 5.17). Male courtship involves raising the wings to an upright position and vibrating them quickly, which produces an ultrasonic pulse. Females did not produce any sounds, suggesting that females may use male ultrasonic courtship songs to select a mate.

To examine the role of male ultrasound in courtship, the research team conducted an experiment in which they deafened one group of females by puncturing their tympanal membranes. In control females, no manipulation was done. A sham treatment was carried out on another group of females, in which non-sound-detecting membranes on the leg were punctured. Sham treatments allow individuals to experience the same treatment but without the intended consequences; in this case, the females were not deafened. Individual males in the laboratory were then released in a small flight tunnel 1 m downwind from a small group of either deafened or hearing females, and ensuing courtship and mating behavior was noted.

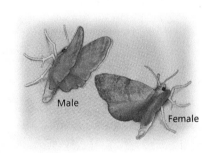

Figure 5.17. **Moth courtship.** A male (left) lands near a female (right) and vibrates his wings to produce ultrasonic pulses during courtship. (*Source:* Nakano et al. 2006)

Hearing females mated within two minutes in 84% of the trials. In contrast, deafened females mated in only 59% of the trials. These data suggest that male ultrasonic pulses are an important component of mate choice by females in this species. Next let's examine mechanoreception of low-frequency sounds and how elephants use this information.

Featured Research

Long-distance communication in elephants

Many animals use contact calls to maintain associations with others in their social group (Kondo & Watanabe 2009). For some animals—especially those that live in large, wide-ranging social groups or in habitats with dense vegetation—maintaining visual contact may not be possible. In such circumstances, sounds can be used to maintain contact. But sounds attenuate (diminish) as they travel through the environment. How can contact be maintained over long distances? One possibility is by using infrasounds, which have very low frequencies and attenuate more slowly. Several species of large-bodied mammals, including elephants, can produce and detect low-frequency infrasonic sounds. Do elephants communicate with one another over long distances using infrasound frequencies?

Karen McComb and colleagues investigated infrasonic social communication in African elephants (*Loxodonta africana*) (McComb et al. 2003). Elephants live in social groups centered on adult females, which are typically related, and their immature offspring (Figure 5.18). Elephants produce unique contact calls that are audible to humans to maintain social cohesion, as well as unique low-frequency (less than 20 Hz) vocalizations that are not audible to humans.

Elephants exhibit characteristic behaviors in response to hearing a contact call of a family member: they produce contact calls of their own and move in the direction of the call. In response to hearing vocalizations of less known, potentially threatening individuals, they behave differently: individuals tend to bunch into a defensive posture and do not move toward the sound. These differences in behavior allowed the research team to determine the distance over which individuals responded to familiar and unfamiliar infrasonic contact calls by playing the calls at different distances.

The study was conducted at Amboseli National Park in Kenya on a population of elephants that had been studied for almost 30 years, so all individuals could be recognized and family relationships were known. The team recorded infrasonic contact calls of different elephants on specialized low-frequency digital audiotape. The researchers took advantage of situations when a particular female was absent from a group and played either her contact call or the call of an elephant that was much less familiar to that group. They did this starting at a distance of 2.5 km and noted the behavior of individuals in the group. If no response was detected, they approached the group at 0.5 km intervals and again played the call until they noted a behavioral response. They also measured several behaviors indicative of call detection, including listening (ears held stiff), smelling, or streaming (the production of a new secretion from glands in the forehead). Data were collected for seven different groups over seven years.

Elephants responded to familiar contact calls by producing their own contact calls or moving in the direction

Figure 5.18. African elephants. Elephants live in social groups.

of the call when it was 1.0 km away or closer. The unfamiliar vocalizations elicited a response when played up to 1.5 km away, but individuals never moved toward or produced contact calls in response to the less familiar vocalizations; instead, they tended to bunch together in a defensive posture. The researchers concluded that infrasounds are used to maintain contact among individuals in their group. The ability to detect long-distance acoustic vocalizations also allows individuals to acquire information about the presence of another, potentially threatening, group while it is still fairly distant.

In the examples above, we have seen how the detection of sounds provides important environmental information about the presence of conspecifics. Next, let's examine mechanoreception in water and see how this allows fish to find their prey.

Catfish track the wake of their prey

Featured Research

Many fish live in habitats where visibility is reduced by cloudy water or dense vegetation. In addition, many fish are nocturnal. Therefore, these animals often rely on nonvisual sensory systems to obtain information about their environment. Fish detect mechanical waves that travel through water both in their otolith organs (inner ear structures) and via their **lateral line system** (Coombs & Montgomery 1999). This latter mechanoreceptor system consists of pits or tubes that run along the side of the fish's body and head (Figure 5.19). These pits contain sensory receptors called neuromasts, which provide information about water velocity and acceleration, as well as the direction of water movements. Fish use mechanoreceptors both for hearing and for detecting their own body position and orientation in dim light environments, as well as for getting information about water pressure or movement that can indicate the presence of other organisms (such as predators or prey).

lateral line system In fish, a network of mechanoreceptors that provide hydrodynamic information.

Kirsten Pohlmann and her colleagues investigated how nocturnal, fish-eating catfish find food in the dark. They studied European catfish (*Silurus glanis*), which are 20 to 25 cm long and feed on small fish (Pohlmann, Grasso, & Breithaupt 2001). The research team used guppies (*Poecilia reticulata*) as prey fish because this species creates a swimming wake that has been well described. Pohlmann's

Figure 5.19. Lateral line system. The lateral line system is composed of many neuromasts that detect vibrations.

team aimed to determine whether catfish can use cues provided only from the wake of individual fish to stalk and attack their prey. An infrared video system tracked the movement paths of both the predator and the prey in an environment of complete darkness. The researchers used four catfish and observed each one stalking up to ten individual guppies. Each catfish was allowed to acclimate to the experimental tank for one hour in the dark, after which the first guppy was added to the tank. After a guppy was consumed, another was added every 20 minutes to replace the one that had been eaten. The researchers ended a trial when ten guppies had been consumed or when no guppy had been consumed in 20 minutes. They used software to digitize the movement sequences of both the catfish and the guppies and classified them in one of three ways: (1) path following (the catfish followed the guppy wake), (2) head-on encounters (the catfish encountered the guppy head-on without a previous encounter), or (3) attack on a stationary guppy (Figure 5.20).

Pohlmann's team found that 80% of the attacks occurred on moving guppies. In the majority of all attacks, the catfish were following the same path as the guppy before the attack. The researchers concluded that the catfish appeared to be following the wake of the guppies to find and attack their prey in the dark.

However, the wake of a fish contains two cues that a predator might use to track its prey: hydrodynamic signals and chemical signals. To tease apart which sensory cue catfish use, Pohlmann's team conducted a second experiment using the same setup (Pohlmann, Atema, & Breithaupt 2004). In this experiment, they manipulated either the lateral line or the gustation system of the catfish and compared individuals' behavior to that of controls. The lateral line provides hydrodynamic information and can be rendered temporarily nonfunctional by immersing a fish in a $CoCl_2$ (cobalt chloride) solution for six hours prior to the experiment. In fish, olfactory receptors are located in the nose while taste receptors are present in the mouth; but in some fish, including catfish, taste receptors are also found on parts of the body surface (the external gustation system). In catfish, the external gustation is the major chemical sense used to find food and

Figure 5.20. Wake following. Two examples (a and b) of wake following, leading to an attack. The blue lines represent the catfish predator and the orange lines the guppy prey. (*Source:* Pohlmann, Grasso, & Breithaupt 2001)

TABLE 5.1 Catfish attack mode and success. Intact catfish relied heavily on the wake-following attack mode, whereas lateral line–ablated fish most often used a head-on attack mode. Taste–ablated fish used both head-on and wake-following attacks. (*Source:* Pohlmann, Grasso, & Breithaupt 2001)

TREATMENT	CAPTURE SUCCESS RATE (% CAPTURE)	PERCENT OF ATTACKS		
		WAKE FOLLOWING	HEAD-ON	STATIONARY
Intact	65	55	30	15
Nonfunctioning lateral line	17	6	88	6
External taste ablated	60	27	58	15

so it was manipulated by surgically removing (ablating) an area of the dorsal medulla oblongata, which controls this system.

Using identical procedures to record the attack behavior of the catfish, the researchers found that individuals with a nonfunctioning lateral line attacked guppies head-on 88% of the time. Most of the gustation-ablated fish also attacked guppies head-on, though less often (58% of attacks), with only 27% of their attacks characterized as path following (Table 5.1). The capture success rates also differed across treatments. Control and gustation-ablated fish exhibited a similar, relatively high, capture success (approximately 60%). In contrast, the success rate for individuals with a nonfunctioning lateral line system was only 17%. Pohlmann and colleagues concluded that the lateral line is a very important factor in catfish's ability to track the wake of their prey; in contrast, the external gustation system does not appear to be as important in facilitating this mode of attack behavior. The lateral line system may be key not only because it provides constant updated information about the movement of the prey but also because it provides feedback to the predator about its own movements and position.

For animals that live in environments with reduced visibility, the ability to acquire such sensory information is an important adaptation that helps them capture prey. Next, we turn our attention to mechanical waves that move through substrates.

Substrate-borne vibrations

The examples above focused on the detection of mechanical waves that travel through air and water. Mechanical waves also travel through a variety of substrates. For spiders, vibrations in their web enable the detection of prey. Other species find potential mates or predators by detecting vibrations that travel through stems or the ground. A growing body of literature suggests that substrate-born vibrations are a rich source of environmental information for thousands of insect and spider species, as well as for many species of crustaceans, amphibians, reptiles, and even mammals (Hill 2009). Again, we see that the mechanoreceptors detecting these vibrations are not restricted to the head of an animal. In many species, receptors are found on the legs or feet, body regions that are frequently in contact with the substrate.

Next, let's examine a recent study of mechanoreception of substrate-borne vibrations in antlion larvae.

Antlions detect substrate-borne vibrations

Featured Research

Arnold Fertin and Jérôme Casas conducted an experiment to determine whether insect larval antlions can detect their ant prey using substrate-borne vibrations

SCIENTIFIC PROCESS 5.1

Antlion mechanoreception

RESEARCH QUESTION: *How do antlions detect prey in their sand pits?*

Hypothesis: Antlions detect prey via substrate-borne vibrations.

Prediction: Antlions will toss sand in response to vibrations.

Methods: The researchers:

- Recorded the substrate-borne vibrations made by walking ants (Figure 1)
- Created ant-walking vibrations in pits
- Recorded antlion behavior in response to the vibrations

Figure 1. Vibration recording. Mechanical waves produced by a walking ant are recorded by a microphone buried in the sand. (*Source*: Fertin & Casas 2007)

Results:

- 26 of 30 antlions tossed sand in the direction of the vibration.
- 13 of 30 antlions attacked the needle.

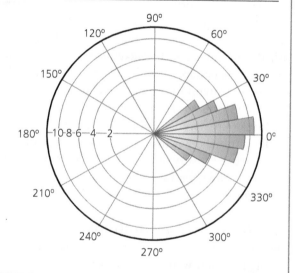

Figure 2. Antlion sand-tossing behavior. Distribution and direction of sand tossing in response to vibrations caused by the needle simulating ant footfalls. The needle's position relative to the antlion is standardized here at 0 degrees. The distance of the needle from the antlion is indicated by the numbers in the left half of the circle. (*Source*: Fertin & Casas 2007)

Conclusion: Antlions detect substrate-borne vibrations in their prey in their sand pit.

EVALUATE

How could you determine how the antlions detect vibrations from prey?

(Fertin & Casas 2007). Antlions (family Myrmeleontidae) are found worldwide, often in arid environments. Their common name is a reference to their larvae's feeding behavior: they feed on ants. Species that live in sandy habitats typically dig small, funnel-shaped pits (5 cm deep, 7.5 cm in diameter) at the steepest angle that the sand can maintain (Scientific Process 5.1). These funnels function as a trap for prey because the slope is steep and unstable: any ant stepping into a pit will begin to slide down and then attempt to climb out. Antlion larvae wait buried in the sand at the base of the funnel. Once an ant enters a pit, the antlion throws sand in its direction, causing small avalanches on the sides of the pit. These bring the prey farther down the slope until it can be attacked. Given that the antlion larvae are buried and cannot visually detect their prey, how do they detect those prey that fall into their traps?

Fertin and Casas tested the hypothesis that they do so via substrate-borne vibrations. The researchers first recorded the substrate waves made by an ant walking on 5 mm of sand placed over a condenser microphone. They could then mimic the sounds of a walking ant by using an electromagnetic shaker with a 5 mm needle in contact with the surface of the sand. This method allowed Fertin and Casas to experimentally introduce ant-walking vibrations in an individual antlion's pit and observe its response. Individual antlion larvae (*Euroleon nostras*) were collected from the field and placed in small sand-filled arenas; individuals readily constructed sand pits. The researchers then placed the tip of the shaker needle 10 mm behind the larvae's head at different positions and simulated the vibrations of a walking ant.

None of the 30 antlions responded to the needle before it produced vibrations. However, once the needle began to produce vibrations, 26 larvae responded by tossing sand in the direction of the needle. In addition, 13 antlions moved up and bit the tip of the needle, much as they would in response to the presence of actual struggling prey in their pit. Fertin and Casas concluded that such behavioral responses provide clear evidence that antlion larvae detect their prey using substrate-borne vibrations. Antlion larvae possess mechanoreceptor sensilla on their head, legs, and abdomen; further work is required to determine the specific sensory receptors that aid in detecting vibrations made by their prey.

The above examples represent different types of information animals gather about their environment by detecting mechanical waves that travel through air, water, and substrates. Next, we examine two less well-known sensory systems: the detection of electric and magnetic fields.

5.5 Some animals can detect electric or magnetic fields

Some aquatic organisms, such as skates and rays, often need to find buried prey or potential mates. As these buried organisms respire, muscles contract, resulting in the movement of ions, which creates a weak electric field. This electric field can be detected in water or moist environments due to the density of the medium (Ruxton 2009).

Other organisms need to determine the appropriate direction for their long-distance movements. The earth's geomagnetic field is made up of electric currents and magnetic materials, and animals can use this field to determine directions. We examine these sensory systems next, focusing first on sharks and then on birds, taxa in which electroreception and magnetoreception, respectively, have been the focus of much research.

Electroreception

Some species, particularly aquatic ones or animals in moist environments, have evolved **electroreception**, the ability to detect weak electric fields. These species can electrolocate individuals in dark or murky environments or find buried prey. Electroreception is most widespread in elasmobranchs (sharks and rays) but has

electroreception The ability to detect weak electric fields.

also been identified in some bony fishes and amphibians, a few mammals (e.g., duck-billed platypuses, echidnas), and some insects.

Sharks detect electric fields

In sharks and rays, electric fields are detected in structures known as ampullae of Lorenzini, a set of pores mostly located on the head that contain a crystalline gel and electrosensitive neurons (Figure 5.21). The presence of an external electric field stimulates these neurons, which then send an electrical impulse to the central nervous system. Electric fields generated by organisms attenuate rapidly in aquatic environments and so can be detected only over distances of a few meters. How have researchers studied electroreception? A typical method is to create weak electric fields and observe the behavior they elicit.

Stephen Kajiura and Kim Holland examined the orientation pathway and attack behavior of naïve juvenile scalloped hammerhead (*Sphyrna lewini*) and sandbar sharks (*Carcharhinus plumbeus*) in response to electric fields that simulated the presence of prey (Kajiura & Holland 2002). The research team investigated the hypothesis that hammerhead sharks have greater electrodetection ability due to the enlarged lateral expansion of their head (Figure 5.22). The ability to detect electric fields produced by other organisms would enhance the ability of these predators to locate benthic prey (fishes and crustaceans) buried in the sea floor.

The researchers created a large experimental pen (10 m × 20 m) that enclosed part of a natural reef in Kaneohe Bay, Hawaii. In the shallow part of the pen, the researchers buried a 1 m² acrylic plate that contained four dipoles, positive and negative charges separated by a small distance, producing a weak electric field in a specific location. A single shark was introduced into the shallow portion of the pen, and prey odor was released in the center of the dipole array to stimulate food-searching behavior. An individual dipole was then activated, creating a weak electric field that simulated the type of electric field produced by a buried prey species. Different dipoles were activated across different trials, and a video camera recorded the sharks' behavior.

Both species oriented toward the active dipole by changing their swimming direction to move closer to it. These changes in orientation occurred when both species were less than 30 cm from the active dipole, and both then attacked it with a bite (Figure 5.23). None of the sharks attacked a nonactive dipole, and all individuals stopped biting an active electrode when the current to it was turned off. These data provide strong evidence that scalloped hammerhead and sandbar sharks orient to weak electric fields and that such fields elicit feeding behavior. However, the researchers found no support for the hypothesis tested because there was no significant difference between the species in the behavioral response thresholds for the weak electric fields used in the experiment. Perhaps hammerheads can detect weaker electric fields than sandbar sharks, but that question requires further study.

The ability to detect electric fields is a relatively rare sensory ability, but it is an important source of information for a variety of species searching for hidden prey. Next, we examine a sensory system found in a wide variety of taxa that is used to determine direction: the ability to detect magnetic fields.

Figure 5.21. Ampullae of Lorenzini on shark head. These structures detect electric fields.

Figure 5.22. Shark heads.
(a) Hammerhead sharks have an enlarged lateral expansion of their head relative to (b) sandbar sharks.

Magnetoreception

Many animals migrate long distances, making two-way movements from one location and back each year. To accomplish these movements, individuals need to determine the appropriate direction to move in different seasons. The earth's magnetic field can provide such directional information (Figure 5.24). **Magnetoreception**, the ability to detect the earth's magnetic field, provides animals with directional information in three ways. First, the magnetic field flows from south to north—it has polarity. Second, magnetic field lines are oriented vertically with respect to the earth's surface at magnetic north and south; at these poles, we can say that the lines have a 90° angle of inclination. These lines become more parallel to the ground as they move toward the magnetic equator, where they have a 0° angle of inclination to the earth's surface. Thus, their angle of inclination with respect to the horizontal varies with latitude. Third, the intensity of the field varies predictably: it tends to be strongest at the poles and weakest at the magnetic equator. Use of a geomagnetic compass has been documented in dozens of species in many taxa, including snails, crustaceans, insects, fish, amphibians, reptiles, birds, and mammals. These behavioral studies have typically examined how manipulations of field direction affect the movement of individuals (e.g., Wiltschko & Wiltschko 2005).

It is less clear how animals detect the earth's magnetic field. Research has focused on two mechanisms. In many animals, certain cells contain crystals of magnetite, a form of iron oxide whose magnetic properties depend on its size and shape. Because crystal magnetite is a permanent magnet, its magnetic field interacts with the earth's magnetic field, creating a force (pressure) that can be detected. Experiments have shown that exposure to external magnetic fields can temporarily disrupt an animal's orientation ability (Wiltschko et al. 1998). More recent work has demonstrated that birds can detect both variation in intensity of an external magnetic field and its direction (Stapput et al. 2008; Wu & Dickman 2012). But where is the magnetic sense located?

Experiments as well as neurophysiological studies on a variety of vertebrates have revealed that magnetite appears to be frequently located in the front of the

Figure 5.23. Shark response to weak electric field. (a) An individual is swimming before the dipole is active. (b) The dipole is activated (indicated by the vertical line) to produce a weak electric field, and the shark initiates orientation to it. (c) The shark makes a rapid 90° turn to move toward the active dipole. (*Source:* Kajiura & Holland 2002)

magnetoreception The ability to detect and orient using magnetic fields.

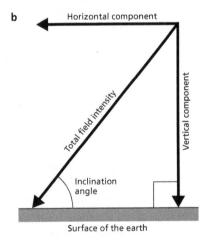

Figure 5.24. Magnetic field of the earth. (a) The magnetic vectors around the earth. Arrows indicate the local vector and angle of inclination; their length indicates the intensity (strength) of the field. (*Source:* Wiltschko & Wiltschko 2005) (b) The total field intensity is composed of a horizontal component and a vertical component. (*Source:* Lohmann, Putnam, & Lohmann 2008)

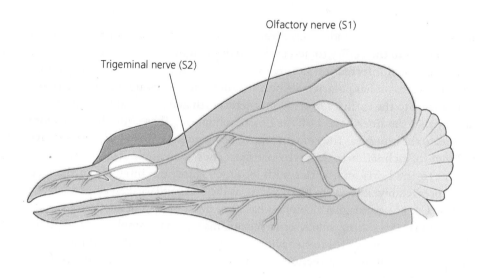

Figure 5.25. Lateral view of pigeon head. The olfactory nerve S1 carries nerve impulses from the olfactory mucosa to the brain, while the trigeminal nerve S2 carries nerve impulses from the upper beak to the brain. S1 and S2 were sectioned in the two treatments. (*Source:* Mora et al. 2004)

Figure 5.26. Percentage of correct choices in four pigeons. Individuals with a sectioned olfactory nerve (triangles) continued to make a high level of correct choices. Those whose trigeminal nerve was sectioned (circles) exhibited a significant decline in their performance. (*Source:* Mora et al. 2004)

head (Kirschvink, Walker, & Diebel 2001), particularly in the olfactory lamellae of fish and in the upper beak of birds (Diebel et al. 2000; Fleissner et al. 2003). Cordula Mora, Michael Davidson, Martin Wild, and Michael Walker conducted a series of experiments to identify the nerve involved in transmitting magnetic information to the brain in birds (Mora et al. 2004). They focused on two candidate nerves in the upper beak, the olfactory and ophthalmic branch of the trigeminal (Figure 5.25).

Mora and her colleagues trained captive pigeons to discriminate the presence or absence of an induced magnetic anomaly in an experimental chamber. The researchers used a magnetic coil to create a magnetic field that was substantially different in intensity and angle of inclination from the normal background field. Four pigeons were trained to move to one feeder if the anomalous magnetic field was present but to move to the other feeder if the background magnetic field was present. If birds made the correct choice, they were rewarded with food. After 24 consecutive training sessions, individuals selected the correct feeder almost 70% of the time, indicating the ability to detect the anomalous magnetic field. Next, the research team cut the ophthalmic branch of the trigeminal nerve in half the subjects and the olfactory nerve in the others. Those with impaired olfactory nerves continued to make the correct choice about 70% of the time, whereas those with impaired trigeminal nerves made the correct choice only 50% of the time (Figure 5.26). Thus, magnetite in the upper beak appears to provide information about magnetic field intensity and inclination that is then transmitted to the brain via the ophthalmic branch of the trigeminal nerve. These results provide strong support for the hypothesis that magnetite in the upper beak is involved in magnetoreception in pigeons.

However, magnetite-based magnetoreception is not the whole story. Some birds and amphibians cannot orient properly in captivity unless they are exposed to short wavelengths of light, which suggests that the visual system may also play a crucial role in magnetoreception (e.g., Wiltschko & Wiltschko 1981; Deutschlander, Phillips, & Borland 1999). This finding has led to the idea that magnetoreception might also be accomplished when a receptor molecule absorbs photons (light particles). The energy absorbed temporarily changes the chemical property of the molecule in a manner that makes it responsive to external magnetic fields (Ritz, Adem, & Schulten 2000). Photopigments known as cryptochromes are present in the retina of vertebrates; they possess such chemical properties and so are considered to be possible receptor molecules (Gegear et al. 2010). There is still much to learn about the biophysical basis of magnetoreception in animals, and this area of research remains active (Wiltschko & Wiltschko 2005; Gould 2010).

We have seen how researchers have discovered the diverse set of sensory systems that animals use to acquire information from the environment. This information helps direct animals' behavior in the search for food, mates, and members of a social group. This is by no means a complete accounting of sensory capabilities, however: for instance, many species can acquire environmental information by detecting polarized light (Horváth & Varjú 2004). For now, it is important to realize that understanding how animals perceive their world provides greater insight into understanding their behavior.

So far, we have examined sensory systems of individual species. In the next section, we examine how adaptations in sensory systems evolve in response to selection pressures from other species.

5.6 Predator and prey sensory systems co-evolve

Interactions between two species can affect evolution when the fitness of each species is affected by the outcome of the interaction. For example, in a predator–prey encounter, successful prey capture enhances predator fitness, while prey that escape have higher fitness than those that do not. Selection will act on each species, but in opposite directions with respect to the interaction: it will favor predator adaptations that enhance capture efficiency while simultaneously favoring adaptations in prey that enhance escape. This common scenario can lead to what is often referred to as a co-evolutionary arms race that in turn leads to changes in the traits of predators and their prey over time. Let's examine one such arms race that involves sensory systems.

Insect tympanal organs: an evolved antipredator adaptation

Many insects have receptor organs that can detect sound frequencies ranging from 20 to 100 kHz. The basic structure, the tympanal organ, has a vibrating membrane called the tympanum, which is associated with an internal airspace. In addition, the tympanal organ has from one to several thousand receptors called chordotonal sensilla that transmit electrical impulses to the brain when stimulated by high-frequency sound waves. These structures are found on the thorax, legs, abdomen, or base of the wing, and they range from a single organ in praying mantises (Figure 5.27) to six pairs of organs in some grasshoppers (Yager 2012).

Phylogenetic analyses suggest that ultrasonic-detecting insect organs evolved independently 18 different times. The presence of ultrasonic-detecting tympanal organs in so many diverse insect taxa may indicate that they evolved as an antipredatory response to their main predator, echolocating bats (Yack & Dawson 2008). Further support for this hypothesis comes from the fact that while insect species differ in the range of frequencies their chordotonal sensilla detect, most exhibit

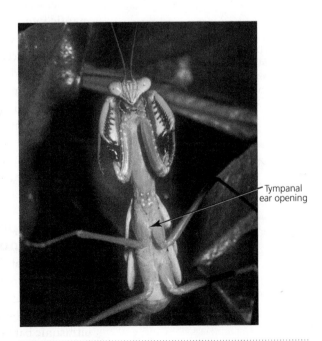

Tympanal
ear opening

Figure 5.27. Tympanate ear. Praying mantis, showing the location of the tympanal ear opening.

Figure 5.28. Evasive flight in mantids. Change in flight paths of four different mantids (different colors) after playing an ultrasonic click similar to one produced by a bat predator. The sequence of colored arrows represents the successive locations of a moth in two-dimensional space. Note that all individuals drop quickly (move lower on the Y-axis) after detecting the bat click (black arrows). (*Source:* Yager 2012)

Figure 5.29. Percentage of mantid escapes. Deafened mantids had a significantly lower escape frequency than individuals that could hear. (*Source:* Triblehorn et al. 2008)

peak sensitivity between 20 and 60 kHz, the range of frequencies typically used by echolocating bats when they are hunting prey.

Echolocating bats vary the rate at which they emit ultrasonic vocal pulses. When bats search their environment, they produce pulses at a low rate (fewer than 10 pulses per second). When prey has been detected, they increase the pulse rate, presumably to facilitate orientation toward the moving object. Then, just prior to attack, they increase the pulse rate even more, to about 200 pulses per second as the attack is conducted. This variation in pulse rate provides important information to a potential victim that can detect it, regarding its distance from an approaching predator and the onset of an attack.

Once ultrasonic frequencies are detected, insects with tympanal organs display a variety of responses (Yager 2012). When low–pulse rate sounds are detected, individuals respond by flying away from the source of the sound, presumably to reduce probability of detection. However, when high–pulse rate sounds are detected, individuals exhibit rapid, erratic changes in their flight pattern and will often initiate a strong dive to avert capture (Figure 5.28). How effective are these antipredator behaviors?

Jeffrey Triblehorn and his colleagues tested the hypothesis that the ability to detect ultrasonic frequencies provides a strong survival advantage to species hunted by bats (Triblehorn et al. 2008). The research team manipulated the ability of praying mantids (*Parasphendale agrionia*) to detect ultrasound by filling their single tympanal organ with Vaseline. This method eliminated mantid responses to frequencies between 20 and 60 kHz, effectively deafening them to the ultrasonic frequencies used by hunting bats. The researchers examined bat–mantid interactions in free-flight encounters in a flight room. One mantid, either a control or a deafened individual, was released along with a single big brown bat (*Eptesicus fuscus*). The researchers recorded all bat vocalizations, the flight behavior of both individuals, and the outcome of the interaction (capture or escape).

Control mantids that could detect ultrasonic frequencies escaped in 74% of trials, typically by initiating a power dive just before the attack (Figure 5.29). In contrast, deafened mantids escaped in only 35% of trials, and none exhibited a power dive in response to an attack. These data support the hypothesis that the ability to detect ultrasonic frequencies is an effective antipredator adaptation. With this background, we now examine co-evolution between bats and their prey.

Featured Research

Predator–prey sensory system co-evolution: bats and moths

Nocturnal bats around the world have used echolocation to detect and pursue their prey for over 50 million years (Figure 5.30) (Springer et al. 2001). Predation using ultrasonic vocalizations created strong selection on their prey to detect an approaching bat, and an effective antipredator response evolved. Ears sensitive to ultrasonic bat vocalizations have evolved independently in six orders of insects: Lepidoptera (moths), Neuroptera (lacewings), Coleoptera (beetles), Dictyoptera Mantodea (praying mantids), Orthoptera (crickets and grasshoppers), and Diptera (parasitic flies). As we saw above, this attribute allows individuals to take evasive action just prior to attack. But a second counteradaptation has also evolved in many species: the ability to produce ultrasonic sounds that can deter bat attacks. Two types of such adaptations are known to exist.

First, some moths, particularly those in the family Arctiidae, contain toxic compounds that are distasteful to predators, including bats. Many species acquire these compounds from the plants they consume as caterpillars. In this large family, some species have evolved the ability to produce short, repetitive ultrasonic clicks using modified corrugated exoskeletal structures called tymbals. Bats can detect these clicks and learn to avoid the unpalatable species making them. The production of signals that indicate noxiousness is a phenomenon known as aposematism. (We will examine aposematism in more detail as an important adaptation in predator–prey communication in Chapter 6.)

Figure 5.30. Bat attacking a moth. Bats use ultrasonic vocalizations to hunt moths.

In a groundbreaking study, Dorothy Dunning obtained the first evidence of the antipredator nature of acoustic aposematism. She presented caged bats with unpalatable, ultrasound-producing arctiid moths and palatable, silent non-arctiid moths and found that the bats avoided the sound-producing moths more often than they avoided silent moths (Dunning 1968).

More recently, Nikolay Hristov and William Conner tested the acoustic aposematism hypothesis using naïve bats that had never been exposed to moth sounds (Hristov and Conner 2005). The hypothesis assumes that predators learn to avoid sound-producing moths because they are unpalatable. It predicts that bats will at first attack noxious, sound-producing moths but will quickly learn to associate the sound with bad taste and will attack sound-producing moths less frequently than silent moths.

Hristov and Conner tested the hypothesis by conducting a simple experiment in a flight cage in which they presented naïve big brown bats (*Eptesicus fuscus*) with moth species that differed in sound-production (S+ or S–) and chemical defense (C+ or C–) capabilities. When none of the moths were chemically defended, the bats attacked sound-producing and silent moths with a similar high frequency. However, when the sound-producing moths were noxious and the silent moths were palatable, bats quickly learned to avoid the sound-producing moths as predicted (Figure 5.31). Thus, the ability to produce ultrasonic sounds, combined with a chemical defense, enhances the survivorship of arctiid moths that have evolved the ability to produce sounds that bats can detect.

The second hypothesized advantage of the production of ultrasonic sounds by arctiid moths is that it interferes with a bat's ability to echolocate by jamming the bat sonar. Many species of tiger moth in the genus *Bertholdia* produce complex acoustic sounds and increase the frequency of sound production just prior to a bat's attack. This suggests that the sounds produced by the moth might function to interfere with a bat's ability to localize and capture its prey.

Aaron Corcoran and his colleagues tested this hypothesis by examining interactions between big brown bats and one palatable species of tiger moth, *Bertholdia trigona* (Corcoran, Barber, & Conner 2009). The research team trained bats to

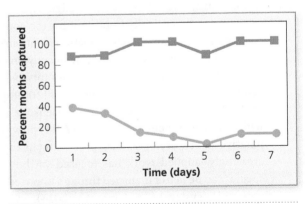

Figure 5.31. Percentage of moths captured as a function of moth traits. Control moths (blue) were palatable and could not produce ultrasound. Experimental moths (orange) could produce ultrasonic clicks and were chemically defended (noxious). Bats quickly learned to avoid noxious moths that produced ultrasonic clicks. (*Source:* Hristov & Conner 2005)

attack tethered moths in a flight-room. Moths were presented individually, and attack success was recorded. In treatment trials, a single *B. trigona* that was fully capable of producing sound was tethered in the room. Control trials included individual moths that were similar in size to *B. trigona* but were silent. In a second control, the researchers presented individual *B. trigona* that could not produce sounds because their sound-producing tymbals were ablated (removed).

Control moths (both silenced *B. trigona* or other silent species) were contacted over 400% more often than sound-producing tiger moths and this difference in successful attack rate existed from the first trial and was constant over the course of the experiment. These data support the hypothesis that the ultrasonic sounds produced by palatable *B. trigona* are an effective antipredator adaptation that interferes with a bat's ability to successfully attack its flying prey.

In bat–moth predator–prey interactions, we see how differing fitness interests of species can lead to a co-evolutionary arms race involving sensory system adaptations. Like all co-evolutionary arms races, there is no ultimate winner: we expect ongoing evolutionary changes (novel adaptations) and counteradaptations in such species' interactions. For example, the European bat (*Barbastella barbastellus*) has evolved a novel counteradaptation: it produces ultrasonic pulses that are 10–100 times quieter than those produced by other aerial-hawking bats, which presumably makes it more difficult for moths to detect them (Goerlitz et al. 2010).

Chapter Summary and Beyond

In a heterogeneous and dynamic world, animals benefit by acquiring information about food, mates, and the presence of rivals and predators. This information is acquired via their sensory systems. Species vary in their sensory capabilities and so differ in their perception of the environment: they each possess a different "umwelt." Van Dyck (2012) suggests that knowledge of the "umwelt" concept can be critically important for the conservation of species. Environmental information is gathered in a variety of ways, but most animals rely heavily on chemoreception, the detection of chemical odorants, perceived as tastes and odors. Touhara and Vosshall (2009) provide a review of recent research on chemosensory receptors, while de Brito Sanchez and Giurfa (2011) provide a comparative review of neural taste processing in animals. Increasingly, applied work has focused on how invasive species can be better managed through knowledge of chemosensory systems (e.g., Choe et al. 2014).

Visual sensory systems provide information through the detection of electromagnetic radiation. Species differ greatly in the wavelengths they can detect. Many animals, including insects and birds, can detect ultraviolet wavelengths that humans cannot perceive, and an appreciation of these differences in perceptual abilities is important for understanding the behavior of these species (e.g., Martin 2012). For reviews on the evolution of animal photoreceptor sensitivities and color vision, see Rosenthal (2007) and Osorio and Vorobyev (2008).

Animals detect mechanical vibrations that travel through air, water, and substrates via their mechanoreceptors. Species vary in the frequencies of sounds traveling through air or water that they can detect. Large-bodied mammals such as elephants can detect low-frequency infrasounds, while bats, toothed whales, and some insects can detect high-frequency ultrasounds. Many invertebrates also detect mechanical waves that travel through substrates such as the ground. Hill (2009) provides a comprehensive review of this ability to detect substrate-borne vibrations. Some animals have evolved sensory capabilities that allow them to detect electric and magnetic fields. The former allows individuals to detect hidden prey, while the latter provides information about directionality that can facilitate long-distance movements. Finally, sensory systems can evolve in response to adaptations in other species. Such co-evolutionary arms races are common in predator–prey interactions and can also occur between the sexes (e.g., Rowe, Cameron, & Day 2005). Conner and Corcoran (2012) provide a comprehensive review of insect–bat sensory system co-evolution, and Arnqvist (2006) reviews male–female co-evolution involving sensory systems.

CHAPTER QUESTIONS

1. Based on the results of Zhao and colleagues' (2003) study of sweet and umami taste receptors in mice, identify which of these tastes each knockout mouse type (KO-1, KO-2, KO-3) can detect.

2. Imagine you trained your dog to recognize differently colored blocks. You successfully trained him to touch either a yellow block or a green one on command using food rewards. However you were not able to train him to touch a red or an orange block on command. Propose a hypothesis to explain this observation.

3. In what ways are ultraviolet wavelengths and infrasounds similar?

4. Giraffes live in social groups, often in dense forests or in more open savannahs. A researcher predicts that they might use infrasounds for communication. Why might infrasounds be effective for long-distance communication?

5. Pohlmann and her colleagues evaluated the hypothesis that the lateral line system provides information to predatory catfish about their prey. What data supported the hypothesis?

6. Describe the basic methodological approach adopted by both Fertin and Casas (2007) and Kajiura and Holland (2002) to identify the sensory capabilities of antlion larvae and sharks.

7. What prediction can be made regarding the acoustic aposematism hypothesis that explains the evolution of ultrasound production in insects?

Chapter 6

Communication

If you live in the eastern United States near a moist habitat, you may be treated to a remarkable sight. Just after dark each summer evening, many backyards come alive, as fireflies (or lightning bugs) flash on and off. Fireflies are beetles in the order Coleoptera and are bioluminescent—that is, they produce light. If you watch carefully, you can observe differences in the duration and rate of flashes among individuals. You are in fact watching different species, each with its own distinctive flash pattern for courting conspecifics. The flying males of each *Photinus* species display a species-specific flash pattern to females sitting on vegetation, who respond with their own distinctive flash pattern. Males will land nearby, while continuing to flash, and then mate. The field also contains predatory fireflies in the genus *Photuris*. These species mimic the flashes of *Photinus* females to lure males. Once a *Photinus* male lands nearby, the predatory *Photuris* attacks and eats it (Figure 6.1) (Lloyd 1975; Champion de Crespigny & Hosken 2007).

This amazing display is an example of animal communication. As we see in this chapter, animal communication involves an individual's use of signals to influence the behavior of others. Signals are generally believed to be accurate indicators of conditions: the courtship flash displays of male and female *Photinus*, for example, indicate species and sex (Figure 6.2). However, signals are not always reliable, as the predatory *Photuris* demonstrates. The evolution and accuracy of signals will depend on the benefits and costs of signal production and the fitness interests of those involved. And, while the simplest communication involves two individuals, third-party eavesdroppers can intercept signals, which can affect the behavior of both eavesdroppers and signalers.

Figure 6.1. Firefly deception. Predatory female *Photuris* firefly eating a male *Photinus* firefly.

6.1 Communication occurs when a specialized signal from one individual influences the behavior of another

Figure 6.2. Firefly. Fireflies communicate using a bioluminescent flash display.

communication The process in which a specialized signal produced by one individual affects the behavior of another individual.

signal An evolved trait that is selected for its effect on the behavior of a receiver.

Communication is an interaction in which one individual, a signaler, produces a signal that affects the behavior of another, a receiver. An animal **signal** is an evolved trait that is selected for its effect on the behavior of a receiver and is adaptive for the signaler and, typically, the receiver (Bro-Jorgensen 2010). Signals can take many forms, including the patterns of firefly flashes, chemical pheromones (Applying the Concepts 6.1), morphological traits like body colors or structures (e.g., horns and antlers), electrical discharges, substrate-borne vibrations, and behaviors such as mating displays and vocalizations. Note that some of these, like vocalizations and behaviors, can be rapidly turned on and off, while others, like morphological structures or chemicals, can persist for longer durations.

Because signals are evolved traits, they differ from cues, which are consistent aspects of the environment that can guide behavior in a way that enhances fitness. For example, a group of vultures on the ground can provide a cue to other scavengers, such as hyenas, about the location of a carcass. Similarly, elevated concentrations of carbon dioxide provide a cue to mosquitoes searching for a vertebrate blood meal. But selection has not acted on the vulture group to communicate the location of a carcass to competitors, nor has it acted on mosquito victims to communicate their location to blood-seeking insects.

Much research on communication has identified different signals and their effect on the behavior of receivers. We begin by focusing on two detailed examples of different types of signals: a behavioral signal in honeybees and a vocal signal in mammals and birds.

Honeybees and the waggle dance

Karl von Frisch was awarded the Nobel Prize in 1973, in large part for his work on communication in honeybees (*Apis mellifera*). Honeybees live in colonies, and

APPLYING THE CONCEPTS 6.1

Pheromones and pest control

Up to one-third of food production worldwide each year is destroyed by insect pests (Witzgall, Kirsch, & Cork 2010). One response to this problem has been the widespread use of chemical insecticides. However, these chemicals are not species-specific, and many kill nontarget species as well. Insecticides can also facilitate the evolution of resistance by the same pests they are used to kill (Witzgall, Kirsch, & Cork 2010). Increasing attention to this issue has produced an alternative solution: the development and use of pheromones as a tool in pest management. Many insects use volatile sex pheromone signals to find a mate. Pheromones can be effective in minute amounts, are species-specific, and are environmentally benign. One technique is to release female pheromones throughout a field or orchard to disrupt mating, because males cannot then locate females.

Jianhuamo Mo and colleagues studied the effectiveness of pheromones in controlling damage by light-brown apple moths

(*Epiphyas postvittana*) in citrus orchards (Mo et al. 2006). They compared two orange groves: one control and one into which they released small quantities of an apple moth female sex pheromone. Each orchard contained dozens of individual constrained test females. After one week, females were examined to determine whether they had been inseminated by a male. The researchers found that females in the treatment orchard were almost never inseminated, compared to approximately 50% of those in the control orchard. In addition, examination of orchard fruits revealed that 20% of control orchard fruits had evidence of insect damage, whereas less than 10% of the treatment orchard fruits were damaged. Results like these are encouraging the development of pheromone-based management for a diversity of insect pests. ∎

individual workers leave the hive in search of food. Like many other insects, honeybees use their visual and chemosensory systems to find nectar and pollen. When an individual scout finds a rich food source, it flies back to the hive and recruits others to help exploit the food. Such communication allows the colony to rapidly exploit the food resource before competitors do. What signal is used to recruit others? In a series of studies, von Frisch manipulated the location of a distant food source and observed the behavior of both scouts and new recruits. He determined that the scout performs a specific behavioral signal that he called the waggle dance (von Frisch 1967).

During a **waggle dance**, the scout moves in a figure-eight pattern on a vertical wall of the honeycomb. In the linear movement of the dance, the scout vigorously wags its body, and the duration of the wagging, von Frisch argued, indicates the distance to the food. Subsequent work indicates that every 75 milliseconds of waggling translates into a distance of approximately 100 m from the hive (Figure 6.3) (Seeley 1985). The waggle dance also describes the direction of the food source, relative to an imaginary line that runs from the hive to the sun. For instance, if the sun is on the horizon (as at dawn) and the scout's linear movement is 30° left of vertical, then the food is 30° left of the sun (Figure 6.4).

But do honeybees also use odors to find food? There has been some debate about the relative importance of odor versus the waggle dance signal used by recruits to find a food source (e.g., Vadas 1994). Von Frisch recognized that odor is used to locate food resources that are close to the hive. He proposed, however, that the waggle dance provides the crucial signal, especially for food located long distances away—distances of more than, say, 500 m (von Frisch 1971). Yet most recruits take much longer to fly to the food source than would be expected based on a straight-line route (e.g., von Frisch, Wenner, & Johnson 1967; Gould & Gould 1988). If the waggle dance provides an accurate signal, travel times should be much shorter. What is causing this extra time? As an alternative explanation, Adrian Wenner and Dennis Johnson proposed that the bees doing the waggle

waggle dance Behavior performed by a honeybee scout that recruits workers to exploit a food source.

Figure 6.3. Waggle dance. The waggle dance conveys the direction of and distance to a food source to attending bees. (*Source:* Grütter & Farina 2009)

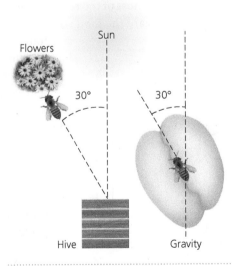

Figure 6.4. Linear movement of the waggle dance. The bees' linear movement in the vertical plane of the hive (right) indicates the direction of the food relative to the location of the sun (left).

dance present the odor of the food source, not a precise direction and distance, to nest mates; this behavior then stimulates recruits to search for food on their own (von Frisch, Wenner, & Johnson 1967; Wenner, Well, & Johnson 1969). They suggested that new recruits fly downwind a few hundred meters before starting their search and use olfaction as a method of locating food by flying upwind, in a zigzag pattern, back toward the hive (von Frisch, Wenner, & Johnson 1967; Wenner 2002). If so, this could account for the longer travel time than expected. In fact, many insects that rely only on odor to find food do fly back and forth into the prevailing wind (Kennedy 1983). The next study examines these two possibilities.

Odor or the waggle dance in bees

Featured Research

Which is more important, odor or the waggle dance signal? One way to find out is to track the actual movements of new recruits. Joe Riley and colleagues examined the flight paths of recruits that viewed the waggle dance from a hive placed in a large, flat, mowed field (Riley et al. 2005). The field contained few natural sources of nectar or pollen, so the only food available was located at a feeding station, which was placed 200 m due east of the hive and contained a sucrose solution with no artificial odors. The researchers recorded average wind speed and direction during the experiment to determine their effect on the flight of bees. Bees were required to move through a clear plastic tube to enter and leave the hive, which facilitated bee capture. This set-up allowed the researchers to mark most of the bees in the colony with small numbered tags on their backs. They could then record the identity and movement of individual bees into and out of the colony. At the start of the experiment, individual scouts were allowed to visit the feeder, and their identity was recorded. Their waggle dance indicated that food was located due east of the hive. The researchers then captured recruits who had never previously visited the feeding station and placed small transponders on their backs that transmitted continuous information about their location (Figure 6.5). These individuals were released either at the hive or at one of three locations about 200 m southwest of the hive, and their movements were tracked.

Figure 6.5. Bee with transponder. A honeybee with a transponder attached to its back can be tracked using radar.

Most of the recruits released at the hive almost immediately began flying east toward the food station (Figure 6.6). However, only two of the 19 recruits actually found the feeding station. After 200 m, most individuals began to follow a more circuitous flight path, which the researchers interpreted as searching behavior for the food. In addition, the bees that were transported southwest of the hive and then released similarly tended to fly due east for 200 m before also adopting a circuitous flight path. The wind data indicated that no odor from the feeding station was available to the bees at their release sites. There was no evidence that individuals flew downwind and then returned in a zigzag pattern to find the food. Instead, the bees had to fly *across* the prevailing wind to maintain a due east heading (Figure 6.7).

The flight paths of new recruits confirmed von Frisch's hypothesis that the waggle dance appears to signal the distance and direction of the food. In addition, the data demonstrate that recruits can travel to the area of a novel food source without odor cues. However, they also show that the dance signal is not sufficient to allow a recruit to locate a food source precisely.

Figure 6.6. Bee flight paths. Each line represents the flight path of an individual. (*Source:* Riley et al. 2005)

Additional work has demonstrated that new recruits use odors to pinpoint the exact location of the food (Grütter, Balbuena, & Farina 2008; Grütter & Farina 2009). In sum, the waggle dance is a behavioral signal that allows individuals to travel to the area of a distant food source, but bees then use odors to find the food itself. Next let's examine a vocal signal produced by many species.

Auditory signals: alarm calls

Many birds and mammals vocalize when they spot a predator. These **alarm calls** are unique vocalizations produced when a predator is nearby. In a set of groundbreaking studies, Dorothy Cheney and Robert Seyfarth found that vervet monkeys (*Chlorocebus pygerythrus*) (Figure 6.8) produce different alarm call vocal signals that affect receivers differently (Cheney & Seyfarth 1980; Cheney & Seyfarth 1985; Seyfarth, Cheney, & Marler 1980). Vervet monkeys are attacked by a variety of predators, including leopards (*Panthera pardus*), eagles, and snakes; when a vervet spots one of these animals, it gives an alarm call that alerts others in its group. However, individuals produce different alarm call signals for different types of predator (Seyfarth, Cheney, & Marler 1980). Why?

Each predator represents a threat that requires a different behavioral response (Figure 6.9). Leopards, for example, attack from the ground. When a leopard is spotted, a "bark" call is given, and vervets escape attack by moving up into trees. "Cough" calls are given when an eagle is sighted. Because eagles attack from above, avoiding their attack requires vervets to move down from treetops and into dense bushes. Snakes, in contrast, often hide in dense grass. When a python or cobra is observed, vervets give a "chutter" call that causes individuals to stand erect

Figure 6.7. Flight track in wind. Flight track (blue) of an individual bee during a southeast wind (gray arrows). The bee did not simply fly downwind and in fact needed to adjust its flight direction to fly due east across the wind (gray arrows). (*Source:* Riley et al. 2005)

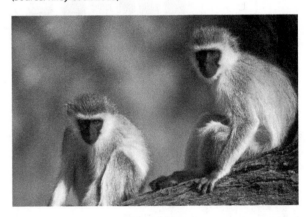

Figure 6.8. Vervet monkeys. These African primates live in social groups.

■ **alarm call** A unique vocalization produced by social animals when a predator is nearby.

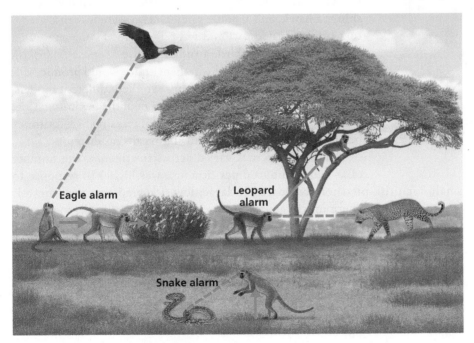

Figure 6.9. Vervet alarm calls. Vervets produce different alarm calls and then subsequent behavioral responses to different types of predators.

and look into grass clumps. Cheney and Seyfarth's work stimulated research on the alarm call signal systems of other species, as we see next.

Titmouse alarm calls

Jason Courter and Gary Ritchison examined alarm calls in the tufted titmouse (*Baeolophus bicolor*), a small songbird (Courter & Ritchison 2010). Titmice produce an alarm call in response to a perched avian predator. Vocalizations can be visualized in a sonogram, which characterizes frequencies of sound as a function of time (Figure 6.10). The titmouse call is composed of three basic notes, called Z, A, and D, which are typically produced in sequence. However, individuals often vary the number of D notes. This call recruits other birds (both con- and heterospecifics) to approach and mob the predator. Avian mobbing behavior, in which birds produce loud vocalizations, often serves to harass a predator and can drive it away from an area (see Chapter 9).

Titmice co-occur with several potential avian predators. Small raptors like Eastern screech owls (*Megascops asio*), sharp-shinned hawks (*Accipiter striatus*), and Cooper's hawks (*A. cooperii*) are known predators of titmice, while large raptors like great horned owls (*Bubo virginianus*) and red-tailed hawks (*Buteo jamaicensis*) rarely attack titmice. The size of a predator thus correlates negatively with risk to a titmouse. Courter and Ritchison conducted a study to determine whether titmouse alarm call signals differ with the size and degree of threat posed by a perched avian predator (Courter & Ritchison 2010).

The experiment consisted of six treatments, each presenting a different predator, and a control. At several sites, the researchers established a feeding station. Lifelike models of different birds were placed on platforms 1 m from the feeding station; the control presented nothing on the platform. Three models were high-risk predators (Eastern screech owl, sharp-shinned hawk, and Cooper's hawk), two were low-risk predators (great horned owl and red-tailed hawk), and one was a nonpredator control bird, a ruffed grouse (*Bonasa umbellus*). The researchers examined whether titmice would produce different alarm call signals in response to these different models. As soon as the titmice were within 25 m of the feeding station, a model was uncovered, and the researchers recorded titmice behavior for six minutes and all alarm calls produced.

Figure 6.10. Titmouse alarm call. Sonogram of the tufted titmouse alarm call showing the structure of Z, A, and D notes. Inset: tufted titmouse. (*Source:* Courter & Ritchison 2010)

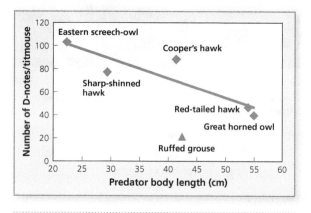

Figure 6.11. Alarm calls vary with predator size. Titmice produced more D notes per minute when presented with models of a small hawk compared to larger models or controls. (*Source:* Courter & Ritchison 2010)

Alarm call signals varied across treatments. The number of D notes produced per titmouse was higher in response to small, high-risk predators compared to large, low-risk predators and controls (Figure 6.11). Furthermore, in general, the smaller the predator, the longer the mobbing response (Figure 6.12): mobbing behavior lasted for approximately 250 seconds for the small-bodied, high-risk predators, compared to only 150 seconds for the larger, low-risk predators and less than 50 seconds for the controls. These results indicate that titmice produce different alarm call signals that correlate with varying levels of threat and lead to differences in the behavior of receivers. This finding illustrates the rapid speed and flexibility of auditory signals in response to changes in the environment.

Information or influence?

There is an important caveat here. In these and other examples, we do not know whether the signal actually encodes information about conditions, like the presence of a specific predator, although this metaphor has previously dominated work in animal communication (e.g., Otte 1974). For example, vervet alarm call signals may not mean "leopard present" or "snake present," but rather "climb a tree" or "look down" (Scott-Phillips 2008; Owen, Rendell, & Ryan 2010). Drew Rendall, Michael Owen, and Mike Ryan contend that the assumption that signals encode specific information is problematic for two reasons (Rendell, Owen, & Ryan 2009; Owen, Rendell, & Ryan 2010). First, it can imply a language-like meaning of communication that is challenging to document. Second, it can encourage attempts to characterize the information encoded in a signal, rather than focusing on factors that shape signal properties.

Others argue that removing the concept of information from the study of communication is unwarranted (Carazo & Font 2010; Seyfarth et al. 2010), and this discussion continues to stimulate further work. What we can say is that signals influence behavior *as if* they contained information about the location of a food source, the threat of a predator, or the phenotype of a signaler.

Signals vary widely in their physical properties. Next we examine how these differences and the environments through which they travel affect their evolution.

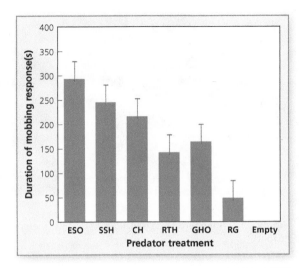

Figure 6.12. Duration of titmouse mobbing behavior. Mean (+ SE) duration of mobbing behavior for each treatment. Titmice spent more time mobbing models of small predators than they did mobbing larger ones or controls. ESO = Eastern screech owl; SSH = sharp-shinned hawk; CH = Cooper's hawk; RTH = red-tailed hawk; GHO = great horned owl; RG = ruffed grouse. (*Source:* Courter & Ritchison 2010)

6.2 The environment influences the evolution of signals

For effective communication, a signal must travel from one individual to another, sometimes over long distances. Because chemical, morphological (visual), and mechanical (auditory and tactile) signals differ in their physical properties and in how they are produced, they also differ in many other aspects, such as how rapidly they move through the environment, how quickly they can be modified by the signaler, their persistence once produced, and whether they can be blocked by physical objects such as vegetation or rocks.

Chemical signals can be transmitted readily through water or air (when they are volatile), are relatively long lasting, and can travel great distances. They can also be deposited on a substrate, as illustrated by the territorial scent marking of mammals (Figure 6.13) and the foraging trails of ants. Individuals can control the strength of these signals by controlling the amount deposited. In addition, chemical signals can travel around environmental barriers such as dense vegetation. Once transmitted, however, chemical signals cannot be modified, and their effectiveness is greatly impacted by conditions (e.g., temperature, wind, rain).

Visual signals, like behavioral displays, move rapidly through the environment and so can be detected quickly, allowing a more rapid response. In clear skies or clear water, they can also be perceived over fairly large distances. However, their perception requires sufficient ambient

Figure 6.13. Scent marking. A wolf marks his territory with urine.

Figure 6.14. Ant pheromone trail. Recruits follow a pheromone trail to food.

light levels, and they can be blocked by obstacles. Thus, in deep ocean or murky water and at night, visual signals are not effective, nor do they transmit well in dense vegetation.

Auditory signals are mechanical waves that can bypass obstacles: sound waves can travel around solid objects. That means that auditory signals can sometimes be perceived when visual signals cannot. In addition, auditory signals can be modified rapidly by a sender: they can be turned "on" or "off" depending on conditions (say, when a predator is nearby), and they can be produced at different amplitudes (decibel level) to overcome background noise. However, auditory signals can attenuate rapidly (lose energy) as they move through an environment, limiting their effective range. In addition, auditory signals are broadcast widely and so can be detected by many individuals, including predators.

With this background, let's examine three studies that have focused on how environmental conditions have affected the evolution of chemical, visual, and auditory signals.

Featured Research

Temperature affects ant chemical signals

Ants live in colonies where offspring are reared. Scouts travel away from the colony to find and retrieve food resources for nest mates. Across species, two foraging strategies have been observed. In some species, individual workers search for and bring back food items. In other species, foraging is done socially: a scout that finds a rich food source deposits a volatile chemical pheromone on the surface as it returns to the colony, making a pheromone trail (Figure 6.14). Such signals can persist for 30 or more minutes and allow new recruits to travel directly to the food source. After finding food, new recruits also deposit chemical pheromones as they return to the colony, which maintains the trail until the food is depleted.

Chemical pheromones in ant food trails are volatile. Therefore, the temperature of the surface should affect the duration of the signal: the signal should disappear more rapidly in warmer conditions as a result of evaporation. Louise van Oudenhove and her colleagues tested the hypothesis that the persistence of a chemical food trail signal will be negatively affected by surface temperature. They studied the foraging behavior of *Tapinoma nigerrimum*, a social foraging ant species found throughout the Mediterranean region (van Oudenhove et al. 2011). This species feeds on nectar, and workers deposit volatile chemical trails between a food source and the nest.

The research team first quantified the activity of free-living colonies by counting the number of workers on feeding trails (measured as the number of individuals that crossed a fixed mark over three minutes) once each hour over the course of several days. They simultaneously recorded the surface temperature at the colony entrance. Worker activity was highest for surface temperatures between 20 and 30°C.

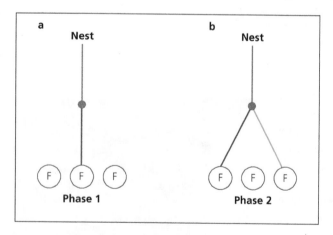

Figure 6.15. Ant food trail experimental design. (a) In Phase 1, ants deposited pheromones between the food (labeled F) and the nest on the original bridge (gray and blue). (b) In Phase 2, ants could select either the original bridge (blue) or a new bridge (orange) to move toward a new food location (labeled F) after both bridges had been warmed and then cooled to room temperature.

Activity declined sharply for surface temperatures above 30°C, and they recorded no worker activity for surface temperatures above 48°C. Clearly, ambient temperature affected foraging behavior in a manner consistent with the hypothesis, but why?

To determine the effect of temperature on pheromone persistence, the research team trained laboratory-housed colonies to feed on a solution of honey and water. Workers were required to cross two sections of a glass bridge to travel from the nest to the food. In the initial phase of the experiment, the colony was allowed to exploit the food patch to establish a pheromone trail across both sections of the bridge. In Phase 2, the second section of the bridge was removed for several minutes and heated to a fixed temperature (25, 30, 35, 40, 50, or 60°C) for ten minutes and then allowed to cool to room temperature (25°C) (Figure 6.15). Simultaneously, a novel section of bridge underwent the same heating and cooling procedure. Once both sections were cool, they were placed at the end of the fixed bridge section in a new Y-shaped orientation. Food was placed at the end of each bridge section, and ants were allowed to travel on either the original or novel section to move to a food patch. The researchers recorded the number of ants using each bridge for ten minutes. If the pheromone persisted after the heating treatment, ants would be expected to preferentially use the original bridge.

For the three low-temperature treatments (less than 40°C), workers preferentially used the original bridge. However, for the three high-temperature treatments (40, 50, and 60°C), there was no significant difference in bridge use (Figure 6.16). These results suggest that surface temperature plays a critical role in food trail persistence and use. At high surface temperatures, chemical pheromone signals are an ineffective communication mode. This may explain why ants that use chemical food trails are restricted to foraging at lower surface temperatures. Ambient temperature clearly influences the effectiveness of these chemical pheromones. In a similar manner, visual signals are affected by light level and the visual background environment, as we see next.

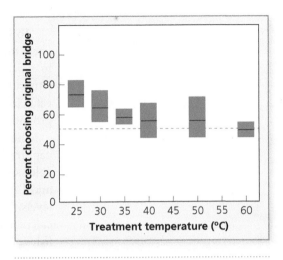

Figure 6.16. Experimental bridge choice. Median percent of workers crossing the original bridge. Each box represents the middle 50% of the data for each treatment temperature. The median is the horizontal line within each box. The dashed line represents 50%. (*Source:* Oudenhove et al. 2011)

Habitat light environment affects fish visual signals

Featured Research

Visual signals often must be detected against a potentially complex visual environment. For effective communication, selection may often favor visual signals that contrast most strongly with the background environment (Endler 1992; Marchetti 1993). Environments can differ dramatically in both their light level and their spectral properties (dominant wavelengths): consider the difference in light level between a dense forest and an open desert. Environmental light level and spectral properties can also differ dramatically in the same habitat, as occurs in shady and sunny locations within a stream or around a lake.

Suzanne Gray and her colleagues examined the effect of different light environments on the effectiveness of the visual signal in male *Telmatherina sarasinorum*, a small fish found in Lake Matano, Indonesia (Gray et al. 2008). Males exhibit a discrete color polymorphism: they are either all blue or all yellow. Females are cryptically colored (gray), and so male body color is likely a signal used in communication with females during courtship. Males court females in two distinct habitats: shallow beach sites composed of sand and cobble, and deepwater sites that contain algae-covered roots. These "beach" and "root" sites are interspersed, and all males and females visit both habitats to reproduce.

To humans, these habitats appear to present different light levels and visual backgrounds. Gray and her colleagues wondered whether females preferentially mate with males based on the effectiveness of their color display in each habitat. In particular, they tested the prediction that a male's reproductive success will be highest in habitats in which his body color contrasts most strongly with the visual background, because in such conditions his visual signal should be most effective.

The researchers established 12 transect sites (six in each habitat). They first captured males along each transect and used a radiometer to measure (1) the color spectrum (wavelengths of visible light) reflected from the sides of his body and (2) the associated habitat background color spectrum under ambient light levels. The level of chromatic contrast between a male and his environment was calculated using spectral sensitivities of fish retinal photopigments to quantify a fish's perception of the contrast (Chapter 5). Analyses of the spectral properties of each morph in each habitat revealed that the blue morph contrasted most strongly with the beach habitat color spectrum, while the yellow morph contrasted most strongly with the root habitat color spectrum (Figure 6.17a). Does male morph color affect reproductive success differently in each habitat, as predicted?

To answer this question, the research team needed to characterize the reproductive success of males of each morph in each habitat. In this species, two male mating tactics exist. Some males perform active mating displays to attract females, while other males do not display and are known as "sneakers." Females can mate with one displaying male that fertilizes all her eggs. However, if one or more sneaker males are present at the mating, they can fertilize eggs by releasing sperm near the spawning site of the pair. In this case, all males that release sperm fertilize eggs. Both male morphs use both strategies in both habitats studied.

To quantify the reproductive success of males in each habitat, observers conducted focal observations of individual males and females for up to ten minutes, recording all spawning events, the presence of all sneaker males, and the color morph of each male present during a spawning event. To quantify the reproductive success of each male morph in each habitat, the researchers assumed that eggs were fertilized equally by all males present (e.g., when only one male was present, he fertilized 100% of the eggs; when two males were present, each fertilized 50% of the eggs, and so forth). These data allowed the research team to characterize the mating success of both male morphs in each habitat.

As predicted, each male morph had its highest reproductive success in the habitat in which its morphology contrasted most strongly with the background environment: blue morphs had higher reproductive success in the beach habitat, while yellow morphs had higher reproductive success in the root habitat (Figure 6.17b). These data support the prediction that visual signal effectiveness is affected by background environmental properties and may help to explain the evolution of different male coloration signals in this species.

Finally, let's consider how the environment affects auditory signals.

Figure 6.17. Chromatic contrast and reproductive fitness. (a) Mean ($\pm$SE) chromatic contrast. Blue males had high chromatic contrast in the beach habitat (circles), while yellow males had high contrast in the root habitat (triangles). (b) Mean ($\pm$SE) reproductive fitness. Blue males had the highest fitness in the beach habitat (circles), while yellow males had the highest fitness in the root habitat (triangles). (*Source:* Gray et al. 2008)

Habitat structure affects bowerbird auditory signals

To be effective, auditory signals must be detected over background noise: they must stand out from their environment. Auditory signals can travel over long distances, but they degrade (lose quality) and attenuate (lose intensity) as they travel; furthermore, the rates of degradation and attenuation are affected by habitat structure. For example, higher frequencies attenuate more rapidly in dense vegetation (Blumenrath & Dabelsteen 2004), which should favor the use of lower frequencies, because selection should favor individuals whose vocalizations are best transmitted through their particular habitat (Morton 1975) (Applying the Concepts 6.2). James Nicholls and Anne Goldizen tested this hypothesis by comparing the vocal signals (advertisement calls) of bowerbirds living in different environments (Nicholls & Goldizen 2006).

Satin bowerbirds (*Ptilonorhynchus violaceus*) live in a variety of habitats in eastern Australia ranging from dense rainforests to open woodlands. Males construct a stick bower on the ground, from which they display to attract mates; we will examine these structures in more detail later in this chapter. Males also produce a loud call to attract females, and previous work noted geographical differences in these advertisement calls (Tack et al. 2005). Nicholls and Goldizen wondered whether such variation was due to differences in the vegetation structure across populations.

To test the prediction that call frequency should be lower in more densely vegetated habitats, the researchers recorded male advertisement calls from 18 locations that varied in habitat and vegetation structure. They used sonograms to characterize the average call structure at each site, recording the minimum

APPLYING THE CONCEPTS 6.2

Urban sounds affect signal production

Birds use auditory signals to defend territories and attract mates. Historically, these vocalizations occurred in their native rural habitat, but today, many birds live in noisy urban environments. Hans Slabbekoorn and Ardie den Boer-Visser investigated whether urban environments affect the songs of great tits (*Parus major*) (Slabbekoorn & den Boer-Visser 2006). Great tit songs are composed of two- to four-note phrases of different frequencies. The researchers recorded the songs of great tits in ten large European cities and compared them to the songs of great tits in matching forests outside each city.

Slabbekoorn and den Boer-Visser found that urban birds produced songs that had a higher minimum frequency and were shorter and faster than the songs of birds living in forests (Figure 1). The change in songs should allow for better transmission of the acoustic signal. The researchers suggest that the most likely explanation for the difference in minimum song frequency is learning: birds learn to modify their songs to compensate for low-frequency urban sounds, such as the noise of cars and trucks. This behavioral plasticity may be a critical element that allows species to adapt to urban life. These results add to a growing body of evidence that human alteration of the environment affects animal communication (Laiolo 2010). ■

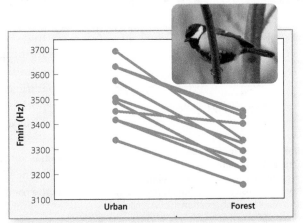

Figure 1. Great tit song. Minimum frequency in songs of ten urban and rural matched populations. Each line connects urban–forest population pairs. Inset: great tit. (*Source:* Slabbekoorn & den Boer-Visser 2006)

Figure 6.18. Satin bowerbird call sonograms. (a) Satin bowerbird. (b) Sonograms differ by location and habitat. Note differences in frequencies within the calls at different sites. (*Source:* Nicholls & Goldizen 2006)

Minimum and dominant frequency of calls is lower in high tree density habitats.

Figure 6.19. Bowerbird call variation and habitat structure. The minimum and dominant frequencies of calls are lower in habitats with higher tree density. Each point represents the mean values for each location. (*Source:* Nicholls & Goldizen 2006)

frequency, maximum frequency, dominant (or loudest) frequency, and call duration, and found substantial differences in the calls (Figure 6.18). To characterize vegetation, they counted the number of tree stems wider than 5 cm at 1 m above the ground that occurred along four 40 m transects. Bowerbirds vocalize at a height of a few meters off the ground, and stems of this diameter will scatter sounds to attenuate a call (Bradbury & Vehrencamp 1998).

As predicted, call structure varied across sites and was related to habitat type. Minimum frequency and dominant frequency were negatively correlated with tree density—that is, they were lower in sites with more trees (Figure 6.19). What produces these population differences? One possibility was that morphological differences across populations, such as body size, could account for this correlation, but after examining the data, the researchers ruled this explanation out. Another possibility concerns learning: juveniles may simply learn those habitat-specific calls that they hear best. Further work is required to test this hypothesis.

These examples show how the environment can affect signal use and evolution. Recall that the function of communication is to influence the behavior of another individual. As such, the evolution of signals is strongly influenced by the fitness interests of both signaler and receiver. This aspect is the focus of the rest of the chapter.

6.3 Signals often accurately indicate signaler phenotype and environmental conditions

For signals to evolve, they must increase the fitness of the signaler. Because signals affect the behavior of another individual, selection on signals will also be affected by receivers. Receiver behavior should not be influenced by signals that do not enhance their own fitness. As a result, the fitness interests of signalers and receivers may differ, which will affect the evolution of a signal. In particular, the potentially divergent interests can lead to two outcomes. In the first, the signal will evolve to become an accurate indicator of signaler phenotype or the environment. In the second, the signal will evolve to become an inaccurate indicator of conditions. We start by examining the former outcome and then address the latter in the next section.

Signals as accurate indicators: theory

Three conditions favor the evolution of signals as accurate, or "honest," indicators (Maynard Smith & Harper 2003). First, if the fitness interests of the signaler and the receiver are similar, both parties will benefit from an accurate relationship between the signal and the signaler phenotype or environmental conditions. Second, we can expect signals to be accurate indicators when they cannot be faked. Suppose, for example, that a signal is a function of body size. In contests over resources or territories, males often produce vocal signals. The frequency of a vocalization, or its pitch, is often negatively related to body size. For example, only the largest males with the longest sound-producing structure—such as the larynx in frogs and mammals or the harp in crickets—can produce the lowest-frequency vocalizations (Clutton-Brock & Albon 1979; Scheuber, Jacot, & Brinkof 2003). In these cases, the ability to produce low-frequency vocalizations is an accurate signal indicator of male size. Finally, and perhaps most commonly, signals may be accurate indicators because they are costly to produce or maintain (Zahavi 1975; Grafen 1990). For example, the long horns of beetles used in aggressive interactions (Emlen 1994), the wide eye span of stalk-eyed flies (David et al. 2000), and the wattle size and color of pheasants used to attract mates (Ohlsson et al. 2002) are physiologically costly to produce: only males with access to abundant resources possess the largest, most extreme form of these traits (Figure 6.20). Other signals used in courtship include behavioral displays and the bright plumage or body coloration of many birds and fish, and these, too, are costly to produce and maintain: they make the male more visible (Magnhagen 1991) and are therefore found in those individuals best able to avoid predators. Because they are accurate indicators of male phenotype, females frequently prefer mates with extreme signals (see Chapter 12).

Many signals are considered to be accurate indicators because one of the above three conditions exist. Next, we illustrate examples of accurate signals with

Figure 6.20. Costly traits. (a) Horn length in beetles, (b) eye span in stalk-eyed flies, and (c) wattle size and red color in pheasants are examples of traits that are costly to produce: each is larger and/or more intense when individuals have access to better diets.

studies of predation, courtship, and aggression. In the first example, the fitness interests of predator and prey are similar. The latter two examples illustrate how the high fitness cost of a behavioral signal can lead to accurate indicators of signaler phenotype.

Aposematic coloration in frogs

Some species possess **aposematic coloration**—bright coloring on their bodies that makes them stand out from the environment and indicates that they contain noxious chemicals or poisons that make them unpalatable or dangerous prey. Such coloration is common in insects but is also found in fishes, amphibians, and snakes (Figure 6.21) (Ruxton, Sherratt, & Speed 2005). These species benefit if predators learn to avoid attacking them. Predators, too, benefit by learning such an association, because they will then not waste time hunting unpalatable prey.

Ralph Saporito and colleagues conducted an experiment to investigate whether the bright body coloration in the dendrobatid frog (*Oophaga pumilio*) functions as an aposematic signal. This frog commonly lives in leaf litter in the tropics and has a bright reddish-orange dorsal color with blue appendages. Dendrobatid frogs contain skin alkaloids that predators find distasteful (Daly & Myers 1967; Daly, Spande, & Garraffo 2005).

Saporito's team examined predation attempts on clay models molded to resemble the toxic *O. pumilio* and a nontoxic, brown leaf-litter frog similar to frogs of the genus *Craugastor* (Figure 6.22) (Saporito et al. 2007). Each model measured approximately 20 mm in length, the average size of *O. pumilio* in the study

Featured Research

■ **aposematic coloration** Brightly colored morphology in a species that stands out from the environment and is associated with noxious chemicals or poisons that make individuals unpalatable or dangerous prey.

Figure 6.21. Aposematic coloration. (a) Coral snake, (b) poison dart frog, and (c) cinnebar moth caterpillar. All of these species are unpalatable to predators.

Figure 6.22. Clay models and frogs. (a) Toxic *Oophaga pumilio* and (b) its clay model. (c) Nontoxic *Craugastor fitzingeri* and (d) its clay model.

region in Costa Rica. To quantify predation, the researchers placed 800 frog models on either the forest floor or a white piece of paper at 5 m intervals along 40 transects. The white paper was used to take into account the effect of cryptic coloration, since half the brown models would then be obvious to predators. After 48 hours, the models were collected. The soft modeling clay retained impressions from predation attempts. Birds left U- or V-shaped attack marks, while mammals left teeth marks.

Over 12% of the models were attacked, mostly by birds. Although more models placed on the forest floor were attacked than those placed on white paper, the difference was not significant. However, the brown models were attacked by birds at almost twice the rate of the brightly colored models (Figure 6.23). These data support the hypothesis that bright coloration is a signal to predators that prey are unpalatable. Bright body coloration benefits the signaler when predators associate the colors with distasteful prey, allowing them to learn to avoid such items. Next, we examine accurate signaling in courtship behavior.

Figure 6.23. Predator attacks on frog models. Brown cryptic models were attacked more often than red (aposematic) models. (*Source:* Saporito et al. 2007)

Courtship signaling in spiders

Featured Research

Chad Hoefler and colleagues studied the mating behavior of wolf spiders (*Pardosa milvina*) to determine if males display accurate signals of their quality to females (Hoefler, Persons, & Rypstra 2008; Hoefler et al. 2009). In the presence of a female (or silk containing female pheromones), males engage in active displays that involve leg raises. Females prefer to mate with those males that display the most intense courtship behaviors (Rypstra et al. 2003). Hoefler and his research team hypothesized that males in good condition should perform leg raises at a higher rate than those in poor condition (Scientific Process 6.1).

They tested this prediction by conducting two experiments that manipulated the condition of males and females and tested their behavior over three days. As predicted, a male's condition affected his behavior. Males in good condition displayed leg raises at higher rates than males in poor condition. In addition, male behavior was consistent over all three days in three of the four test groups. In the group that matched males in good condition with females in poor condition, display rate declined over the three days.

In addition, females that mated with males with high display rates had higher fitness: they had more spiderlings than females that mated with males with low display rates, and their spiderlings lived significantly longer. Taken together, these results indicate that male display is an accurate signal of male quality: only males in good condition were able to maintain a high display rate, and display rate was generally consistent over time. Females obtained fitness benefits by mating with males that displayed at high rates. How can we interpret the behavior of good-condition males that decreased their display rate over time when courting females in poor condition? All males that displayed to a female whose condition was similar to or higher than their own displayed highly repeatable courtship levels over each day, indicating that males are capable of continuing with the high rate of leg raises. Perhaps it was not beneficial for males that perceived a female to be in poor condition to expend the energy to continue a high-intensity courtship. This hypothesis remains to be tested.

Accurate signaling is also common in a third context as well: aggressive interactions. As we see next, the cost of signaling also plays a role in selecting for accuracy of signals used in aggressive interactions.

SCIENTIFIC PROCESS 6.1

Signaling in male wolf spiders
RESEARCH QUESTION: *Do male wolf spider leg raises provide an accurate signal to females?*

Hypothesis (1): Male condition affects leg-raising rate.

Prediction (1): Good-condition males will display at a higher rate than poor-condition males.

Hypothesis (2): High display rate in males indicates high quality.

Prediction (2): Females mated to males with high leg display rates will have higher reproductive success.

Methods (1): The researchers:

- Collected 60 juvenile male and 60 juvenile female spiders and raised them in the lab for three weeks. Poor-condition treatment (PC) spiders were fed one cricket twice per week. Good-condition treatment (GC) spiders were fed four crickets twice per week.
- Created four mate-pairing test groups:
  (1) GC male with GC female
  (2) GC male with PC female
  (3) PC male with GC female
  (4) PC male with PC female
- Measured male courtship rate in a cylindrical arena (8 cm high by 19 cm diameter)
- Repeated the trial for the next three days using the same individuals to test for consistency

Methods (2): The researchers:

- Mated females with males in two groups:
  (1) Males (*n* = 21) with high leg display rates (5.9 leg raises/min)
  (2) Males (*n* = 36) with low leg display rates (3.0 leg raises/min)
- Measured the number of spiderlings and their survival

Results (2): Females mated to males with a high display rate had more spiderlings and higher offspring survival than females mated to males with a low display rate.

Results (1): GC males displayed at higher rates than PC males. Display rate was consistent across the sample, except for GC males mated to PC females.

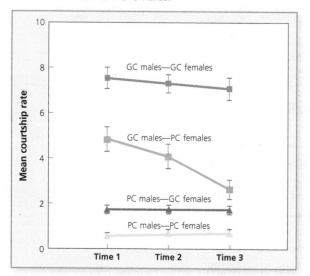

Figure 1. Courtship. Mean (± SE) courtship rate over trials for each treatment. (*Source:* Hoefler et al. 2009)

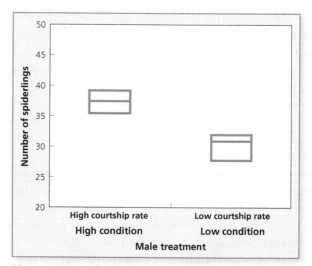

Figure 2. Reproductive success. Reproductive success of females mated to males with different courtship rates. Each box represents the middle 50% of the data. The median is each horizontal line. (*Source:* Hoefler et al. 2009)

Conclusion: Mate condition affects male courtship display rate and is an accurate signal of male quality and fitness.

> **EVALUATE**
> What fitness benefit do females obtain by mating with males that exhibit high display rates?

Aggressive display and male condition in fighting fish

Many animals, including fish, engage in aggressive contests over food and mates (see Chapter 11). When two fish fight, they often begin by engaging in a head-on opercular flare—an encounter in which the opercula, the bony flaps that cover the gills, flare out at the opponent. The frequency and duration of the display increase during a contest, and both are strongly correlated with winning: individuals that have high flare rates and long flare durations typically win.

Mark Abrahams, Tonia Robb, and James Hare hypothesized that the opercular flare display is an accurate signal of fighting condition because it is costly (Abrahams, Robb, & Hare 2005). The opercula play an important role in respiration. For the gills to function properly in gas exchange, they must be adequately ventilated by moving water. In calm water, ventilation is accomplished through opercular movements. Fish engaged in long flare displays in calm water must therefore sacrifice normal flapping movement and consequently water ventilation. Abrahams and colleagues

SCIENTIFIC PROCESS 6.2

Fighting fish opercular display

RESEARCH QUESTION: *Does opercular display behavior vary with the condition of a fish?*

Hypothesis: Opercular displays are difficult to perform at high rates because they reduce water ventilation over the gills.

Prediction: Males in the best physiological condition will perform opercular displays at a higher rate than individuals in poorer condition.

Methods: The researchers:

- Manipulated males by placing individuals in water that differed in oxygen level: normoxic water (at least 4 ppm dissolved oxygen) allowed fish to remain nonstressed; hypoxic water (3 ppm dissolved oxygen), created by bubbling nitrogen into the water, produced stressed individuals.
- Recorded fish display behavior in ten-minute trials for two treatments: with or without an intruder (represented by a mirror image in the test tank)
- Recorded the percentage of time a fish spent displaying (operculum flared out) over the course of 19 trials

Results: Fish in normoxic water displayed significantly more than fish in hypoxic water.

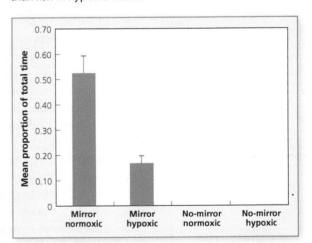

Figure 1. Displays. Mean (+SE) proportion of time displaying across treatments. (*Source:* Abrahams, Robb, & Hare 2005)

Conclusion: Only males in good physiological condition can display at a high rate. Opercular displays are an accurate signal of male condition.

EVALUATE

What is the assumed relationship between opercular flare rate and fighting ability?

APPLYING THE CONCEPTS 6.3

Human luxury brands as costly signals

Luxury products are priced at a substantial premium, but most provide essentially the same function as nonluxury products (think about brands of clothes, handbags, or watches). Why do consumers purchase these costly items? Are these signals that influence the behavior of others (e.g., Bouska & Beatty 1978)? Can we understand such behavior in the context of animal signaling theory?

Ron Nelissen and Matijn Meijers conducted a set of experiments designed to answer these questions (Nelissen & Meijers 2011). They hypothesized that because displays of luxury brands are costly, these signals should be perceived as accurate indicators of status and wealth. Further, the researchers investigated the benefits of producing such a display.

In one experiment, undergraduates viewed one of two videotaped job interviews of a man seeking employment as a laboratory assistant. The two videos differed only in the man's polo shirt logo (either the luxury-brand logo Tommy Hilfiger or no logo). Based on the video, the participants were asked to rate the applicant's suitability for the job, how much he should be paid, his status, his attractiveness, his kindness, and his trustworthiness. When the man was wearing the luxury-brand logo, participants rated him as more suitable for the job, deserving of a higher wage, and of higher status than when he was wearing no logo.

This experiment suggests that luxury-brand labels may act as a signal that affects receiver perception of, and behavior toward, the signaler. Those wearing luxury brands are perceived as being of higher status and wealth and as worthy of higher pay. The result is a benefit for the signaler (e.g., the acquisition of more money). Animal signaling theory can help explain how and why such behavior has evolved. ■

hypothesized that only individuals in the best physiological condition would be able to sustain long opercular displays because of this cost. To test this hypothesis, they measured the ability of fish to maintain long opercular displays in normoxic (normal oxygen level) and hypoxic (low oxygen level) water. They predicted that flare displays would be significantly shorter under hypoxic conditions (Scientific Process 6.2).

The team examined the behavior of a popular freshwater species, Siamese fighting fish (*Betta splendens*), known for its aggressive behavior. Fish were exposed to four experimental treatments that varied in both dissolved oxygen level (normal or hypoxic) and the presence or absence of an intruder. To simulate an intruder, the researchers placed a mirror behind a removable opaque partition at one end of the tank. Fish appear to perceive their image in a mirror as an intruder and will display aggressively toward it.

Fish did not display when there was no "intruder" but readily displayed to the "intruder" that appeared in the mirror. In addition, opercular display behavior was significantly lower in hypoxic than normoxic conditions, as predicted. Abrahams and colleagues concluded that opercular displays are accurate signals of physiological condition because they are costly. They suggest that opercular displays cannot be faked: a fish flaring its operculum is analogous to a terrestrial vertebrate holding its breath. Only individuals in good physiological condition can maintain long opercular flare displays.

These examples show how, and under what conditions, signals can be accurate indicators of signaler phenotype or environmental conditions; recent work has even applied these ideas to humans (Applying the Concepts 6.3). However, not all signals are accurate indicators of phenotypes or conditions, as we see next.

6.4   Signals can be inaccurate indicators when the fitness interests of signaler and receiver differ

Not all communication involves accurate signals. What happens when the fitness interests of signaler and receiver conflict, as in our predator–prey firefly example at the beginning of this chapter? In such cases, the signaler can benefit by producing an inaccurate, or "dishonest," signal. But when inaccurate signals exist, receivers that are able to discriminate accurate from inaccurate signals will also benefit. This produces a **co-evolutionary arms race** (Dawkins & Krebs 1979) between signalers and receivers: signalers are selected to produce signals that affect the behavior of receivers in ways that enhance signaler fitness (e.g., by producing inaccurate signals that are difficult for receivers to discern), and receivers are selected for better discrimination abilities so that they are not influenced by inaccurate signals (Rice & Holland 1997; Holland & Rice 1998; Lozano 2009). Who wins in such a "race"? Neither. The evolution of the signal trait and its effect on receivers is an ongoing process: it is influenced by the fitness benefits and costs of signal production and accuracy in the signaler and discrimination ability in the receiver (Krebs & Dawkins 1984; Dawkins & Guilford 1991; Johnstone & Grafen 1993; Rowell, Ellner, & Reeve 2006).

Inaccurate signals are common in interspecific signaling, such as between prey and predators. Here, the fitness interests of the species differ radically: prey fitness is enhanced by avoiding predation, while the predator benefits from consuming the prey. One well-known case of inaccurate signaling is **mimicry**—the adaptive resemblance of one species (the mimic) to another (the model) so that a third species (the receiver) is duped. In **Batesian mimicry** (Bates 1862), a palatable mimic resembles an unpalatable model that predators have learned to avoid. The predator perceives an inaccurate signal from the mimic and so does not attack it. In this case, the relative fitness benefit of the signaler favors the evolution of a signal that is difficult to discern. A poor signal that is easy to discern will lead to the death of the signaler, but an effective signal will only result in the loss of a meal for the predator. A second example is **aggressive mimicry**, in which a predator mimics a nonthreatening model species in order to gain access to food. Again, the receiver is duped, but in this case, it is attacked by the aggressive mimic (as we saw in the *Photuris* and *Photinus* fireflies). Let's look at one detailed example of each.

Batesian mimicry and Ensatina salamanders

As we saw earlier, aposematism, or bright warning coloration, is commonly found in species with toxins. These visual signals evolved to convey accurate information to predators about the unpalatability of a prey species, and both signaler and receiver benefit from it. However, this signaling system is also ripe for "deceit," or inaccurate signaling. Individuals of a third species that does not produce a toxin, the Batesian mimic, can benefit if they resemble the toxic model species.

Batesian mimicry is found in many animal taxa. Shawn Kuchta, Alan Krakauer, and Barry Sinervo examined mimicry in Ensatina salamanders (*Ensatina eschscholtzii*), which live throughout central California (Kuchta, Krakauer, & Sinervo 2008). Most subspecies are cryptically colored, but one, the yellow-eyed salamander (*Ensatina eschscholtzii xanthoptica*), is brightly colored. As its name suggests, it has yellow eyes, and its legs and ventral region are bright orange. Is it a Batesian mimic? One hypothesis is that it mimics aposematic toxic newts in the genus *Taricha* (Figure 6.24) (Stebbins 1949). *Taricha* newts and the yellow-eyed salamander co-occur geographically, and both possess an orange ventral region and bold yellow patches on the iris.

To test this hypothesis, the research team conducted a series of feeding trials with Western scrub jays (*Aphelocoma californica*). Scrub jays feed on a variety of insects and small terrestrial vertebrates, including salamanders. In the experiment, captive jays were housed individually in a large aviary. On the first day, all jays were presented with an edible salamander (*Batrachoseps*

co-evolutionary arms race The back-and-forth process of adaptation in one species favoring counteradaptation in another.

mimicry The adaptive resemblance of one species to another.

Batesian mimicry The resemblance of a palatable mimic to an unpalatable model that predators have learned to avoid.

aggressive mimicry A situation in which a predator mimics a nonthreatening model.

Featured Research

Figure 6.24. Salamander mimicry.
Ensatina salamander mimic (right) and its toxic model (left). Note their yellow eye coloration and orange ventral surface.

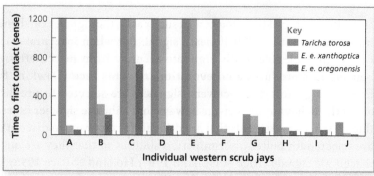

Figure 6.25. Jay attacks on salamander models and mimics. Time to first contact with models for ten birds (A–J). Scrub jays rarely contacted the toxic model (blue). Jays took longer to contact the mimic Ensatina salamander (orange) than the cryptic salamander (gray). (*Source:* Kuchta, Krakauer, & Sinervo 2008)

attenuatus or *B. luciae*) to ensure that they would eat salamanders. The next day, all jays were presented with a toxic newt (*Taricha torosa*). On the third day, each jay was presented with either a yellow-eyed salamander (mimic) or a palatable Ensatina subspecies (*E. e. oregonensis*). On the final day of the experiment, each jay was given the other subspecies of Ensatina. All salamander prey offered to the jays were first euthanized to eliminate behavioral differences and thus force jays to make feeding decisions based on morphology alone. Prey were presented in a lifelike posture in the middle of the aviary.

The researchers found that seven of the ten jays did not touch the toxic model. Three jays briefly poked at it but did not attempt to consume it. These ten jays had apparently learned to avoid the toxic prey in the wild. However, the behavior of the jays differed with respect to the two salamander subspecies. All ten birds contacted the palatable Ensatina subspecies (*E. e. oregonensis*) more quickly than they contacted the yellow-eyed salamander. The yellow-eyed salamanders were not even touched by three of the jays (Figure 6.25). They were consumed in only half the trials, whereas the palatable Ensatina subspecies (*E. e. oregonensis*) was consumed in nine out of ten trials. The researchers concluded that the yellow-eyed salamander effectively mimics the coloration of the toxic newt *T. torosa*, thereby obtaining a fitness benefit. The mimicry was not perfect: jays consumed more salamander mimics than they did toxic models. Nonetheless, attack intensity on mimics was less than on nonmimics, so we can see how selection would favor the evolution of Batesian mimicry as an inaccurate morphological signal.

Featured Research

Aggressive mimicry in fangblenny fish

Recall that in aggressive mimicry, a predator mimics a nonthreatening model so that it can closely approach its prey. Karen Cheney and Isabelle Côté examined aggressive mimicry in a coral reef system in Australia (Cheney & Côté 2007). Ectoparasites, such as gnathiid isopod larvae, infest many reef fish (Grutter 1994). These ectoparasites are also a food source for **cleaner fish**—small fish that feed on parasites and the dead skin of larger fish. In essence, the cleaner fish performs a service to the parasitized "client" fish, creating a mutualistic relationship. However, when client fish have few parasites, cleaner fish may feed on healthy tissue and can harm clients (Cheney & Côté 2005), so the mutualism depends on how heavily the client fish is parasitized. The juvenile bluestreak cleaner wrasse (*Labroides dimidiatus*) is a common cleaner fish found throughout the Indo-Pacific. Another species, the bluestriped fangblenny (*Plagiotremus rhinorhyncos*), closely resembles the bluestreak wrasse but consumes only the healthy scales, mucus, and dermal tissue of client

cleaner fish Fish that feed on ectoparasites and the dead skin of other fish.

fish. Both species have black bodies with neon blue stripes (Figure 6.26). Given the close morphological resemblance between the two, Cheney and Côté tested the hypothesis that fangblennies are aggressive mimics of juvenile cleaner fish. Clients with high parasite loads, they predicted, should seek interactions with cleaners and so may not be as vigilant about discriminating a cleaner from a fangblenny. On the other hand, clients with low parasite loads have less to gain by interacting with cleaners and so should be more vigilant to deception.

Staghorn damselfish (*Amblyglyphidodon curacao*) (Figure 6.27), a common reef client fish, were collected from a site in the Great Barrier Reef along with juvenile bluestreak cleaner fish and fangblennies. The experiment consisted of a treatment (damselfish parasitized) and a control (damselfish not parasitized). To parasitize damselfish, a fish was confined to a small section of a tank, where approximately ten unfed gnathiid isopods were released. On average, approximately six isopods attached themselves to the fish, a typical parasite load for *A. curacao*. Unparasitized fish were left free of isopods. During each trial, all three fish were allowed to interact for 15 minutes, and the researchers recorded the amount of time damselfish spent being cleaned by the cleaner fish, which fish (cleaner or client) terminated the interaction, the number of times the damselfish avoided an interaction with the cleaner fish, the number of successful attacks by the fangblenny on the damselfish, and whether the damselfish aggressively chased the fangblenny or cleaner fish.

Cheney and Côté found that parasitized damselfish were cleaned for a longer period of time than were unparasitized individuals. In addition, parasitized individuals terminated fewer cleaner fish interactions and avoided cleaner fish less than did unparasitized fish (Figure 6.28). All these findings indicate that parasitized damselfish benefited from interactions with cleaner fish, just as expected. Fangblennies also attacked parasitized and unparasitized fish at a similar average rate of 4.5 times per 15-minute observation. However, the proportion of successful attacks was higher on parasitized damselfish. In addition, parasitized damselfish retaliated less often to fangblenny attack than did unparasitized fish.

These data suggest that fangblennies in this system are aggressive mimics of juvenile cleaner fish, and that their success depends on the state of a client fish. Cheney and Côté suggest that when client fish have higher rates of parasitism, they are more likely to seek out interactions with cleaner fish. Because of the benefits of

a

b

Figure 6.26. Cleaner and mimic. There is a close resemblance between (a) the fangblenny mimic and (b) the bluestreak cleaner model.

Figure 6.27. Staghorn damselfish. A common reef fish in the western Pacific.

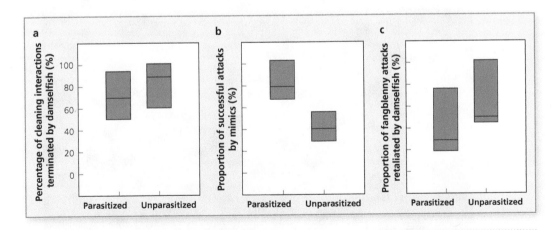

Figure 6.28. Interaction between damselfish and cleaner fish. Parasitized damselfish (a) terminated fewer interactions with cleaner fish, (b) were attacked more successfully by fangblennies, and (c) initiated fewer retaliatory chases than did unparasitized damselfish. In each plot, the median is represented by the horizontal line, and the box represents the middle 50% of the data. (*Source:* Cheney & Côté 2007)

parasite removal, client fish are willing to incur the costs of mimics and so are less vigilant. This allows aggressive mimics like fangblennies to be more successful.

Intraspecific deception: false alarm calls

So far, our examples have involved mimicry and interspecific "deception," or inaccurate signaling between species. Within-species, or intraspecific, deception also occurs but is less common for two reasons. First, intraspecific interactions are more common, and so individuals can learn to ignore inaccurate signals faster (Dawkins & Guilford 1991; Johnstone & Grafen 1993). We can say that inaccurate signals are subject to negative frequency-dependent selection, which keeps them rare, as we discussed in Chapter 3. Second, recall that natural selection will favor receivers that can discriminate accurate from inaccurate signals. For instance, we have seen that male quality is often associated with a morphological or behavioral signal. Since poor-quality males can benefit by providing an inaccurate signal, how do females avoid deception? They do so by focusing on signals in males that are costly to produce or difficult to fake (Zahavi 1975; Grafen 1990). Inaccurate signals will then yield low fitness, because they will be ignored.

However, one type of signal that involves deception of conspecifics is the production of false alarm calls—alarm calls that are produced when no threat is nearby. Let's look at two examples. In each case, the caller produces an inexpensive inaccurate signal and then takes advantage of its effect on receiver behavior.

Featured Research

Topi antelope false alarm calls

Topi (*Damaliscus lunatus*) are large antelope (weighing 75–150 kg) that live in the open savannahs of sub-Saharan Africa. When topi spot a predator—such as a lion or even humans—they emit snort vocalizations and then stare in the direction of the predator with their ears pricked up (Figure 6.29). Alarm snorts likely function to inform a predator that it has been spotted, because many predators can successfully capture prey only when they have the element of surprise (see Chapter 9).

During the breeding season, males defend individual mating territories from other males. Herds of nonterritorial females roam through male territories

Figure 6.29. Topi. A male with a female in his territory gives an alarm snort and looks off to the distance.

selecting mates during their one-day estrus (the sexually receptive, fertile period). A female will visit many territories and can mate multiple times with several males—a behavior that can be used to determine when a female is fertile.

Jakob Bro-Jorgensen and Wiline Pangle noticed that males frequently gave alarm calls during visits by estrus females, even when no predators appeared to be nearby (Bro-Jorgensen & Pangle 2010). They proposed that these false alarms "deceive" a female into staying on a male's territory so that he can obtain additional matings. This **sexual deception hypothesis** predicts that males will give false alarm calls only when an estrus female is in their territory and only when she is trying to leave.

To test this hypothesis, the researchers observed the behavior of 53 estrus and 20 non-estrus females between 2005 and 2009 in Masai Mara National Reserve in Kenya. Individuals were identified by their coat color and marking patterns. The researchers recorded the territorial location of the female, all matings, alarm snorts, and departure attempts (i.e., when a female attempted to leave a male's territory). Following each snort, the researchers scanned the area for predators. Predators, if present, were easy to locate because the habitat was open and topi tend to stare at a predator following a snort alarm. The researchers also recorded all alarm snorts to determine whether true alarm snorts—those produced when a predator was actually present—and false alarm snorts differed in acoustic structure.

They found that alarm snorts given in the presence of predators were acoustically identical to false alarm snorts given in the absence of predators. The researchers observed that false alarms were most likely to be given when an estrus rather than a non-estrus female was in a male's territory, and especially when she was trying to depart the territory (Figure 6.30). Males often gave false snorts from the edge of their territory. After hearing a snort, females moved away from the male, which in turn tended to move her back toward the center of his territory. Following false alarm snorts, a male secured an average of 2.8 additional matings before the female left his territory.

From these results, the researchers concluded that male topi give inaccurate alarm signals to keep estrus females from leaving their territory. Females apparently cannot distinguish true from false alarms and so move away from the apparent danger—which leads them back to the center of the deceiver's territory. The male benefits by acquiring additional matings and siring more offspring (e.g., Dziuk 1996). Why do females respond to this inaccurate signal? Although the cost of producing the inaccurate signal is quite low for males, the cost to the female of ignoring any alarm signal can be much higher.

Figure 6.30. Male alarm call rate. Mean (+ SE) alarm call rates. Males give more false alarms when visited by estrus females. EF—DEP = estrus female attempting departure; EF—No DEP = estrus female not attempting departure; NEF = non-estrus female; SOL = solitary female. (*Source:* Bro-Jorgensen & Pangle 2010)

sexual deception hypothesis
The hypothesis that males will produce deceptive signals to females in order to enhance their own reproduction.

Capuchin monkeys and inaccurate signals

Featured Research

In many social groups, dominant individuals—which may be older or larger than others—have preferential access to resources (see Chapter 11). Dominant individuals will often feed first, and if food is limited and easily defendable, they may prevent subordinates from feeding. Can inaccurate alarm calls allow subordinates to gain access to food? If so, then we can make several predictions:

1. These calls should be produced by subordinates and not by dominant individuals, who already have free access to resources.

2. The calls should be given when food is clumped rather than dispersed, because it is more difficult for subordinates to gain access to clumped resources.

3. The calls should be given when a caller is close enough to food to take advantage of the distraction.

4. The calls should benefit the caller by eliciting antipredator behavior in dominants, such as moving to a safer location (and most likely away from the food).

Brandon Wheeler tested these predictions on a population of tufted capuchin monkeys (*Cebus apella nigritus*) in Iguazú National Park in Argentina (Wheeler 2009). Capuchins are medium-sized frugivorous (fruit-eating) primates that live in social groups of up to several dozen individuals (Figure 6.31). Within a group, dominant and subordinate individuals can be identified by their interactions (see Chapter 15). Wheeler studied a single group of about two dozen members over a period of 18 months. Individuals were identified by unique facial marks. Wheeler focused on the production of antipredator "hiccup" calls. Two or more of these indicate a high level of danger and often elicit antipredator behavior, especially "look" and "escape" reactions, in which receivers scan the environment and flee to a safer location.

Wheeler established food patches consisting of one to six platforms suspended in tree branches several meters off the ground—a typical feeding location for capuchins (Figure 6.32). The amount and distribution of food was manipulated. When only a few platforms were present, the food was clumped and thus easier to defend; when more platforms were included, the food was more dispersed and thus harder to defend.

At the start of each day, bananas, a preferred food, were placed on the platforms. Upon arrival of the group, the researchers selected a focal individual and

Figure 6.31. Tufted capuchin. Capuchins live in social groups.

Figure 6.32. Feeding platform. The feeding platform used to examine capuchin behavior.

Figure 6.33. Alarm calls vary with food distribution. False alarms were given more often when food was clumped in one or two platforms rather than more evenly distributed across five or six platforms. (*Source:* Wheeler 2009)

Figure 6.34. Alarm calls vary with distance to food. False alarm calls were most often given when a capuchin was less than 2 m from food. (*Source:* Wheeler 2009)

collected data using continuous sampling. The observers recorded all hiccup calls produced by the focal animal as well as its location with respect to the food platforms when the call was produced. In addition, the observer recorded the position of the focal individual every 30 seconds (scored as less or farther than 2 m from the platform). Hiccup calls were classified as accurate if a predator was observed nearby and inaccurate if no predator was observed.

Wheeler observed the behavior of capuchins at 321 feeding platform trials and collected data from 499 individual focal animals. While only 14 individuals produced 25 inaccurate alarm calls, 24 of those were produced by a subordinate. Most were produced when the food patch consisted of no more than three platforms (i.e., when food was clumped); few were given when the patch consisted of five or six platforms (Figure 6.33). Individuals were almost always within 2 m of the food patch when an inaccurate alarm call was emitted (Figure 6.34), and the false alarms produced antipredator behavior in 40% of receivers. The caller often benefited from the "deception," because in most cases the receiver abandoned the food platform, allowing the caller to feed.

These results are consistent with each of Wheeler's predictions and imply that subordinate capuchins produce inaccurate alarm calls to distract others so that they can gain access to clumped food. Note, however, that such calls were given sparingly, probably because of frequency-dependent selection: subordinates that produce them too often are likely to be ignored.

6.5 Communication can involve extended phenotype signals

As we've seen, most communication involves signals associated with an organism's morphology, behavior, or the chemical pheromones it produces. Another class of signals can also be defined—**extended phenotype signals** are those that are expressed beyond the body of an individual and often include modification of

extended phenotype signal A signal expressed beyond the body of an individual that often includes modification of the environment.

Figure 6.35. Extended phenotype. A sand pillar constructed by male fiddler crabs to attract females.

the environment. Many extended phenotype signals are involved in mate choice and include the construction of mud pillars by crabs (Figure 6.35) (Matsumasa et al. 2013), the creation of sand craters by fish (Schaedlin & Taborsky 2006), and the manipulation of pebbles and plant material by birds (Soler et al. 1996).

Extended phenotype signals persist longer than behavioral signals and continue to function in the absence of the signaler (Schaedlin & Taborsky 2009). They are also more flexible than morphological signals because they can be more readily modified with changes in environmental conditions. Most are costly to produce and maintain and so are considered to be reliable indicators of the phenotype of the signaler. As such, they provide an additional signal to others about the condition of a signaler.

Let's examine two species, bowerbirds and sticklebacks, that illustrate extended phenotype signaling.

Featured Research

Bowerbirds construct and decorate bowers

As noted earlier, male bowerbirds construct elaborate stick structures, bowers, to attract a mate. Males construct and decorate bowers with colorful objects from their environment such as flowers and snail shells, as well as plastic objects discarded by humans (Figure 6.36). They defend their bower from other males and will steal decorations from a rival, even destroying another male's bower if it is undefended. In addition, males display in front of their bower to attract females, and mating occurs in the bower structure.

Gerald Borgia tested the hypothesis that a bower functions as a signal of male quality (Borgia 1985). He assumed that the quantity of decorations at a bower correlates with male quality. If so, then male mating success would be expected to correlate positively with the quantity of bower decorations.

Figure 6.36. Bowerbird bower. Males construct stick bowers and decorate them with colorful objects.

This prediction was tested by observing a population of 22 male satin bowerbirds (*Ptilorhynchus violaceus*) over two years in New South Wales, Australia. Borgia quantified the size of each bower, the number of decorations on each male's bower, and the number of copulations the male obtained in his bower.

Male mating success was highly skewed, with the top-five males obtaining over 50% of the copulations. There were significant positive correlations between mating success and the size of a bower as well as the number of snail shells, blue feathers, and yellow leaves on a bower. These data suggest that bowers function as signals of male quality. Bowers require great effort to build, decorate, and defend. Only males in the best condition can elaborately decorate and defend large bowers, and such males appear to be the most attractive to females.

Bowerbirds represent perhaps the most conspicuous example of species with extended phenotype signaling. Many male fish also build nests, and some species add decorations to them to attract females. This is particularly common among sticklebacks, a group of small freshwater fish found throughout the world. Let's look at one example of work that investigated the signaling function of such nest decorations.

Featured Research

Sticklebacks decorate their nests

Sara Östlund-Nilsson and Mikael Holmlund investigated the function of nest decoration behavior in three-spined sticklebacks (*Gasterosteus aculeatus*) (Östlund-Nilsson & Holmlund 2003). Three-spined stickleback males construct nests of soft algae within a shallow depression in sand. Females lay eggs in a nest, and only the male cares for the eggs. Some males decorate the entrance to the nest with materials that provide a striking color contrast: for instance, males that build a nest with green algae will place a piece of red algae at the entrance, and vice versa. Östlund-Nilsson and Holmlund investigated male preference for objects of different color and whether nest decorations affect female mate choice; in other words, they asked, "Are nest decorations an extended phenotype signal?"

The research team captured individuals from the wild and housed them in individual aquaria. Breeding males were offered green algae and, when they began to construct a nest, an equal number of potential decorations that differed in color and material: red, green, blue, and silver foil sticks as well as red and blue shiny spangles. All males readily added these objects as decorations near the entrance of their nest and showed a strong preference for red sticks.

Given that males used the offered materials as decorations, the researchers conducted a mate choice experiment in which they manipulated the quantity of decorations at a nest. They first matched pairs of males for size, weight, and coloration. Each male was allowed to construct a nest using green algae. Males were then captured and moved to the other nest to control for differences in nest-building behavior. One male was randomly assigned to the "nest decorated" treatment, and the researchers added five red sticks, two red spangles, and two blue spangles to the nest entrance. For the control male, the researchers added one piece of green algae at the nest entrance. Males were then leashed to restrict their movement but could still perform mating displays. Females were allowed to view each male for ten minutes and then interact with the males. The researchers recorded the time spent near each male and whether she mated with a male.

All eight females preferred the male whose nest had been decorated. On average, they spent over 80% of the time near the male with the decorated nest (Figure 6.37); the three females that mated did so with the male that had the decorated nest. This study supports the hypothesis that males use contrasting colored objects as extended phenotype signals to attract females. Males will decorate their nest entrance with colored objects that contrast strongly with the color of the nest, and females prefer to mate with males that have decorated nests. Why do females prefer to mate with males that elaborately decorate their nest? If nest decorations are costly to produce and maintain, then male quality may correlate with the extended phenotype signal that he constructs.

Our discussion of signals to this point has largely assumed that communication occurs between one signaler and one receiver. In many situations, however, other individuals are nearby and so can intercept signals that are produced. We examine the implications of this in the last section of the chapter.

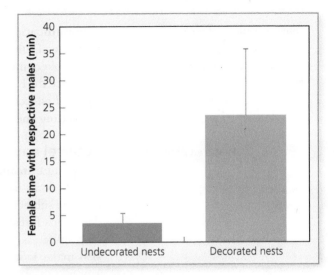

Figure 6.37. Female choice of males. Mean (+SE) time spent with respective males. Females spent more time in front of males with nests containing colored sticks and spangles. (*Source:* Östlund-Nilsson & Holmlund 2003)

6.6   Communication networks affect signaler and receiver behavior

While the simplest communication system involves two individuals, signals can also be intercepted by others. These **bystanders**, or **eavesdroppers**, are present but do not take part in the signaling (Matos & Schlupp 2005). In such

bystander A third-party individual that detects a signal transmitted between a signaler and a receiver. Also called an **eavesdropper**.

communication network
A communication system involving a signaler, a receiver, and at least one bystander.

communication networks, eavesdroppers can benefit by learning about the presence of competitors or predators. Not surprisingly, eavesdropping is a widespread phenomenon (Valone 2007). In turn, the presence of bystanders can also influence the behavior of signalers, a phenomenon known as an audience effect.

We first examine two studies of eavesdropping and then end the chapter by illustrating one example of an audience effect.

Featured Research

Squirrel eavesdropping

In the autumn and early winter, Eastern gray squirrels (*Sciurus carolinensis*) frequently cache (or store) food items like nuts for later retrieval. Food caches, however, are subject to pilfering: conspecifics or heterospecifics can steal food. In the deciduous forests of eastern North America, squirrels live with many potential cache pilferers, including other squirrels and birds. Blue jays (*Cyanocitta cristata*) also make food caches and retrieve them by remembering visual landmarks around the cache location (Bednekoff & Kotrschal 2002). Given that these species both cache and live in the same area, Kenneth Schmidt and Richard Ostfeld wondered whether squirrels might eavesdrop on the vocal communication signals of potential cache pilferers to learn about their proximity. A jay that observed a squirrel caching food could easily remember its location and return later to pilfer the contents. The researchers hypothesized that squirrels might alter their feeding preferences for different food items depending on whether jays were nearby or too far away to watch. They predicted that when jays are nearby, squirrels should reduce their preference for feeding on cacheable food items that could be stolen after being cached.

To test this prediction, Schmidt and Ostfeld established pairs of food patches in eastern New York in early winter (Schmidt & Ostfeld 2008). Each food patch was filled with pea gravel and 15 hazelnuts. One patch contained unshelled hazelnuts, while the adjacent tray contained shelled hazelnuts. Squirrels frequently cache intact hazelnuts but rarely cache nuts without shells, presumably because unshelled nuts cached in the ground rot quickly. Food items were mixed into the pea gravel, forcing squirrels to dig through the gravel to find them. The first few nuts were relatively easy to find, but food discovery then became more and more difficult. This design ensured that a squirrel would leave a patch before all the food was found, because the last few items would be very difficult to acquire. The number of food items left, called the giving-up density, indicated how much effort the squirrel allocated to feeding in that patch (see Chapter 8).

Jays use vocalizations to communicate with conspecifics to defend territories and attract mates. Can squirrels use jay vocalizations to assess the threat of pilfering? Schmidt and Ostfeld recorded the number of nuts left in each patch for three experimental treatments: (1) blue jay vocalizations from nearby speakers (less than 25 m away), (2) blue jay vocalizations from far-away speakers (more than 125 m away), and (3) vocalizations of other common nonpilfering birds, such as northern cardinals (*Cardinalis cardinalis*) (control group). Food patches were available to the squirrels all day. At the end of each day, the researchers collected all the nuts that remained in each tray, the giving-up density (GUD).

In each case, squirrels consumed the same number of noncacheable nuts per food patch. However, significantly more cacheable nuts were left in the patch when jay vocalizations were played nearby (Figure 6.38). These results suggest that squirrels eavesdrop on heterospecific blue jay vocalizations to learn blue jays' presence and location. When jays appear to be nearby, squirrels reduce the amount of effort they devote to acquiring cacheable food, presumably because it is more likely to be pilfered if cached.

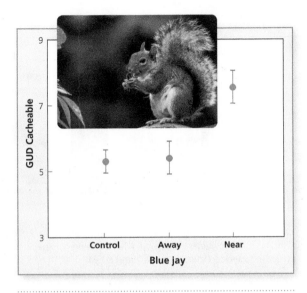

Figure 6.38. Giving-up densities of cacheable food. Mean (± SE) giving-up density (GUD) of cacheable food items for each treatment. Squirrels left more cacheable nuts (higher GUDs) when jay vocalizations were played nearby than when vocalizations were played far away or for controls. Inset: gray squirrel. (*Source:* Schmidt & Ostfeld 2008)

Eavesdropping in túngara frogs

Featured Research

During the breeding season, several species of frogs produce vocalizations to attract mates. However, predators can also more easily locate a calling male. For example, in the tropics, predatory bats are known to identify the location of túngara frogs (*Physalaemus pustulosus*) (Figure 6.39) using their vocalizations. In a large pond, multiple frog species often start and stop calling synchronously. It is relatively easy to understand why they all might stop simultaneously: a predator is nearby. But why should they all start simultaneously?

Steven Phelps, A. Stanley Rand, and Michael Ryan proposed that males eavesdrop on one another to assess predation risk (Phelps, Rand, & Ryan 2007). Túngara frogs frequently live sympatrically (co-occur) with *Leptodactylus labialis* frogs and together produce a mixed-species chorus at breeding sites. Can males use the vocalizations of others to assess the level of predation risk?

Phelps and colleagues captured male frogs in Panama and housed them in the laboratory for two days before testing. On the evening of testing, a male was placed in a small container. The test began with a ten-minute acclimation period during which conspecific calls were played. Playbacks continued for the next five minutes, and the number of calls produced by the subject was recorded. At Minute 16, the playback stopped, and a small disc similar in size and shape to that of a predatory bat was passed over the subject to elicit antipredator behavior (flattening the body, freezing). From Minutes 17 to 25, one of four treatments was presented: (1) silence; (2) a repeated conspecific call; (3) a repeated call of sympatric *L. labialis*; or (4) a repeated call of an allopatric (geographically isolated) frog, *Physalaemus enesefae*, that has a similar call to that of the túngara frog. If túngara frogs eavesdrop on calls to assess predation risk after a

Figure 6.39. Túngara frog. A male calls to attract females. These frogs are common in Central America.

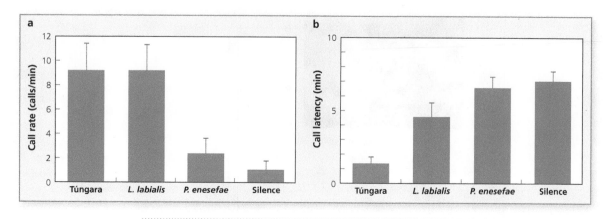

Figure 6.40. Frog calls. Mean (+ SE) calling behavior. After a simulated attack, frogs (a) produced calls at a higher rate and (b) had a lower latency to call when hearing conspecific calls and the calls of a heterospecific sympatric species. (*Source:* Phelps, Rand & Ryan 2007)

simulated attack, the researchers predicted that they should respond most readily to conspecific calls, followed in order by the calls of sympatric *L. labialis*, the calls of allopatric *P. enesefae,* and silence. The researchers measured both the latency to begin calling and the call rate.

Male túngara frogs responded as predicted. They had the shortest latency to calling and the highest call rate in response to túngara frog calls, followed by the calls of *L. labialis*, the calls of *P. enesefae*, and silence (Figure 6.40). Indeed, there was no significant difference in their call rate or latency to call in response to either the call of *P. enesefae* or silence, indicating that the calls of *P. enesefae* were essentially ignored. The researchers concluded that túngara frogs eavesdrop on both conspecific and sympatric heterospecific vocal signals to assess predation risk. They propose that this behavior may be one reason why many frogs participate in mixed-species choruses when courting females.

Featured Research

Audience effects in fighting fish

In the first two cases, we saw how receiver behavior was affected by eavesdropping within a communication network. Signalers are also affected by the presence of bystanders, as demonstrated in the audience effect we see next.

Claire Doutrelant, Peter McGregor, and Rui Oliveira investigated how a bystander affects aggressive behavior in male fighting fish (*Betta splendens*) (Doutrelant, McGregor, & Oliveira 2001). As we discussed earlier in this chapter, male fighting fish are highly aggressive and compete for territories and mates. Aggressive signal displays include flaring of the gill covers, as well as tail beats and bites directed at an opponent.

The researchers designed a simple experiment that examined the aggressive behaviors of a focal male as it interacted with a rival male when a bystander was either present or absent. The focal male and its rival, which were closely matched for size and color, were separated by a removable opaque partition. A second tank was placed next to the focal tank that contained a female bystander, a male bystander, or no bystander (Figure 6.41). When the partition was raised, the focal male was allowed to interact with the rival male for a ten-minute session, after which the partition was replaced. After two hours, the partition was raised again for another ten-minute interaction session. In one session, a bystander was present, while in the second, none was present, with the order randomized across

Figure 6.41. Experimental design. Ma and Mb represent the locations of a focal male and a rival male (position randomized across trials). F and M represent the location of a (a) female and (b) male bystander, respectively. The male bystander was placed farther away than the female bystander to prevent interactions. (*Source:* Doutrelant, McGregor, & Oliveira 2001)

focal males. For trials with a bystander, the focal male was allowed to view the bystander for three minutes before the partition was raised.

Doutrelant and colleagues found that focal males displayed more tail beats and spent more time with erect gill covers but directed fewer bites at their opponent when a female bystander was nearby, compared to when there was no bystander (Figure 6.42). In contrast, the presence of a male bystander did not affect male aggressive display behavior.

These results demonstrate that signaling behavior can be affected by the presence of an audience. Males used more conspicuous displays (erect gill covers, tail beats) but fewer overt aggressive behavioral signals (bites) in a contest with a rival male when a female was present. The researchers suggest that males were attempting to simultaneously display to the female while interacting with a rival and so used more conspicuous behavior. In addition, they speculate that females may also be less attracted to highly aggressive males. Interestingly, no such audience effect was observed when the bystander was a male, suggesting that third-party males are ignored.

Many animals live in complex social environments surrounded by both conspecifics and heterospecifics. This variation creates the opportunity for individuals to acquire information from others by intercepting their signals but also can affect the production of signals. Much work has focused on understanding how and when such audience effects influence signaling behavior.

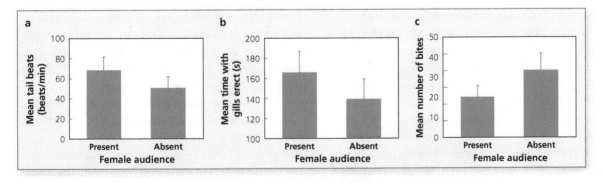

Figure 6.42. Female audience affects male behavior. Mean (+ SE) aggressive behaviors of treatment males. Male aggressive behavior was affected by the presence of a female audience. (*Source:* Doutrelant, McGregor, & Oliveira 2001)

Chapter Summary and Beyond

Communication occurs when an individual uses a specialized signal to influence the behavior of another. Signals affect behavior as if they contain information about signaler phenotype or the environment and include specialized morphology, vocalizations, behavior, physiological traits, and chemical pheromones. Recent work has demonstrated that some signals affect behavior as if they conveyed information about individual identity (e.g., Tibbetts & Dale 2007). To be effective, a signal must stand out from background noise so that it can be perceived. Therefore, selection favors signal characteristics that contrast with the environmental background and minimize degradation and attenuation in different habitats (e.g., Peters, Hemmi, & Zeil 2007; How et al. 2008). Attention is increasingly being paid to examining how and when animals use multiple types of signals in communication (Bro-Jorgensen 2010).

The fitness interests of signalers and receivers affect the evolution of signals. Signals will be accurate indicators of conditions when the fitness interests of signaler and receiver coincide, as occurs in aposematic coloration, or when signals, such as those used in courtship and aggressive interactions, are difficult to fake or costly. On the other hand, selection can favor the evolution of signals that are inaccurate indicators of conditions when the fitness interests of two individuals differ. This creates a co-evolutionary process that affects the evolution of both the signaler and the receiver. Mimicry in predator–prey systems is a common example of inaccurate signaling. In Batesian mimicry, a palatable species resembles an unpalatable species, and a predator is duped. Ruxton, Sherratt, and Speed (2005) provide a comprehensive review of Batesian mimicry. In aggressive mimicry, a predator resembles a nonthreatening species to dupe a prey species, as occurs with cleaner fish and their aggressive mimics. Recent research has examined how client fish find and recognize cleaners. For example, Cheney et al. (2009) showed that the typical blue and yellow color patterns of cleaner fish contrast sharply with a background of coral reef and blue water, providing a conspicuous signal to a wide variety of client fish species. Inaccurate signaling within a species is relatively rare, because such signals must be cheap to produce and are subject to negative frequency-dependent selection. False alarm calls are one example of inaccurate signaling within a species. Individuals that produce a false alarm call can benefit from increased matings or feeding.

Some species use physical signals that involve active modification of the environment. These extended phenotype signals tend to be honest indicators of the phenotype of an individual. Schaedelin and Taborsky (2009) provide a review of empirical examples, and Bailey (2012) summarizes models to explain the evolution of such traits. Recent work has investigated the use of body ornaments such as tattoos and piercings as potential examples of such signaling in humans (Carmen et al. 2012).

In communication networks, signals can be intercepted by bystanders. Such eavesdroppers learn about others and can modify their behavior accordingly. In addition, when bystanders are present, audience effects, in which individual signalers modify their behavior, are common. Recently, Semple, Gerald, and Suggs (2009) observed how bystanders affect mother–infant interactions in rhesus macaques (*Macaca mulatta*), and Rowell, Ellner, and Reeve (2006) examined how communication networks can affect the evolution of accurate and inaccurate signals. McGregor (2005) provides a comprehensive summary of work on both eavesdropping and audience effects in animal communication networks.

CHAPTER QUESTIONS

1. Fireflies that produce flash displays are rather unique among nocturnal insects in that they use a visual rather than a chemical signal (pheromones) to communicate. Provide a hypothesis to explain why visual signals are rare among nocturnal insects.

2. Male bison produce snort vocalizations during dominance contests. Vocalization frequency is affected by body size: the larger the individual, the lower the frequency of sounds it can produce. Form a hypothesis to predict how female mating behavior might be affected by male vocalization frequency.

3. In the cleaner fish example, three species were involved in the system of aggressive mimicry: the client, the cleaner, and the aggressive mimic. However, predatory *Photuris* fireflies attract and eat male *Photinus* fireflies by producing

a flash display that resembles the one produced by *Photinus* females. Explain how this is an example of aggressive mimicry when only two species are involved.

4. Female crayfish can realize higher fitness by mating with dominant males. Two hypotheses describe how a female can recognize a dominant individual. One is that dominant individuals produce unique pheromones after winning a contest with a rival male. The other hypothesis is that females can learn about relative fighting ability by eavesdropping on fights. Design an experiment to test these alternatives.

5. Explain how frequency-dependent selection would act on a Batesian mimic.

6. Compare the communication networks in the squirrel–blue jay system studied by Schmidt and Ostfeld (2008) and the fish system studied by Doutrelant, McGregor, and Oliveira (2001). In each case, identify the third-party eavesdropper and that individual's effects on behavior.

Chapter 7

Learning, Neuroethology, and Cognition

Many years ago, during an ornithology class field trip to the Chiricahua Mountains of southeastern Arizona, we heard a loud police whistle nearby. Almost instantly, we began to hear vocalizations of a flock of Mexican jays (*Aphelocoma wollweberi*) that appeared to be moving toward the location of the whistle (Figure 7.1). We followed the jays to find a researcher who was busy collecting data on aggressive behavior.

This flock was being studied by Jerram Brown and his students. They had uniquely color-banded individuals and were investigating the behavior and ecology of these social birds. In Brown's early work, researchers placed small piles of food in the center of a flock's territory to attract many individuals, who then competed for access to the food. One year the researchers blew a police whistle a few times just before putting out the food, and soon the jays learned to associate this novel sound with the presence of food at a specific location. Over time, the researchers noticed that yearlings learned the association from older birds, with unmarked young birds quickly responding to the sound even though no whistle had been blown in the area for almost a year (Brown 1997b).

Figure 7.1. Mexican jay. These birds learned to associate a whistle with the presence of food.

In this chapter we examine the broad array of ways that animals learn, ranging from simple habituation to complex problem solving. We also examine the physical changes in the brain that are associated with learning. We then discuss individual and social learning and see how the latter can lead to the development of animal traditions and even culture. We end by discussing animal cognition.

7.1 Learning allows animals to adapt to their environment

learning A relatively permanent change in behavior due to experience.

Learning is a relatively permanent change in behavior as a result of experience. It is a process by which animals modify their behavior, or adapt to their environment, in ways that allow them to experience increased fitness. For example, the ability to successfully capture prey is a foraging skill that can affect fitness and improve with experience, as we see next.

Featured Research

Improved foraging efficiency in salamanders

Red-backed salamanders (*Plethodon cinereus*) are common in the eastern deciduous forests of North America (Figure 7.2). They feed under rocks and logs on a wide variety of invertebrate prey. Salamander neonates can recognize, attack, and capture prey within one week of hatching, without prior experience. Does their foraging efficiency improve with experience? Megan Gibbons, Arianne Ferguson, and Danielle Lee conducted an experiment to investigate the hypothesis that learning would improve salamanders' skill as predators (Gibbons et al. 2005).

Gibbons and her colleagues examined the predatory hunting skill of lab-reared two-week-old individuals that had no prior opportunity to forage. To standardize hunger levels, three feeding trials were separated by about 14 days, the digestion period for these neonates.

To characterize feeding efficiency, the researchers presented four termites (*Reticulitermes flavipes*) to an individual and recorded capture attempts (a jaw snap at one termite), successful captures, and the number of prey that escaped after capture in each ten-minute trial. From these data, they quantified the number of prey captured and the salamanders' foraging accuracy (measured as the number of prey eaten, divided by the number of attacks, captures, and escapes).

Neonates captured few prey during the first trial. The proportion captured improved significantly, however, in Trials 2 and 3 (Figure 7.3). These data suggest that individuals rapidly learned to forage in a more efficient manner as they gained experience. In this example, only one learning experience was required for a large gain in foraging efficiency. Such improved performance can enhance survivorship in the wild.

Evolution of learning

In the last example, individuals learned to hunt more effectively, a skill that should improve their fitness. This finding raises the question, Is learning always adaptive? A long-standing answer was that evolution should always favor improved learning ability in dynamic environments. Recent work, however, suggests a more complex answer.

Theory indicates that two factors affect the evolution of learning: environmental stability and the usefulness of past experience (e.g., Mery & Kawecki 2002; Dunlap & Stephens 2009). Consider two possible types

Figure 7.2. Red-backed salamander. This salamander feeds on small invertebrates.

of worlds: one is fixed and experiences no change (e.g., no new predators or parasites ever appear), while the other is dynamic and changes unpredictably (e.g., new predators and parasites may appear in the middle of the breeding season). Now imagine two habitats (A and B) that differ in quality for reproduction: breeding in one leads to high fitness, while breeding in the other leads to reproductive failure. In the fixed world, the high-fitness habitat (A) will always be the best place to reproduce. In the dynamic world, both habitats have the possibility of leading to high-fitness due to the unpredictable nature of predators and parasites. Sometimes A will be best and at other times B will be best.

Now let's assume that in this changing world, each habitat has a 50% chance each breeding season of being high quality or causing reproductive failure (Figure 7.4). In which world should learning evolve? Neither. This answer may not be intuitive, so let's examine why. In the fixed world, individuals that breed in the high-fitness habitat will quickly outcompete those that reproduce in the other habitat. If habitat choice is influenced by genes, the world will soon be full of individuals that selected only the high-fitness habitat, and there would be no need to learn about the other habitat or the differences between them. On the other hand, in the dynamic world, there is nothing to learn, because learning is useful only if individuals can benefit from their experience. In other words, there must be a predictable relationship between experience and the best option available at a given moment. In the dynamic world, the habitats change unpredictably, and so both habitats always have a 50% chance of being the best option, no matter what happened in the past. Again, there is no benefit to learning because information about the environment has no value.

Learning theory thus demonstrates that two factors affect the evolution of learning: the regularity of the environment and the reliability of past experience. These factors are independent of each other but are not mutually exclusive. As environmental regularity increases, learning will become less favored, because in a completely regular world, evolution will fix behavior. As the reliability of experience increases, learning will be strongly favored, because individuals that learn will have higher fitness than those that do not (Dunlap & Stephens 2009). The real world usually falls between these two extremes, and so we expect learning to evolve. Indeed, most—if not all—animals that have been studied show the capacity to learn. Next we examine a fairly simple form of learning—learning not to respond to a nonthreatening stimulus.

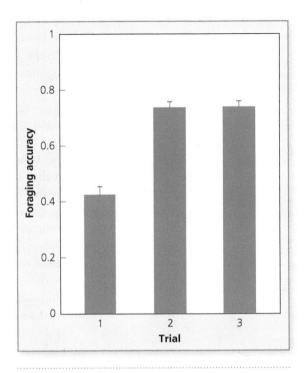

Figure 7.3. Salamander foraging accuracy. Mean (+SE) foraging accuracy of prey capture increased after one trial. (*Source:* Gibbons, Ferguson, & Lee 2005)

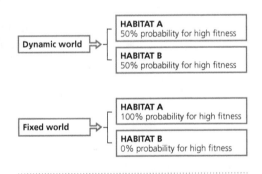

Figure 7.4. Dynamic and fixed worlds. In a dynamic world, reproduction in each habitat can lead to high fitness, but in a fixed world, only one habitat leads to high fitness.

Green frog habituation to intruder vocalizations

Perhaps the simplest form of learning is **habituation**, the reduction and then lack of response to a stimulus over time. An environmental stimulus is anything in the environment (biotic or abiotic) that an individual can perceive, and any reaction to that stimulus is a response. Habituation to stimuli that are irrelevant to survival and reproduction, such as leaves blowing in the wind, is useful because it allows time and effort to be spent in more productive ways. For example, the **dear enemy hypothesis** (Temeles 1994) predicts that individuals will show reduced aggressive interactions with territorial neighbors, compared to strangers who represent a greater threat, as a territory owner becomes more familiar with its

Featured Research

■ **habituation** The reduction and then lack of response to a stimulus over time.

■ **dear enemy hypothesis** Territory owners will show reduced aggressive interactions toward territorial neighbors, compared to strangers.

neighbor. The mechanism for this reduction in aggression is habituation. Strangers represent potential threats as new competitors for territorial space, while a neighbor represents a much lower threat because it has already established a nearby territory. A territory owner can save time and effort by essentially ignoring a neighbor while aggressively interacting with a stranger.

Patrick Owen and Stephen Perrill investigated whether the dear enemy hypothesis explains variation in aggression in territorial green frogs (*Rana clamitans*) (Owen & Perrill 1998). Males defend territories around ponds, where they vocalize to attract females. Territorial males respond to strangers with both aggressive vocalizations and physical attacks. Aggressive vocalizations can be distinguished from advertisement calls used to attract females by their lower dominant frequency. Owen and Perrill predicted that if males habituate to familiar stimuli, there should be a decrease in their response to a new rival's vocalization after an initial aggressive response.

The researchers studied males at four ponds near Indianapolis, Indiana. They simulated intruders by synthesizing the calls of two males to standardize their length and intensity and playing them from a speaker placed 1 to 2 m away from a focal calling male. For each focal calling male, the researchers played one of the synthetic intruder calls at 5- to 15-second intervals for up to an hour. During playback, the researchers recorded the focal frog's movements toward the speaker and its aggressive vocalizations. When the focal frog stopped moving and began to produce advertisement calls again, the researchers stopped the playback and started a 15-minute rest period, after which they initiated the same intruder call once more. The researchers again recorded the focal male's movement toward the speaker. When movement again ceased and advertisement calls were being produced, the researchers played the second synthetic call to simulate a new intruder and recorded the vocalizations of the focal frog.

Habituation to the intruder calls was observed in both the focal frog's movement and its vocalizations. At first, focal frogs made several movements toward the speaker, but this behavior declined over time (Figure 7.5). Movement toward the speaker increased following the rest period and re-initiation of the synthetic call, but at a lower level.

The intruder calls also affected the focal frogs' call production. Prior to the playback of the synthetic intruder, the focal males produced advertisement calls but switched to aggressive calls when the synthetic call began. Eventually, focal frogs switched back to advertisement calls, but only until a new synthetic call was played. These results support the dear enemy hypothesis and demonstrate habituation to a stranger: both physical and vocal responses to the simulated intruder were initially high, but over time, both responses declined as the intruder became more familiar and was perhaps perceived to be less of a threat.

The above examples illustrate how learning allows individuals to modify their behavior in adaptive ways. But *how* do animals learn? We examine that question in the next two sections. First, we look inside the brain to see proximate changes in neurons associated with learning. Then, we examine how animals learn by making associations.

Figure 7.5. Movement in response to calls. Mean number of movements over time as conspecific synthetic intruder calls are broadcast at different dominant frequencies. The rest period between broadcasts begins at 65 minutes, and the second broadcast begins at 80 minutes. Individuals habituate to each broadcast call and show less movement over time. Inset: green tree frog. (*Source:* Owen & Perrill 1998)

Learning is associated with neurological changes

How are experiences translated into changes in behavior? In Chapter 5, we saw how information from experiences and environmental stimuli is perceived via sensory receptors and relayed to the central nervous system through nerves. Recall that nerves are composed of neurons, cells that receive and transfer electrical and chemical signals. The junction between two neurons, the synapse, is believed to play an important role in learning and memory. Two aspects of the nervous system have garnered much attention in research on learning: changes in neurotransmitters and the number of synapses between neurons. One neuroethological approach to understanding such proximate mechanisms of learning is to characterize synapse characteristics before and after a learned experience to determine what changes occur. This is the approach used in the next two studies.

Neurotransmitters and learning in chicks

One important form of learning is **imprinting**, rapid learning that occurs in young animals during a short, sensitive period and has long-lasting effects. For example, offspring often learn the phenotype and identity of their parents through filial imprinting, because their parents are the first objects they encounter. Vulnerable offspring benefit from this learning by being able to quickly identify their parents and remain close to them. Konrad Lorenz made imprinting famous when he showed how graylag geese (*Anser anser*) hatchlings would imprint on his boots when these were the first objects they saw. In the absence of their parents, they simply followed him (his boots) around (Figure 7.6) (Lorenz 1935). More recently, this behavior has been used for the reintroduction of endangered birds to their former habitats (Applying the Concepts 7.1).

Many birds, like domestic chickens (*Gallus gallus domesticus*), visually imprint on a stimulus when they hatch. In their brain, the intermediate and medial parts of the hyperstriatum ventrale (IMHV) appear to play an important role in memory related to imprinting; earlier work demonstrated that lesions in this area of the brain prevent imprinting, as does the blocking of postsynaptic neurotransmitter receptors (Horn 2004).

Rhiannon Meredith and her colleagues studied another imprinting mechanism in chicks by investigating whether the release of neurotransmitters from the presynaptic neuron is also associated with such learning (Meredith et al. 2004). The researchers divided chicks into two groups. Half were trained by exposure to a visual imprinting stimulus (a red box or a blue cylinder). The other birds were used as a control group and were given no visual stimulus for imprinting. During imprinting training, birds were placed on a running wheel; typically, they attempted to move toward the stimulus, and the wheel recorded these movements. Less than ten minutes after training, the researchers measured the strength of imprinting, or preference score, by sequentially placing two objects in front of the chick while it was on the running wheel. One was the imprinted object, while the other was a novel object. The preference score was calculated by dividing the amount of running the chick did toward the imprinted object during a fixed time period by the total amount of running in response to both objects. A score of 50 showed no preference for the imprinted object, while higher scores showed greater preference and a higher strength of imprinting. At the end of the experiment, chicks were sacrificed, the IMHV tissue

Featured Research

> **imprinting** Rapid learning in young animals through observation of parents or other objects. Individuals are typically attracted to objects they imprint on.

Figure 7.6. Lorenz and geese. A clutch of geese that imprinted on Lorenz's boots.

APPLYING THE CONCEPTS 7.1

Operation Migration and imprinting

Many birds follow long migration routes from their summer breeding areas to their winter feeding grounds. Birds typically learn these routes from their parents. However, the hand-reared birds of endangered species that are being reintroduced into the wild do not have this option. How can they learn their natural migration routes? William Sladen and his colleagues hypothesized that imprinting could be used to train young birds to follow a human flying an ultralight aircraft (Sladen et al. 2002).

Work began with a common species, the Canada goose (*Branta canadensis*). The researchers took eggs from parents, hatched them in an incubator, and hand-reared them. Birds did indeed imprint on the human handlers and followed them as they flew the aircraft, first over short distances, but eventually over longer flights. The same procedure was then used to train endangered species raised in captivity, such as trumpeter swans (*Cygnus buccinator*) and whooping cranes (*Grus americana*) (Figure 1) (Ellis et al. 2003). This success is particularly impressive given the cranes' 1,800 km migration. This

Figure 1. Imprinting on aircraft. Whooping cranes following an ultralight aircraft.

research inspired the movie *Fly Away Home*, in which captive Canada geese imprint on a young girl. She trains the birds to follow their historic migration route all the way from Ontario, Canada, to North Carolina. ∎

a

The birds imprint on a red box or blue cylinder.

was dissected, and assays were conducted to measure the release of several amino acid neurotransmitters, including glutamate and gamma-aminobutyric acid (GABA).

The research team found higher glutamate in the IMHV tissue of trained chicks compared to that of the controls (Figure 7.7), but no difference in GABA. However, among the trained birds, chicks with the highest preference score—those that most strongly imprinted on the test object—had higher levels of GABA in their brain (Figure 7.8).

These results suggest that neurotransmitters play a role in imprinting. One neurotransmitter, glutamate, was more strongly released in

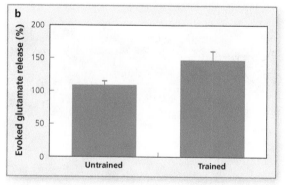

b

Figure 7.7. Glutamate release and training. (a) Training protocol. (*Source:* Horn 1998) (b) Mean (+ SE) glutamate release in different chicks. More glutamate was detected in trained than in untrained birds. (*Source:* Meredith et al. 2004)

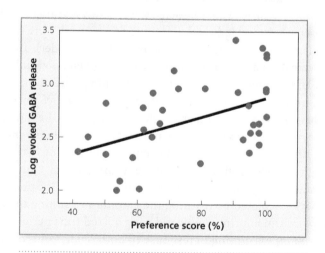

Figure 7.8. GABA release. Increased GABA was associated with higher preference scores (strength of imprinting). (*Source:* Meredith et al. 2004)

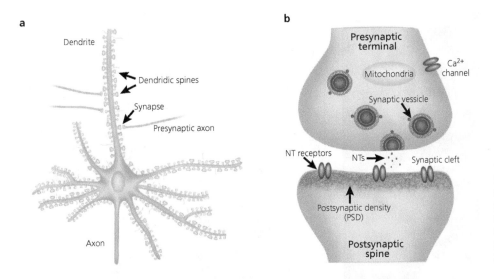

Figure 7.9. Neuron and dendritic spines. (a) Dendritic spines are small protuberances on the dendrites of neurons in the brain. (b) Each spine forms a synapse with another neuron. NT = neurotransmitter. (*Source:* Smrt & Zhou 2010)

the brain of birds that visually imprinted on an object, while another, GABA, was correlated with the strength of imprinting. This work illustrates that neurotransmitter release in presynaptic neurons played a role in imprinting over the short timescale of the experiment. Next, we examine recent research that investigates long-lasting memory.

> **dendritic spines** Small protuberances on a dendrite that typically receive synaptic inputs.

Dendritic spines and learning in mice

Featured Research

Memory is the retention of a learned experience and is a critical factor in animals' ability to utilize their prior experiences. How is memory stored in the brain? Although we are just beginning to understand the answer to this question, changes in the brain appear to be involved.

Structural change in the brain, especially in the number of synapses and in the strength of chemical synapses between neurons, is a phenomenon known as neural plasticity (Hering & Sheng 2001). Synaptic connections in the brain are dynamic, owing in part to the formation and elimination of **dendritic spines**, small protuberances on a dendrite that typically receive synaptic inputs (Figure 7.9). The dynamic nature of dendritic spines means that the number of synaptic connections between neurons can vary. Recent work suggests that the dynamic nature of spines plays a role in learning.

Guang Yang, Feng Pan, and Wen-Biao Gan examined spine formation associated with learning in mice (Yang, Pan, & Gan 2009). One group of mice was trained to learn a new motor skill: running on a rotating rod (a rotorod) suspended above the cage floor (Figure 7.10), an activity that is analogous to humans "logrolling" in water. The mice had to learn novel motor coordination skills and balance in order to stay on the rod as its speed increased. Another group of mice received no training and served as controls. The researchers used transcranial two-photon

Figure 7.10. Rotorod. Mice are placed on a rotating rod that increases in speed to develop a novel motor skill. Researchers measure the ability of mice to stay on the rod. (*Source:* Carter, Morton, & Dunnett 2001)

Figure 7.11. Dendritic spine formation. Image of dendrite (a) before and (b) after training, showing new spine formation (red arrows). (c) Mean (+SE) spine formation. Both young and adult animals that were trained and learned a new behavior had more spine formation. (*Source:* Yang, Pan, & Gan 2009)

microscopy to examine the fluorescent-labeled dendritic spines of living subjects. This technique involves surgically thinning the cranium to allow for high-resolution imaging of dendritic spines.

After one to two days of training, both young and old mice that learned the new motor skill showed significantly higher levels of dendritic spine formation compared to controls (Figure 7.11). In addition, the performance of mice on the rotorod (the revolutions per minute they could achieve) was associated with the number of new spines formed: mice that had developed more new dendritic spines performed the task better. Interestingly, over the course of the following two weeks, most of the new spines disappeared, so that the net result was essentially no difference in the total number of spines between trained and control mice. In fact, the researchers estimated that less than 0.1% of the new spines would persist for life.

How do these results relate to lifelong memory? One possible explanation is that lifelong memory may be associated with both the formation of new spines during learning and the pruning of synapses that, in essence, "remodel" the brain following learning. Further studies are required to examine this hypothesis.

Such proximate studies of brain structure allow us to understand how the brain changes with learning. We can also ask evolutionary questions about learning and memory. Are there brain differences between species that are associated with differences in learning and memory? The next study provides an answer.

Featured Research

Avian memory of stored food

Many animals **cache** (or store) food such as hard nuts for later use. Caching provides food for the future but is only useful if that food can be relocated. What is the mechanism that allows animals to find their caches? Some rely on local landmarks, like rocks and logs (Jones & Kamil 2001), but most use their memory.

cache Food stored in a hidden location for later retrieval.

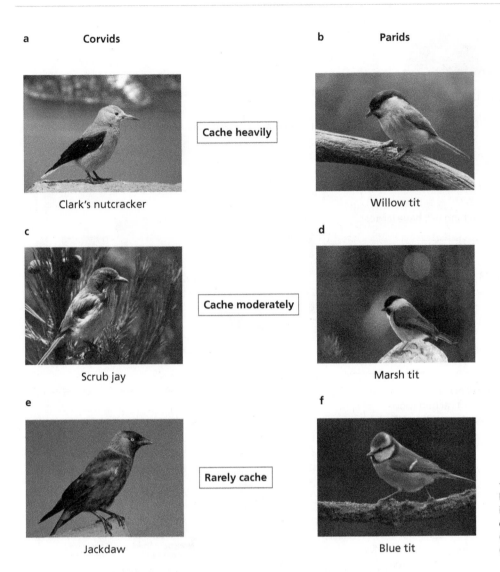

a   **Corvids**

Cache heavily

Clark's nutcracker

b   **Parids**

Willow tit

c

Cache moderately

Scrub jay

d

Marsh tit

e

Rarely cache

Jackdaw

f

Blue tit

Figure 7.12. Corvids and parids. There is variation in the level of caching within each group. (a) Clark's nutcracker, (b) willow tit, (c) scrub jay, (d) marsh tit, (e) jackdaw, and (f) blue tit.

Caching is particularly common in two families of birds, Corvidae (crows and jays) and Paridae (chickadees and titmice), but caching propensity varies across species (Figure 7.12). Some, such as Clark's nutcrackers (*Nucifraga columbiana*) and willow tits (*Poecile montanus*), cache heavily and rely on stored food for survival. Others, such as scrub jays (*Aphelocoma coerulescens*) and marsh tits (*Parus palustris*), cache occasionally, while species such as jackdaws (*Corvus monedula*) and blue tits (*Parus caeruleus*) rarely, if ever, cache food.

If cache recovery is accomplished through memory, then species that cache heavily should rely more on a well-developed memory. Examining the brain structure of such species could help confirm this hypothesis. In many vertebrates, spatial memory is associated with the hippocampus, known in birds as the hippocampal

Figure 7.13. Avian brain. The hippocampal formation (HF) is located in the temporal medial lobe.

formation, which is located in the medial temporal lobe (Figure 7.13). Does increased caching behavior correlate with increased hippocampal formation size? Early work found that across families, the hippocampal formation is larger in

SCIENTIFIC PROCESS 7.1

Brain structure and food hoarding

RESEARCH QUESTION: *Does brain structure differ between birds that do and do not need to remember the location of cached food?*

Hypothesis: Hippocampal formation (HF) size in the brain correlates with spatial memory capability.

Prediction: Species that rely more on caching will have larger HF size.

Methods: The researchers:

- Assembled data on HF size for 13 species of corvids and ten species of parids, adjusted for differences in body and brain size
- Characterized each species as: (1) a nonhoarder, (2) a nonspecialized hoarder that caches food occasionally, or (3) a specialized hoarder that relies heavily on cached food for survival

Results: In both families, specialized food hoarders had larger HF size than species that rely less heavily on cached food. Species that do not cache food exhibited the smallest HF.

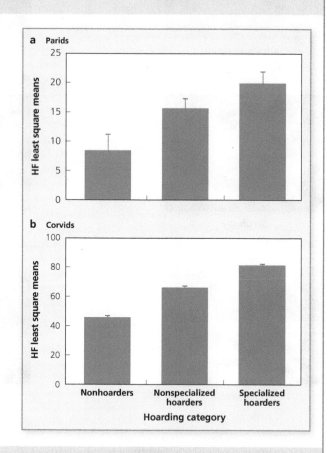

Figure 1. Hippocampal formation size. Mean (+SE) HF size for species of (a) parids and (b) corvids that differ in food-hoarding behavior. (*Source:* Lucas et al. 2004)

Conclusion: HF appears to play an important role in facilitating the memory of cache locations for future food retrieval.

EVALUATE

Propose a hypothesis to explain why all corvids and parids do not have a large HF size.

species of birds that often cache food (Krebs et al. 1989). However, species in different families may differ in brain structure for a variety of reasons. Jeff Lucas and colleagues (Lucas et al. 2004) examined how species *within* the same family vary in food-hoarding propensity and brain structure (Scientific Process 7.1). They gathered data from studies on 23 corvid and parid species. The degree of food caching ranged from a complete absence of caching to a strong reliance on cached food for survival. They found a strong association across species within both families: those that relied heavily on food caches indeed had a larger hippocampal formation than those that rarely cached food, suggesting a link between hippocampal formation size and function (memory of cache locations) in these species.

Each of these studies illustrates how learning is associated with changes in the brain, ranging from rapid changes in dendritic spine formation associated with learning a new skill to evolutionary changes in brain structure associated with long-term memory. Next, we examine how animals learn by making associations.

7.3 Animals learn associations between stimuli and responses

Learning often requires making associations. For example, when we pick up a leash, our dogs associate that gesture with going for a walk. One common type of learning involves exhibiting a particular behavior when a certain stimulus is present—a stimulus–response association. Many animals make such associations, allowing them to learn about their environment, prepare for any similar future events, and respond accordingly.

Classical conditioning

The researcher most often associated with stimulus–response learning is Ivan Pavlov. Pavlov was studying the physiology of digestion when he observed that dogs often began salivating when they could smell or see food. More interestingly, they would also begin salivating when they saw the technician who normally fed them. They had formed an association between the technician and the upcoming delivery of food (Pavlov 1927).

This ability to learn new associations between a stimulus and an innate, or unlearned, response is called **classical conditioning** (or **Pavlovian conditioning**). Classical conditioning begins with an innate response to a stimulus, such as salivating in response to the sight of food. Pavlov called the food the unconditional stimulus (US) and salivating the unconditional response (UR). Next, Pavlov paired a neutral stimulus, such as the ringing of a bell, with the presentation of food. Such "conditioning" was done repeatedly until the animal learned to associate the bell with the arrival of food. In Pavlov's terms, the bell was a conditional stimulus (CS), and salivation in response to the bell was a conditional response (CR). Subsequently, a dog would salivate after simply hearing the bell, even in the absence of food: a new predictable relationship had developed between a stimulus (bell) and a response (salivating). Often the word "conditioned" is used instead of "conditional" when referring to a stimulus or response. Next, we examine how such conditioning can affect fitness.

> **classical conditioning**   A type of learning in which a novel stimulus is paired with an existing stimulus and elicits a particular innate response. Eventually the novel stimulus alone elicits the same response as the existing stimulus. Also called **Pavlovian conditioning**.

Pavlovian conditioning for mating opportunities in Japanese quail

Classical conditioning allows individuals to become better prepared for future events by learning new associations. How does such learning affect fitness? Elizabeth Adkins-Regan and Emiko MacKillop studied how classical conditioning affects mating behavior in Japanese quail (*Coturnix japonica*), medium-sized Asian birds commonly raised in captivity for egg production.

The researchers conditioned adult males and females in the laboratory to two different mating situations using cages that differed in size, location, and appearance (wire or Plexiglas construction) (Adkins-Regan & MacKillop 2003). Each bird learned that mating always occurred in one cage condition (conditioned stimulus, or CS+) but not the other (CS−) (Figure 7.14). For half the birds, a mate was added after two minutes only when they were placed in the Plexiglas cage, while for the other half, a mate was added only when they were in the wire cage. For the focal males, once the female was

Featured Research

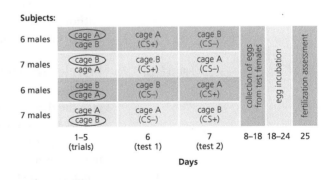

Figure 7.14. Conditioning design for males. Cage A is made of wire and Cage B of Plexiglas. For training trials 1–5, mating only occurred in the CS+ cage (circled). During Tests 1 and 2, mating occurred in both cages. Females were given similar training but were not allowed to mate during training. (*Source:* Adkin-Regan & MacKillop 2003)

Figure 7.15. **Percentage of eggs fertilized.** (a) A male copulating with a female. Mean (+ SE) percentage of eggs fertilized by (b) males and (c) females in the CS+ and CS− cages. (*Source:* Adkin-Regan & MacKillop 2003)

added, mating ensued. For females, no mating was allowed during conditioning, because female Japanese quails can store sperm, which could have biased the results. The male was thus kept behind a wire mesh screen. Training occurred twice a day for five days.

After training, the focal individuals were tested in both cages: researchers presented these individuals with a mate and then counted the number of inseminations that occurred. Eggs were collected, incubated, and allowed to develop for one week before being examined for embryos.

Males fertilized more eggs in the cage where they had been conditioned to expect a mate (the CS+ condition). Females also had more eggs fertilized in the CS+ condition (Figure 7.15). Both sexes thus achieved greater reproductive success in the condition where they had learned that mating opportunities occur.

These results show that Pavlovian conditioning can affect fitness. What caused these differences in fertilization success? The researchers hypothesized that males transferred larger or more effective ejaculates and that females behaved differently to allow more sperm transfer under the conditioned stimulus. Although the exact mechanism by which this occurred is not yet understood, it is clear that fertilization success was affected by the learning of conditions.

Fish learn novel predators

Featured Research

Another important component of fitness is predator avoidance. Many animals face the threat of predation, but how do they know to avoid predators? One possibility is that a prey can innately identify its predators. However, many animals move large distances when migrating or searching for appropriate habitats and so may encounter many different predators. Can they identify all possible predators innately, or do they learn about them?

Matthew Mitchell and colleagues examined the ability of juvenile lemon damselfish (*Pomacentrus moluccensis*) to learn about novel predators (Mitchell et al. 2011) (Scientific Process 7.2). Lemon damselfish larvae develop in open water for about a month, after which they settle on the Great Barrier Reef off the coast of Australia. Settlers are small (1 cm in length) and suffer high predation.

SCIENTIFIC PROCESS 7.2

Fish learn predators

RESEARCH QUESTION: *How do fish learn about predators?*

Hypothesis: Damselfish make associations between conspecific chemical alarm cues and the odor of a heterospecific fish to learn their predators.

Prediction: Individuals will reduce feeding and increase antipredator behavior when they detect odors that have been associated with chemical alarm cues.

Methods: The researchers:

- Created two cocktails of odors: one from four species of known fish, A, B, C, and D (A = *S. dermatogenys*, a predator; B = *P. fuscus*, a predator; C = *C. plebeius*; D = *R. aculeatus*); and one from two species of novel fish, E and F (E = *C. batuensis*, a predator; F = *A. steinitzi*).
- Paired the cocktail with a conspecific chemical alarm substance (treatment)
- Paired the cocktail with seawater (control)
- Tested individuals by placing them in a 13 L flow-through tank with a shelter (a pot)
- Added food and recorded behavior for five minutes
- Added a stimulus odor (A–F) and recorded the amount of time test fish spent feeding, their distance from the shelter, and the time spent in the shelter for five minutes

Results: Fish exposed to the cocktail plus the chemical alarm substance exhibited less feeding behavior than did control fish and fish exposed to a novel odor.

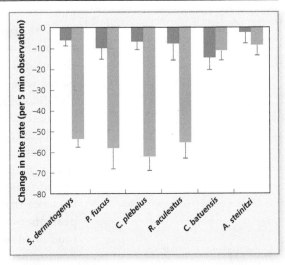

Figure 1. Feeding behavior. Mean (+ SE) change in feeding behavior for treatment fish (orange) and controls (blue) when exposed to the odor of six fish species. From left to right, these represent fish A, B, C, D, E, and F. (*Source:* Mitchell et al. 2011)

Conclusion: Fish learn associations between the odor of a fish and chemical alarm substances.

EVALUATE

What data suggest that damselfish do not have an innate ability to detect a predator?

Many fish, including damselfish, respond to conspecific chemical alarm substances released from damaged epidermal cells when, for example, a fish is wounded or killed by a predator. Fish that perceive chemical alarm substances respond with innate antipredator behaviors: they reduce feeding, increase vigilance behavior (scanning the environment for predators), and may spend more time in shelter. The researchers investigated whether fish learn their predators by association with the presence of the chemical alarm.

The experimenters created a "cocktail" of chemical odors using adults of four fish species caught from the reef: two fish predators and two nonpredators. They placed individuals of each species in separate tanks for six hours—enough time for the tank water to accumulate species-specific odors. The researchers then mixed water from each of the species' separate tanks to form the "cocktail" of chemical odors.

Mitchell's team next collected damselfish recruits that had not yet settled using traps set 50 to 100 m from the reef. Since predators of these fish are only found on reefs and not in open waters, the damselfish recruits were assumed to be naïve in regard to these predators. On Day 1, half the damselfish were conditioned to the cocktail odors plus a conspecific alarm substance. The other half were conditioned to the cocktail odors plus seawater as a control. The next day, damselfish were tested by exposing them either to odors from one of the cocktail species or to the odors of two novel species, only one of which was a fish predator. The researchers recorded the amount of time the test fish spent feeding, their distance from the shelter, and the time spent in the shelter.

When the odor of each species was added, damselfish conditioned with the cocktail of species and alarm substance reduced their feeding time significantly more than did the controls, although there was no difference in shelter use between the two groups. The test fish had apparently learned a new association between the odor of each cocktail fish and predation risk, as indicated by the chemical alarm substance. Individuals tested with the odor of the cocktail fish paired with seawater did not respond with antipredator behaviors. Individuals also did not respond to the odors of the two novel fish species, even though one was a predator.

These results demonstrate that lemon damselfish can rapidly learn associations between fish odors and the risk from unfamiliar predators, and can respond in an appropriate manner by reducing their feeding time. In addition, because damselfish did not respond to the odor of a novel predator, there was no evidence that they have innate knowledge of predatory species; instead, it appears they must learn to identify predators. The ability to make these important (and rapid) associations allows damselfish to minimize predation risk in new environments.

Operant conditioning

operant conditioning A process in which an animal learns to associate a behavior with a particular consequence.

As we just saw, classical conditioning involves learning novel associations that affect innate behavior. In contrast, **operant conditioning** involves learning associations between learned behaviors and outcomes. Edward Thorndike and B. F. Skinner pioneered the study of operant conditioning, also known as instrumental conditioning (see, e.g., Thorndike 1911; Skinner 1938). From Skinner's perspective, much behavior is learned from the consequences of this conditioning. Any of four types of operant conditioning can occur, depending on whether a behavior becomes more or less likely as a result of the presentation or removal of a stimulus as follows:

1. In positive reinforcement, a behavior becomes more likely due to the presentation of a stimulus, such as food.
2. In negative reinforcement, a behavior becomes more likely due to the removal of a stimulus, such as pain.
3. In positive punishment, a behavior becomes less likely due to the presentation of a stimulus.
4. In negative punishment, a behavior becomes less likely due to the removal of a stimulus.

Skinner revolutionized the study of behavior in the mid-twentieth century, and his work is still used extensively to train animals (Applying the Concepts 7.2).

APPLYING THE CONCEPTS 7.2

Dog training

Have you ever rung a friend's doorbell only to hear uncontrolled barking? Have you had a dog run full force at you through the front door? Dog owners face behavioral problems like these all the time. Usually the owner simply yells at the dog: "Get away from the door," "Sit down," "No," or some other command that may be ineffective. Sophia Yin studied domestic animal behavior and used operant conditioning with positive reinforcement to train dogs with problem behaviors.

To address dogs' problems with greeting people, Yin and colleagues turned first to professional dog trainers and guided them to train dogs to remain on a small mat for one minute, despite loud distractions (Yin et al. 2008). The trainers began with two exercises. First, the six dogs learned to eat from a remote-controlled food-reward dispenser. Once the dogs ate readily from the food dispenser, the trainers played a tone, which they followed by giving each dog a food reward. After the dogs learned to look for food after hearing the tone, the researchers added a condition: the dog had to look at the trainer before the food would be dispensed. In the second exercise, the dogs learned to walk toward a "target" and touch it

with their nose, which again resulted in the tone and a food reward. Once the dogs had mastered these two exercises, the real training protocol began.

The protocol had three stages: (1) a "down-stay" on a rug-covered platform; (2) running to the platform and doing a down-stay; and (3) the down-stay with distractions (e.g., a doorbell, loud knocking, or people walking). Training sessions lasted up to 30 minutes, depending on the dog's motivation. The trainers taught the dogs to complete the behaviors for each stage separately and then put them together in sequence. After eight days in the laboratory, dogs remained in a down-stay for a full minute despite distractions, whereas before training, they did so for only five seconds.

Dog owners used these same techniques with similar success: barking decreased from 19 to two barks per minute, while jumping decreased from eight to almost zero jumps per minute after training. The authors concluded that positive reinforcement is a powerful technique for modifying dog behavior. Dog trainers and owners are using these same techniques all over the world (Miller 2001). ■

Much of Skinner's work involved understanding how and when each operant condition should be applied to produce the desired behavioral outcome. He also invented novel research protocols and apparatuses to study operant conditioning. A case in point is the operant chamber, or "Skinner box," which typically contains a lever that a rat or pigeon can press.

Operant chambers can be very useful tools to study animal problem solving. Operant conditioning is also known as **trial-and-error learning**, because behavior often changes incrementally as the animal "makes progress" toward solving a problem. Skinner developed a graphical representation of such progress, such as a decline in errors over time, called the individual's **learning curve**. Learning curves allow researchers to quantify learning rate and characterize differences in learning ability, as we see next.

trial-and-error learning Learning to solve a problem through repetition.

learning curve A graphical representation of a change in learning over time.

Learning curves in macaques

Featured Research

One way to understand how quickly animals learn is to train them to overcome an innate preference. For example, when given a choice between a small and a large quantity of food items, many primates, including humans, strongly prefer the larger quantity (Boysen, Berntson, & Mukobi 2001). This innate preference makes sense because food intake affects fitness, and so more is usually better. Elisabeth Murray, Jerald Kralik, and Steven Wise studied the strength of this preference in rhesus macaques (*Macaca mulatta*) (Murray, Kralik, & Wise 2005). They offered six subjects a choice between one or four peanut halves that were placed in the experimenter's open hands. If a test subject reached for the hand with one food item, it received four; when it reached for the hand with four food items, it received one. Here, one action provided four times more rewards than the other

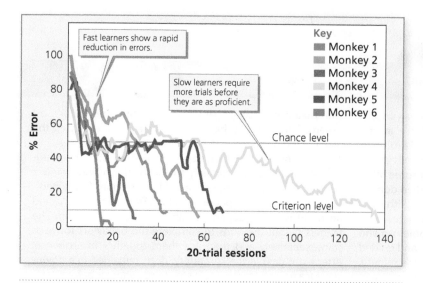

Figure 7.16. Macaque learning curves. Individual macaques (different lines) learned to select the hand with one food item instead of the hand with four at different rates. Those macaques that learned the fastest had the fewest errors in the least number of trials. (*Source:* Murray, Kralik, & Wise 2005)

and therefore represented a more positive reinforcing outcome. Twenty trials were conducted each day, with a 20-second delay between them. How quickly did individuals learn to select the hand that contained only one food item?

As expected, each subject showed a strong initial preference for the hand with four food items (the "incorrect" but innate choice). Macaques did learn to select the hand with one food item in order to receive four, demonstrating trial-and-error learning, but there was tremendous variation in their learning curves (Figure 7.16). One macaque learned rather quickly, attaining a low error rate after about 340 trials, but another macaque took over 2,700 trials to attain a low error rate. In fact, a wide variation in learning curves is common in many species (Boutin 2007).

Featured Research ▶ ## Trial-and-error learning in bees

Following Skinner's work and the development of operant chambers, many learning studies, like the previously described work on macaques, were conducted in laboratory settings designed to control as many factors as possible. Can operant learning be studied in more natural settings? And, given that individuals differ in their learning rate, does learning ability affect fitness?

Nigel Raine and Lars Chittka set out to answer those questions (Raine & Chittka 2008). Bumblebees (*Bombus terrestris*) use a variety of cues to learn which flowers provide nectar and pollen food rewards. Bees feed on many types of flowers but have an innate preference for the color blue, because blue flowers tend to have more nectar (Raine & Chittka 2007). Using operant conditioning, Raine and Chittka examined the learning curves of bees trained to associate the color yellow with a food reward in the laboratory.

The researchers worked with 12 colonies that contained uniquely marked workers. All bees were first allowed to feed from artificial flowers that contained sugar water. Each flower was multicolored—both blue *and* yellow. Once bees were used to feeding from artificial flowers, the researchers began training trials. Bees were placed in a flight arena where half the artificial flowers were all blue and half were all yellow. Only the yellow flowers contained sugar water, whereas the blue flowers were empty. Thus, bees had to learn that they would be rewarded for probing yellow flowers and not rewarded for probing blue flowers.

The researchers recorded over 100 flower visits per bee as bees searched for food. At first, bees in all colonies made mostly incorrect choices because of their innate preference for blue flowers, but all eventually learned to select mostly yellow flowers. However, the colonies differed in their learning rate: some learned quickly, while others did so much more slowly (Figure 7.17).

Figure 7.17. Bumblebee colony learning curves. Average learning curves for bees from three colonies. Note the variation in learning rate among colonies. Inset: bumblebee. (*Source:* Raine & Chittka 2008)

Next, the researchers investigated whether this variation in learning was associated with differences in fitness in a natural environment. They allowed bees from each colony to forage on wildflowers around Queen Mary's College in London. They then measured foraging success based on a bee's weight when departing and its weight when returning to the hive, with the difference in weight being the mass of nectar that the bee was bringing to the hive. Bees were allowed to forage naturally for six days, providing data from thousands of foraging bouts.

Colonies differed in food delivery: those quick to learn in the laboratory delivered more food per hour to the colony than did slow learners, which researchers hypothesized would lead to enhanced survivorship (Figure 7.18). From these experiments, the researchers concluded that bees can learn where to feed by trial and error and that learning ability can affect fitness.

These examples demonstrate that animals often learn by making associations. Many individuals may learn in isolation, as we saw in the work on Japanese quail. But in many species, individuals live with conspecifics, a situation that provides a rich social environment and enhanced learning opportunities, as we see next.

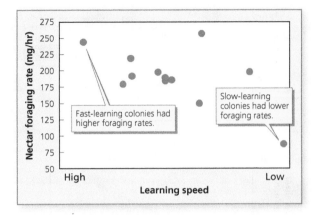

Figure 7.18. Learning speed and feeding rate. Mean nectar foraging rate for the 12 colonies. Colonies with a high learning speed in the experiment had higher mean feeding rates on wildflowers. (*Source:* Raine & Chittka 2008)

7.4 Social interactions facilitate learning

In social species, other individuals are a source of information for learning, a phenomenon known as social learning. In individual learning, such as trial-and-error learning, information is acquired through an individual's own activities, while in social learning, individuals learn by observing others. Social learning can be favored by selection when it reduces the time and energy costs of learning (Rieucau & Giraldeau 2011).

Young animals have much to learn about their environment, and for many species, one of the most important lessons is how to avoid being eaten. We begin with an example illustrating the benefits of social learning in this context.

Learned antipredator behaviors in prairie dogs

Featured Research

Many animals are highly vulnerable to predators. For example, black-tailed prairie dogs (*Cynomys ludovicianus*) are large social rodents that live in arid grassland habitats in North America (Figure 7.19) that also contain predators such as black-footed ferrets (*Mustela nigripes*), coyotes (*Canis latrans*), bobcats (*Lynx rufus*), and a variety of hawks and rattlesnakes. How do young prairie dogs learn to avoid them?

Debra Shier and Donald Owings tested the hypothesis that young prairie dogs learn antipredator behaviors from adults (Shier & Owings 2007). They captured 36 juveniles with their mothers within two days of emergence from their burrows. Each mother and her litter were housed in a separate enclosure with an artificial underground burrow at one end. Each juvenile first went through a pretraining assessment of antipredator behavior with predator test stimuli: a black-footed ferret, a prairie rattlesnake (*Crotalus viridis*), a moving hawk model, or a cottontail rabbit (*Sylvilagus audubonii*), which served as a nonpredator control. The ferret and rabbit were presented within a mesh box, while the snake was presented behind a mesh barrier. The model hawk was flown over the enclosure. The researchers videotaped each juvenile for ten minutes to record its antipredator response to the test stimulus.

Figure 7.19. Prairie dogs. Highly social rodents that live in large colonies.

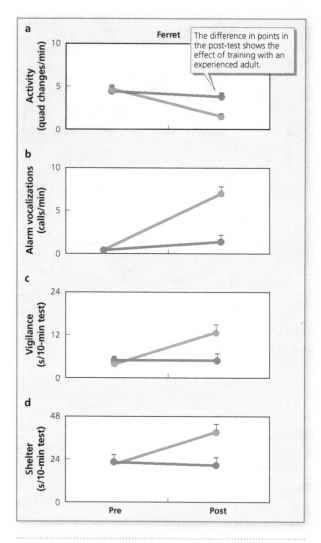

Figure 7.20. Juvenile antipredator behavior. Mean (+ SE) antipredator behavior of juveniles trained with an experienced adult (orange) and those not trained with an experienced adult (blue) for exposure to one predator type, a ferret. (a) Activity level, (b) alarm vocalizations, (c) vigilance rate, and (d) time in shelter before (pre) and after (post) training. Note large differences in behavior in the post-training data. (*Source:* Shier & Owings 2007)

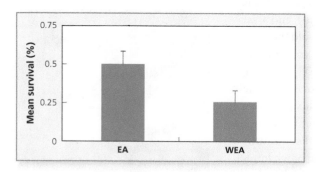

Figure 7.21. Survivorship differences. Mean (+ SE) survival probabilities. Juveniles trained with an experienced adult (EA) had higher survivorship in the wild after one year than juveniles trained without an experienced adult (WEA). (*Source:* Shier & Owings 2007)

Next the young pups underwent a five-week training period in which they were randomly divided into two groups and exposed to the predators. One group spent five weeks with an experienced adult that had lived in the wild and had previous exposure to the chosen predators. The other group either trained alone or with an inexperienced sibling. Following training, juveniles were given a post-training test in which their response to each test stimulus was recorded when they were alone. The researchers measured the juveniles' activity level, frequency of fleeing, antipredator vocalizations, and vigilance behavior.

The post-training tests indicated that the treatments strongly affected antipredator behavior. Prairie dog juvenile pups trained with an adult were less active, produced more alarm calls, exhibited greater levels of vigilance, and more often fled to the burrow when exposed to a predator (Figure 7.20).

Following these tests, all juveniles and adults were marked and released back into the wild in a new colony to assess their survivorship. One year after testing, the juveniles reared with experienced adults had higher survivorship (Figure 7.21) than those reared alone or with inexperienced siblings. Shier and Owings concluded that juveniles learn antipredator behaviors from experienced adults and that learning helps them survive. Interestingly, social learning may play a role in how human children learn about dangerous animals (Applying the Concepts 7.3).

In addition to avoiding predators, animals need to find food to survive and reproduce. In the next example, we see how social learning facilitates learning about food patches.

Learning about food patches

Birds such as terns feed on fish near the water's surface by diving into the water. Their food patches—schools of fish—are always moving, which can make locating food especially challenging. Individuals will often use a strategy known as **local enhancement**, in which they look to other foraging individuals as a cue to the location of food. This phenomenon is an important aspect of social learning.

Location, however, is only one aspect of a food patch. Perhaps even more important is the quality of that patch, or how much food it contains. How can an individual learn about food patch quality? One possibility is again local enhancement. The more food a patch contains, the more foragers will typically be present at that patch. An even better strategy for determining patch quality is to observe individual feeding rates in a patch. Feeding rates will be high in rich patches and low in poor patches. Knowledge obtained from others about the quality of a resource is called **public information**, which can improve an individual's assessment of resource quality (Valone 1989).

APPLYING THE CONCEPTS 7.3

Human social learning about dangerous animals

Prairie dogs and many other animals learn about predators through social learning (Griffin 2004). H. Clark Barrett and James Broesch tested the hypothesis that human children also learn socially to remember the relative danger posed by different species (Barrett & Broesch 2012).

The researchers conducted a simple experiment using children under the age of 12 from one urban area (Los Angeles) and one rural area (a Shuar village in the Amazon region of Ecuador). In each societal group, half the children were assigned to the social learning treatment and half were assigned to the control (no learning). Each child was shown individual images of 16 different animals on cards. The species depicted came from several different continents and were largely unfamiliar to most of the children prior to the study.

To gauge the ability to infer level of danger or an animal's diet from the images, the control children were asked a series of questions about each card. In the social learning treatment group, the experimenter told the children each animal's name,
whether it ate plants or animals, and whether it was "dangerous" or "safe." The cards were then shuffled, and each image was shown again, but this time the experimenter asked the children to repeat the information. To assess longer-term learning, this procedure was repeated one week later.

Few of the children could correctly name any of the animals depicted on the cards, and the treatment children performed poorly on this task both immediately after learning the names and one week later. In addition, both control and treatment children performed poorly when asked about a species' diet, answering correctly only about 50% of the time. However, children who had been informed about the level of danger of each animal performed significantly better on this question than control children, both immediately after learning this information and one week later. Shuar and U.S. children performed similarly on the task. These data suggest that human children can learn socially about animals and the level of danger they present. ■

Social information use in sticklebacks

Do animals use public information to assess patch quality? Isabelle Coolen and her colleagues examined this question in nine-spined sticklebacks (*Pungitius pungitius*), small fish that live in lakes and rivers throughout the Northern Hemisphere (Coolen et al. 2005). They predicted that if only local enhancement information is available, individuals will use it, because high numbers of fish should indicate the best patch. If fish can observe the feeding success of others, however, they should use that public information to select a food patch—regardless of the number of fish at the patch.

To test this prediction, the researchers divided a tank into three areas (Figure 7.22). Each side compartment contained a feeder, creating a "food patch." The center area contained a compartment with a single observer, the test fish. In a first experiment, a test fish could see conspecific "demonstrators" at each food patch. Six demonstrators were at one food patch, and two were at the other. Across from the test fish were six conspecifics viewed through a one-way mirror. The conspecifics served as a "companion shoal" so that the test fish was never isolated, even when the demonstrators were removed. No food was provided from either feeder, so only local enhancement information was available—the number of demonstrators. After ten minutes, the barriers and all demonstrators were removed and the observer was allowed to swim freely in the entire tank, although no food was actually present. The researchers recorded the mean percentage of time the test fish spent in each compartment and at the center near the companion shoal.

The test fish spent significantly more time at the patch that had previously contained six demonstrators and near the central compartment than in the compartment with only two demonstrators (Figure 7.23). Sticklebacks were apparently using the number of fish at each food patch as a cue to its quality.

Featured Research

local enhancement A strategy in which an individual uses the presence of another as a cue for focusing on a particular part of the environment.

public information Information obtained from the activity or performance of others about the quality of an environmental parameter or resource.

Figure 7.22. Stickleback experimental tank. (a) Stickleback. (b) Test aquarium. Individual test fish were placed in the observer compartment. Two or six demonstrators were placed in each side compartment, which contained a food patch, the feeder. The companion shoal of six individuals did not have access to food during demonstrations. (*Source:* Coolen et al. 2005)

To determine whether sticklebacks also use public information to estimate patch quality, the researchers conducted a second experiment in which they manipulated the feeding rates of demonstrators at each patch. The experimental tank was the same as before, but now the test fish could watch demonstrators feed on bloodworms from each patch. The patch with two demonstrators was designated the "rich" patch, and food was released from the feeder there six times during the ten-minute trial. The patch with six fish was designated the "poor" patch, and food was released there only twice during the trial. Feeding rates therefore differed at the two patches.

After all demonstrators and food were removed, the test fish exhibited a strong preference for the rich patch, even though only two fish had been there (Figure 7.24).

Figure 7.23. Percentage of time in each patch when seeing demonstrators that do not feed. Mean (+ SE) percentage of time spent in different locations. When demonstrators did not feed during the observation period, test fish spent more time near the patch that had six demonstrators compared to the patch that had only two demonstrators. (*Source:* Coolen et al. 2005)

Figure 7.24. Percentage of time in each patch when seeing demonstrators that feed. Mean (+ SE) percentage of time spent in different locations. When feeding rates of demonstrators differed during the observation period, test fish spent more time at the rich patch that had two demonstrators compared to the poor patch that had six demonstrators. (*Source:* Coolen et al. 2005)

Coolen's team concluded that fish do use public information to select a patch. When local enhancement and public information provide contradictory information, as they did in the second experiment, fish rely on public information to select a food patch, likely because it provides more accurate information about patch quality.

In the above cases, individuals learned how to avoid predators or about the quality of food patches by observing others. Such learning by observation is common and requires no special effort on the part of demonstrators. Recently, however, evidence suggests that some demonstrators play a more active role in the social learning process—a phenomenon known as teaching.

Teaching

The hallmark of **teaching** is the active participation of an experienced individual in facilitating learning by a naïve conspecific. Animal teaching has been defined in many ways, but three widely accepted criteria for evidence of teaching are (1) one experienced individual, the teacher, modifies its behavior only while in the presence of a naïve pupil; (2) the behavior is costly for the teacher; and (3) the pupil acquires knowledge or a skill more rapidly through the actions of the teacher than it would otherwise (Caro & Hauser 1992). Let's examine two cases that satisfy these criteria.

teaching The active participation of an experienced individual in facilitating learning by a naïve conspecific.

Ptarmigan hens teach chicks their diet

Some birds, like chickens, are precocial, or highly developed and mobile when they hatch, and so quickly need to learn what to eat. Jennifer Clarke examined how precocial white-tailed ptarmigan (*Lagopus leucura*) (Figure 7.25) chicks learn their diet (Clarke 2010). In particular, she investigated whether hens taught their chicks what to eat.

Ptarmigans feed on a variety of leaves, flowers and buds, and berries of shrubs and wildflowers. Their natural habitat contains many dozens of plants, but the diet of ptarmigans is composed of just a small subset of available plants. How do chicks learn the plants (and plant parts) to eat? Chicks follow hens for several weeks after hatching. Occasionally, when one or more chicks are nearby, a hen will drop bits of food it is eating, produce a unique vocalization (a food call), and bob its head up and down. In response to this food call display, nearby chicks will run to the hen and feed. Are these displays a form of teaching?

Clarke observed seven marked hens and their chicks in over 100 feeding bouts and noted all food call displays that occurred. In these bouts, she recorded the food items eaten by hens and chicks, the food plant associated with each food call display, and the behavior of the chicks in response to the call.

While hens fed on over a dozen species of plants, they produced food calls in association with only six. Food call displays always resulted in chicks feeding on the plant the hen had been feeding on, and there was a positive correlation between the proportion of food calls associated with those species and the proportion of food in the diet of the chicks (Figure 7.26). Biochemical analyses of the plants consumed by the chicks indicated that they were significantly higher in protein content than a random mixture of plants available in the environment. Protein is a limited but important nutrient for growing chicks (Scott, Holm, & Reynolds 1963).

Clarke also noted variation in the food calling behavior of the hens studied and in the food consumed by their chicks. Chicks of hens that produced the greatest number of food call displays consumed significantly more protein in their diet

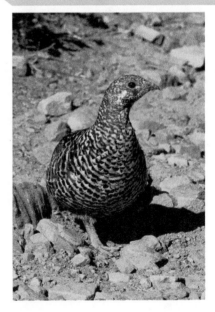

Figure 7.25. White-tailed ptarmigan. A common bird in alpine meadows in western North America.

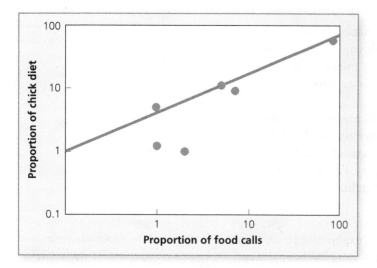

Figure 7.26. Hen food call frequency and chick diets. The proportion of food calls given by hens to each of six plant species and the proportion of those plants in the diet of chicks. The axes are logarithmically scaled. (*Source:* Clarke 2010)

than chicks of hens that produced the fewest food call displays. Clearly, hens were biasing food call production to plants high in protein, and chicks learned to consume a diet high in protein. These data are consistent with the definition of teaching given above, although the cost of teaching to the parent was only inferred (a reduced feeding rate).

Some argue that clear evidence of teaching requires an additional criterion: a modification in behavior by the teacher based on variation in learning by the pupil. In other words, evidence that the teacher modifies its behavior when the pupil is not learning. Let's examine one such case in ants.

Tandem running in ants

Social insect colonies provide much opportunity for social learning by individuals. *Temnothroax albipennis*, an ant common throughout Europe, lives in small colonies that often contain fewer than 100 workers (Dornhaus et al. 2008). Colonies are located in loose rubble or small rock crevices, and when these sites become unsuitable, colonies often move to more stable, higher-quality locations. A small number of scouts are always searching the environment for food or better nest locations. When a valuable resource is located, the scout returns to the nest to recruit additional workers. Recruitment takes the form of a tandem run (two individuals moving together) in which the experienced scout physically leads the naïve worker to the resource. Is this behavior a form of teaching?

Nigel Franks and Tom Richardson studied tandem running in ants housed in artificial nests in the laboratory (Franks & Richardson 2006). They used 1 m arenas with the nest at one end and a food source located at the other. Franks and Richardson videotaped the movements of eight independent tandem runs, including the movement of the first scout to find the food, its behavior while engaged in tandem running with a naïve recruit, and the movement of the recruit back to the nest.

Frame-by-frame analysis indicated that leaders moved toward the food when the follower maintained antennal contact by tapping on the legs and body of the leader. Occasionally, the recruit would break away from the leader and engage in a circuitous movement, perhaps acquiring visual information about the environment to learn the route it was traveling. When physical contact was broken, the leader would slow down or stop moving until the follower was again close enough to maintain contact (Figure 7.27). Slowing down to wait for the recruit meant that tandem runs were costly for the leader—it took four times longer for the leader to move between the nest and food when engaged in a tandem run. However, recruits engaged in a tandem run located the food significantly faster than naïve scouts: they benefited from the behavior of the leader. In addition, Franks and Richardson found that followers would subsequently become leaders for new recruits.

Tandem running behavior satisfies all three primary criteria for teaching. Furthermore, leaders modify their behavior in response to the behavior of followers in such a way that facilitates learning, and the information learned by the

a

1

3

Food

2

Nest

b

7
6
5
4
3
2
1
0

Running speed (mm/s)

1

2

3

— Leader
— Follower

Time

Figure 7.27. Tandem running ants.
(a) Schematic paths of leader (orange) and follower (blue) from the nest to food. (b) Associated running speeds of each ant making each path. The leader occasionally stops (Points 1 and 3) or slows (Point 2) to wait for the follower. (*Source:* Leadbeater, Raine, & Chittka 2006)

followers allows them to later become tandem leaders/teachers themselves. This demonstrates that teaching behavior is not limited to vertebrates.

The above examples illustrate how social interactions between individuals can promote learning of a diverse array of behavior in both juveniles and adults. Next, we examine one outcome of social learning: the development of animal traditions and culture.

behavioral traditions Differences in behavior among populations that are transmitted between generations through social learning.

7.5　Social learning can lead to the development of animal traditions and culture

We have seen that conspecifics can facilitate learning because they act as a source of information, and that such learning is beneficial, so social learning is widespread. Social learning can also lead to variation among populations. For instance, some populations of primates use stones to crack open nuts (Figure 7.28), while others do not. Why do these differences among populations exist? One explanation might involve genetic differences. Alternatively, the variation across populations might correspond to ecological differences, because behavior that is adaptive in one environment might not be adaptive in another. Or differences among populations might exist due to ecological constraints. For instance, the use of stones to crack nuts requires the presence of both nuts and appropriate stones. If either is absent, the behavior will also be absent. A final possibility involves social learning, which could allow a trait to spread rapidly through one particular population, resulting in populations that differ in their behavioral traits. Among populations, differences in behavior that are transmitted between generations through social learning are called **behavioral traditions**.

One type of behavioral tradition commonly observed in populations of birds and cetaceans (whales, porpoises, and dolphins) are local song dialects, characteristic differences in songs that vary geographically.

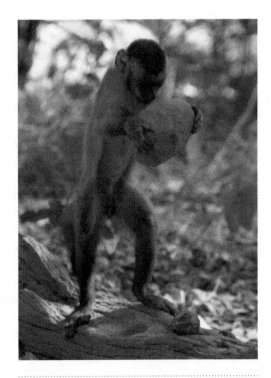

Figure 7.28. Tool use. Wild bearded capuchin monkeys use stone tools to obtain food.

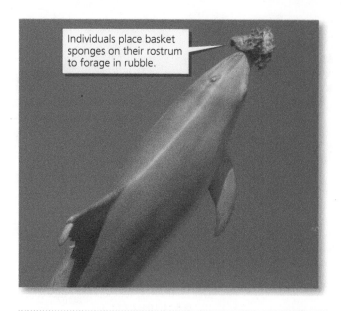

Individuals place basket sponges on their rostrum to forage in rubble.

Figure 7.29. Dolphin tool use. Dolphins use marine sponges to dig up hidden prey on the seafloor, which is strewn with sharp rubble.

animal culture Differences in multiple traditions among populations.

Figure 7.30. Bluehead wrasse. A species that exhibits a behavioral tradition in spawning site use.

Song is learned from an adult tutor (as we saw in Chapter 4), and so changes can spread rapidly through different populations (e.g., Noad et al. 2000; Wright, Dahlin, & Salinas-Melgoza 2008).

Another example of a behavioral tradition involves tool use in dolphins. Bottlenose dolphins (*Tursiops* sp.) off the coast of Australia remove marine sponges from the substrate and wear them on their rostrum or beak as they probe the seafloor to locate submerged prey in rough rubble (Figure 7.29). Sponge tool-use behavior is found only in one population of dolphins near Shark Bay and tends to be most commonly observed in females. Do calves learn this behavior through social learning?

Janet Mann and her colleagues used 14 years of data on the behavior of mothers and their calves to examine this question. They found that this behavior appears to be transmitted primarily from mothers to their offspring, suggesting that social learning plays a role (Sargent & Mann 2009). Further work indicates that sponge tool-use behavior may function to minimize damage to the skin while probing the rocky substrate for prey (Patterson & Mann 2011).

Chimpanzees also exhibit behavioral traditions involving tool use. For example, populations vary not only in their use of stones to crack nuts, but also in their use of twigs to fish for termites, and in specific grooming and courtship behaviors (Whiten et al. 1999). Differences in multiple traditions may be evidence of broader differences among populations, or **animal culture** (Whiten & van Schaik 2007).

One way to test genetic, ecological, and social learning explanations for animal traditions is through experimental manipulation of populations. Manipulating populations can be difficult or impossible with primates and cetaceans, but it is feasible with smaller vertebrates, like fish. Such an approach was used to study a behavioral tradition in coral reef fish.

Featured Research

Behavioral tradition in wrasse

Robert Warner studied the mating behavior of bluehead wrasse (*Thalassoma bifasciatum*) off the coast of Panama (Warner 1988; Warner 1990). These small fish live year-round in coral reefs and spawn each day (Figure 7.30). Males defend territories along the vertical face of the reef (Figure 7.31), where they display to attract females. Females move to these territories and deposit their eggs within them for fertilization. Individuals live for a maximum of three years, but over the course of 12 years, Warner noticed remarkable consistency in the location of spawning sites: despite great variation in population size and many available sites, he never observed the use of a new spawning site. Why did this population use the same sites repeatedly? One explanation is that females simply used the best sites for reproduction. Alternatively, spawning sites might be learned from adults as part of a behavioral tradition.

To test these alternatives, Warner conducted a clever experiment. He first recorded the spawning sites of individual fish in six reef patches. He then removed all adults from these sites and replaced them with adults from the same population that had used a different reef for spawning. Warner predicted that if fish assess a spawning site based on its quality, such as resource availability or predation risk, then the new fish should use the same sites used by the original fish. In contrast, the new fish might establish their own traditions by choosing new sites, and subsequent generations would stick to these choices. As a control, Warner briefly removed all adults from three other reef spawning sites and returned them after a few hours.

Warner found that new individuals used some of the same sites as fish that were removed but also many new sites (Figure 7.32). Furthermore, their use of the sites was remarkably consistent over three more years—a new behavioral tradition had begun. In the control sites, the exact same spawning sites were still used.

How do traditions develop in these fish? Warner observed that females frequently arrived at the reef to spawn in small groups and that they often varied in size and thus age (Warner 1990). He suggested that young females learn the traditional spawning sites from older females. Warner further proposed that all spawning sites are essentially equivalent in quality for reproduction, because sites do not directly affect the fitness of offspring: eggs develop into pelagic young far from the reef itself. As such, spawning sites can then be maintained by tradition.

We now understand that social learning can help explain variation in behavior among populations. Additional studies like Warner's are needed to understand how often animal culture underlies this variation and under what circumstances we expect behavioral traditions to arise (Boyd & Richerson 1996). In the last section of the chapter, we examine variation in the cognitive abilities of animals.

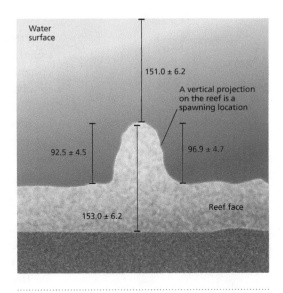

Figure 7.31. Bluehead wrasse spawning sites. The physical characteristics of a typical spawning site. Measurements in centimeters. (*Source:* Warner 1988)

Figure 7.32. Bluehead wrasse spawning site use. The proportion of total spawning rates on one reef during the experiment, showing that new wrasse used different breeding sites. The letters represent different spawning sites used by old (pre) and new (post) fish. (*Source:* Warner 1990)

7.6 Animals vary in their cognitive abilities

There is a long history of interest in understanding the relative cognitive abilities of species. **Cognition** is broadly defined as the acquisition, storage, retrieval, and use of information for decision-making (Shettleworth 2001) and usually is applied to learning that involves more than basic associations. Cognitive ethologists have used many approaches to study the cognitive abilities of animals, including those that focus on tool use, problem solving, numerical abilities, spatial memory, and self-recognition (Pearce 2008). Such studies have documented variation in performance on different cognitive tasks among individuals and across species (e.g., Thornton & Lucas 2012). Next, we examine how researchers have studied tool use, problem solving, and numerical abilities to understand the cognitive abilities of different species.

cognition The ability to generate and store mental representations of the physical and social environment to motivate behavior or solve problems.

Tool use in capuchin monkeys

Featured Research

The ability to use tools is one aspect of cognition. As discussed earlier, many species use tools to acquire food that otherwise would be unavailable. Examples include sticks used by chimpanzees, finches, and crows to probe holes to feed on hidden insects; stones used by primates to crack open nuts; and, as we saw earlier, even sponges used by dolphins to feed on fish hidden in sharp rubble on the ocean floor (Seed & Byrne 2010). Selection can favor tool use when it enhances foraging ability.

In many cases, individuals learn to use a tool by trial and error after having observed others (e.g., Lonsdorf 2005). But some species appear capable of understanding the problem to be solved by the tool and then discover a functional solution without trial-and-error learning. One approach to understand this enhanced cognitive aspect of tool use is to examine how individuals select a tool to solve a problem.

Elisabetta Visalberghi and her colleagues experimentally examined how bearded capuchin monkeys (*Cebus libidinosus*) selected stones to use as hammers to open nuts (Visalberghi et al. 2009). Wild capuchins at a study site in Brazil naturally use stones to open palm nuts. Typically a nut is placed on a log and a stone is dropped on the nut to open it. Success in obtaining food requires the use of heavy stones that will not fracture during use. At the study site, all stones made of sandstone are too fragile while stones composed of harder quartzite less than 200 g in size are too light to be used successfully: only heavy quartzite stones make effective hammers to open palm nuts. Can capuchins select the correct stones to open palm nuts?

The experiment consisted of providing eight different individuals with a palm nut at a log and a choice of stones supplied by the researcher after all natural stones were removed from an area. The first two experiments (Experiments 1 and 2) offered natural stones that differed in type (sandstone or quartzite) or varied in size and weight (large and heavy versus small and light quartzite). The research team recorded the first stone touched, the stone transported to the log and used to hit the nut, and the individual's success at obtaining food. All individuals were tested ten times in each experiment. Capuchins overwhelmingly selected the functional stone: the quartzite and heavier stone were selected in over 90% of the trials (Figure 7.33).

In the next three experiments, the research team offered capuchins novel artificial stones whose weight did not correlate with size (a condition not typical in nature). Individuals were offered two stones that were the same size but differed in weight (Experiment 3); two stones that were either light and large or heavy and small (Experiment 4); and three stones that were light and small, light and large, or heavy and large (Experiment 5). All of these novel stones were made of the same material and so only differed in the properties described.

For the novel stones in which size and weight were uncorrelated, capuchins again overwhelmingly selected the functional stone (Figure 7.33), but did so after more manipulations of the stones. In these experiments, individuals more often tapped, lifted, or rolled the stones prior to selecting one for use. For instance, tapping of a stone occurred in 12%, 26%, and 48% of trials in Experiments 3, 4, and 5, respectively. This work demonstrates that wild capuchins do not simply rely on past experience such as trial-and-error learning to select functional tools to obtain food. It implies a cognitive ability that suggests an understanding of the relevant feature of a functional tool—its weight, independent of its size.

Problem solving and insight learning

A second aspect of cognition is known as **insight learning**—spontaneous problem solving without the benefit of trial-and-error learning. One of the first researchers to study insight learning was Wolfgang Köhler. As we have seen, trial-and-error learning is one way to solve problems. Köhler was interested in

insight learning Spontaneous problem solving without the benefit of trial-and-error learning.

Figure 7.33. Capuchin stone tool choice. In five experiments (1–5) that provided one functional and one nonfunctional stone tool, capuchins overwhelmingly selected the functional stone. (*Source:* Visalberghi et al. 2009)

whether animals could learn in other ways. We have all experienced that "aha" moment when "the light bulb goes on" and we suddenly find a solution to a problem. Köhler wondered whether animals could learn to solve problems in the same way.

He worked with a small captive colony of chimpanzees at the Canary Island Anthropoid Station, where he developed a series of problems for the chimps to solve using food as motivation. In one experiment, he placed bananas outside of the chimp's enclosure and placed several sticks in the enclosure that they could use to reach the food. One chimp, Tschego, spent half an hour trying to reach the food with her arms before she quit. As a group of younger individuals approached the food, Kohler wrote, "suddenly, Tschego leap[ed] to her feet, seize[d] a stick, and quite adroitly pull[ed] the bananas till they [were] within reach" (Köhler 1925, p.32). In another experiment, a chimp named Sultan quickly figured out how to put two sticks together so that they were long enough to reach the food. In these experiments, the animals had no prior experience with the problem and yet appeared to solve it rather quickly, as one would expect from individuals that possessed the capacity for insight learning.

Building on Köhler's work with chimpanzees, Preston Foerder and his colleagues examined the capacity for insight learning in elephants. Elephants are highly social and thought to be intelligent, and so are good candidates for insight learning. The research team worked with Asian elephants (*Elephas maximus*) at the Smithsonian National Zoological Park (Foerder et al. 2011).

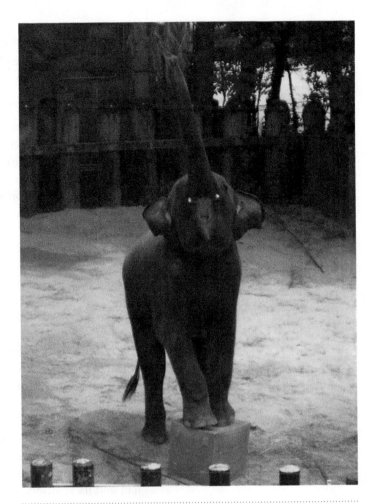

Figure 7.34. Elephant insight. This elephant, Kandula, moved the cube and stood on it to reach food.

In the first experiment, the researchers placed food out of the elephants' reach but left sticks in the enclosure as possible tools. They then tested whether any of the elephants would use the sticks to reach the food. Unlike Köhler's chimps, none did.

In a second experiment, the researchers hung fruit over the animals' heads, out of reach, and left a large moveable cube in the enclosure. The youngest elephant, Kandula, quickly learned to maneuver the cube to the proper location so that he could stand on it and get the food (Figure 7.34). Because Kandula did not use trial-and-error learning, the researchers concluded that the solution he devised was indeed an example of insight learning. Examples of insight learning are not only restricted to mammals, as we see next.

Insight learning in keas

Featured Research

The above examples illustrate one common approach in studies of insight learning and cognition: presenting animals with a problem that involves obtaining food that is out of reach. Such work on birds often involves a string-pulling task, in which food is typically suspended from a string attached to a perch (Dücker & Rensch 1977; Heinrich & Bugnyar 2005). The food is out of reach, and so to obtain it, a particular sequence of behaviors is required: reach down from the

Figure 7.35. Kea. This parrot is native to New Zealand.

perch, pull up on the string with the beak, place a foot on the pulled-up loop of string, release the string, and repeat until the food is brought within reach.

Solving this problem requires understanding both "connectivity" (the fact that food is attached to the string) and a cause–effect relationship among physical objects (pulling up on the string will bring the food closer). Such understanding might be achieved through trial-and-error learning or through insight. The former predicts slow improvement in the problem-solving task over time, whereas the latter predicts instantaneous solution with no further improvement over time and would indicate higher cognitive function.

Dagmar Werdenich and Ludwig Huber presented a string-pulling task to seven hand-raised keas (*Nestor notabilis*), a large parrot (Figure 7.35) native to New Zealand (Werdenich & Huber 2006). All individuals were born in captivity and ranged in age from seven months (a juvenile) to six years. The birds had no prior experience with strings. Each bird was first familiarized with the wooden perch and then tested alone, twice each day. On the first day of testing, a familiar toy was attached to the string and baited with food. All six adult birds solved the problem in the first trial, and four obtained the food within 15 seconds. The other two adults needed 120 and 330 seconds, respectively, to obtain the food. The juvenile solved the problem on the second trial, needing only 24 seconds to obtain the food. For all birds, there was no improvement in performance (time to obtain the food) after the task was solved.

These data suggest that adult keas can solve a complex problem without trial-and-error learning and provide support for the notion that, for the task presented, they have an understanding of the cause–effect relationship of the problem and its solution. Interestingly, the juvenile and one adult apparently relied more on trial-and-error learning as they solved the problem, both demonstrating individual differences in cognitive function and suggesting the importance of studying the development of cognitive ability with age.

In addition to tool use and problem solving, a third aspect of animal cognition involves numbers. Because "number" is an abstract concept rather than an inherent property of an object, the ability to count and assess "more" rather than "less" is used to characterize cognitive ability. As we saw earlier in macaques, many primates exhibit **numerical competency**: they can recognize relatively small numerical quantities. But this ability is not restricted to primates. Irene Pepperberg demonstrated quite sophisticated numerical competency in an African gray parrot (*Pasittacus erithacus*) named Alex that was trained to vocally label objects and their color. Alex could correctly answer questions such as "How many blue keys?" and discriminate numbers up to nine (Pepperberg 2006). Pepperberg's findings have stimulated additional work on other species, as we see next.

numerical competency The ability to recognize numerical quantities.

Featured Research

Numerical competency in New Zealand robins

As was the case with Alex the parrot, much of the research on numerical competency has been conducted in the laboratory. More recently, Simon Hunt, Jason Low, and Kevin Burns investigated whether species in the wild could exhibit numerical competency and whether the development of this capability required extensive training. They studied a population of color-banded wild New Zealand robins (*Petroica australis*) in the Karori Wildlife Sanctuary in New Zealand (Figure 7.36). Because of their remote location, these birds are not wary of humans and so are easy to observe up close. The birds live in forests and regularly

Figure 7.36. New Zealand robin. These birds display numerical competency.

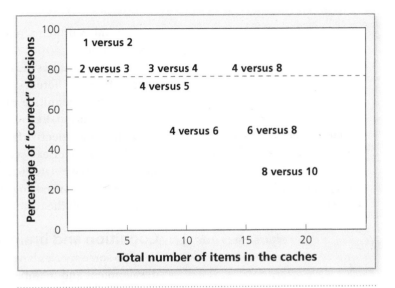

Figure 7.37. Percent correct choices. The percentage of time birds chose the cache with more food items in each of the treatments. Treatments are labeled with the number of food items hidden in each location. Treatments above the dashed line differ significantly from random choice. (*Source:* Hunt, Low, & Burns 2008)

capture large invertebrates that they dismember into smaller pieces before eating. These food items are frequently cached for later consumption in depressions in tree branches. Birds rely on cached food to survive the winter, but caches may be pilfered by others. It may therefore be important for an individual to assign the caches relative importance: a cache with many food items should be more valuable than one with a single item. Such an assessment requires numerical competency.

The research team tested whether birds used numerical judgments when retrieving cached food (Hunt, Low, & Burns 2008). They presented mealworms (*Tenebrio molitor* larvae) to 14 subjects on a tree branch that contained two cache sites. While a subject was observing, the researcher placed a number of mealworms in one cache location and then covered it with a piece of leather. They then placed a different number of mealworms in the other cache location and covered it with another piece of leather. Items were placed in caches one at a time, at five-second intervals, so that the birds could observe each addition of an item to the cache. The researchers then recorded which cache the robin selected. Robins readily learned to remove the leather lids to obtain cached items. Birds were allowed to consume the contents of only one cache, so there was incentive to select the larger cache with more food.

The research team varied both the total number of items in the two caches and the difference in food items between the caches. In general, the birds preferred the larger cache when the total number of food items was fewer than ten (Figure 7.37). When the caches contained more than ten food items, there was no significant difference in cache selection. Additional experiments ruled out the use of scent, the volume of food, the time needed to add to a cache, or cache location bias as explanations of the results. Because the caches were covered, the robins could not visually compare the number of items in each cache. They must have counted items as they were placed in each of the two locations, compared the numbers, and then selected the larger cache. These results indicate that wild New Zealand robins have a sophisticated numerical sense when dealing with a small number of items, and that this ability is not developed through extensive prior training.

The above studies illustrate how researchers have examined different aspects of cognition of different species. Increasingly, attention has turned toward understanding how cognitive differences evolve and the relationship between brain size and cognition. Examination of the evolution of cognitive ability requires that we understand selective factors that have favored its development (Whiten & van Schaik 2007; MacLean et al. 2012). One might simply assume that enhanced cognitive ability correlates with enhanced fitness, but there are energetic costs associated with enhanced cognitive function. Let's examine two recent studies that have focused on the relationship between cognition and brain size.

cognitive buffer hypothesis The hypothesis that large brains, despite their costs, provide survival benefits through flexible behaviors that are learned.

Featured Research

Cognition and brain architecture in birds

Why do some species have larger brains than others? Given the high costs of their development and maintenance, one might ask, "Why have big brains evolved?"

One answer is articulated in the **cognitive buffer hypothesis**: large brains, despite their costs, provide survival benefits through flexible behaviors that are learned (Allman, McLaughlin, & Hakeem 1993). This hypothesis states that large brains increase individuals' capacity to learn and allow them to survive and reproduce better in unpredictable, changing environments.

Daniel Sol, Tamás Székely, András Liker, and Louis Lefebvre evaluated this hypothesis by testing one of its predictions: species with relatively smaller brains should have lower survival rates than species with relatively larger brains (Sol et al. 2007). From a variety of literature sources, they collected data on relative brain size (brain mass/body mass) and adult mortality rates (adjusted for differences in body mass) for 236 species of birds in 43 different families. Many ecological and life history factors are known to affect adult mortality, including whether a species is migratory or not, its reproductive strategy, and its level of parental care, and so these were factored into the analysis as well. After controlling for differences in ecology and life history, the research team still found significant negative relationships between relative brain mass and adult mortality rates when analyzing both the species and the family data (Figure 7.38). These data suggest that big-brained species have lower mortality rates (higher adult survival), a finding that is consistent with the cognitive buffer hypothesis and has prompted greater focus on understanding the benefits and costs of evolving larger brains (e.g., Healy & Rowe 2013).

Figure 7.38. Brain size and adult mortality in birds. Annual adult mortality correlates negatively with relative brain size. (*Source:* Sol et al. 2007)

Featured Research

Brain size and cognition in guppies

The analysis by Sol and his colleagues provides intriguing support for the cognitive buffer hypothesis, but it relies on a correlation. Can we find additional support from an experiment?

Alexander Kotrschal and his colleagues experimentally evaluated the relationship between relative brain size and cognitive ability in guppies (*Poecilia reticulata*) (Kotrschal et al. 2013). The research team created six replicate laboratory populations from guppies collected in Trinidad. They selected for relatively large brains in three populations and for relatively small brains in the other three populations. To accomplish this, they randomly paired single males and females and

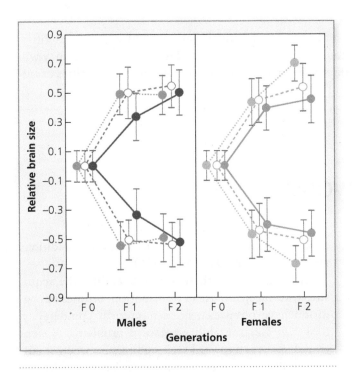

Figure 7.39. Selection on brain size. Mean (±SE) relative brain size in both sexes changed over three generations in each population. (*Source:* Kotrschal et al. 2013)

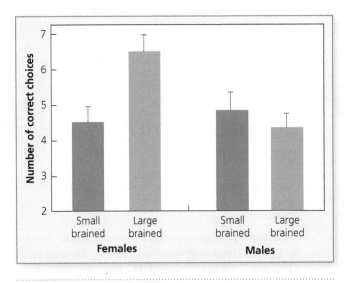

Figure 7.40. Foraging performance and brain size. Mean (+SE) number of correct choices. Females with larger brains made more correct choices than females with smaller brains. Male performance did not differ for large- and small-brained individuals. (*Source:* Kotrschal et al. 2013)

allowed them to reproduce. After females gave birth, the parents were sacrificed to measure their brain and body mass. The offspring of parents with the largest and smallest relative brains were used as breeders for the next two generations in each population.

Relative brain size evolved rapidly in the replicate populations over three generations. Both sexes exhibited an increase in relative brain size ranging from 5%–8% (Figure 7.39). A total of 48 F_2 generation individuals were then tested for their cognitive ability to discriminate between two or four symbols. During training trials conducted twice daily, white cards with two or four symbols (black squares and circles) were placed at either end of the aquaria. Food was provided only at the end containing the card with four symbols. During a test trial, both types of cards were present but no food was provided. Each individual was exposed to three training trials followed by a test trial repeatedly until it had completed 48 training trials and eight test trials.

The researchers compared the number of correct choices (selecting the side of the tank with four symbols) for each fish for each of the eight test trials. The four- and two-symbol sides were randomly altered to control for a potential side bias in the fish, and there was no evidence that guppies have an innate preference for four symbols over two.

Large-brained females made significantly more correct choices than small-brained females; there was no difference in performance for males (Figure 7.40). The result for females supports the cognitive buffer hypothesis: large-brained individuals apparently learned to associate the four black symbols with the presence of food more rapidly than did small-brained females. For males, however, brain size did not affect learning ability. The researchers suggest that food is a more important resource for females than males (whose fitness is more strongly affected by finding mates), and so this may explain the difference observed across the sexes.

The examples above illustrate that animals vary in their cognitive abilities. Continued studies of tool use, insight learning, and numerical competency will continue to provide different avenues to understand the mental capacities of animals and the evolution of cognition.

Chapter Summary and Beyond

Learning allows animals to modify their behavior to adapt to their environment. The simplest forms of learning involve imprinting, which allows animals to learn about their parents, and habituation, which allows them to lessen their response to nonthreatening stimuli. Recent work has examined imprinting involved in species recognition in birds in the wild (e.g., Hansen, Erik, & Slagsvold 2008). Learning is associated with changes in the brain, including increased neurotransmitter release and the formation and elimination of dendritic spines. Further research has examined how spine size and membrane neurotransmitter receptors are modified with learning (Kasai et al. 2010). Across species of birds, we also see a correlation between the size of the hippocampal formation and the propensity to rely on memory to find cached food. Roth et al. (2010) have examined variation in this relationship in rodents and note that a similar pattern occurs with respect to the size of the hippocampus.

More sophisticated learning involves the formation of associations between a stimulus and a response. In classical conditioning, a new association is formed between a previously neutral stimulus and an innate behavioral response. In operant conditioning (trial-and-error learning), learned behaviors change based on positive and negative reinforcement or punishment. The formation of these associations involves trial-and-error learning, and learning curves can be used to characterize the rate and proficiency of learning. While much work has focused on vertebrates, sophisticated learning abilities are also found in many invertebrates (Perry, Barron, & Cheng 2013).

Social learning occurs when animals observe and learn from others. This type of learning is widespread across taxa (e.g., Valone 2007) and can enhance the speed of learning. Through social learning, populations can acquire traditions and culture—learned behaviors transmitted across generations. Researchers are actively

debating definitions and examples of animal traditions and culture (Laland & Janik 2006).

Animals vary in their cognitive ability: the acquisition, storage, and use of information. Studies of cognition have often examined numerical competency, tool use, and the ability to solve problems. Recent work has turned toward seeking links between cognitive ability and fitness (e.g., Keagy, Savard, & Borgia 2009; Cole et al. 2012; Thornton, Isden, & Madden 2014) and examining whether individuals assess the cognitive abilities of potential mates (Boogert, Fawcet, & Lefebvre 2011).

Work on animal cognition is particularly intense in cetaceans, primates, and dogs (e.g., Pak 2010; Seed & Tomasello 2010; Howell & Bennett 2011) and has led to ethical debates about the treatment of captive animals (e.g., Allen & Bekoff 2007; Grimm 2011). This area of research has also led to a discussion of animal consciousness, or whether or not animals "think" about objects or events (Griffin 1984; Griffin & Speck 2004). Ongoing work examines animals' neural capacity for consciousness and focuses on behavior that suggests animal self-awareness (Bekoff & Sherman 2004).

CHAPTER QUESTIONS

1. In one experiment, juvenile chipmunks were housed in two large outdoor aviaries that contained large trees at one end and underground burrows at the other. In Aviary A, the researchers shook the tree branches violently for three seconds, which caused a loud sound twice a day. In Aviary B, the researchers also shook the tree branches violently for three seconds twice a day, making a loud sound—but in this treatment, several acorns, a favorite food of chipmunks, would fall to the ground five seconds later. In both aviaries, chipmunks were initially startled by the loud sound and ran to underground burrows. Predict

how the behavior of chipmunks in the two aviaries might change after repeated trials.

2. You discover a fruit fly mutant that has no memory. How will this mutation affect its ability to learn?

3. Explain how Warner's experiments (1988; 1990) with wrasse demonstrated a behavioral tradition of spawning site use. How could he rule out genetic or ecological factors to explain the observed behavior?

4. Local enhancement and public information are two forms of social learning. How do they differ?

5. Imagine two groups of primates. One group is known to learn socially by observation, while the other exhibits only individual trial-and-error learning. Imagine that one individual in each group learns a new foraging skill, such as using a rock to crack open hard nuts. What prediction can you make about the number of individuals in each group that will display this skill over time?

6. You have taught your puppy Max to lift a front paw when you say the word "shake." As an adult, Max still lifts a paw when you give this command. Describe the changes that likely occurred in the puppy's brain associated with this learned behavior.

Chapter 8

Foraging Behavior

One day we noticed quite a mess at one of our bird feeders. We had filled it full of mixed seeds, and now it was almost empty. Most of the seeds were scattered all over the ground. What had happened? After refilling the feeder, we kept careful watch and got our explanation. Carolina chickadees (*Parus carolinensis*) kept knocking smaller seeds out of the feeder until they found a larger sunflower seed, which they ate. They continued to push the small seeds out, ignoring them in their search for those larger seeds. Why were they passing up all those perfectly good seeds, spending extra time and energy to get the sunflower seeds? Animals selectively choose where, on what, and how long to forage, often passing up some food items in favor of others.

Animals eat to acquire energy and nutrients for survival and reproduction. How they feed, however, reveals a surprising degree of complexity and a large number of decisions. First, a forager needs to find food. Some animals, like many rodents, feed on small, mobile prey or food hidden by sand, sediments, or bark, which can be challenging to find (Figure 8.1). After finding a potential food item, a forager is faced with a decision: Should the item be eaten, or should it be ignored and the search continued? For example, some prey items, such as large nuts or hard-shelled invertebrates, may require the feeding animal to spend extensive time chewing and crushing the exterior to acquire the food inside. These items may not be worth the time and effort needed to consume them. And when a forager like a bee or a hummingbird finds itself in a patch of flowers, it must decide how long to stay, visiting flowers to extract nectar, before moving on to another patch (Figure 8.2).

Figure 8.1. Chipmunk feeding. A chipmunk eats a food item that was hidden under a log.

Figure 8.2. Bee and flower diversity. (a) Bees have many flowers to choose among. (b) Each flower differs in the amount of nectar it contains and that nectar's ease of removal.

In this chapter, we first examine how animals use their sensory systems to locate food. Then we focus on animals that use vision to find food and consider how they can more effectively find cryptic prey. Next, we show how mathematical models help us understand how animals select their diet and exploit food patches. Finally, we explore situations in which some individuals usurp the food discoveries of others.

8.1   Animals find food using a variety of sensory modalities

Before an animal can eat, it must find food. Recall from Chapter 5 that a variety of sensory systems, or modalities, can be used when animals search for food. Natural selection favors those modalities that most efficiently and accurately provide information about the location of food, but the efficiency of a modality can vary with environmental conditions. For instance, visually searching for food in murky water may be difficult. Given the diversity of sensory modalities that an animal possesses, what sensory systems do foragers use to find food? And are some sensory modalities favored over others? Let's examine two examples.

Featured Research

Bees use multiple senses to enhance foraging efficiency

All animals possess multiple sensory systems. This leads to a natural question: Is foraging more efficient when multiple senses are used? Ipek Kulahci, Ana Dornhaus, and Daniel Papaj examined this question in bumblebees (*Bombus impatiens*) (Kulahci, Dornhaus, & Papaj 2008). Bees feed on nectar and pollen in flowers that can differ in color and shape (visual cues), as well as in odor (olfactory cues). Kulahci and colleagues studied the foraging behavior of individually marked bees to determine how feeding efficiency varies with the number of sensory cues that can be used to find flowers containing food.

In the laboratory, a box was attached to the colony container. The box contained a manual gate that allowed the researchers to release one individual at a time into an experimental arena. The arena contained eight artificial flowers—small circular wells that contained either 30% sucrose solution (a food reward) or water (no reward). During training, bees were exposed to two yellow flower shapes, circles or crosses, which surrounded the well (Figure 8.3). Bees were also exposed to two odors: 2 μL of either diluted peppermint or clove essential oil that was placed behind the perforated flowers.

Individual bees were exposed to one of three training treatments that varied the cues associated with reward flowers. In each treatment, half the flowers (randomly selected) contained sucrose and half contained water. In one treatment, only flower shape (a visual cue) indicated the reward (e.g., crosses indicated reward flowers and circles nonreward flowers, or vice versa). In the second treatment, only odor (an olfactory cue) indicated a reward flower (e.g., peppermint indicated reward flowers and clove nonreward flowers, or vice versa). Finally, in the last treatment, both shape and odor (visual cue plus olfactory cue) indicated reward flowers (e.g., clove plus circle indicated reward flowers and peppermint plus cross indicated nonreward flowers).

During training, bees visited all flowers and eventually learned which ones contained rewards. Training was deemed complete when an individual achieved 80% correct flower choices (reward flowers) over its last ten choices. Once trained, a bee was immediately tested by allowing it to enter the arena, where all the flowers were now nonrewarding. The researchers recorded foraging time spent on each flower and the number of "correct" flower choices the bee made (flowers that the bee had learned contained rewards).

A total of 31 bees were tested. Bees trained on two sensory modalities had significantly higher feeding performance (calculated as correct flower visits divided by total decision time) than bees trained using only a single sensory modality (Figure 8.4); two-modality bees made more correct choices and spent less time deciding where to feed. These results suggest that the ability to use multiple sensory modes (vision and odor) facilitates efficient foraging in bumblebees. How this increased efficiency occurs in nature requires further research.

a

36 cm

75 cm

b

Figure 8.3. Bumblebee foraging experiment. (a) An experimental patch with eight artificial flowers. (b) A bee feeds on a "cross" flower. (*Source:* Kulahci, Dornhaus, & Papaj 2008)

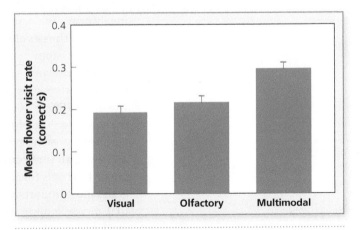

Figure 8.4. Flower visit rates. Mean (+ SE) flower visit rates. Bees trained to use both visual and odor cues (multimodal) had higher correct flower visit rates than those trained on just a single cue. (*Source:* Kulahci, Dornhaus, & Papaj 2008)

Given that many foragers, like Kulahci's bumblebees, likely use multiple sensory modalities to find food, can we determine the relative importance of each modality? Recent work on a small nocturnal primate shows one approach to this question.

Gray mouse lemurs use multiple senses to find food

Marcus Piep and colleagues (Piep et al. 2008) studied how gray mouse lemurs (*Microcebus murinus*) find their prey. Gray mouse lemurs are very small (weighing about 60 g) nocturnal primates that live in Madagascar, where they feed on a variety of arthropods and fruits. Many small nocturnal mammals, like mice, rely on olfaction to locate prey, while diurnal primates rely heavily on vision. Although gray mouse lemurs are primates, they are ecologically similar to mice because of their size and nocturnal habit. Observations of wild mouse lemurs suggest that individuals might find insect arthropod prey both visually and by hearing their rustling movements in the vegetation (Goerlitz & Siemers 2007). Other work suggests that individuals may use olfaction to find fruits (Siemers et al. 2007). Given the diversity of sensory systems that gray mouse lemurs appear to use, Piep and colleagues sought to determine the relative importance of visual, acoustic, and olfactory cues in their foraging behavior (Scientific Process 8.1).

To answer this question, they created a "choice test" experiment in an arena containing two dishes. All tests were conducted in low light that simulated nocturnal conditions. In each trial, one of the dishes contained a single mealworm (a *Tenebrio molitor* larvae) and the other was empty. The researchers manipulated the number and type of sensory cues present at the dish with the mealworm. Each dish was covered by a cone-shaped plastic lid that contained slits that allowed the passage of smells and sounds.

To manipulate auditory cues, the researchers used either live mealworms, which crawl and produce sounds that mouse lemurs can hear, or freshly killed mealworms, which produce no sound. To manipulate olfactory cues, the researchers either placed the mealworms in a sealed bag (no olfactory cue) or did not seal the mealworms in a bag (olfactory cue present). To manipulate visual cues, they used a cone lid that was either black (mealworm not visible) or transparent (mealworm visible).

In the trials, individuals were presented with different combinations of these cues so that one, two, or all three cues were present at the dish that contained the mealworm. The research team recorded the lid that the mouse lemur opened first—and thus its choice of the two dishes—and then analyzed the proportion of correct decisions.

Piep's research team found that when only olfactory or auditory cues were present, individuals selected the correct dish (the one that contained the mealworm) in 70% to 80% of the trials. When only a visual cue was present, that rate increased to 90% of the time, despite the low light level. However, when any two cues were present, mouse lemurs selected the correct dish over 90% of the time, and when all three cues were present, they correctly found the mealworm in 159 out of 160 trials (over 99% of the time). Piep and colleagues concluded that vision is the most important sensory modality for mouse lemurs. They also concluded that mouse lemurs use all three sensory modalities to find prey, and the more modalities they use, the better their hunting success. Why? The use of all three sensory modalities provides complementary sensory information under a variety of environmental conditions. Visual cues may not be available, say, when mouse lemurs forage in dense vegetation in the wild.

SCIENTIFIC PROCESS 8.1

Prey detection by gray mouse lemurs

RESEARCH QUESTION: *How important are vision, hearing, and smell in prey detection by nocturnally foraging gray mouse lemurs?*

Hypothesis: Lemurs will detect prey more accurately with the use of multiple sensory modalities.

Predictions: Lemur prey detection rate will be lowest with the use of one sensory modality and highest with the use of multiple sensory modalities.

Methods: The researchers:

- Used a choice test experimental design in which a single mealworm was placed under one of two dishes, 10 cm apart, covered with cone lids
- Manipulated sensory cues so that one, two, or three cues were present at the dish with food
- Conducted experiments under low light levels, testing six lemurs on each treatment 14 times

Results: Individuals used all three senses to locate prey successfully. Food detection probability increased with the number of sensory modalities used.

Figure 1. Correct choices. Mean (+ SE) percent correct choices for different treatments (V = visual; A = auditory; O = olfaction). (*Source*: Piep et al. 2008)

Conclusions: Gray mouse lemurs use vision, olfaction, and hearing to locate prey. Detection increases when more than one sensory modality is available. Detection is highest when a visual cue is present.

EVALUATE
What is the relative importance of visual, acoustic, and olfactory cues for successful foraging by mouse lemurs?

These studies illustrate how different species find prey using a variety of sensory modalities. This is only half the story, however. Selection should favor prey that can avoid detection. We examine the consequences for visual foragers of prey adaptations to minimize predation in the next section.

8.2 Visual predators find cryptic prey more effectively by learning a search image

Many predators, especially birds and mammals like the gray mouse lemur, rely heavily on vision to find prey. This tendency has favored the evolution of behavioral and morphological adaptations in prey species that allow them to blend into their background to avoid being detected by predators. Here, we first examine the effectiveness of such cryptic coloration. We then consider a counteradaptation found in predators that improves their ability to hunt for cryptic prey.

Cryptic coloration reduces predator efficiency in trout

Jörgen Johnsson and Karin Kjallman-Eriksson examined the ability of brown trout (*Salmo trutta*) juveniles (known as parr) to find cryptic or conspicuous prey in a simple laboratory experiment (Johnsson & Kjallman-Eriksson 2008). Brown trout often hunt diurnally and therefore rely more heavily on visual cues for foraging than do nocturnal fish. These animals live in streams and are opportunistic feeders that eat a wide variety of invertebrate prey. Since aquatic invertebrates have different emergence times over the year, trout often have to learn to search for new types of prey as they become available. In addition, trout parr possess tetrachromatic color vision—that is, they have four different types of cone cells, which allow them to see in the red, green, blue, and ultraviolet spectra. This ability to distinguish colors is thought to be an adaptation that allows animals to find cryptic prey.

Johnsson and Kjallman-Eriksson asked two research questions. First, do predators find conspicuous prey more quickly than cryptic prey? Second, does hunting ability improve with experience? They predicted that trout would find conspicuous prey in less time than they would cryptic prey. They also predicted that predators would learn to find a cryptic prey item more effectively as they gained more experience searching for it (Scientific Process 8.2).

The researchers established two test aquaria: one had a bottom covered with brown grains, while the other had a bottom covered with green aquarium grains. Before each trial, a single, small brown maggot (*Calliphoridae* sp.) prey was placed on the aquarium bottom while the test fish was kept behind an opaque partition. The maggot was cryptic on the brown background and conspicuous on the green background. After the maggot was put in place, the partition was lifted and the researchers recorded the amount of time it took the test fish to find the food item.

The trout always found the conspicuous prey faster than the cryptic prey, indicating that foraging efficiency is reduced when predators search for cryptic prey. This means that background color matching can provide significant survival benefits for prey. However, search times decreased at a similar rate over the six trials both when fish searched for cryptic prey and when they searched for conspicuous prey, indicating that trout parr can improve their ability to find cryptic prey with only a moderate amount of experience.

Blue jays use a search image to find prey

Why do predators like the trout parr have less difficulty finding cryptic prey over time? They are thought to have learned a **search image**, or the distinctive visual features of a single prey type (Tinbergen 1960). Of course, a predator may pay a price for narrowing its focus, as its probability of detecting other prey is thus likely to decrease. You may have experienced this effect yourself if you have ever tried to find an important school paper in your dorm room. You form a mental image of its size, shape, and unique markings. You probably do not pay attention to much else around you as you go through your room looking for the paper.

In a set of classic laboratory experiments, Alexandra Pietrewicz and Alan Kamil studied visual prey detection in blue jays (*Cyanocitta cristata*) to determine whether they use a search image to find cryptic prey (Pietrewicz & Kamil 1977; Pietrewicz & Kamil 1979). Blue jays commonly feed on moths, so Pietrewicz and Kamil trained individual birds to search for images of moths that were displayed

search image The distinctive visual features of a single prey type, which, once learned, can enhance prey detection.

SCIENTIFIC PROCESS 8.2

Cryptic prey reduces predator efficiency

RESEARCH QUESTION: *How does cryptic prey coloration affect trout predator foraging efficiency?*

Hypotheses: Prey that match their background will be harder for predators to detect, and predator hunting efficiency will increase with experience.

Predictions: Trout will find noncryptic prey faster than cryptic prey. Trout will find cryptic prey faster with experience.

Methods: The researchers:

- Established two test aquaria, identical except for the color of the aquarium bottom: one was brown plastic covered with brown grains, and the other was green plastic covered with green grains.
- Placed a single prey (a maggot) in one of six different locations on the aquarium bottom. The maggot was cryptic on the brown background and conspicuous on the green background.
- Recorded the amount of time until the test fish found the food item
- Tested 42 parr, half with cryptic prey and half with conspicuous prey

Figure 1. Background. The two experimental tanks with differently colored bottoms and scattered prey. (*Source*: Johnsson & Kjällman-Eriksson 2008)

Results: Individuals found noncryptic prey faster than cryptic prey. Search times to find prey decreased with experience.

Figure 2. Search time. Median search time for conspicuous prey (orange) and cryptic prey (blue). Error bars denote interquartile range, which represents one-quarter of the range of the data. (*Source*: Johnsson & Kjällman-Eriksson 2008)

Conclusions: Background color matching can benefit prey by reducing predator hunting efficiency. Predator search efficiency for cryptic prey can increase over time.

---

as video images on tree trunks. The moths and tree trunks were similar in color, and so the moths were cryptic when on this background.

Pietrewicz and Kamil designed a very clever apparatus to examine whether blue jays use search images to find prey. They used two different species of bark-like *Catocala* (Noctuidae) moths, which are active during the night and rest during the day on cryptic substrate. *Catocala relicta* rests on white birch tree trunks, whereas *Catocala retecta* rests on oak tree trunks. The researchers took

> **EVALUATE**
> Explain how you would determine whether trout hunting ability improves faster with experience when they hunt for cryptic or conspicuous prey.

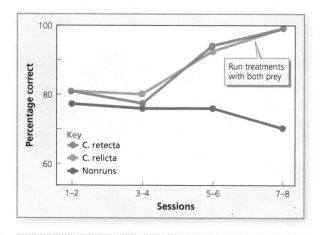

Figure 8.5. Search image results. Percent correct behavior over sessions. During positive trials, jays became more accurate within runs of either species compared to nonruns in which both species were present. (*Source:* Pietrewicz & Kamil 1997)

matched pictures of birch and oak tree trunks with and without the moths on them. They trained jays to peck ten times at a positive image (a moth on a cryptic tree trunk) and to peck one time at a negative image (no moth on a tree trunk). Both these behaviors resulted in positive reinforcement: a food reward of half a mealworm.

To examine the jays' use of search images, Pietrewicz and Kamil created two treatments. In the "run" treatment, the jays were shown eight negative images and eight positive images of one of the moth species so that they could learn a search image for that species. In the "nonrun" treatment, the jays were shown a randomly intermixed group of four positive images of one moth species, four positive images of the other moth species, and eight negative images of no moths. The researchers assumed that jays in the latter treatment did not have an opportunity to learn a search image for either species.

Over eight sessions, the percentage of correct responses increased in the run condition but not in the nonrun condition (Figure 8.5). In fact, the increase occurred quite rapidly. Pietrewicz and Kamil concluded that jays were using a search image of the moth species, which resulted in the increased success rate observed in the run condition. This test provided the first experimental evidence of animals using a search image. It also showed that short-term changes in the behavior of a predator can result in intense predation. In subsequent experiments, Bond and Kamil (1999) confirmed that when jays form a search image of one moth type, their ability to find other moth types is reduced.

Finding prey is just one aspect of foraging behavior. Animals must also decide, first, which prey types to include (and which not to include) in their diet and, second, how long to spend feeding in one food patch before moving on to another. To address these questions, researchers have developed models based on the assumption that animals attempt to maximize their fitness while feeding. We examine this approach next.

8.3 The optimal diet model predicts the food types an animal should include in its diet

Energy and nutrients are obtained from food, which is limited in most environments. In addition, food is typically heterogeneously distributed in the environment: most places contain little or no food, while other locations, known as food patches, may contain abundant food. For example, a dense aggregation of flowers is a food patch for a bumblebee, and a tree infested with insect larvae is a food patch for a woodpecker.

Optimal foraging theory (OFT) assumes that natural selection has favored feeding behaviors that maximize fitness. Many OFT models assume that fitness while feeding is a positive function of an individual's energy intake rate—the energy acquired while feeding divided by total feeding time. OFT models describe the relationship between possible behaviors and the fitness they produce. The behavior that produces the highest fitness is called the **optimal behavior** (Maynard Smith 1978). OFT models provide ultimate (i.e., evolutionary)

optimal foraging theory An approach to studying feeding behavior that assumes that natural selection has favored feeding behaviors that maximize fitness.

optimal behavior The behavior that maximizes fitness in an optimal foraging theory model.

explanations about the function of behavior. Given a set of possible behaviors and assumed environmental conditions, the models predict the behavior that maximizes fitness and will thus occur in a population (Mitchell & Valone 1990). They are widely used because they produce testable predictions and can be applied to many species and behaviors. Two classic OFT models examine diet and food patch use. Let's start with the diet model, which predicts what food types an animal will consume.

The diet model

Animals encounter food types that differ in size, energy content, and handling time—the amount of time it takes to manipulate a food item so that it is ready to eat. Given this diversity, individuals need to decide which food items to eat (i.e., include in their diet) and which to reject. The optimal diet model (Pulliam 1974) is based on three assumptions:

1. Foragers maximize fitness by maximizing energy intake rate.
2. Food items are encountered one at a time and in proportion to their abundance.
3. All food items in the environment can be ranked according to their profitability.

The **profitability** of a food item is the energy it contains divided by its handling time. For example, consider a squirrel feeding on a sunflower seed and a walnut. Sunflower seeds have soft shells and small handling times but also contain relatively little food, or energy, compared to that found in a walnut. The walnut has a hard seed coat and thus a much longer handling time but contains more food, or energy. If the walnut contains 100 joules (J) of energy and takes 25 seconds to consume, its profitability is 4 J/s. If the sunflower seed contains 10 J of energy and has a handling time of one second, its profitability is 10 J/s. In this example, the sunflower seed has higher profitability and so is a higher-ranked food item.

profitability For a food item, the energy it contains divided by the handling time.

A graphical solution

To see how the optimal diet model works, let's first assume that all food types have the same amount of energy (E), say, 10 J, but differ in handling time (h). Table 8.1 shows the handling times for each of five hypothetical food types along with their profitabilities. We see that Type A has the highest profitability, whereas Type E has the lowest. Finally, let's assume that all food types are equally abundant and randomly scattered throughout the environment.

TABLE 8.1 Food item characteristics. Food types with their handling times and profitabilities.

FOOD TYPE	ENERGY (J)	HANDLING TIME (S)	PROFITABILITY (J/S)
A	10	1	10.0
B	10	3	3.3
C	10	8	1.2
D	10	24	0.4
E	10	29	0.3

TABLE 8.2 Possible forager diets. Five hypothetical diets.

DIET	EAT	REJECT	DESCRIPTION
1	Only Type A	Types B, C, D, and E	Extreme specialist
2	Types A and B	Types C, D, and E	
3	Types A, B, and C	Types D and E	
4	Types A, B, C, and D	Type E	
5	All types	Nothing	Extreme generalist

generalist A forager that consumes a wide variety of items in its diet.

specialist A forager that has a narrow diet.

What diet maximizes the energy intake rate? Think about this question and write down your answer. You might begin by thinking that a forager should eat all food items it finds because all food items provide it with energy. A forager that eats a wide variety of food types is called a **generalist**, because it has a broad diet and consumes food items in proportion to their availability in the environment. Alternatively, you might think that a forager should maximize its energy intake rate by eating only Type A items because they have the highest profitability. Foragers with restricted diets are called **specialists**, because they consume a small subset of potential food types. Or maybe you think that a diet somewhere in between these two extremes should maximize an animal's energy intake rate. At this point, it is not obvious which diet maximizes fitness.

To determine the optimal diet, let's compare the profitabilities of each food type. Food Type A is the most profitable, so a forager should always include Type A in its diet. What other food types should be added to the diet to maximize energy intake rate (fitness)? You might imagine the five possible diets described in Table 8.2.

The first diet is the most specialized and the last the most general. Note that as we move down the list, each diet adds only the next most profitable food item. There are other possible diets, but it makes no sense for a forager to eat a low-ranking food item like Type E while ignoring one of higher profitability like Type A.

We can solve for the optimal diet by computing a forager's average energy intake rate per item in the diet. This is the average energy obtained per item divided by the sum of the average time it takes to find an item and the average time to handle an item in the diet:

$$\frac{\text{Average energy}}{\text{intake rate/item}} = \frac{(\text{average energy obtained/item})}{\left[\left(\text{average search time/item}\right) + \left(\text{average handling time/item}\right)\right]}$$

TABLE 8.3 Mean search time per item for each diet.

DIET	AVERAGE SEARCH TIME PER ITEM (S)
1	24
2	12
3	8
4	6
5	5

Because we assumed that all food items contain the same energy, 10 J, the numerator is the same for all diets. We therefore need to know only how the average search time and handling time per item vary with the possible diets. The average search time per food item is the rate at which an individual encounters food in the diet. An individual with a more general diet will have a high encounter rate with food—and thus a lower search time per item—because it eats almost every food item it encounters. An individual with a narrow diet, on the other hand, will have a longer search time per item. Table 8.3 shows how the average search time per item eaten varies for each diet listed in Table 8.2. Note that as more food types are added to the diet, the average search time per item declines.

TABLE 8.4 Mean handling time per item for each diet.

DIET	EAT	AVERAGE HANDLING TIME PER ITEM (S)
1	Only Type A	1
2	Types A and B	2 = (1 + 3)/2
3	Types A, B, and C	4 = (1 + 3 + 8)/3
4	Types A, B, C, and D	9 = (1 + 3 + 8 + 24)/4
5	All types	13 = (1 + 3 + 8 + 24 + 29)/5

Average handling time per item also varies with diet breadth. From Table 8.1, we can see that if the forager eats only Type A food items, its average handling time per item will be one second. If the diet includes both Type A and Type B items, the average handling time per item will be two seconds (half the items will be Type A, with a handling time of one second, and half will be Type B, with a handling time of three seconds). Table 8.4 shows how the average handling time per item varies for the five possible diets.

We can graphically display the average search time per item and average handling time per item for the five possible diets (Figure 8.6). Our graph also shows the combination of search time *plus* handling time per item, which we need to determine the optimal diet. In our example, Diet 3 has the lowest combined search and handling time per item, and so it maximizes the energy intake rate. The model predicts that a forager should always eat Types A, B, and C when encountered and should always reject Types D and E. Why should a forager reject Types D and E? Because the cost in handling time of including them in the diet is greater than the benefit derived from reduced search time; including these food types in the diet lowers an individual's energy intake rate. The model also predicts that each food type should always be accepted or always be rejected. There is no middle ground. This prediction is called the zero-one rule.

Now look at the answer you wrote down. Is it the same as the one produced by the model? What logic did you use to generate your prediction? This simple example illustrates the strength of the OFT approach to studying behavior. The OFT approach makes explicit assumptions and generates predictions that can be tested.

In our example, all food items have the same energy content. We can make the model more general and more realistic by allowing each food item to differ in both energy content and handling time. We can use a mathematical solution to show how this is done (Toolbox 8.1). In general, we find that diet breadth is affected by the abundance of the most profitable prey items and not by the abundance of the least profitable items.

The optimal diet model makes many simplifying assumptions about food types. It also assumes that there are no competitors or predators in the environment and that the forager does not need to search for mates. But remember that the model is a hypothesis that makes a testable prediction: animals will select a diet that maximizes their energy intake rate. Can it predict the diet of real animals? Let's look at one classic field study.

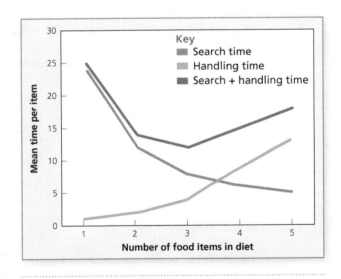

Figure 8.6. Diet model solution. A graphical solution for the optimal diet, with lines showing how mean search time per item (blue) and mean handling time per item (orange) vary with each diet. The optimal diet minimizes mean search time per item plus mean handling time per item (gray).

Figure 8.7. Northwestern crow. These birds feed on a variety of foods, including clams.

Diet choice in northwestern crows

Howard Richardson and Nicholas Verbeek used the optimal diet model to understand the feeding behavior of northwestern crows (*Corvus caurinus*) on Mitlenatch Island off the coast of British Columbia (Figure 8.7) (Richardson & Verbeek 1986). Crows on this island eat mainly Japanese littleneck clams (*Venerupis japonica*) but often reject clams that they have found. Richardson and Verbeek used the optimal diet model to understand why.

Clams on the island range in size from 10 mm to almost 40 mm in length. Crows open clams by dropping them onto rocks repeatedly to break them and then prying open the shells with their bills. To test the model, Richardson and Verbeek needed to measure handling times, the energy content of the clams, and encounter rates (search times for each item). They measured handling times for clams of different sizes in two ways. First, they observed crows feeding on clams of different sizes (ranging from 10 mm to 40 mm) and recorded the time from the first drop until the clam was eaten, which they defined as the handling time. They then collected the empty shell to determine its size so they could see how handling times varied with clam size. Second, in order to increase their sample size and obtain handling time data on all sizes of clams, they dropped over 500 clams of different sizes onto rocks from heights similar to those used by the crows themselves. They measured the number of drops it took (and the corresponding time) until the clam could be pried open with a fingernail.

To measure energy content, they collected and dried the soft tissue from 100 clams of different sizes. Previous work allowed them to determine the energy content of a clam based on the amount of dry tissue collected (James & Verbeek 1984). The researchers found that both energy content and handling time increased with clam size. Because both energy content and handling time varied with size, the researchers assumed that clams of different sizes represent different food types. They determined that the largest clam size was the most profitable and that profitability declined as size declined. Finally, they measured the availability of differently sized clams in 25 × 25 cm square plots of land called quadrats. In each of over 200 quadrats, they measured and counted all clams in the top 15 cm of substrate, which was the depth available to crows. These data provided information about the abundance of—and thus encounter rate for—each clam size.

Using the above information, the model predicted that crows should eat all clams greater than 29 mm in size and reject all clams less than 29 mm in size to maximize their energy intake rate. To test the model, Richardson and Verbeek collected data on the behavior of crows, noting the size of both those clams that were eaten and those that were rejected (clams that were picked up and then ignored). They found that the crows almost always ate clams that were larger than 30 mm and always rejected clams that were less than 26 mm (Figure 8.8). However, the crows did not follow the zero-one rule. Instead, they exhibited a partial preference: they sometimes ate clams of intermediate size. For example, clams that were 29 mm in size were eaten about 50% of the time. Although the diet model does not predict partial preferences, overall, the behavior of the crows was quite similar to the predicted behavior. Richardson and Verbeek concluded that the observed diet yields an energy intake rate very close to that of the optimal diet.

Many experiments using a variety of animals have produced results that closely match the predictions of the optimal

Figure 8.8. Crow diet choice. Each data point shows the percentage of clams eaten as a function of clam size. (*Source:* Richardson & Verbeek 1986)

diet model. However, animals often exhibit partial preferences for some food types. Why? The model assumes that foragers have perfect knowledge of the abundance of food types, as well as the energy content and handling times of each type. Of course, real foragers can only estimate these parameters. According to one explanation, the uncertainty results in partial preferences. For instance, the crows probably did not know for sure when they found a clam of size 28 mm (which the model predicts should be rejected) rather than one that was 29 mm in size (which should be eaten). Given this uncertainty, it is not surprising that clams of these sizes were sometimes eaten and sometimes rejected. In addition, the diet model assumes that there are no competitors and no predation risk. These factors, too, can lead animals to modify their diet in ways that include exhibiting partial preferences (Bell et al. 1984; Berec & Křivan 2000; Lima, Mitchell, & Roth 2003).

Ant foraging: the effect of nutrients

The optimal diet model assumes that fitness is only a function of energy intake rate. But animals require more than just energy—they also need about two-dozen elements, including carbon, nitrogen, phosphorus, and potassium (Fausto da Silva & Williams 2001). Such dietary needs can also affect feeding behavior and diet choice. For example, herbivores often face a situation of sodium limitation because plants contain low sodium concentrations. Thus, many herbivores, ranging from butterflies to large ungulates, are attracted to natural or artificial mineral licks that contain sodium (e.g., Boggs & Dau 2004; Tracy & McNaughton 1995).

Michael Kaspari, Charlotte Chang, and Johanna Weaver tested the hypothesis that sodium limitation affects the feeding behavior of ants (Kaspari, Chang, & Weaver 2010). They examined the recruitment of ants (attraction to a food source) to two food baits: sucrose (a high-energy food) and sodium chloride (NaCl, or salt). The researchers hypothesized that to maximize their energy acquisition, ants would recruit most heavily to sucrose. However, if sodium is limited and essential, ants should also recruit heavily to it. To examine the relative strength of ant recruitment to the two food sources, the researchers focused on areas that differ in the availability of salt in the environment. Road salt, containing NaCl, is commonly applied to roads in winter. Residual salt can be long lasting and can affect ecosystems near these roads. The researchers predicted that ants living far from salted roads (where sodium is rare) would exhibit a more pronounced preference for NaCl baits than would ants living nearer to salted roads. In other words, environmental availability of sodium should affect the diet selection of ants.

Kaspari and colleagues tested this prediction by establishing two sets of transects in Massachusetts that ran parallel to a salted road that was treated with NaCl during winter. Each set contained four transects located 1, 10, 100, and 1,000 m from the road. At 1 m intervals along each transect, the researchers placed a single 2 mL vial on the ground that was half-filled with cotton soaked with either sucrose or NaCl. A total of 90 vials were used in each transect: 45 with sucrose and 45 with NaCl, randomly selected for placement at each point along the transect. One hour later, the researchers collected each vial and recorded any ants inside. At each transect, the researchers also collected soil samples to calculate soil NaCl concentration and thus estimate sodium availability at different distances from the road.

NaCl concentrations were about ten times higher 1 m from the road compared to all other distances, indicating that road salt application does increase sodium availability near roads. About one-third of all ants recorded were *Tapinoma sessile*, a small omnivorous ant that consumes large amounts of vegetation. Near the road, these ants strongly recruited to vials containing sucrose: over 90%

Figure 8.9. Ant NaCl preference. A greater ratio of NaCl to sucrose vials was used by ants as distance to the salted road increased. (*Source:* Kaspari, Chang, & Weaver 2010) *Inset: Tapinoma sessile.*

diminishing returns A decline in instantaneous harvest rate as a food patch is depleted.

marginal value theorem An optimal foraging model that predicts how long an individual should exploit a food patch

of the vials visited contained sucrose, and so the NaCl vials were largely ignored. However, as distance from the road increased, ant use of NaCl vials also increased, so that at 1,000 m from the road, there was greater recruitment to the NaCl vials (Figure 8.9). These data demonstrate that sodium limitation affects the feeding behavior of these ants and illustrate how nutrient requirements can affect diet choice. In areas where sodium is limited, animals will often modify their behavior and diet to increase their ingestion of sodium.

Diet choice is one important aspect of a forager's behavior. However, foragers also need to decide how long to exploit one food patch before departing for another. We examine this behavior next.

8.4 The optimal patch-use model predicts how long a forager should exploit a food patch

After animals exploit a patch, it can take some time before food is replenished in that patch. For example, after a bee removes nectar from a flower, it can take 24 hours for the nectar to be fully replenished. After a woodpecker removes insects from a tree trunk, it may take days to months before more insect larvae are deposited there. Therefore, when a forager enters a patch with abundant food, it initially harvests food at a high rate, but as the patch becomes depleted, the harvest rate declines: we say that the forager experiences **diminishing returns** as it exploits a food patch. Given diminishing returns, the forager must decide how long to exploit the present (depleting) food patch before abandoning it to search for a richer patch.

To determine the optimal level of patch use, we need to examine the benefit a forager can gain by spending just a bit more time in a patch. This marginal benefit of feeding is the harvest rate at any point in time (or the instantaneous harvest rate). When a forager enters a patch with abundant food, the marginal benefit of feeding is high, because the instantaneous harvest rate is high. Because of diminishing returns, the marginal benefit declines as the patch becomes depleted. When the patch contains little food, the marginal benefit of feeding is very low, because it takes a long time to find the next food item.

The optimal patch-use model

Eric Charnov created an OFT model to predict how long a forager should stay in a food patch to maximize its fitness (Charnov 1976). Charnov's model, the optimal patch-use model, is based on four assumptions:

1. Foragers attempt to maximize energy intake rate.
2. All patches are identical (contain the same kind and amount of food).
3. Travel time between patches is constant.
4. As a forager depletes a patch, its instantaneous harvest rate declines—that is, it experiences diminishing returns.

Charnov's model, called the **marginal value theorem**, predicts that a forager should stay in a patch until its marginal benefit of feeding declines to equal the average energy intake rate from the environment. The average energy intake rate from the environment is calculated by dividing the total energy acquired from all patches by the total time to travel to and then exploit the patches (travel times plus patch times). In Charnov's original model, the travel time between patches is held constant, and all patches are identical.

How much time should a forager spend in a patch to maximize energy intake rate? To answer this question, we must first determine how much total energy the animal obtains when it stays in a food patch. Initially, there is abundant food in a patch, so the cumulative gain curve begins with a steep slope (Figure 8.10). However, the longer the time spent in a patch, the less food is left (i.e., diminishing returns), and so the slope of the curve becomes shallower as more time passes. Next, we need to include the amount of time spent traveling to a patch (T_t) in addition to the time spent in the patch (T_p). The energy intake rate can now be calculated for any T_p: it is the energy accumulated, divided by T_t plus T_p. We can solve for this relationship graphically, because the energy intake rate for any T_p is the slope of the line that runs from the start of the travel time to the cumulative gain curve at each T_p. The patch time that maximizes energy intake rate is the T_p where this line is tangent to (i.e., just touches) the cumulative energy gain curve (Figure 8.11).

We see that the model predicts the amount of time a forager should spend in a patch for any fixed travel time. This number depends on the travel time: as travel time increases, foragers should spend longer in each patch (Figure 8.12). This prediction can be tested, as we see next.

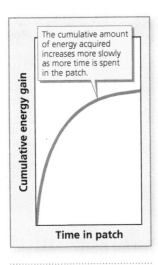

Figure 8.10. **Cumulative gain curve.** The graph shows how cumulative energy acquired by the forager increases with time spent in the patch.

The cumulative amount of energy acquired increases more slowly as more time is spent in the patch.

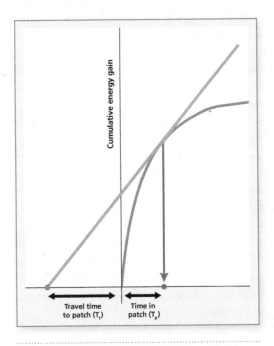

Figure 8.11. **Marginal value theorem.** For a fixed travel time, T_t, energy intake rate is maximized when the forager spends T_p time in a patch.

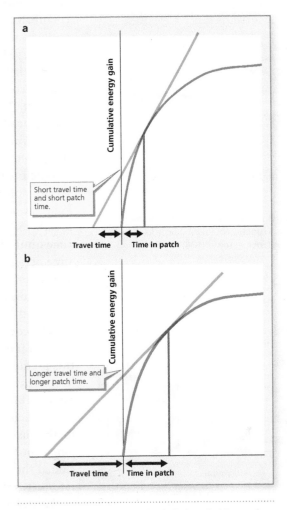

Figure 8.12. **Travel time and patch time.** In (a), travel time between patches is short, and so the optimal patch time is short. In (b), travel time is longer, and so the optimal patch time increases.

Patch use by ruddy ducks

Michael Tome tested the predictions of the marginal value theorem with ruddy ducks (*Oxyura jamaicensis*) (Tome 1988). Ruddy ducks feed on aquatic invertebrates and vegetation on muddy lake bottoms. They repeatedly dive from the water's surface to the lake bottom to find food. A short time later, they emerge to take a breath before diving again for food.

Tome created a large concrete and glass aquarium (5 m long × 2 m wide × 2 m deep). The bottom of the aquarium contained 16 wooden trays filled with sand, and the sides had glass windows for observations. Each tray constituted a potential food patch: Tome buried wheat grains in some of the trays during his experiments and trained birds to search the sand for food. The model assumes that ducks would experience diminishing harvest rates while feeding on wheat grains in the artificial food patches. To test this assumption, Tome buried 150 wheat grains in a patch whose location was fixed, which allowed the ducks to learn which patch contained food. He then released a duck to feed from the patch for a fixed amount of time that varied from five to 200 seconds. Afterward, he collected all the wheat grains left in the patch, which told him the amount of food eaten by the duck in that time. He then plotted the number of wheat grains harvested from a patch against time spent in the patch for six ducks (Figure 8.13). He found that the harvest rate did indeed decline as ducks depleted the patch: the cumulative gain (i.e., harvest rate) curve is steep when a bird enters the patch and levels off as the patch becomes depleted. These data match the model's assumption.

Figure 8.13. Ruddy duck cumulative gain curve. Number of food items harvested as a function of feeding time in a food patch. Inset: ruddy duck. (*Source:* Tome 1988)

Next, Tome established two identical patches on wooden trays placed on the aquarium bottom that each contained 150 buried wheat grains. The locations of each patch were kept constant from trial to trial, and birds were allowed to feed for seven days before the start of trials to learn their location and quality. When a bird first exploited a patch, it dove from the surface. When it abandoned the patch, it returned to the surface to breathe before diving to the second patch. Tome measured the time until a bird abandoned the first patch, as well as the travel time between patches (in other words, the time to travel to the surface to breathe and then dive back down to the second patch).

Tome already knew the relationship between time spent in a patch and the number of wheat grains eaten. From this information, he could calculate the optimal number of food items a bird should eat in a patch to maximize its energy intake rate. To test this prediction, he collected data on the number of food items left in a patch (to determine how many grains were eaten) for the same six birds, conducting ten trials for each.

Tome found that the model accurately predicted patch use for the majority of ducks (although two ducks ate more and spent longer in the patches than predicted) (Figure 8.14). He concluded that most of the ducks were maximizing energy intake rate, just as the model assumes.

Next, he tested the prediction that a forager should remain longer in a patch as travel time between patches increases. In the first experiment, the travel time between patches was small, because birds learned the location of the two patches and dove

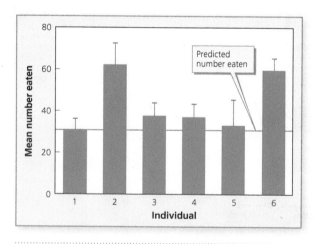

Figure 8.14. Ruddy duck patch data and prediction. Mean (+ SE) number of food items eaten by each duck. The line shows the predicted value based on the model. (*Source:* Tome 1988)

directly to them. In a second experiment, Tome increased travel time by randomly changing the location of a single food patch on the aquarium bottom in each trial. This required the ducks to search for the patch, and they needed about five search dives (rather than one) before they found it. Tome stopped the trial when a duck abandoned the patch, surfaced, and then dove again to search for other food patches. He again collected the number of food items left in the patch, allowing him to calculate the number eaten. This time, five of the six birds ate significantly more food from the food patch and thus spent significantly more time there, just as the model predicts (Figure 8.15). Tome concluded that the ducks were indeed attempting to maximize their energy intake rate, as assumed by the marginal value theorem.

The optimal patch-use model is sufficiently general that it can be applied to a variety of species, including humans (Applying the Concepts 8.1). But this model, like the diet model, makes several simplifying assumptions. It assumes that there are no foraging costs other than time spent traveling between and exploiting patches. It also assumes that a forager has perfect information about the environment: it knows that all patches are identical and that the travel time between all patches is the same. Let's see how these assumptions have been modified to better understand the feeding behavior of animals.

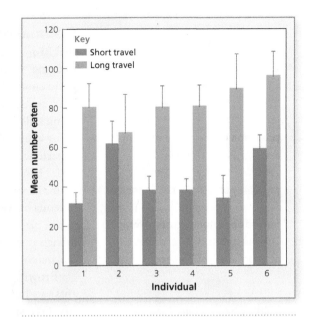

Figure 8.15. Travel time affects ruddy duck patch-use time. Mean (+ SE) number of items eaten increased with travel time. (*Source:* Tome 1988)

APPLYING THE CONCEPTS 8.1

Human patch-leaving decisions

Humans are often faced with deciding how long to spend exploiting one resource "patch" before moving to another. For instance, in hunter-gatherer societies, individuals must decide how long to spend searching for food in one area before moving to another (Smith & Wishnie 2000). As another example, you may often need to decide how long to hunt for information on one website before changing to another (Pirolli 2005).

John Hutchinson, Andreas Wilke, and Peter Todd evaluated the applicability of the optimal patch-use model to human behavior by presenting 100 subjects with the opportunity to acquire money on a computer fishing game (Hutchinson, Wilke, & Todd 2008). Individuals observed a screen that displayed a cartoon character with a fishing pole (the cursor) at a fishing pond. Fish popped up randomly for two seconds and could be caught by the observer by clicking on them with the mouse. The rate at which fish appeared varied positively with their density in a pond, and harvesting of fish reduced the fish remaining. Individuals could switch to a new pond at any time but could not return to an abandoned one. Moving to a new pond required some travel time during which no fish could be captured. The final payout to the subject depended on the total number of fish captured, so spending too much time at a heavily depleted patch or switching too often (i.e., spending too much time moving and not fishing) would result in poor payoffs. Thus, individuals sought to maximize their gain by determining how long to stay at each pond (patch).

The research team tested individuals on several conditions that varied the distribution of fish in the ponds and the travel time between ponds. Of particular interest were two conditions that had identical resource distributions (a high variance in the number of fish in ponds) but different travel times between each pond (15 seconds or 25 seconds). The model predicts the optimal average time individuals should spend at each patch and that more time should be spent at each patch when travel times are longer. How did the subjects do?

Individuals spent more time than was optimal exploiting the ponds. When travel time was short, they spent an average of 170 seconds at each pond, whereas the optimal average time was 47 seconds. For the longer travel time, individuals spent on average 207 seconds at each pond, while the optimal time was 65 seconds. Note, however, that individuals did increase their patch-use time as travel time increased, as predicted by the model. The researchers suggest that individuals may have spent more time than was optimal exploiting these patches because they needed to learn about the environment, specifically the distribution of fish in the ponds. ■

energetic costs of foraging The energy used to exploit a patch and the metabolic costs incurred while feeding.

predation risk cost The fitness cost associated with being killed by a predator. Often equated with the probability of being killed by a predator.

missed opportunity costs of foraging The fitness costs of not engaging in other activities while feeding.

quitting harvest rate The forager's instantaneous harvest rate when it leaves a food patch.

Featured Research

giving-up density (GUD) The density of food items in a food patch after being exploited by a forager.

Optimal patch model with multiple costs

Joel Brown developed a new patch-use model that includes the energetic costs of foraging, predation risk costs, and missed opportunity costs (Brown 1988). **Energetic costs of foraging** include the energy used to exploit a patch and the metabolic costs incurred while feeding. **Predation risk costs** involve the probability of being killed while feeding. Finally, **missed opportunity costs** are the costs of forgoing other activities that might yield even higher fitness. For example, a forager experiences high missed opportunity costs if by feeding it misses opportunities to mate. Brown's model predicts that a forager will maximize its fitness when it remains in a patch until its marginal benefit of staying in the patch (again, the instantaneous harvest rate) declines to equal the marginal (instantaneous) cost of being in the patch. Since the marginal cost now includes energetic, predation risk, and missed opportunity costs, Brown's model is more general than the original optimal patch-use model, and it is applicable to a wider range of foraging situations.

Brown's model predicts that if two patches have the same food (identical benefits) and the same costs, they should be harvested down to the same food density. (We say they should have the same **quitting harvest rate**.) This prediction holds even if patches differ in their initial food density. Remember that the marginal value theorem model assumes that all patches are identical. Brown's patch-use model is more general because it applies even when patches differ in food amount. Let's examine one test of this prediction.

Fruit bat foraging on heterogeneous patches

A simple way to test Brown's model is to provide foragers with patches that have identical benefits and costs but differ in initial food abundance. Recall that Tome collected the amount of food left in a patch when a duck abandoned it, obtaining what is called the **giving-up density (GUD)** of food. Brown's model predicts that the patches will be harvested down to the same GUD if foragers are attempting to maximize their energy intake rate.

Francisco Sánchez tested this prediction using Egyptian fruit bats (*Rousettus aegyptiacus*) (Sánchez 2006). Fruit bats are large, nocturnal bats that live in tropical Asia and Africa. They obtain juices and pulp from a variety of fruits and also feed on nectar.

Sánchez examined the behavior of captive bats that exploited artificial feeders in a large outdoor flight cage (Scientific Process 8.3). He wanted to determine whether bats attempt to maximize their energy intake rate while feeding on patches that differ in initial resource amount. He trained bats to feed from artificial feeders that consisted of a large cylinder with one opening that contained a mix of sugar water and protein. Brown's model, like the marginal value theorem, requires that a forager experience a decline in its harvest rate while feeding from a food patch. Bats often face diminishing returns in nature because they often feed on fibrous fruits: juice and pulp are easy to extract at first, but extraction becomes progressively more difficult. To create diminishing returns in the artificial feeders, Sánchez placed several small pieces of rubber tubing inside the food solution. As a patch was depleted, the rubber tubing made it progressively more difficult for the bats to extract the sugar water. Before conducting his experiment, Sánchez examined the harvest rates of bats in these feeders by allowing them to feed for several nights. Each night he recorded the total time a bat spent at a particular feeder and the amount of food remaining at the end of the night—the GUD. When he plotted the corresponding harvest rate data, he found that bats did experience diminishing returns in the artificial feeders, just like the ruddy ducks in Tome's experiment.

SCIENTIFIC PROCESS 8.3

Patch use by fruit bats
RESEARCH QUESTION: *How do fruit bats exploit food patches that differ in food amount?*

Hypothesis: Bats will attempt to maximize their energy intake rate.
Prediction: Bats will equalize the giving-up density across the patches.

Methods: The researchers:
- Housed bats in an outdoor flight cage that contained feeding stations. Each station contained three artificial feeders that differed only in the amount of food they contained.
- Recorded giving-up densities in all feeders each day for three days

Results: Bats experienced diminishing returns when using the feeders. Bats equalized the giving-up densities of the three feeders at each station.

Figure 1. Feeder. Schematic drawing of artificial feeder. The cubes are the rubber tubing, and the bar represents the feeder opening. (*Source:* Sánchez 2006)

Figure 2. Food harvested. Cumulative gain curve for the feeders. (*Source:* Sánchez 2006)

Figure 3. GUD. Giving-up densities of the three feeders (patches) on each day. (*Source:* Sánchez 2006)

Conclusions: Bats maximize energy intake rate when feeding on patches that differ in initial food amount.

EVALUATE
What data indicate that the bats were attempting to maximize their energy intake rate while feeding on the three types of artificial feeders?

To test the prediction that bats would equalize GUDs across patches that differed in initial resource amount, Sánchez provided bats with three identical feeders that were placed very close together to create similar energetic, predation, and missed opportunity costs. No predators were introduced into the flight cage, and the captive bats could not engage in other activities. However, the feeders contained different amounts of food: 42, 84, or 168 g of sugar per feeder in the water-protein mix. Sánchez established five feeding stations in the aviary in different locations and allowed 12 bats to feed from them all night. Each morning, he collected the amount of food remaining in each feeder to determine how much had been consumed. He repeated this procedure for three nights.

Sánchez found that the bats did equalize GUDs as predicted by the model. He concluded that the bats maximized their energy intake rate while feeding.

His work and others have shown that Brown's patch-use model can explain the feeding behavior of many organisms (Brown & Kotler 2004). Brown's model can be applied to a wider range of ecological conditions than Charnov's original optimal patch-use model because it allows for differences in patch resource amounts and foraging costs. Sánchez's work shows how the model can predict feeding behavior when the initial resource amounts of food patches differ. Next, we examine how Brown's patch-use model has been tested when patches differ in their foraging costs.

Kangaroo rat foraging with variable predation costs

Brown's model predicts that if foraging benefits are the same but costs differ between two patches, foragers should spend less time at the patch with higher costs, leaving it with a higher GUD. This predicted relationship can be used to investigate costs experienced by individuals as they forage.

Cathy Herman and Thomas Valone conducted an experiment to investigate how variation in predation risk affected patch use (Herman & Valone 2000). They studied nocturnal, seed-eating Merriam's kangaroo rats (*Dipodomys merriami*) in the Mojave Desert. The vegetation structure of this desert is characterized by a mosaic of shrubs and open areas that create two microenvironments for kangaroo rats: individuals can search for seeds under shrub canopies or in the open between shrubs (Figure 8.16).

Herman and Valone investigated how perceived predation risk affects foraging behavior. One way for a rodent to assess recent activity of a predator is by detecting its urine (Kats & Dill 1998). Therefore, the researchers examined how fresh urine scent affected patch use by kangaroo rats feeding either under shrubs or in the open.

They established 48 pairs of standardized food patches filled with sand and a fixed amount of seed. In each pair, one tray was placed under the canopy of a creosote bush (*Larrea tridentata*) and another tray was placed 4 m away in the open. Next to each tray was a small wooden dowel. The researchers used these dowels to add the scent of a common mammalian predator at the study site, urine of gray fox (*Urocyon cinereoargenteus*). The experiment consisted of four treatments: urine present at both trays, urine only next to the bush tray, urine only next to the open tray, or urine at neither tray. Because the benefits of foraging were identical at the two trays, any differences in the GUDs at the paired trays should reflect differences in perceived cost for the foragers. Trays were established at dusk, and the following morning at dawn the remaining seeds in each tray were collected, providing the GUD.

Over the course of the winter, there was a strong effect of microhabitat and predator scent on the patch-use behavior of kangaroo rats. Kangaroo rats exhibited a preference (lower GUD) for feeding at the shrub tray when predator scent was found at both trays, at neither tray, or only at the open tray. However, when predator scent was present only at the shrub tray, kangaroo rats no longer preferred to feed there (Figure 8.17). Kangaroo rats preferred to spend more time feeding in the safety of shrub canopies, except when evidence of recent mammalian activity was present only at the shrub tray. This indicates that kangaroo rats perceive lower foraging costs (predation risk) under shrub canopies, except when mammalian urine is present only at the shrub microhabitat.

These data show how kangaroo rat foraging behavior is affected by perceived predation risk costs that vary spatially (shrub versus open microhabitat) and temporally (nightly absence or presence of predator urine). Recently, conservation biologists have begun to use GUDs to understand how human activities can

Figure 8.16. Desert habitat and kangaroo rat. (a) Mojave Desert habitat with widely spaced shrubs and open bare ground. (b) Merriam's kangaroo rat.

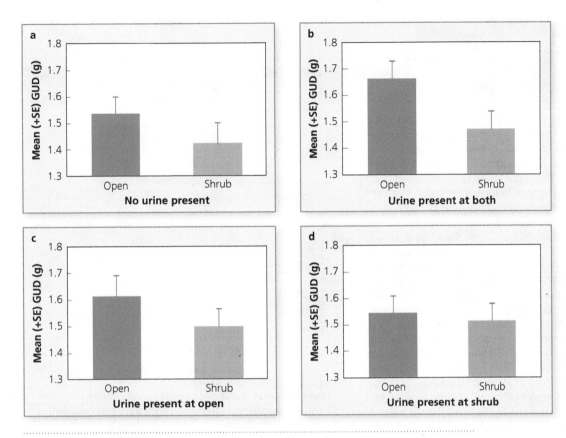

Figure 8.17. Giving-up densities for the open and shrub food patches. Mean (+ SE) GUDs for open and shrub trays for each experimental treatment. (*Source:* Herman & Valone 2000)

increase perceived costs for nocturnal animals and negatively affect such species (Applying the Concepts 8.2).

The marginal value theorem is a simple model, but it captures essential features that have shaped the evolution of feeding behavior. Brown's model is more general and has allowed a greater understanding of patch-use feeding behavior. There remains one more limiting assumption, however: individuals have perfect information about their environment.

Incomplete information and food patch estimation

Recall that the marginal value theorem assumes that an individual knows the exact number of food items in each patch. However, in nature, most animals can only estimate the number and quality of food items in each patch they encounter. This incomplete information might explain why some foragers spend slightly longer in patches than predicted, as we saw for two of the ducks in Tome's experiment.

How do animals estimate the quality of a patch? Individuals gain sample information about a patch from the quantity and quality of food they find. In addition, they may have prior knowledge about the frequency of food patch types (their quality) in the environment from previous foraging events. Combining sample information with prior knowledge allows individuals to obtain a more accurate estimate of food patch quality. How do they do this?

You've made such estimates yourself if you have ever played blackjack. The object of this game is to acquire a set of cards that adds up to 21 without going over that value—or at least beats the dealer's hand. You can ask for any number of cards, but if your total exceeds 21, you "bust" and lose the game. How do you decide how many additional cards to obtain? You use two types of information: prior knowledge and sample information.

APPLYING THE CONCEPTS 8.2

GUDs and conservation

Conservation biologists are studying patch use and using GUDs to examine how humans affect other species. As humans expand into previously uninhabited areas, for example, they modify the environment by adding artificial lights. Many nocturnal species perceive high light levels to be risky and in response, reduce their feeding behavior. Brittany Bird, Lyn Branch, and Deborah Miller studied how different light levels change the feeding behavior of nocturnal beach mice (*Peromyscus polionotus leucocephalus*) on Santa Rosa Island, Florida (Bird, Branch, & Miller 2004). Beach mice live in sand dune habitats and feed on seeds. At four sites, Bird and colleagues established two linear transects of food trays that contained 5 g of seed mixed into beach sand. One transect was established with a light source (two 40 W bug lights) at one end (light transect), while the other transect was placed more than 20 m away from a light source (dark transect). Each transect consisted of nine patches placed at 2 m intervals. Trials were conducted around the time of the new moon (i.e., dark nights), and GUDs were collected for two or three nights for each transect.

The research team found that mice GUDs were nearly two times higher at patches near the light source than at patches away from the light source. In the light transects, GUDs decreased as distance from the light increased, so that at 14 to 16 m from the light source, GUDs were similar between light and dark transects. Because human dwellings often have high levels of nighttime illumination, urbanization can negatively affect nocturnal species. ◼

Figure 1. Mouse GUDs. Mean (± SE) giving-up density. Giving-up densities are higher on light transects (orange) compared to dark transects (blue) from zero to 12 m away from a light source (light source is at distance 0). Inset: beach mouse. (*Source:* Bird, Branch, & Miller 2004)

You already know the cards in a standard playing deck. The values range from one to 11 but are not equally represented. There are four cards of each value from two through nine, 16 cards of value ten (the tens, jacks, queens, and kings), and four cards (the aces) of a value of either one or 11. This information constitutes your prior knowledge, and it allows you to determine the probability that you will go over 21 with the next card. For instance, if your first two cards total seven, there is zero probability that the next card will result in a total that exceeds 21, so you will always take another card. On the other hand, if your first two cards total 14, there is a relatively high probability that the next card obtained will give you a total that exceeds 21.

What is that probability? You can determine this number by observing the cards that have already been played. This is the sample information, and it allows you to estimate the frequency distribution of the cards remaining in the deck. If many cards of low value (say, less than eight) have already been played, then the probability of drawing a high card is higher, and you are more likely to bust. If many cards of high value (say, greater than seven) have already been played, the probability that the next card will make you bust is much lower. If you did not know the cards in a standard playing deck, blackjack would be much more difficult. It is the ability to track sample information that so often separates winners from losers at the blackjack table.

Combining prior knowledge with sample information is a powerful way to estimate unknown parameters. How does this skill relate to animal behavior? Modifications of the optimal patch-use model assume that animals estimate food patches much as blackjack players estimate the likelihood that one more card will result in a set of cards totaling over 21. This process of combining sample information and prior knowledge is called Bayesian estimation because it is done mathematically using Bayes's theorem. Let's examine one recent test of this idea.

Bayesian foraging bumblebees

Featured Research

To answer the question of whether bumblebees estimate food patches in a Bayesian manner, Jay Biernaskie, Steven Walker, and Robert Gegear created an enclosure that allowed a single bee to search artificial flowers for sugar water (Biernaskie, Walker, & Gegear 2009). They made artificial flowers using microcentrifuge tubes that contained either sugar water (a food reward) or plain water (no food). They then created a foraging arena that contained ten food patches comprising 12 artificial flowers each (Figure 8.18). Using this design, they trained individual bees to learn different distributions of patch types. This information became the bee's prior knowledge. One set of bees learned that all patches were the same: five flowers contained rewards and seven flowers did not; hence, the mean rewards per patch were five. The location of the rewards was randomly assigned within each patch, so the bees could not memorize the exact location of rewards within patches. Such a condition is called a uniform environment because there is no variation in the number of rewards across patches. More formally, patch variance in this distribution of patch types was zero. The researchers predicted that the patch quality estimate of bees trained in this uniform environment should decline each time the bees found a food reward, because this would indicate the presence of one less food reward in that patch. Finding rewards represents the bees' sampling information.

This leads to our first prediction:

Prediction 1: Bees trained in the uniform environment should have a lower probability of staying in a patch after each reward is found, and they should always immediately leave after finding the fifth reward.

A second set of bees was trained in a high-variance environment. Here, half the patches contained nine rewards, and half the patches contained one reward. The mean patch in this environment contained five rewards (as in the uniform environment), but the variation in rewards among patches was much higher. For bees trained in the high-variance environment, there was a different relationship

Figure 8.18. Foraging arena. (a) Food patches with 12 artificial flowers each. (b) Bee exploiting a flower.

between the number of rewards found and the expected number of rewards left in the patch. A bee using prior knowledge of the high-variance environment could now learn about the quality of a patch from sampling information. After finding one reward, it would know that there was a 50% chance the patch contained eight more rewards, and so the estimated patch quality would be relatively high. Once a bee found a second reward, it would know with 100% certainty that the patch contained seven more rewards, and so the estimate of the patch would be very high. After that, each reward found would mean that one less was left, and so the estimate of the patch quality would decline. This generates our second prediction:

Prediction 2: If bees use prior knowledge, individuals trained in the high-variance environment should have a high probability of staying in the patch after finding the second reward. The probability should decline thereafter.

The research team tested these predictions using 20 bees, half of which were trained in each environment, that fed on patches in a test environment. For each bee, the researchers recorded the number of flowers visited and how often the bee left a patch after finding a reward. This procedure was repeated for ten test patches for each bee.

As predicted, bees trained in the uniform environment had a lower probability of staying in the patch for each reward found. Bees in the high-variance environment, on the other hand, showed a much higher probability of remaining in the patch after finding the second reward compared to bees trained in the uniform environment. The likelihood of patch departure, however, did not decline for each reward found thereafter (Figure 8.19). These results are consistent with

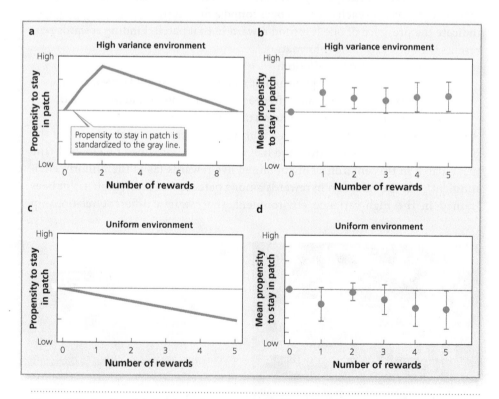

Figure 8.19. Predicted and observed patch-leaving behavior. (a and c) Predicted propensity to remain in a patch as a function of the number of rewards found in each environment. (b and d) Observed (± error bar) propensity to stay in a patch as a function of the number of rewards found in each environment. (*Source:* Biernaskie, Walker, & Gegear 2009)

the idea that bees can learn different prior distributions. The data also suggest that bees combine information from prior distributions with sample information. It is important to note that these animals are not calculating probabilities. Rather, they display adaptive learning regarding the distribution of resources in their environment. Such estimation allows them to exploit food patches more efficiently. A wide variety of animals, ranging from insects to mammals, do the same (Valone 2006).

The original optimal foraging models are based on many simplifying assumptions. Recent work has incorporated more realistic assumptions about the environment (e.g., patches have different initial amounts of food) and the forager (e.g., multiple foraging costs and incomplete knowledge about food patch quality). Although these models may still seem somewhat simplistic, they have been remarkably successful at predicting complex foraging behaviors in a wide variety of animals.

8.5 Some animals obtain food from the discoveries of others

Throughout this chapter, we have examined the foraging behavior of individuals, independent of the behavior of others. But some foragers are kleptoparasites: they obtain food from other individuals. Heterospecific kleptoparasites usurp food from a different species (Iyengar 2008). Such behavior not only reduces search costs but can also increase diet breadth. For example, Stéphane Morissette and John Himmelman quantified the success of kleptoparasites that steal food from the sea star (*Leptasterias polaris*) (Morissette & Himmelman 2000). This echinoderm extracts large buried clams from the sediment to feed on them, a process that can last for several days (Figure 8.20). A variety of opportunistic carnivores that cannot extract and feed on large buried clams rely on kleptoparasitism to obtain such food. These kleptoparasites include whelks, crabs, and another sea star, *Asterias vulgaris*.

Figure 8.20. Sea star eating clam.
Leptasterias polaris opening a clam to eat.

To document interactions among these species, Morissette and Himmelman used underwater cameras at a study site in the Gulf of Saint Lawrence to film 20 *L. polaris* individuals feeding on a large clam provided by the researchers. The researchers estimated the mass of the clam before and after it was eaten by *L. polaris*, both when kleptoparasites were excluded by a cage and when they were present. This allowed the research team to determine how much food mass, on average, was lost to kleptoparasites.

After the initiation of feeding, kleptoparasites arrived within a few hours and frequently contacted the feeding sea star when not excluded by a cage. Such interactions sometimes lasted several hours, and all kleptoparasites successfully obtained food from the clam prey: the sea star lost 6%–20% of clam prey mass to the heterospecific kleptoparasites.

In this marine system, a variety of species gain access to a unique resource via kleptoparasitism of food captured by the heterospecific *L. polaris*. Perhaps even more often, however, individuals obtain food from the food discoveries of conspecifics, as we see next.

Spice finch producer-scrounger game

Featured Research

producer An individual in a group that searchers for food.

scrounger Individuals in a group that exploit the food discoveries of others.

Some individuals within stable social groups often usurp food from conspecifics. In such cases, we classify individuals that search for and find food as **producers**, while those that usurp food from producers are called **scroungers**.

The existence of conspecific producers and scroungers within a foraging group is particularly interesting because successful scrounging requires at least one producer: no one obtains food if all group members are scroungers. Furthermore, as the proportion of scroungers increases in a group, fewer scrounging opportunities will exist. One might correctly conclude that the foraging success (fitness) of scroungers should be high when they are rare relative to producers in a group but should decline as the frequency of scroungers in the group increases. Given this relationship, what frequency of scroungers might we expect in a foraging group?

The producer-scrounger game theory model was developed to answer this question (Barnard & Sibley 1981; Vickery et al. 1991). This model assumes that searching for (and finding) food is incompatible with searching for scrounging opportunities. This might often be true when finding food requires searching the ground (head down), whereas searching for successful scrounging opportunities requires scanning the environment to observe producers (head up). The model also assumes that the feeding rate (fitness) of scroungers decreases as the frequency of scroungers in the group increases, so that the fitness of scroungers is higher than that of producers when scroungers are rare but lower than that of producers when scroungers are common. The model predicts that there is an equilibrium frequency of scroungers and producers at which their fitness is equal (Figure 8.21): both behavioral strategies can persist in a population at this stable equilibrium.

Kiernon Mottley and Luc-Alain Giraldeau tested the producer-scrounger model using captive flocks of spice

Figure 8.21. Producer-scrounger model. Payoffs (fitness) of the scrounger (blue line) and producer (orange line) strategies as a function of scrounger frequency in a group. The intersection denotes the predicted frequency of scroungers in groups. (*Source:* Mottley & Giraldeau 2000)

finches (*Lonchura punctulata*), a granivorous bird native to Southeast Asia (Mottley & Giraldeau 2000). These birds exhibit little aggression, and in captivity, individuals can behave as both producers and scroungers (Giraldeau, Soos, & Beauchamp 1994). Mottley and Giraldeau trained birds to feed from two sides of a set of artificial patches that contained seeds. On one side of the patches, individuals were producers, while on the other side, they were scroungers. Birds were trained to pull a string on the producer side of the patch, which caused seeds to fall into a collecting dish. Seeds in the dish were accessible to the producer that pulled the string and to all scroungers on the other side.

The researchers then formed multiple flocks of six birds. Each flock had differing numbers of producers and scroungers because each individual was restricted to just one side of the food patch (producers on one side, scroungers on the other) (Figure 8.22). This allowed the research team to create six treatments that manipulated the proportion of producers and scroungers in a flock, ranging from one producer and five scroungers to six producers and zero scroungers. This procedure was replicated for three different flocks. One final manipulation affected the benefit of scrounging. When the food patches were uncovered, scroungers had ready access to seeds that fell into the collecting dish. However, when the patches were covered, scrounger access was more limited, slowing their ability to obtain seeds and so reducing the benefit of scrounging. In each trial, a focal individual was observed to determine its feeding rate.

The feeding rate for scroungers was indeed lower when patches were covered for all flock types. As expected, for both uncovered and covered patches, the feeding rate of scroungers declined as the proportion of scroungers increased

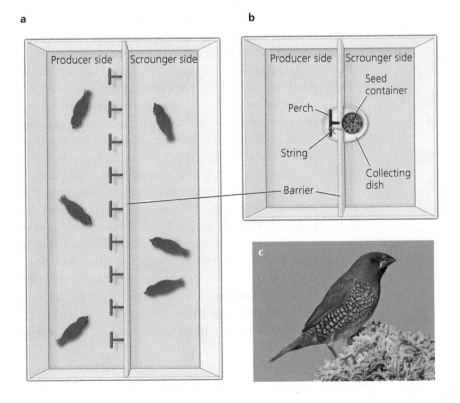

Figure 8.22. Producer-scrounger experimental set up. (a) Producers and scroungers were restricted to separate sides of the arena. (b) When producers pulled the string, seeds fell into a container from which both producers and scroungers could feed. (c) Spice finch. (*Source:* Mottley & Giraldeau 2000)

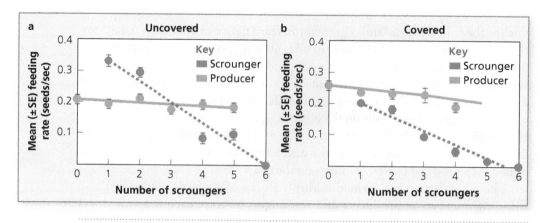

Figure 8.23. Producer and scrounger feeding rates. (a) For uncovered patches, producer (orange circles and line) and scrounger (blue circles and line) feeding rates were equal when flocks contained three scroungers. (b) For covered patches, producers (orange) and scroungers (blue) had equal feeding rates only when there was a single scrounger. (*Source:* Mottley & Giraldeau 2000)

(Figure 8.23). In contrast, the feeding rate of producers was largely unaffected by the proportion of scroungers for both types of patches. For uncovered patches, producers and scroungers had equal fitness when they were equally common; for covered patches, scroungers and producers had equal fitness when a flock had one scrounger and five producers.

Next, Mottley and Giraldeau examined how birds would behave when all were free to use either side of the patch. Would flocks of birds conform to the predicted proportions of producers and scroungers? In the second experiment, the patch was covered for Days 1–8 and uncovered for Days 9–16. The researchers again measured the behavior of focal individuals in three different flocks, noting their foraging strategy (producer or scrounger) and their feeding rate.

Individual birds tended to adopt one or the other foraging strategy, although there was some switching of patch sides at the start of a trial. In just a few days, the flocks adjusted their scrounger frequency to the predicted values: for covered patches, flocks contained about one scrounger, while for uncovered patches, there were about three scroungers per flock (Figure 8.24).

These data are consistent with the predictions of the producer-scrounger model, which assumes that individuals attempt to maximize feeding rate. This suggests that foraging spice finches are capable of fine-scale adjustments in their behavior based on current environmental condition (the benefit of scrounging) and the behavior of others. Within stable groups of conspecifics, some individuals find food while others usurp those discoveries, but within the group, producers and scroungers obtain similar feeding rates.

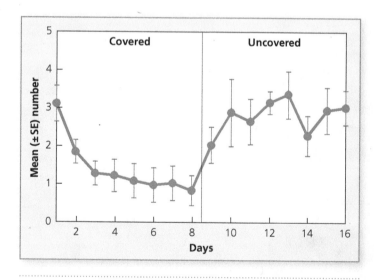

Figure 8.24. Number of scroungers. For covered patches (Days 1–8), flocks tended to have a single scrounger. For uncovered patches (Days 9–16), flocks contained about three scroungers. (*Source:* Mottley & Giraldeau 2000)

Mathematical solution to the optimal diet model

For simplicity, assume that there are only two food types in the environment, which we will call Type 1 and Type 2:

1. Type 1 contains an amount of energy E_1 and requires handling time h_1.
2. Type 2 contains an amount of energy E_2 and requires handling time h_2.
3. Type 1 is more profitable because $E_1/h_1 > E_2/h_2$.
4. Foragers spend time T_s searching for food.
5. The food types differ in their abundance in the environment. A forager encounters food Type 1 at a rate of λ_1 items per minute searching. Its encounter rate with food Type 2 is λ_2 items per minute searching. The inverse of the encounter rate $(1/\lambda)$ is the mean search time to find a particular food type.

Since we have only two food types, there are only two possible diets. A forager can specialize on Type 1, the more profitable type, or it can have a general diet and eat both food types. (A third possibility, specializing on Type 2 items, makes no sense, because that is the lower-ranked food type.) Which diet leads to a higher energy intake rate? To answer this question, we compare the energy intake rates of the two possible diets for a forager that searches for time T_s. We begin with the specialist diet and calculate the total energy acquired and the total handling time. The total energy acquired is simply the number of Type 1 items encountered during search time T_s—or λ_1 multiplied by T_s—multiplied by the energy, E_1:

$$\text{Total energy acquired} = (\lambda_1 \times T_s)E_1$$

By the same reasoning, the total handling time of the number of Type 1 items encountered during search time T_s is the number found $(\lambda_1 \times T_s)$ multiplied by their handling time h_1:

$$\text{Total handling time} = (\lambda_1 \times T_s)h_1$$

and again we know that the total search time is T_s.

The energy intake rate R_1 for the specialist diet is the total energy acquired by feeding on Type 1 items divided by the total search time for those items plus the total time spent handling those items:

$$R_1 = \frac{(\lambda_1 \times T_s) \times E_1}{T_s + \left[(\lambda_1 \times T_s) \times h_1\right]}$$

$$= \frac{\lambda_1 \times T_s \times E_1}{T_s \times [1 + \lambda_1 \times h_1]}$$

which simplifies to

$$R_1 = \frac{\lambda_1 \times E_1}{1 + (\lambda_1 \times h_1)}$$

Now we consider the generalist diet. We go through the same steps to calculate the energy intake rate for the generalist diet, but this time we have two terms, one for each food type:

$$\text{Total energy acquired} = \left[(\lambda_1 \times T_s)E_1\right] + \left[(\lambda_2 \times T_s)E_2\right]$$

$$\text{Total handling time} = \left[(\lambda_1 \times T_s)h_1\right] + \left[(\lambda_2 \times T_s)h_2\right]$$

where the total search time is again T_s.

The energy intake rate for this diet, $R_{1 \text{and} 2}$, is again the total energy acquired divided by the total search time plus the total time spent handling those items:

$$R_{1 \text{and} 2} = \frac{\left[(\lambda_1 \times T_s) \times E_1\right] + \left[(\lambda_2 \times T_s) \times E_2\right]}{T_s + \left[(\lambda_1 \times T_s) \times h_1\right] + \left[(\lambda_2 \times T_s) \times h_2\right]}$$

$$= \frac{T_s\left[(\lambda_1 \times E_1)\right] + \left[(\lambda_2 \times E_2)\right]}{T_s + \left[(\lambda_1 \times T_s) \times h_1\right] + \left[(\lambda_2 \times T_s) \times h_2\right]}$$

which simplifies to

$$R_{1 \text{ and } 2} = \frac{\left[(\lambda_1 \times E_1) + (\lambda_2 \times E_2)\right]}{1 + \left[(\lambda_1 \times h_1) + (\lambda_2 \times h_2)\right]}$$

A forager should generalize if the energy intake rate from eating both food types is greater than the energy intake rate from specializing on Type 1 only. That means that a forager should be a generalist if $R_{1 \text{ and } 2} > R_1$, or

$$\frac{\left[(\lambda_1 \times E_1) + (\lambda_2 \times E_2)\right]}{1 + \left[(\lambda_1 \times h_1) + (\lambda_2 \times h_2)\right]} > \frac{\lambda_1 \times E_1}{1 + (\lambda_1 \times h_1)}$$

Rearranging produces this inequality:

$$\frac{E_2}{h_2} > \frac{E_1}{\dfrac{1}{\lambda_1} + h_1}$$

The model predicts that Type 2 should always either be rejected (if the inequality is false) or eaten (if the inequality is true): foragers should always either eat or reject a food type. This prediction is the zero-one rule.

The model also predicts that the decision to specialize or generalize is not affected by the encounter rate with the lower-ranked food item. The term λ_2 does not appear at all in the final inequality. Type 2 could be infinitely abundant, and the forager's diet would still not change. Why not? Look again at the inequality. Just one behavioral difference distinguishes the generalist from the specialist: when an individual finds a Type 2 food item, that item is eaten by the generalist and rejected by the specialist. The forager should eat the item (and be a generalist) if the energy intake rate for that item (the left-hand side of the inequality) is greater than the energy intake rate of rejecting that item instead of searching for a Type 1 item (the right-hand side of the inequality). Since this judgment does not involve finding the Type 2 item, its encounter rate is irrelevant. Thus, we see that diet selection is really about what *not* to eat.

Chapter Summary and Beyond

Animals find food using a variety of sensory modalities. Many foragers locate food visually and have trouble finding cryptic, as opposed to conspicuous, prey. Animals can increase their hunting success on cryptic prey by forming a search image that allows them to focus on a single prey type. Recent work has examined interactions among prey camouflage, predator search images, and the evolution of prey polymorphisms (Bond 2007). In addition, olfactory search images are being used to train dogs to identify explosive materials (Gazit, Goldblatt, & Terkel 2005).

Optimal foraging theory assumes that natural selection has favored behaviors that maximize fitness largely by maximizing energy intake rate, and it has become a powerful approach to studying the behavior of animals (Rosenzweig 2001). The optimal diet model is used to understand the diet breadth of foragers. This model predicts that animals should adopt a zero-one rule for each food type and that diet breadth should be affected by the abundance of the most profitable food items and not by the abundance of the least profitable items. Sih and Christensen (2001) provide a comprehensive review of studies that have tested this model. Their review finds good support for the model, especially for foragers that consume immobile prey items.

The optimal patch-use model predicts how long a forager should remain in a patch. It predicts that foragers should spend longer in patches as travel time increases, and empirical tests show that it does often successfully predict behavior. Modification of the patch-use model includes the addition of energetic, predation risk, and missed opportunity costs of foraging. Brown and Kotler (2004) show how incorporation of these costs provides even greater understanding of feeding behavior.

While the optimal patch-use model assumes that animals have perfect knowledge of the food patches they encounter, most animals must estimate the quality of a food patch. Many appear to do so by combining prior knowledge about the distribution of patch types in the environment with current sampling information from a patch in a manner similar to Bayesian updating. Valone's (2006) review shows that many animals, including mammals, birds, fish, and insects, are capable of making such estimates.

Many animals usurp food that has been discovered by others. Such kleptoparasitism can enhance diet breadth and reduce search time for food. Within a social group of conspecifics, there often will exist variation in feeding behavior: producers search for food, while scroungers obtain food discovered by producers. The producer-scrounger model explains this diversity and predicts the equilibrium frequency of each strategy in a group. Recent work has investigated how often these strategies are fixed or flexible (Morand-Ferron, Varennes, & Giraldeau 2011) and how individual condition affects the adoption of these strategies (David & Giraldeau 2011).

CHAPTER QUESTIONS

1. In a review of empirical work, Sih and Christensen (2001) found that the optimal diet model better predicted the diets of animals that searched for immobile prey like seeds or clams compared to the diets of animals that hunted mobile prey. What assumptions of the model might best explain this finding?

2. Imagine the following experiments designed to understand the patch-use behavior of foragers. As in the experiment by Herman and Valone (2000), you set out two seed trays 1 m apart that contain the same amount and kind of food buried in the patch.

   Experiment A: One patch contains heavy soil, while the other contains light sifted sand. Which patch do you predict will have the higher GUD for kangaroo rats?

   Experiment B: On a cold morning, blue jays are foraging for sunflower seeds in two nearby seed trays filled with sand. One tray is in a warm, sunny location, and the other is in a cold, shady location. What prediction can you make about the relative GUDs of these two patches?

3. You record the diet selection of black-chinned hummingbirds in a meadow for several weeks. Ten species of flowers are common in this field, but hummingbirds only feed on seven of them and ignore the other three. After a rainstorm, the abundance of one of the ignored flowers increases dramatically, but the abundance of all the other flowers is unaffected. Using the optimal diet model, how do you think this change will affect the diet selection of the hummingbirds?

4. Imagine an experimental arena in which a shrew (a small insectivorous mammal) must search food patches for buried invertebrates. The arena contains ten identical patches that are located 5 m apart, and each contains 17 prey items. You record the patch-use behavior of the individuals and determine that, on average, they spend one minute in each patch and eat nine items before departing. What would you predict about the patch-use behavior of shrews if you increased the size of the arena so that the patches were 25 m apart?

5. Consider two different groups of spice finches with fixed strategies exploiting uncovered patches in Mottley and Giraldeau's first environment. Group A has one scrounger and five producers, while Group B has five scroungers and one producer. Identify the forager(s) with the highest fitness.

6. What assumption is shared by the optimal diet and patch-use models?

Chapter 9

Antipredator Behavior

Our college campus, like many others, has a large population of gray squirrels (*Sciurus carolinensis*). They nest in trees and feed on the nuts and seeds the trees produce. Gray squirrels are typically found in dense woodlands, where many mature trees are close together, but they have also adapted well to urban environments such as our campus. In addition to their usual diet of nuts, our urban squirrels scavenge food from garbage cans and people. They will visit one garbage can after another, sometimes diving to the bottom to find food. Squirrels will walk right up to a table in a large, open plaza and sit on their hind legs, a behavior that often entices students to give them cookies or French fries.

However, a couple of years ago, a pair of red-tailed hawks (*Buteo jamaicensis*) colonized our campus, made a nest in one of the light stands high above the soccer field, and raised a pair of young. Ever since, the hawks have been regulars on campus, and students and others have enjoyed watching them soar overhead and deliver food to their nestlings. The squirrels on campus have also noticed the hawks, because squirrels are one of these birds' favorite foods (Figure 9.1).

Since the hawks have been present, we have observed significant changes in the behavior of the campus squirrels. Now we rarely see squirrels looking for food far from the safety of trees, and when they do, they often dart rapidly from one place to another, rarely stopping in

Figure 9.1. Red-tailed hawk. Red-tailed hawks are frequent predators of squirrels.

wide-open places to solicit food. A squirrel will often interrupt a search for food on the ground and stand up on its hind legs, presumably to look for an approaching hawk. We have even seen a few attacks. Hawks typically attack when a squirrel is on the ground and far from a tree. Once the squirrel sees the hawk, it flees to the nearest tree, where it sometimes produces loud "chuck" calls and vigorously wags its tail back and forth. Invariably, this behavior causes other squirrels in the area to do the same, which produces quite a racket. These squirrel behaviors are examples of antipredator adaptations.

In this chapter, we examine how animals reduce the likelihood that they will be killed by a predator—that is, their predation risk. We begin by examining ways prey can reduce the probability of being detected by a predator. We also examine behaviors that enhance survival once an individual has been attacked. Next, we focus on two common antipredator adaptations that involve modifying behavior: reducing overall activity level and increasing vigilance—being alert for predators. Both, however, involve trade-offs that affect behavior, as we will see. Next, we consider how animals reduce predation risk by living in groups. We conclude by focusing on counterintuitive behaviors that involve interactions with predators in order to deter an attack.

9.1 Animals reduce predation risk by avoiding detection

cryptic coloration Morphological coloration that matches the color of the environment to reduce detection by predators.

Obviously, individuals cannot be killed by a predator if predators do not detect them. As we saw in Chapter 8, one common morphological adaptation is to blend into the background through **cryptic coloration**—body coloration that matches the color of the environment. When that fails, individuals can modify their behavior to reduce risk. They might try to flee or simply lie low in order to be overlooked. Experiments have found evidence for each of these behavioral adaptations.

Featured Research

Predator avoidance by cryptic coloration in crabs

Karen Manríquez studied the effectiveness of cryptic coloration in the crab *Paraxanthus barbiger* (Manríquez et al. 2008). These South American crabs live and feed on the benthic substrate, or bottom, of marine coastal habitats, where they are eaten by many species of fish. The benthic habitat can consist of a uniform sandy color or a more heterogeneous mixture, thanks to small shell fragments called shell-hash. Young juveniles display color variation that ranges from brown to tan to white, with many spots and stripes (Figure 9.2). As juveniles grow, their body color changes to a more uniform light purple color. Juveniles experience much higher predation than adults, and so Manríquez and colleagues hypothesized that the complex coloration of juveniles is an adaptation to minimize detection by predators, as the color of the juveniles is more cryptic on the heterogeneous shell-hash background. The team predicted that this coloration reduces predation.

To test this prediction, the researchers tethered individual juvenile crabs (with a mean carapace width of 11 mm) to 20 cm × 20 cm ceramic tiles of different colors. One set of tiles had a uniform white surface, while the other had a heterogeneous shell-hash surface. The tether was 10 cm long, which kept the crabs on the tile for the duration of the field experiment. All the plates were randomly placed in the ocean for a fixed period of time, and the researchers recorded the number of crabs alive at the end of the experiment. As predicted, they found very different survival rates for the crabs on the two different surfaces. Only 30% of the crabs on the white surface tiles survived, while more than 60% of the crabs on the shell-hash-colored tiles survived. From these results, the researchers concluded that juvenile crabs are indeed more cryptic when on the heterogeneous shell-hash background.

The effectiveness of cryptic coloration requires that animals live in, or use, environments that match their body color. Manríquez and colleagues next conducted a follow-up experiment to see whether crabs would select a shell-hash-colored surface over a more uniformly colored surface. They predicted that juvenile crabs would prefer a shell-hash background over a uniformly colored background if given a choice, and that this preference would be stronger when predation risk was high.

To test these predictions, they established experimental aquaria that contained ten small plastic trays (20 cm × 40 cm × 5 cm). One-half of each tray was covered by a thin layer of shell-hash, while the other was left empty and thus presented a uniformly white background (Figure 9.3). This set-up gave crabs a choice

Figure 9.2. Crab coloration. Juvenile crabs have a heterogeneous coloration.

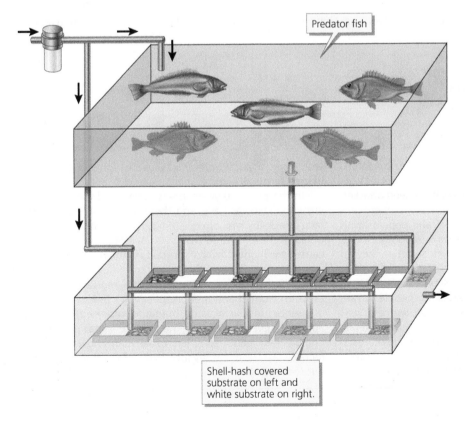

Shell-hash covered substrate on left and white substrate on right.

Figure 9.3. Crab experimental design. Either plain water or water with fish odor could be pumped into the experimental arena, where crabs had the choice of two substrates: plain white or shell-hash covered. (*Source:* Manríquez et al. 2008)

Figure 9.4. Substrate choice results. Mean (+ SE) percentage of individuals occupying heterogeneous substrate with and without fish predator odor. A greater percentage of both small (blue) and large (orange) crabs occupied the substrate when the fish odor was present. (*Source:* Manríquez et al. 2008)

of substrates to use. To manipulate predation risk, the experimenters treated half of the trays with predator odors by pumping seawater from a tank that housed several predatory fish and the other half with seawater that contained no predator odors. In each test aquarium, nine crabs were placed in the center of each tray, and the researchers recorded the proportion of these crabs on each substrate after three hours. To examine how crabs of different size would respond to these manipulations, they tested small crabs (less than 10 mm in length) in one set of experiments and larger crabs (25–30 mm) in another.

Manríquez and colleagues found that almost all of the small crabs preferred the shell-hash background, and that even more crabs used the shell-hash background in the predator odor treatment. Larger crabs showed no preference for either background in the control tanks (perhaps because they experience lower levels of predation risk than do smaller crabs). However, the large crabs strongly preferred the shell-hash surface in the predator odor treatment (Figure 9.4). From these experiments, Manríquez and colleagues concluded that juvenile crabs can select an appropriate cryptic background when given a choice, and that the complex color of juvenile crabs is an adaptation to minimize detection by predators.

This study implies that juvenile crabs can detect differences in surface coloration and select a background that allows them to be more cryptic. How animals assess their environment to match it has yet to be determined, but certainly this behavior is an effective antipredator adaptation. Assessment of predation risk has even been studied in humans (Applying the Concepts 9.1).

Featured Research

Predators and reduced activity in lizards

All animals, even those that lack cryptic coloration, can reduce detection by predators by reducing their overall level of activity. Predators often use motion

APPLYING THE CONCEPTS 9.1

Human fear of predators

Humans have a long evolutionary history with large predators, including several species of large felids, bears, and wolves (e.g., Brain 1970). Even in recent times, humans have been attacked by large predators. For instance, in Uganda in the twentieth century, over 300 people were attacked by lions and leopards, with over 60% of the attacks being fatal (Treves & Naughton-Treves 1999). While such attacks are rare in Western societies, they do occur (Beier 1991), and many people still fear large predators.

Eivin Røskaft and his colleagues assessed fear of predators in Norwegians (Røskaft et al. 2003) by asking people to rate their level of fear of brown bears (*Ursus arctos*), wolves (*Canis lupus*), lynx (*Lynx lynx*), and wolverines (*Gulo gulo*). Overall, 57% of people responded that they were very afraid of bears, while 48% reported a similarly high level of fear of wolves. Less than 30% reported that they were very afraid of lynx or

wolverines. In addition, fear levels were higher in women than in men across all ages.

In the past 300 years, bears have killed more Scandinavians than wolves have, and there have been no reports of fatal attacks by lynx or wolverines. As such, the highest levels of fear were associated with species that in recent times had caused the highest level of mortality in the region. In addition, women suffer significantly higher rates of mortality following an attack than do men (62% versus 29% mortality for bear attacks on women and men, respectively). In Norwegians, the differences in fear of large predators reported by each sex appear to correspond with the level of threat posed. These findings may lead researchers to ask if and how such fear affects behavioral decisions such as the amount of time spent in outdoor activities. ■

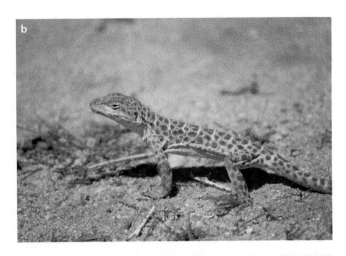

Figure 9.5. Lizards. (a) A marked whiptail lizard. These small insectivorous lizards are common in southwestern U.S. deserts. (b) A predatory leopard lizard.

detection to find their prey and so are less likely to notice slow-moving or stationary prey (e.g., Tarsitano & Jackson 1992; Catania, Hare, & Campbell 2008).

A simple experiment demonstrates how whiptail lizards (*Aspidoscelis uniparens*) reduce their activity level when predators are nearby. Whiptail lizards (which are about 7 cm long) live in deserts in southwestern North America and feed on insects and other arthropods (Figure 9.5a). While searching for food, these lizards can be killed by a variety of predators, including roadrunners (*Geococcyx californianus*), snakes, and larger lizards.

Douglas Eifler and colleagues designed an experiment to determine how whiptail lizards modify their behavior in the presence of predators (Eifler, Eifler, & Harris 2008). They established six 15 m × 15 m pens in the Arizona desert. These pens had 40 cm tall walls that prevented lizards from either leaving or entering. The pens also contained natural desert vegetation. Three of the pens were designated the predator-present experimental treatment and contained two large, adult leopard lizards (*Gambelia wislizenii*), a common predator of whiptails (Figure 9.5b). The other three pens contained no predators and served as the control group. In each pen, the researchers placed six adult whiptail lizards and recorded their behavior over two weeks using focal animal sampling. Each focal observation period lasted for 15 minutes. To determine how the presence of predators affected the behavior of the whiptails, the researchers compared the average behavior of all individuals in each pen.

Eifler and colleagues noticed that the leopard lizards did hunt the whiptail lizards (two were killed) and so presented a high risk of predation for the whiptails. During the experiment, the lizards in the different pens exhibited significant differences in behavior. Whiptail lizards in the predator-present pens were less active: they spent less time moving and moved more slowly than did lizards in the control pens (Figure 9.6). This lower level of activity likely made them less noticeable to predators.

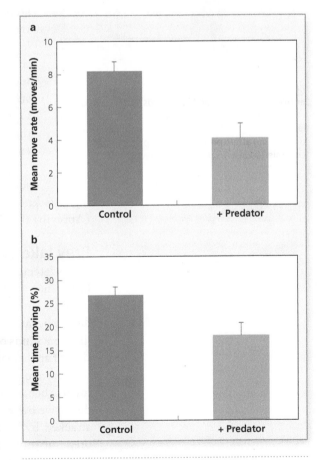

Figure 9.6. Lizard movement. Mean (+ SE) movement behavior. Lizards moved less when predators were present. (*Source:* Eifler, Eifler, & Harris 2008)

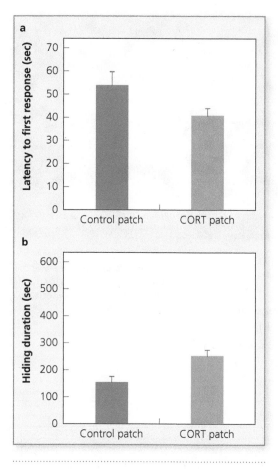

Figure 9.7. Corticosterone experiment and antipredator behavior. Male lizards with elevated corticosterone (CORT patch) (a) had a lower mean (+ SE) latency to respond and (b) spent more mean (+ SE) time hiding after spotting a predator compared to controls. (*Source:* Thaker, Lima, & Hews 2009)

What is the proximate mechanism for these behavioral changes? In vertebrates, exposure to a stressor, such as the sight of a predator, often causes an increase in the plasma concentration of corticosterone (e.g., Hubbs, Millar, & Wiebe 2000; Cockrem & Silverin 2002). This stress hormone affects both physiology and behavior in ways that can reduce predation risk (Sapolsky, Romero, & Munck 2000).

For instance, Maria Thaker, Steve Lima, and Diana Hews manipulated corticosteroid hormone levels in male ornate tree lizards (*Urosaurus ornatus*) and then observed their behavioral response to the sight of a predator (Thaker, Lima, & Hews 2009). Free-living lizards were captured and placed in an outdoor enclosure that provided locations for individuals to bask as well as large rocks under which they could hide. Plasma corticosterone levels were experimentally elevated for half of the males through application of a dermal patch placed on the back for 12 hours. The patch contained 4 µL of corticosterone dissolved in sesame oil. Control males received a patch with oil only. For 20 minutes, each subject was exposed to a caged collard lizard (*Crotaphytus nebrius*), a common predator of tree lizards. The researchers measured latency to first response, time spent hiding, and plasma corticosterone level after the exposure.

The experimental treatment successfully elevated plasma corticosterone levels: they were twice as high in treatment lizards than in control lizards. Additionally, individuals with high plasma corticosterone levels reacted more quickly (lower latency to respond) and spent significantly more time hiding after seeing a predator than did the control individuals (Figure 9.7). In free-living animals, exposure to acute stress, such as the sight of a predator, can elevate corticosteroid levels for hours to days (Cordero et al. 2003). Such a hormonal response can then facilitate more rapid antipredator behaviors if and when a future encounter with a predator occurs.

Behavioral modifications in response to predators like those described above are common among animals, and researchers are applying this knowledge to the challenge of minimizing crop damage from pests (Applying the Concepts 9.2).

Prey take evasive or aggressive action when detected

Once detected by a predator, many prey flee to avoid being killed. Some can move faster than their attacker and so run, swim, or fly to safety. Others flee into an underground burrow or dense vegetation. But others cannot avoid predators by fleeing because they move more slowly than their predators. In such cases, we see different kinds of antipredator adaptations. In Chapter 6, we examined the evolution of aposematic coloration in prey: a bright morphological signal that predators learn to avoid because such prey are chemically defended. Other adaptations involve behavior. For instance, larval lepidopteran caterpillars illustrate common defensive behaviors observed in relatively sedentary prey. These caterpillars are attacked by a variety of fast-moving predators, including insectivorous birds, predatory ants, and insect parasitoids. Most caterpillars employ a variety of defenses to avoid predation that can be categorized as either aggressive or evasive behaviors. Aggressive antipredatory behaviors include biting and regurgitation, while evasive behaviors include thrashing and twisting, head flicking, and dropping from an elevated perch (Greeney, Dyer, & Smilanich 2012).

APPLYING THE CONCEPTS 9.2

Mitigating crop damage by manipulating predation risk

Many vertebrate herbivores are crop pests, resulting in losses of millions of dollars for farmers. Mice, rats, deer, and even kangaroos consume large quantities of grains such as wheat, soybeans, corn, and rice and can also damage fruit trees. Scientists are applying research about antipredator behavior to reduce such crop damage.

To manipulate predation risk for a crop pest, a farmer can apply a predator odor near crops. Many prey use chemical cues such as predator odors to assess whether a predator is nearby and thus the level of predation risk. Prey—in this case, crop pests—often modify their behavior by decreasing the amount of time they spend foraging on crops that are near the predator odor.

Michael Parsons and colleagues conducted an experiment that applied this concept (Parsons et al. 2007). In Australia, Western gray kangaroos (*Macropus fuliginosus*) can be a source of significant crop damage (Figure 1). Parsons and colleagues tested the hypothesis that kangaroos will decrease their feeding on crops if they perceive a high level of predation risk. The researchers manipulated predation risk by applying predator urine near a food source. They used urine from a native Australian predator, the dingo (*Canis lupus dingo*); a North American predator, the coyote (*Canis latrans*); humans (*Homo sapiens*); and water as a control. They conducted their research on a population of 28 gray kangaroos that had free range on a 36 hectare property on a wildlife sanctuary.

Each afternoon, when the kangaroos would normally be feeding, the researchers set up four feeding stations that each contained 400 g of grain. Next they selected one type of odor to use for the afternoon trial (one of the three urines or the water control). The four feeding stations were located at distances of 0, 6, 12, and 18 m from the odor source. The researchers then let the kangaroos feed from all of the stations for two hours. At the end of each trial, they collected all of the food containers and measured how much food was left to calculate how much food was eaten from each station. As we saw in Chapter 8, the amount left in each food station is known as the giving-up density, or GUD (Brown 1988), and is a useful measure of predation risk for foraging animals. If animals leave a lot of food in a food patch (a high GUD), that means they did not eat much because they perceived a high level of predation risk. On the other hand, if an animal leaves very little food (a low GUD), the animal ate a lot because it perceived a low level of predation risk.

The researchers found that when human urine or water was the odor selected, the kangaroos ate most of the food at all four stations. Similarly, kangaroos again ate most of the

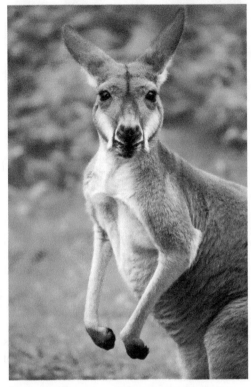

Figure 1. Western gray kangaroo. A species that can be a crop pest in Australia.

food from the stations that were 12 or 18 m away from dingo and coyote urine. However, they ate very little grain (producing a very high GUD) from the feeding stations that were 0 and 6 m from the dingo and 0 m from the coyote odor. In fact, kangaroos ate almost no food from the feeding station that was right next to the dingo odor.

The researchers concluded that the urine of dingos and coyotes was effective in elevating the perceived level of predation risk for gray kangaroos because it greatly reduced the amount of food they consumed when it was nearby (less than or equal to a distance of 6 m). Use of predator urine may be an effective tool to mitigate crop damage. Research still needs to be done to determine whether kangaroos acclimate to the predator scent and thus whether its effectiveness will decline over time. Such nonlethal control of pests is one way that animal behavior research is being applied today. ■

Grant Gentry and Lee Dyer investigated the effectiveness of these antipredator behaviors in response to parasitoid attack in tropical caterpillars (Gentry & Dyer 2002). They collected over 1,600 caterpillars from 69 species in lowland wet tropical forest habitat in Costa Rica. They scored each individual for the types of

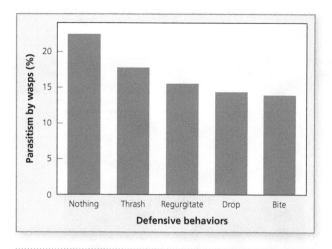

Figure 9.8. Caterpillar defensive behaviors. The percentage of individuals in a species that had been parasitized varied with the types of defensive behaviors exhibited in response to simulated parasite attack.

behavioral defenses it employed during a simulated parasitoid attack, which consisted of lightly brushing a caterpillar with a paintbrush and gently squeezing it with forceps. All caterpillars were then allowed to develop in captivity in order to assess the proportion of individuals that had already been parasitized.

The researchers noted that caterpillar species varied greatly in the number of behavioral defenses they exhibited in response to the simulated attack. Gentry and Dyer found a significant relationship between the types of antipredatory behaviors exhibited by a species and its level of parasitism. Species that had been parasitized the least were those that used aggressive bites and dropped when attacked, while those that exhibited no antipredatory behaviors or only thrashed had high rates of parasitism (Figure 9.8). In this habitat, aggressive defense and dropping when attacked were associated with enhanced survivorship.

Featured Research

Startle display in butterflies

One additional and quite dramatic antipredator behavior seen in some species is a startle display that is typified by abrupt changes in behavior, often involving sudden exposure of conspicuous colors that may be accompanied by a sound (e.g., Schlenoff 1985; Staaterman, Claverie, & Patek 2010). Martin Olofsson and his colleagues examined the effectiveness of such a display in European swallowtail butterflies (*Papilio machaon*) (Figure 9.9) (Olofsson et al. 2012).

Many moths and butterflies rely on cryptic coloration to avoid detection by predators but will suddenly reveal their brightly colored dorsal wing surface when a predator is nearby. The research team conducted an experiment to investigate whether such behavior is an effective startle display that can discourage a common avian predator, the great tit (*Parus major*), from attacking.

The research team captured wild birds and maintained them indoors in individual cages. Butterfly larvae were reared in the laboratory. At eclosion, half the butterflies were euthanized by freezing, while the others were maintained for one week in the laboratory. The experimental trials offered one live and one dead swallowtail to a single bird in an experimental room. Birds were trained to feed on mealworms attached to two small vertical wooden planks, placed 1 m apart, which were mounted on one wall of the room (Figure 9.10). A horizontal board ran along the wall and under both planks to provide a perch from which a bird could feed when at the planks. A 1.5 m oak branch was attached perpendicular to this wall, midway between the feeding planks, and provided an initial perch for the bird to visit the planks.

The butterflies were attached to the wooden planks to appear as if they were resting on a tree trunk. The live butterfly could move its wings to perform a startle display but could not fly from the plank. The experimenters attached a dead mealworm to each plank directly below each butterfly to provide an additional food incentive to the birds. A single bird was then released into the

Figure 9.9. Swallowtail butterfly. This species displays a startle response to bird predators.

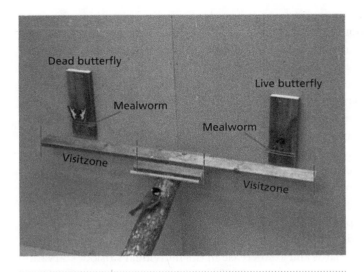

Figure 9.10. Experimental set-up. A bird could attack a live or dead butterfly that was placed above a mealworm food item. (*Source:* Olofsson et al. 2012)

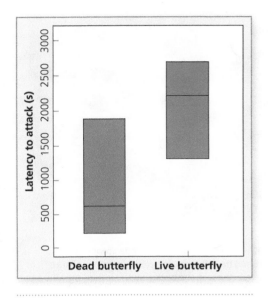

Figure 9.11. Latency to attack. Median (black line) latency to attack live and dead butterflies. Each box represents the middle 50% of the data range.

experimental room for 45 minutes, and its behavior, along with the behavior of the live butterfly, was recorded.

Of the birds that attacked at least one butterfly, significantly more (22 out of 24) attacked the dead butterfly first, and so the latency to attack differed significantly (Figure 9.11). There were a total of 158 visits to the live butterfly, and it responded with a startle display 75% of the time. In 80% of the bird visits that elicited a startle display, the great tit either quickly flew away or hopped back; the birds appeared to be startled and wary to attack the live butterfly. Only three live butterflies were killed during the experiment, suggesting an important antipredator function of the wing-flick startle display by swallowtail butterflies. Additional work is needed to determine why this behavior is such a deterrent to great tits.

Each of these examples illustrates how animals either reduce detection by predators or increase survival once attacked. Next, we examine behavioral trade-offs that reduce predation risk but come at the cost of a reduction in other behaviors.

9.2 Many behaviors represent adaptive trade-offs involving predation risk

A world without predators would allow animals to concentrate their behavior on activities that maximize their fitness, perhaps by increasing time spent searching for mates. In that light, the existence of predators represents a cost to animals: they modify their behavior to reduce the probability that they will be killed, as we saw previously. Sacrificing one activity for another is known as a behavioral trade-off. Such trade-offs are widespread, as we see next.

Increased vigilance decreases feeding time

A widespread trade-off involves **vigilance behavior**, or scanning for predators (Figure 9.12). When many animals search for food, they often lower their head as they focus their attention on finding food items on or in the ground. This head-down

vigilance A behavior in which an individual scans the environment for predators.

Figure 9.12. Vigilance. Flamingos (*Phoenicopterus roseus*) in the foreground have their head up, engaged in vigilance behavior, while those in the background are feeding with their heads down.

position usually results in a reduced visual scanning range. Vegetation, such as grasses or other plants, can also increase the obstruction of an animal's scanning range. Consequently, when an individual has its head down, it experiences a higher risk of predation. To counter this risk, individuals typically raise their head periodically to scan their surroundings for predators. Such vigilance behavior is ubiquitous: you can observe it by watching squirrels, birds, or deer feeding on a lawn.

Because an individual usually cannot simultaneously search the ground for food and scan effectively for predators, vigilance involves a behavioral trade-off. In theory, doubling the amount of time spent scanning will halve an individual's time spent feeding. This will result in both a decrease in the probability of being killed by a predator and an increase in the individual's risk of starvation (McNamara & Houston 1987), illustrating the trade-off. We can predict that animals will adjust their vigilance based on the level of risk in the environment: as predation risk increases, so should vigilance (Brown 1999).

Featured Research

Vigilance and predation risk in elk

One test of this prediction involves elk (*Cervus elaphus*) in Yellowstone National Park. For much of the twentieth century, elk lived in an environment free of wolf (*Canis lupus*) predators, which had been killed by humans. In 1994, wolves were reintroduced into parts of the park, creating areas for elk with and without predators (Figure 9.13). John Laundré, Lucinda Hernández, and Kelly Altendorf examined how elk behavior was affected by the presence of this predator. Over a period of five years, they used focal animal sampling to quantify the time adult elk spent on feeding and vigilance behaviors (Laundré, Hernández, & Altendorf 2001). For each observation, they noted whether the individual was in an area with or without

a

b

Figure 9.13. Wolves in Yellowstone. (a) Wolf pack. (b) Wolf packs occurred in some parts of the park during the study (yellow and orange areas). All other areas were free of wolves. (*Source:* Laundré, Hernández, & Altendorf 2001)

Figure 9.14. Elk behavior. (a) Mean (± SE) vigilance and (b) mean (± SE) feeding behavior of elk. Females with calves (top) and without calves (middle) tended to exhibit higher vigilance in wolf areas (orange) than in nonwolf areas (blue). Males' vigilance level (bottom) did not differ in wolf and nonwolf areas. Females with calves (top) and without calves (middle) tended to spend less time feeding in wolf areas (orange) than in nonwolf areas (blue). Male feeding time (bottom) was not affected by wolves, except in one year. (c) Elk cow and calf. (*Source:* Laundré, Hernández, & Altendorf 2001)

wolves, as well as its sex. Females were further classified as either with or without calves, because females with calves are often more vulnerable to predation.

The presence of wolves clearly affected elk behavior. Females, especially those with calves, spent significantly more time on vigilance behaviors and less time feeding when they were in areas with wolves (Figure 9.14). In general, the presence of wolves had much less of an effect on males. The researchers suggest that reproductive success for males largely depends on maintaining high body mass, and so males have more to lose by reducing feeding time, even if the consequence is higher predation risk.

Rich but risky

In most habitats, there are risky and safe locations. In addition, in any environment, food patches are heterogeneous: some habitats or food patches contain abundant food, and others contain less food. Let's call patches with abundant food *rich patches* and those with less food *poor patches*. Animals obviously should prefer to feed in rich patches because they can obtain higher energy intake rates there. However, many rich food patches also are found in places with higher numbers of predators or higher predator attack rates. That means that in order to feed in a rich patch, individuals must be willing to subject themselves to higher predation risk. We can therefore ask two research questions: How often will animals be willing to accept higher predation risk in order to feed in richer food patches, and what factors might favor animals feeding in richer but riskier food patches? The next two examples illustrate how foragers balance predation risk and food acquisition when rich but risky places to feed are available.

Featured Research

Environmental conditions and predation risk in foraging redshanks

Redshanks (*Tringa totanus*) are medium-sized (approximately 28 cm in length) wading birds in the sandpiper family that live throughout Eurasia. They have a long, straight bill, which they use to find insects, earthworms, mollusks, and crustaceans that are buried in the soil. Redshanks can be found feeding in a variety of habitats, including salt marshes and mudflats near the coast.

Mai Yasué and colleagues (Yasué, Quinn, & Cresswell 2003) examined a trade-off between predation risk and foraging in a population of these birds (Scientific Process 9.1). In particular, they studied how weather conditions and predation risk affected daily feeding decisions during the winter in Scotland. The main predators in this area are sparrowhawks (*Accipter nisus*). The researchers first observed that predators were much more common in the salt marsh habitat than the mudflat habitat, with 20 times more predator attacks on redshanks occurring when they fed in the salt marsh habitat compared to the mudflat habitat. Yasué and colleagues concluded that the salt marsh habitat was riskier for a redshank than the mudflat habitat. Given this level of risk, one might conclude that redshanks should always feed in the mudflat habitat if they want to avoid predators. However, Yasué and colleagues observed that on some winter days, redshanks seemed to feed more often in the riskier salt marsh habitat. Why might redshanks feed more often in salt marshes during the winter?

Birds, like mammals, are endotherms. That means that they generate their own internal body heat, and when it is cold outside, their body temperature is much higher than the ambient temperature. To maintain their high internal body temperature, birds (and mammals) need to consume more food as the temperature drops. This characteristic led Yasué and colleagues to hypothesize that energetic requirements and predation risk affect redshank feeding behavior. That is, when the outside temperature is low, birds cannot find sufficient food by feeding only in the safer mudflat habitat. This hypothesis predicted that the salt marsh habitat contains more food than the mudflat habitat, and that ambient temperature affects where redshanks will feed. On the coldest days of the year, redshanks will prefer to feed in salt marsh habitats.

To test their first prediction, the researchers measured the amount of food available in both the salt marsh and the mudflat habitats every two weeks over the course of the winter. They found that there were roughly twice as many invertebrates in each sample from the salt marsh habitat as there were in those from the mudflat habitat throughout the winter, which supported their first prediction. To test the second prediction, they recorded the number of redshanks that fed in each of the two habitats each day over the course of the winter, as well as the temperature on those observation days. As predicted, environmental temperature

SCIENTIFIC PROCESS 9.1

Feeding trade-off in redshanks

RESEARCH QUESTION: *Why do redshanks sometimes feed in the riskier salt marsh habitat instead of the safer mudflat habitat?*

Hypothesis: Energetic requirements and predation risk affect feeding behavior.

Prediction (1): The salt marsh habitat contains more food than the mudflat habitat.

Prediction (2): As the temperature drops, redshanks will feed more often in the salt marsh habitat to meet their energy needs.

Methods (1): The researchers:

- Collected samples from 2 to 4 cm (depth of bill probe) of the substrate in small plots ($n = 47$) and then counted and identified all invertebrates found there

Results (1): There were twice as many invertebrates per sample in the salt marsh habitat as there were in the mudflat habitat.

Methods (2): The researchers:

- Recorded the number of redshanks feeding in the salt marsh within three hours of low and high tide and the ambient temperature over the course of 91 days.

Results (2): As ambient temperature decreased, more redshanks fed in the salt marsh habitat.

Figure 1. Food abundance. The salt marsh habitat contains more invertebrate food than the mudflat habitat. (*Source:* Yasué, Quinn, & Cresswell 2003)

Figure 2. Redshank behavior. More redshanks fed in the salt marsh habitat when the temperature was low. (*Source:* Yasué, Quinn, & Cresswell 2003)

Conclusions (1): There is twice as much food available for redshanks in the salt marsh habitat as there is in the mudflat habitat.

Conclusions (2): When it is cold, redshanks feed in the risky salt marsh habitat because it contains more food than the mudflat habitat. Feeding there should thus result in higher energy intake rates to meet these individuals' energetic requirements.

> **EVALUATE**
> Describe the adaptive trade-off involving predation risk in this study.

was strongly correlated with the behavior of the birds: the number of redshanks feeding in the salt marsh habitat increased significantly as the temperature declined. Thus, on colder days, most birds fed predominately in salt marshes, where they could obtain more food, but at a cost of suffering higher predator attack rates. From these data, the researchers concluded that the behavior of redshanks was consistent with the hypothesis that individuals trade off higher predation risk in the salt marsh for the higher energy intake rates they obtain there.

This type of foraging–predation risk trade-off is widespread across taxa. While the previous examples examined vertebrates, invertebrates too behave in ways that trade off predation risk and food rewards.

Predation risk and patch quality in ants

Peter Nonacs has studied predation risk–foraging trade-offs in *Lasius pallitarsis*, a common ant in western North America. As in the study on redshanks, Nonacs asked whether ants will trade off higher predation risk to feed in a richer food patch.

An important predator of *Lasius* ants is another ant, *Formica subnuda*, which is a large predatory species. In order to demonstrate that *Lasius* ants respond to the threat of predation by *Formica* ants, Nonacs set up a simple experiment with Larry Dill (Nonacs & Dill 1988). The researchers offered a *Lasius* colony two food patches that contained identical solutions of sugar, proteins, vitamins, and nutrients, a preferred ant food. The two food patches were identical distances away from a colony, but *Lasius* workers had to travel through a small arena that contained a *Formica* predator in order to obtain food from one of the patches. In the control patch, *Lasius* ants did not need to encounter a *Formica* predator to obtain food. Thus, the patches contained identical food, but the control patch was safe (no predators), while the other, the experimental treatment, was risky (*Formica* predator present). Nonacs and Dill found that after some *Lasius* workers were killed, the rest would avoid the risky patch and obtain food only from the safe patch. From this experiment, they concluded that *Lasius* ants prefer to avoid predators.

Once they had demonstrated that *Lasius* ants prefer to avoid *Formica* predators, Nonacs and Dill designed another experiment. They wished to see if these ants would exhibit a behavioral trade-off between predation risk and food rewards. To examine this question, they offered the *Lasius* ants two food patches that differed in both predation risk and food quality (Nonacs & Dill 1990). One food patch was both risky (predator present) and rich with food, and the other was safe (no predator present) but poor in food quality. Both food patches again contained a solution of sugar, protein, vitamins, and nutrients and were placed equidistant from a colony. However, the two food patches now differed in their concentration of food: the rich food patch had a higher concentration of food than the poor food patch.

In this experiment, the *Lasius* ants always had to travel through an arena that contained a *Formica* predator if they wanted to feed from the rich patch. Thus, to feed from the richer patch, workers were subject to higher rates of predation, making it the riskier food patch. Over the course of the experiment, Nonacs and Dill varied the relative difference in quality between the two food patches. Because the food patches offered a food solution, the researchers could manipulate the quality of a food patch by altering the concentration of the food. The safe food patch always had the lowest concentration of food, while the richer food patch contained a concentration of food that was two to 16 times higher than that found in the safe patch.

You can see that the greater the difference in concentration, the greater the potential rewards for feeding at the rich patch would be. Nonacs and Dill predicted that ants would be more willing to feed at the riskier food patch as the difference in quality between the two patches increased. In other words, the ants should be more willing to pay higher predation risk costs (and suffer higher rates of mortality) in order to obtain higher-quality food.

TABLE 9.1 Ant food patch choice. As the relative difference in food concentration increases, ants feed more in the rich, risky food patch (*Source:* Nonacs & Dill 1990).

RELATIVE DIFFERENCE IN CONCENTRATION BETWEEN RICH AND POOR PATCHES	MEAN PERCENT OF FOOD TAKEN FROM RICH, RISKY PATCH
2 times richer	35
4 times richer	39
8 times richer	47
16 times richer	65

Nonacs and Dill recorded the amount of food taken from each patch during a trial and the number of *Lasius* workers that were killed by the *Formica* predator near the richer food patch. The use of the richer patch did increase as the relative difference between patches increased, as predicted (Table 9.1). For instance, when there was only a twofold difference in patch quality between the rich and the poor patches, *Lasius* ants obtained only about 35% of their food from the rich patch. However, when the rich patch was 16 times better than the poor patch, *Lasius* ants obtained approximately 65% of their food from it. Furthermore, feeding from the rich patch was indeed risky for the ants: over 140 *Lasius* workers were killed near the rich patch by *Formica* predators during the experiment. Nonacs and Dill concluded that the ants were using information about both the quality of the food patches and the risk of predation in their foraging decisions. *Lasius* ants were willing to suffer higher predation rates to feed from richer food patches.

Mating near predators in water striders

Featured Research

As these examples show, behavioral trade-offs between predation risk and food rewards can be very sophisticated and are found in diverse taxa. However, predation risk trade-offs are not restricted only to feeding behavior. Predation risk also affects mating behaviors and results in different behavioral trade-offs.

Water striders (*Gerris remigis*) are insects in the order Hemiptera, which also includes cicadas and aphids. Water striders live on the surface of streams and ponds in North America and can walk or "skate" over the top of the water by trapping a layer of air between their feet and the water's surface. They feed on smaller aquatic insects by capturing individuals that are trapped on the surface of the water. During the spring and summer, males mate by attaching themselves to the backs of females. Water striders are eaten by sunfish (*Lepomis cyanellus*) and other fish, but they coexist with their predators. They reduce their probability of being killed both by reducing their overall activity level and by moving to the edge of a pond or stream, where the water is too shallow for fish predators.

Andy Sih and colleagues have been studying the behavior of water striders for many years. In one classic study, they examined the effects of predators on water strider mating behavior (Sih, Krupa, & Travers 1990). The researchers asked a basic research question that had rarely been examined: Do animals reduce their mating behavior in the presence of a predator? They predicted that water striders would exhibit a predation risk trade-off: they would reduce mating behavior when predation risk was high but engage in high levels of mating behavior when predation risk was low.

SCIENTIFIC PROCESS 9.2

Mating behavior trade-off in water striders

RESEARCH QUESTION: *How does the presence of a predator affect the mating behavior of water striders?*

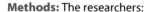

Hypothesis: Water striders modify their behavior in response to variation in predation risk to avoid being killed.

Prediction: Water striders will reduce their mating behavior in the presence of a predator.

Methods: The researchers:

- Created artificial pools in the lab: treatment pools contained predators, and the control pools did not
- Marked individual sunfish predators (three in each predator treatment pool) and water striders (four males and four females in each pool) for identification
- Observed water strider activity every 30 minutes for six hours per trial (a total of 12 observations for each treatment)

a Predator present   **b Predator absent**

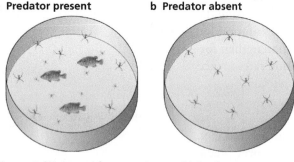

Figure 1. Water strider experiment. (a) Predator present and (b) predator absent. (*Source:* Sih, Krupa, & Travers 1990)

Results: In the pools with no fish (controls), water striders exhibited high levels of overall activity, with mating observed about 20% of the time. In the pools with fish (treatment), water strider activity was greatly reduced, and mating was observed only 5% of the time.

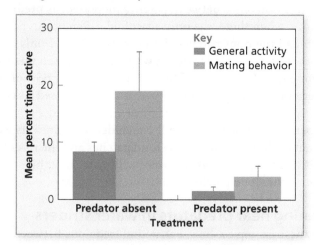

Figure 2. Water strider behavior. Mean (+ SE) activity and mating behavior were more common in the absence of predators. (*Source:* Sih, Krupa, & Travers 1990)

Conclusions: Predation risk can affect the mating behavior of water striders. There is a behavioral trade-off between mating behavior and predation risk.

EVALUATE

Predict how water strider activity level and mating behavior would change with an increase in the number of fish predators.

To test this prediction, they created a set of four artificial wading pools in the laboratory. Water striders were individually marked and then randomly assigned to two treatments: in one, fish predators were present, and in the other, they were absent. During the study, the researchers collected data on the location of all water striders (in the center or at the edge of a pool), the time spent active, and whether the insects were engaging in mating behavior (Scientific Process 9.2).

Sih and colleagues found that the behavior of male and female water striders was affected strongly by the presence of predators. In the pools with no fish predators, water striders exhibited high levels of activity. Individuals were commonly observed in the center of the pool and were seen mating about 20% of the time. In the pools with predatory fish, water strider activity was much lower. Individuals were rarely seen in the center of the pool and were observed mating less than 5%

of the time. Based on their results, Sih and colleagues concluded that water striders do reduce their mating behavior in the presence of predators. In addition to an overall reduction in mating behavior, the researchers also found that the duration of each mating bout was shorter in the presence of predators, perhaps because a mated pair is less mobile (and so more vulnerable) if attacked.

These experiments were among the first to demonstrate a behavioral trade-off involving predation risk and the mating behavior of animals. When predation risk is low (no predators present), water striders are active and engage in much mating behavior. However, when predation risk is high (predators present), water striders reduce both their overall level of activity and their level of mating behavior. We now know that such trade-offs are quite common. Let's examine a recent study on fiddler crabs as another example.

Mating and refuge use in fiddler crabs

Featured Research

Fiddler crabs (*Uca mjoebergi*) live in intertidal habitats and feed on the surface of the sand during low tide. The species gets its name from the morphology of the males, which possess a single large claw that they wave back and forth in a sexual display to attract females (Figure 9.15). Individuals construct and live in burrows in the sand, which provide a safe refuge from bird, mammal, and reptile predators. To feed or attract mates, crabs need to be active aboveground, but they will quickly retreat to their burrow in response to a predator attack.

Leeann Reaney examined how predators and mating opportunities affect refuge use by male fiddler crabs (Reaney 2007). The benefit of refuge use is that it provides protection from predators. The cost is that a male inside a burrow cannot court a female. Reaney hypothesized that male crabs would adjust their refuge-use behavior based on its costs and benefits. Specifically, she predicted that males would emerge from their refuges more quickly after an attack (and thus expose themselves to risk) if potential mating opportunities were available aboveground.

To test this prediction, Reaney created 30 small plots (35 cm × 35 cm) in a coastal intertidal mudflat in Darwin, Australia. Each plot contained several males and their burrows. In 15 of these plots, she tethered a single, sexually receptive female by gluing a thin piece of string onto its carapace and attaching it to a nail buried in the sand. The other 15 plots served as controls and did not contain a tethered female. Reaney first waited for all the males in a plot to become active aboveground. She then flew an artificial model of a bird predator over the plot and recorded how long the males stayed in their burrows after the "attack." She averaged the measurements for all males within each plot and compared the average values for males in each treatment.

Reaney found that all males in the control plots retreated into their burrows and stayed underground for about 120 seconds. In contrast, in the plots with a receptive female, 12% of males did not retreat into their burrow during the attack. Those males that did flee into their burrow reemerged in less than 15 seconds (Figure 9.16). From these results, Reaney concluded that males adaptively adjust their refuge use based on its benefits and costs. Males trade off the safety of the burrow for increased mating opportunities when receptive females are present. When presented with no mating opportunities, males use their burrows much more.

Figure 9.15. Fiddler crabs. Males use burrows in the sand for protection from predators. They wave their large claw to attract females.

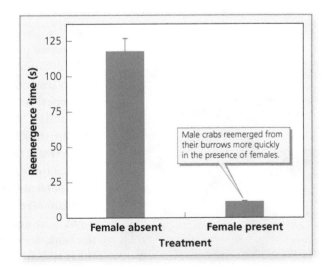

Male crabs reemerged from their burrows more quickly in the presence of females.

Figure 9.16. Male reemergence times. Mean (+ SE) reemergence times of courting male fiddler crabs after a predator attack in plots with and without females present. (*Source:* Reaney 2007)

These examples demonstrate how individuals alter their behavior to reduce predation risk. They also show how this risk reduction comes at the cost of reduced feeding or mating activity. In the next section, we'll look at another common way for individuals to reduce predation risk: association with others.

9.3   Living in groups can reduce predation risk

Many animals live in **social groups**, meaning they live near and interact with other individuals. One important benefit of this behavior for social animals is that living in groups can reduce predation risk. In Chapter 15, we examine social behavior in more detail. Here, we focus on some of the ways that social animals experience lower levels of predation risk than solitary individuals.

social group A set of individuals that live near and associate with one another.

Featured Research

The dilution effect and killifish

Suppose a predator attacks a group of two animals and kills one. For each prey, there was a one-half probability that it would be the one killed. If instead there had been ten prey, there would have been only a one-tenth probability that any one individual would be killed. In a group of 100 individuals, that probability would fall to one-hundredth. The probability of any one individual's dying is diluted by the presence of others. This is called the **dilution effect** (Foster & Treherne 1981).

dilution effect A reduction in the probability of death as a result of associating with others.

In the dilution effect, there is a $1/N$ probability of dying when a predator makes a single kill in a group of N individuals. As group size N increases, the probability that any single individual dies decreases. Clearly, this size increase can greatly reduce an individual's probability of dying as a result of a predator attack, especially in a group of hundreds or thousands of individuals. The only behavior required for the dilution effect to occur is for an individual to join others in a group.

D. J. Hoare and colleagues (Hoare et al. 2004) examined the dilution effect in a simple laboratory experiment using banded killifish (*Fundulus diaphanus*). They predicted that fish should prefer to associate with larger groups rather than smaller groups, particularly when predation risk is high. Banded killifish are small (less than 10 cm standard length) freshwater fish that live in eastern North America. These fish tend to move in schools (or shoals) and can use chemical cues to perceive predation risk. The researchers compared the size of the group that killifish formed in a high–predation risk treatment with the size formed in a control treatment.

Experiments were conducted in a large tank (100 cm × 100 cm × 15 cm). To simulate high predation risk, the researchers added a diluted concentration of killifish skin extract to the tank. Many species of fish respond to the crushed skin extract of conspecifics (individuals of the same species), presumably because it indicates the presence of a predator that has made a kill nearby. In the control treatment, only water was added to the aquarium. For each trial, ten killifish were placed in the experimental tank and allowed to acclimate overnight. The next day, skin extract (treatment) or water (control) was added to the tank, and a 60-minute trial was videotaped. From the videotape footage, the researchers recorded shoal size every 30 seconds. Individuals were considered to be in the same shoal if they were within four body lengths of one another.

The results of the experiment were clear: median shoal size in the control was just two fish, with most fish swimming alone in the aquarium. However, in the

Figure 9.17. Selfish herd. Turkeys (*Meleagris gallopavo*) on the exterior of the flock are more likely to be killed by a predator than are individuals in the center.

high–predation risk treatment, median shoal size was ten fish, the maximum group size. From these results, Hoare and colleagues concluded that banded killifish prefer to associate with more individuals when predation risk is high, as predicted by the dilution effect.

The selfish herd and vigilance behavior

A second way social groups can lower predation risk is for individuals to stay near the group's center. The **selfish herd hypothesis** (Hamilton 1971) assumes that a predator is more likely to kill a member on the outside of a group because it will encounter outside individuals first (Figure 9.17). Individuals can therefore lower their predation risk by placing others between themselves and an approaching predator. In contrast to the dilution effect, the selfish herd hypothesis requires individuals to continually adjust their position in a group.

A third antipredator benefit of associating with others involves vigilance behavior. In social groups, more individuals can be vigilant, and hence "more eyes" can observe potential predators. Many researchers have documented a common pattern known as the **group size effect** (Pulliam 1973), in which the vigilance behavior of each individual decreases as the number of individuals in its group increases (Figure 9.18). The group size effect has been observed in a wide variety of birds, mammals, and even fish, and it is thought to result from two factors. First, as group size increases, each individual is safer because of the dilution effect, and so each individual can reduce its level of vigilance. Second, as group size increases, each individual can afford to scan less frequently because more individuals are scanning. Therefore, the collective vigilance of the entire group can remain high even though each individual scans less often. If one individual in the group sees a predator, it will quickly flee to protective cover. That motion alerts other group members to the imminent danger so that they too can respond appropriately. Next we examine a study of doves that investigated both the group size effect and the selfish herd hypothesis.

Group size effect and the selfish herd hypothesis in doves

Scaled doves (*Columbina squammata*) are small birds that are widely distributed throughout Latin America (Figure 9.19). Raphael Dias studied dove scanning and foraging behavior on the University of Brasília campus (Dias 2006). He investigated two research questions. First, he wondered whether scaled doves exhibit the group size effect. If so, he predicted that they should exhibit reduced scanning rates as their group size increases. Second, he examined whether doves at the edge of a flock are exposed to higher predation risk, as assumed in the selfish herd hypothesis. If so, he predicted they should exhibit higher scanning rates than individuals at the center of a flock.

Dias studied wild flocks all over the campus for two months. When he found a flock of feeding birds, he recorded the group size and then selected one focal animal. He recorded the location of the focal bird in the flock (central or edge position), its scan rate (the number of times it lifted its head to scan the environment over a five-minute period), scan durations, and the total amount of time it spent feeding. He also made note of any position shifts in the focal bird (i.e., any change from a central to an edge position or vice versa). He then moved to a different location on campus to find another flock. In total, he observed 150 flocks that varied in size from one to 19 individuals.

selfish herd hypothesis The hypothesis that a predator is more likely to kill a member on the outside of a group because it will encounter outside individuals first.

group size effect A commonly observed pattern in which the vigilance behavior of an individual declines as group size increases.

Figure 9.18. Group size and vigilance. The relationship between group size and individual vigilance.

Featured Research

Figure 9.19. Scaled dove. These small doves are common throughout Latin America.

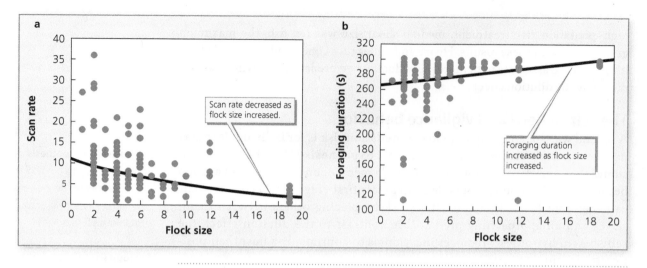

Figure 9.20. Group size effect in doves. The relationship between (a) scan rates and (b) foraging duration for different flock sizes. (*Source:* Dias 2006)

Dias found that individual vigilance behavior decreased significantly as group size increased, indicating that these doves did indeed exhibit the group size effect. In addition, the feeding time of individuals increased as group size increased (Figure 9.20). Furthermore, individuals in the center of a flock behaved differently than did individuals at the edge. Birds in the center exhibited lower scan rates and had higher feeding rates than individuals at the edge of a flock (Figure 9.21). When position shifts did occur, individuals more often moved from

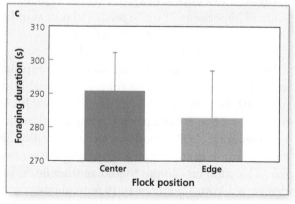

Figure 9.21. Center and edge dove data. Mean (+ SE) (a) scan rate, (b) vigilance, and (c) foraging duration for birds in the center and at the edge of flocks. Doves in the center scanned less, were less vigilant, and also foraged more than doves on the edge. (*Source:* Dias 2006)

the edge of the flock to the center—that is, they exhibited behavior consistent with the selfish herd hypothesis.

Dias concluded that doves exhibit a trade-off between scanning for predators and feeding: scan rate declines and feeding time increases strongly with group size. In addition, the data support an important assumption behind the selfish herd hypothesis: individuals at the edge of a group appear to experience a higher threat of predation than do individuals near the center of a group.

If there are benefits to being near to the center of a group (lower predation risk, higher feeding times), though, then why don't all birds try to move to the center? Dias suggests that a dominance hierarchy maintains group position: dominant individuals most likely can maintain their central position, while subordinate individuals are forced to stay in more peripheral positions (see Chapter 11).

These examples illustrate how individuals in groups can benefit from reduced predation risk. In all of these examples, individuals behave to avoid predators. In the last section, we examine how selection can favor a counterintuitive antipredator behavior: the direct interaction of individuals with predators.

9.4 Some animals interact with predators to deter attack

Predators cannot be entirely avoided. When an encounter occurs, many individuals flee to a safe location, but in some situations, a potential victim does *not* flee. Instead, it actively interacts with a predator. For example, when smaller songbirds observe a hawk or owl, they will often engage in **predator harassment**, behavior that involves rapid movement such as diving at or around a predator and is often coupled with loud vocalizations. Some species, like chimpanzees, will throw objects at a predator. Others, like ground squirrels and kangaroo rats, will kick dirt at a predator. The harassment of a predator by more than one individual is known as **mobbing behavior**.

On the surface, predator harassment behavior appears to increase the likelihood that an individual will be killed, because it means direct interaction with a predator within a short distance. So why would an individual harass a predator? One possible answer is that such behavior can cause a predator to become more defensive, reducing its likelihood of attack. Second, harassment could simply cause the predator to move away. Still other species will engage in behaviors that appear to enhance or "advertise" their presence to a predator. Advertisement behavior might benefit an individual by conveying to a predator that it has lost the element of surprise. Let's see how researchers have tested each of these explanations.

predator harassment Interactions with a predator to deter attack. Such behavior can involve rapid movements, loud vocalizations, and throwing or kicking objects at the predator.

mobbing behavior The harassment of a predator by two or more individuals.

Predator harassment in ground squirrels

Aaron Rundus and colleagues examined interactions between California ground squirrels (*Spermophilus beecheyi*) and rattlesnake predators. Adult squirrels have evolved several behavioral and physiological adaptations to defend their pups from predation by Pacific rattlesnakes (*Crotalus oreganus*). Adults are not susceptible to rattlesnake venom, but pups are vulnerable, and so adults often confront and harass snakes to deter their hunting behavior (Rundus et al. 2007). Squirrels will harass snakes by making rapid movements, engaging in vigorous tail

Featured Research

SCIENTIFIC PROCESS 9.3

Predator harassment by California ground squirrels

RESEARCH QUESTION: *How do California ground squirrels harass rattlesnakes?*

Hypothesis: Ground squirrels use an infrared signal during rattlesnake harassment by increasing blood flow to their extremities.

Prediction: When attacked by a rattlesnake, a ground squirrel will shunt more blood to its tail than it will when being attacked by a gopher snake (*Pituophis melanoleucus*).

Methods: The researchers:

- Conducted four types of randomly ordered trials using 12 adult female squirrels: (1) a baseline trial with no stimuli, (2) a control trial with a conspecific (squirrel) stimuli, (3) an experimental trial with a northern Pacific rattlesnake stimuli, and (4) an experimental trial with a gopher snake stimuli
- Placed a squirrel's home cage next to the opening of the test chamber. Trials began after the squirrel engaged the stimulus animal in the test chamber. Baseline and control trials were compared to experimental trials. Ten-minute trials were spaced two days apart.
- Used infrared video to collect data on squirrel body and tail temperature

Results: The temperature of a ground squirrel's tail was, on average, 2°C higher when it harassed a rattlesnake than when it harassed a gopher snake.

Figure 1. Ground squirrel tail temperature. Note the squirrel's higher tail temperature (indicated by a red tail in the infrared image) when harassing (a) a rattlesnake compared with (b) a gopher snake. (*Source:* Rundus et al. 2007)

Conclusion: California ground squirrels include an infrared component in their harassment of rattlesnakes, presumably by increasing blood flow to the tail, which results in a slowing of the snake attack.

EVALUATE

Why don't ground squirrels increase their body temperature rather than just their tail temperature as an antipredator signal?

flagging (a rapid wagging of the tail), and even kicking dirt and rocks at them. Rattlesnakes are pit vipers and have heat-sensitive pit organs that can perceive infrared radiation (Chapter 5). Other snakes, such as gopher snakes, are insensitive to infrared radiation. Rundus and colleagues investigated whether ground squirrels' harassment of rattlesnakes also had an infrared component and whether such harassment caused snakes to become more defensive (Scientific Process 9.3).

The researchers staged encounters between adult ground squirrels and different stimuli (treatments). One stimulus was an adult Pacific rattlesnake, another was an adult gopher snake, and a third was a control (another ground squirrel).

Each stimulus was housed in a wire mesh cage that prevented direct contact but allowed the passage of visual, auditory, olfactory, and infrared information. All trials were videotaped with an infrared-sensitive video camera, which allowed the researchers to quantify the heat given off by the ground squirrel from different parts of its body during an encounter. The researchers also measured the temperature of both the body and the tail of the ground squirrels.

During encounters with both species of snakes, all ground squirrels exhibited harassment behavior, particularly tail flagging. No harassment behavior was observed in interactions with another ground squirrel. Interestingly, analysis of the videotapes also revealed that the ground squirrel tail temperature was much higher during encounters with the rattlesnake compared to encounters with the gopher snake. Ground squirrels appear to add an infrared component to their harassment behavior, but only when they interact with rattlesnakes that can detect such signals.

How effective is this harassment behavior? Does it indeed put predators on the defensive? To determine how the infrared signal might affect rattlesnakes, another experiment used robotic ground squirrel models based on lifelike taxidermy mounts. The researchers could manipulate the models to exhibit tail-flagging behavior and could alter the temperature of the tail. Rundus and colleagues then trained adult Pacific rattlesnakes to actively search for food in a simulated ground squirrel burrow they created in the laboratory. This set-up required the snakes to travel over 1 m to enter the burrow. After the snakes were trained, a model squirrel was placed next to the burrow entrance, and a snake was introduced into the experimental arena.

The researchers exposed each snake to two treatments. In one, the model ground squirrel tail-flagged with a cold tail (thus producing no infrared signal). In the other, the model tail-flagged with a warm tail (thus producing an infrared signal in the tail). The team found that when a snake interacted with a model that gave off an infrared signal, the snake exhibited more defensive behaviors (e.g., defensive coiling) and moved less (reduced search behavior). This experiment, examining a proximate antipredator mechanism, shows that adding an infrared component to predator harassment behavior is an effective strategy for hindering a snake predator. Why do rattlesnakes exhibit more defensive behaviors in response to the infrared signal? Squirrels often combine tail flagging with substrate throwing, and so snakes may have anticipated additional harassment. Rattlesnakes also often hunt at dusk, when it might be difficult to observe the tail-flagging behavior. The addition of the infrared signal by squirrels may enhance tail-flagging behavior in such low-light conditions. However, this hypothesis remains to be tested.

Mobbing owl predators

A second explanation for predator harassment behavior is that it can cause a predator to move away from an area, the **move-on hypothesis** (Curio, Ernst, & Vieth 1978). Here, the benefit of mobbing is reduced predation risk, because the predator moves away from the mobbers or their offspring. Chris Pavey and Anita Smyth tested this prediction by studying the mobbing behavior of forest birds in response to the presence of powerful owls (*Ninox strenua*) (Pavey & Smyth 1998).

The powerful owl is a large (approximately 1,500 g) nocturnal predator that can capture diurnal birds at dawn and dusk, as well as individuals on their nocturnal roost. The researchers studied the mobbing behavior of birds toward owls at a

Featured Research

move-on hypothesis The hypothesis that predator harassment functions to cause a predator to abandon an area.

Figure 9.22. Owl response to mobbing. On 19 occasions, owls being mobbed mostly watched their mobbers and vocalized. On seven occasions, owls moved to a new perch. On seven other occasions, owls did not respond. Inset: powerful owl. (*Source:* Pavey & Smyth 1998)

Figure 9.23. Advertisement behavior. (a) Impala stotting. (b) Zebra-tailed lizard.

stotting Jumping high into the air with all four feet off the ground and moving forward.

pursuit-deterrence hypothesis The hypothesis that advertisement behavior informs a predator that it has been detected and so pursuit is not likely to be successful.

study site in Brisbane Forest Park, Australia, that comprised both rainforest (12% of the habitat) and open forest (88% of the habitat). Pavey and Smyth noted the location of all roosting owls and all mobbing bouts by birds, as well as the response of the owl to such bouts. To estimate the antipredator benefits of mobbing, the researchers compared the frequency of capture of different species of birds by analyzing owl pellets under roosts to determine the avian species in the owl diets.

Thirty-five mobbing bouts involving seven species and up to 22 individuals were observed. Six of the seven mobbing species resided in open forest. Owls were observed roosting approximately half the time in each habitat, even though the rainforest habitat was relatively sparse. Seven bouts occurred in which the owl did not call or move in response to mobbing. Owls called or watched the mobbers during 54% of the bouts and moved to another perch during 20% of the bouts (Figure 9.22).

The 39 avian prey identified in owl pellets ranged in size from 75 to 800 g. To compare the frequency of capture of mobbing and nonmobbing species, the researchers examined the frequency of capture of the six mobbing and six nonmobbing species that fell in this range. The frequency of predation on mobbing species was nearly an entire order of magnitude lower than that for nonmobbing species. These data suggest a strong association between mobbing and reduced predation. However, five of the six mobbing species were passerines (songbirds), whereas only two of the six nonmobbing species were songbirds. Therefore, another explanation for this pattern is that owls prefer to feed on nonpasserines. The researchers rejected this alternative explanation because previous work indicated that owls feed on passerines almost eight times more often than they feed on nonpasserines (Pavey, Smyth, & Mathieson 1994).

The researchers concluded that the mobbing of powerful owls can lead to the displacement of a predator, as predicted by the move-on hypothesis. They suggest that mobbers (which mostly live in open forest) suffer lower predation than nonmobbers, perhaps because mobbing influences the owls' selection of roosting sites.

Pursuit deterrence and alarm signal hypotheses

Predator harassment is one antipredator behavior exhibited by animals that have spotted a predator. Another involves behaving in a manner that appears to make an individual more obvious to potential predators. For example, after spotting a predator, Thomson's gazelles (*Eudorcas thomsoni*) and impala (*Aepyceros melampus*) will **stot**, a behavior that resembles pogo stick–hopping as they move forward, while zebra-tailed lizards (*Callisaurus draconoides*) will raise their long black-and-white tail and wag it back and forth (Figure 9.23). You may have seen such "advertisement" behavior if you have ever startled a white-tailed deer (*Odocoileus virginianus*) while walking in the woods. Many times, when deer spot a predator (including humans), they lift and wag their tail, revealing the white underside, as they move a short distance away.

At first, these behaviors seem to be at odds with behaving to minimize exposure to predators. Why would a prey advertise itself to a predator? One possible answer is the **pursuit-deterrence hypothesis**: advertisement behavior informs the predator that it has been detected and so pursuit is not likely to be successful. Most predators rely on surprise to successfully capture their prey, as it allows them to close in rapidly before the victim has a chance to flee. If predators know that a prey has already spotted them, the likelihood of successful capture is very low, particularly when the distance between the predator and the prey is large. The predator is then better off searching for another victim.

The pursuit-deterrence hypothesis predicts that prey animals like deer that are far from a predator should exhibit tail-flagging behavior more often than individuals that are close to a predator. Why? Because when an individual is far from a predator, it is relatively safe, and so it can benefit by deterring an attack that has little probability of success. However, if the predator is nearby when spotted, the risk of capture is much higher. It is then unlikely that the predator will not initiate an attack, and so the individual should not advertise itself to the predator.

An alternative explanation for advertisement behavior is that individuals are warning nearby conspecifics, the **alarm signal hypothesis**. This hypothesis makes different predictions than the pursuit-deterrence hypothesis. For example, if the function of tail flagging is to signal others, then tail-flagging behavior should occur very rarely among solitary prey, because there are no other individuals nearby to warn. In contrast, tail-flagging behavior should be commonly seen in groups of prey because of the presence of other individuals nearby. Which hypothesis best explains tail-flagging behavior? White-tailed deer have offered an interesting test case.

Tail-flagging behavior in deer

Featured Research

It is difficult to test the pursuit-deterrence or alarm signal hypothesis using natural predators, because few predator–prey encounters are observed. Most tests of these hypotheses therefore use humans as potential predators. Tim Caro and colleagues (Caro 1995; Caro et al. 1995) examined tail flagging in white-tailed deer in a reserve in southeastern Michigan. The deer population studied was hunted regularly and so treated humans as a threat. Caro tested the predictions of both hypotheses. The alarm signal hypothesis predicts that solitary deer should exhibit tail flagging less than deer in social groups, and the pursuit-deterrence hypothesis predicts that tail-flagging behavior should occur more often as the distance between the predator and the deer increases.

Caro and colleagues recorded deer behavior over the course of a month in late fall. They themselves acted as predators by slowly walking through the reserve. Once they spotted deer, they recorded the number of deer nearby and whether any were aware of their presence. This part was easy, because a deer aware of an observer would stand upright and stare directly at the observer. The researchers then selected a focal animal, took note of its location, and moved slowly toward it as if stalking a prey. During the stalk, the observer would record the number of times the deer exhibited tail-flagging behavior and the distance between the observer and the deer. Once the deer fled, the observer would record its flight distance, flight time, and any tail-flagging behavior.

The researchers found no difference in the incidence of tail-flagging behavior in solitary deer and deer in social groups: both types of deer tail-flagged approximately 85% of the time. The researchers therefore rejected the alarm signal hypothesis. They also found that deer more often exhibited tail flagging as their distance from the human increased, as predicted by the pursuit-deterrence hypothesis (Figure 9.24).

The researchers also made a new discovery. When deer fled into an open area, they continued tail flagging. However, deer that fled into dense woods would often stop tail flagging and become much more difficult to observe. These observations led Caro and colleagues to formulate a new hypothesis. Their **flash disappearance hypothesis** states that the bright white of the underside of the tail gives

alarm signal hypothesis The hypothesis that advertisement behavior functions to warn nearby conspecifics of danger.

flash disappearance hypothesis The hypothesis that the sudden disappearance of the bright white underside of a deer's tail (a target easily observed by predators) against the dark background of dense woods makes it difficult for the predator to locate its prey and will encourage it to give up its attack.

Figure 9.24. Deer tail flagging. (a) White-tailed deer. (b) Percentage of deer that tail-flagged as a function of distance to predator. More deer tail-flagged as the distance to the predator increased. (*Source:* Caro et al. 1995)

the predator an easily observed target. The sudden disappearance of this target against the dark background of dense woods, they predicted, means that the predator will have trouble locating its prey and will give up its attack.

At the conclusion of their study, Caro and colleagues could reject the alarm signal hypothesis. They also found support for the pursuit-deterrence hypothesis and formulated a new hypothesis for tail flagging—it functions to confuse a predator by providing a target that disappears when a deer drops its tail after running into dense vegetation. Additional tests of these hypotheses are needed using real predator–deer interactions.

Chapter Summary and Beyond

Antipredator behavioral adaptations are common because most animals live under the threat of being killed by a predator. Many animals reduce predation risk by selecting appropriate habitats to match background color, reducing their overall level of activity when predators are nearby, and adopting evasive movement behavior to avoid capture. Recent work has begun to examine adaptations of nonvisual crypsis that reduce detection by predators (Ruxton 2009; Booker et al. 2015). Even more remarkable, some species, like octopuses, can rapidly change the color and shape of their body to match their background; recent research is focusing on the mechanisms responsible for these structural coloration changes (Mäthger et al. 2009).

Many behaviors that reduce predation risk also entail a cost in terms of reducing other fitness-enhancing behaviors. Such behavioral trade-offs represent a common

antipredator behavioral adaptation. Several reviews show that such trade-offs affect most behaviors (Lima 1998; Caro 2005; Bednekoff 2007). Recent work has begun to examine how animals respond to longer-term exposure to predators (Ferrari, Sih, & Chivers 2009).

Many animals reduce the risk of predation by associating with others. Antipredator benefits of sociality include the dilution effect, the selfish herd hypothesis, and the effect of group size on vigilance. Recent research has examined whether large groups of individuals can benefit by confusing a predator, how individuals assess group size, and how spacing between individuals affects vigilance (Radford & Ridley 2007; Fernández-Juricic, Beauchamp, & Bastain 2007; Frommen, Hiermes, & Bakker 2009).

Finally, some animals interact directly with predators by harassing them. Such behavior can reduce predation risk by encouraging the predator to leave the area and search for another victim. Other animals appear to advertise themselves to predators, which may then deter an attack because the element of surprise is lost. Recent work has examined human–deer interactions to study additional factors regarding how deer assess the level of predation threat and how they adjust their escape behavior in response (Stankowich & Coss 2007).

CHAPTER QUESTIONS

1. Yasué, Quinn, and Cresswell (2003) based their conclusions on a correlation between ambient temperature and the number of redshanks feeding in the salt marsh habitat. If you could increase predation risk in the mudflat habitat by flying a trained falcon predator over that habitat, what do you predict would happen to the number of redshanks feeding in each habitat?

2. Imagine a fish that defends a territory where it attracts females for mating, and that females of this species prefer males with large territories. Given what you have learned about behavioral trade-offs, what predictions can you make about how the presence or absence of a predator will affect the size of the mating territory defended? Design an experiment to test your predictions.

3. Zebra-tailed lizards wave their black-and-white tails when they spot a potential predator. Could the hypotheses discussed and tested by Tim Caro and colleagues (Caro 1995; Caro 2005; Caro et al. 1995) for tail flagging in deer apply to zebra-tailed lizards? How could you test these hypotheses?

4. The size of a group often increases after a predator is spotted. What antipredator benefit best explains this observation?

5. Design a study to determine whether house sparrows (*Passer domesticus*) exhibit the group size effect.

6. For each hypothesis in the Scientific Process Boxes 9.1, 9.2, and 9.3, formulate the null hypothesis.

7. Consider different factors that might contribute to your individual fear of predators and how this fear might affect your behavior (i.e., How do humans modify their own behavior based on their fear of predators)?

8. We began this chapter by discussing the chuck calls and tail wagging of squirrels as a behavioral response to predators. Which of the hypotheses from this chapter could help us understand this behavior? How would you test these hypotheses?

Chapter 10

Dispersal and Migration

One of our favorite excuses for heading outdoors is bird watching, and Arizona has a wonderful diversity of bird species—one of the greatest in the United States. One day, while in graduate school, we were birding in an arid grassland habitat south of Tucson. Many birders (including ourselves) keep track of the number of species seen in a day, a year, or a lifetime, and we were hoping to spot a Cassin's sparrow (*Aimophila cassinii*) to add to our list.

Suddenly, a rather large bird flew directly overhead, maybe 10 m above us. It sure looked like a jaeger, but no one in the group had ever seen one. Jaegers have a distinct tail morphology, with central tail feathers that project beyond the rest of the tail, making identification fairly easy (Figure 10.1); furthermore, these birds migrate thousands of kilometers. However, jaegers are pelagic, or open-ocean, birds—hardly the kind found in the middle of Arizona. Their long, seasonal movements over the ocean take them from their breeding grounds in the Arctic to their wintering grounds south of the equator. If this *was* a jaeger, it was terribly off course! No wonder other birders and experts didn't believe us.

About a week later, the curator of the bird museum at the University of Arizona called us to see what had just been brought in—a juvenile long-tailed jaeger (*Stercorarius longicaudus*). It had been found dead along a road only a few kilometers from where we had spotted our bird

Figure 10.1. Jaeger. Long-tailed jaeger, a bird that lives in an open-ocean habitat.

and had probably died from stress. Our identification was vindicated. In fact, our jaeger spotting was not the first in the state. A search of museum records revealed that several jaegers had been observed in Arizona over the past 100 years. All were likely lost, blown off course in a storm, or ill. Migrating is serious business, and not all animals make it.

In this chapter, we examine the how and why of animal movements, ranging from short-distance, one-way dispersal movements to round-trip annual migrations of thousands of kilometers. We discuss the mechanisms that allow animals to make these trips, as well as the benefits of such movements.

10.1 Dispersal reduces resource competition and inbreeding

dispersal A relatively short-distance, one-way movement away from a site.

Many animals exhibit **dispersal**, a relatively short-distance, one-way movement away from a site. For dispersal behavior to evolve, the fitness benefit of moving to a new location must exceed the cost. One cost of dispersal involves the time and energy required to move; another is enhanced predation risk while moving through an unfamiliar area. But there is also an important benefit of moving: the new location might result in higher fitness relative to the current site. Locations can differ greatly in their availability of resources and breeding opportunities. As such, the current site might be of low quality as a result of intense competition for resources relative to other locations. Competition for resources is density dependent: it increases with the number of competitors. The **competition hypothesis** states that dispersal functions to reduce competition for resources.

competition hypothesis The hypothesis that dispersal functions to reduce competition for resources.

To test this hypothesis, researchers have manipulated both the density of individuals and the amount of food to which individuals have access, leading to two predictions:

Prediction 1: If two sites have identical amounts of food but differ in their density of individuals, we can expect higher levels of dispersal behavior in the high-density sites.

Prediction 2: If two sites have the same density of individuals but differ in their quantity of food, we can expect higher levels of dispersal behavior in the site with less food.

Studies have tested each prediction using a variety of taxa, as we see in the next two case studies.

Featured Research

Density-dependent dispersal in adult springtails

Collembolans are very small arthropods (most less than 2 mm in length) that live in high densities in the soil, feeding on fungi, algae, and detritus. They are wingless and move by crawling through mulch, leaf litter, and decaying organic matter in moist habitats. When an individual is threatened, it releases its furcula, a tail-like appendage that is held via tension underneath the body. The furcula snaps down onto the substrate, springing the individual into the air and away from danger—hence this species' common name of springtails.

Does the density of springtail individuals affect their dispersal behavior? Göran Bengtsson, Katarina Hedlund, and Sten Rundgren predicted that greater dispersal behavior would occur with higher population density because competition for resources would also be greater. To test this prediction, they established

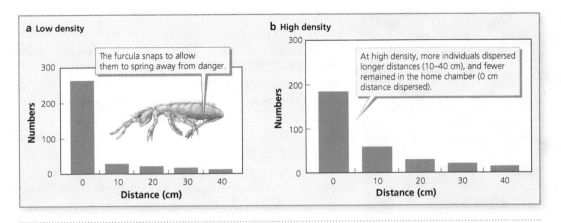

Figure 10.2. Collembola dispersal. On average, (a) individuals in the low-density treatment dispersed shorter distances than (b) those in the high-density treatment. Inset: collembolan. (*Source:* Bengtsson, Hedlund, & Rundgren 1994)

an experimental chamber in the laboratory, consisting of a series of small glass vials filled with soil and linked by rubber tubing that also contained soil (Bengtsson, Hedlund, & Rundgren 1994). Each vial constituted a habitat patch, and 60 adult springtails (*Onychiurus armatus*) were placed in one of the end vials, the home chamber. Individuals could move through the tubing to reach the second, third, fourth, and finally the fifth vial, which was 40 cm from the home vial. Although 40 cm seems like a small distance, for a springtail it represents movement of over 400 body lengths. In order to standardize the physiology and hunger levels of each individual, the researchers maintained the population for one week without food and then for several days with excess food.

The experiment consisted of two treatments, each replicated six times. In the low-density treatment, the home chamber had an area of 20 cm². This equates to 30,000 individuals/m², a typical density for springtails in natural soil. In the high-density treatment, the home chamber had an area of only 7 cm², creating a density of nearly 90,000 individuals/m². A trial began when all individuals were placed in the home vial. The research team quantified dispersal behavior by counting the number of individuals in each of the vials after three days had elapsed.

Bengtsson and colleagues found that more springtails dispersed in the high-density treatment than in the low-density treatment, and those that dispersed moved a greater distance on average (Figure 10.2). For adult springtails, the density of individuals in a habitat and the resulting competition for food strongly affect dispersal behavior. Although these results support the competition hypothesis, they do rely on the common assumption that high density corresponds to intense competition, which drives dispersal behavior.

Bengtsson and colleagues examined the behavior of adults under laboratory conditions. In the next study, researchers examined the dispersal of juvenile birds in the field.

Food-related natal dispersal in northern goshawks

Featured Research

The offspring of many animals aggregate after birth or hatching. These juveniles may face high competition for resources with their kin. **Natal dispersal**, a movement away from an individual's place of birth (natal location) to its first breeding location, is a one-time event that can help alleviate this problem.

natal dispersal A one-time movement away from an individual's place of birth.

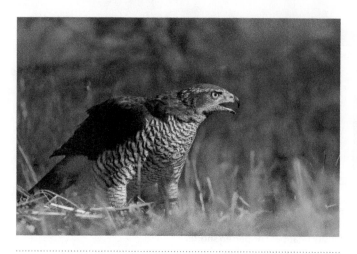

Figure 10.3. Goshawk. Goshawks are large, bird-eating hawks.

One way to test the competition hypothesis is with food supplementation. Juveniles with abundant food should exhibit lower levels of dispersal than those with limited resources. Patricia Kennedy and Johanna Ward tested this prediction on northern goshawks (*Accipiter gentilis*), which are large (approximately 1 kg), bird-eating hawks that live in forests in the Northern Hemisphere (Figure 10.3) (Kennedy & Ward 2003). Goshawks are territorial, and a pair will defend an area of approximately 2 km around the nest. They produce a brood of one to two young each year. Chicks hatch in April and become independent by August, at about 80 days of age.

Kennedy and Ward studied a population of goshawks in the Jemez Mountains of New Mexico for two years. Of the 28 broods studied, the researchers assigned 14 to a food-supplement treatment and designated 14 as controls. For the food-supplement treatment, several dead Japanese quail (*Coturnix japonica*) were provided every other day on a feeding platform located 10 m from the nest. This feeding began at hatching and continued through mid-October, after all juveniles had become independent. Control birds had access to an identical platform located 10 m from the nest, but no food was provided. The researchers visited both platforms equally to standardize human disturbance near a nest. Kennedy and Ward observed that the supplemented broods readily took food from the feeding platforms and so had access to more food than did control broods.

To track the movements of juveniles, Kennedy and Ward placed small radio transmitters on the legs of all juveniles when chicks were 21 days old. These birds were located every other day by car or airplane until mid-October. In total, Kennedy and Ward were able to collect data on the dispersal behavior of 27 juveniles, including 16 that were food supplemented and 11 controls.

Both groups became independent (defined as being away from the nest for more than a week) at about 80 days of age. Dispersal behavior, however, varied greatly between the two groups. Each week following independence, control birds dispersed much greater distances from their natal site than did food-supplemented birds (Figure 10.4). By the end of the study, all control birds had moved out of the study area (more than 25 km away from their natal nest), whereas all the food-supplemented birds had remained. Kennedy and Ward suggested that food-supplemented birds dispersed much shorter distances because they had a much higher estimate of food density in their natal area. That is, those birds estimated the study area to be of higher quality.

These springtail and goshawk studies demonstrate that competition for food or space affects dispersal. Individuals will disperse away from areas of high competition because of either high density or low food availability. Next we examine a second proposed benefit of dispersal behavior— to avoid mating with close relatives.

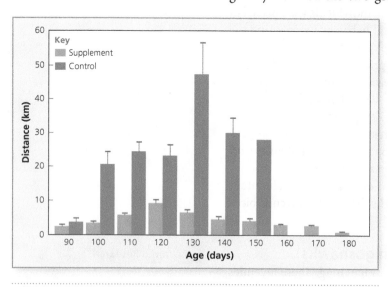

Figure 10.4. Juvenile goshawk dispersal distances. Mean (+ SE) distances traveled from the natal nest for food-supplemented juveniles (orange) and control birds (blue) at each point in time. (*Source:* Kennedy & Ward 2003)

Inbreeding avoidance in voles

Individuals that mate with close relatives tend to have lower reproductive success than individuals that do not—a phenomenon known as **inbreeding depression**. Dispersing away from a natal area can solve this problem. The **inbreeding avoidance hypothesis** posits that one factor driving the evolution of natal dispersal behavior is minimization of inbreeding (Hamilton & May 1977; Bengtsson 1978; Pusey & Wolf 1996). This hypothesis also explains a striking pattern observed in birds and mammals: sex-biased natal dispersal. In birds, juvenile females tend to disperse more often and to greater distances than do juvenile males; in mammals, the opposite pattern tends to occur: males disperse more often and farther (Greenwood 1980; Pusey 1987). When one sex disperses more often and farther than the other, there is less chance that siblings of the opposite sex will settle near each other and mate.

Eric Bollinger, Steven Harper, and Gary Barrett tested the inbreeding avoidance hypothesis using meadow voles (*Microtus pennsylvanicus*) (Bollinger, Harper, & Barrett 1993). These small, herbivorous rodents of North America live in tall, dense grasses, which they feed on and use to hide from predators. They exhibit male-biased natal dispersal: juvenile males are two times more likely to disperse than juvenile females (Boonstra et al. 1987). Bollinger and colleagues asked the research question: Does inbreeding avoidance affect dispersal behavior in juvenile meadow voles? If so, the inbreeding avoidance hypothesis makes the following prediction: juveniles living near their siblings should exhibit greater dispersal behavior than juveniles living near non-kin. Bollinger and colleagues tested this prediction by creating six 8 m × 18 m experimental plots in tall, dense grass in a natural field in Ohio: a perfect habitat for voles. Around each plot, they created a 1 m wide strip of mowed vegetation, surrounded by a 1.5 m wide strip of plowed ground with no vegetation. These two strips formed a buffer area of poor-quality habitat around each plot (Figure 10.5).

The research team placed four voles into each plot: two males and two females. In the experimental treatment, the four voles were siblings, and in the control treatment, they were not. To determine dispersal behavior, Bollinger and colleagues placed four live traps near the center of each plot and 16 live traps in the surrounding

inbreeding depression A reduction in fitness as a result of mating with close relatives.

inbreeding avoidance hypothesis The hypothesis that natal dispersal behavior minimizes the likelihood of inbreeding.

Figure 10.5. Vole dispersal. (a) Experimental plots of tall grass (background), separated by a mowed strip. (b) A captured vole.

plowed strip each day. Any animal captured in the buffer area of poor-quality habitat was considered to be attempting dispersal.

During the ten-day experiment, Bollinger and colleagues found that, as expected for this species, males were more likely to attempt dispersal than females: more males than females were captured in the plowed strip. More importantly, the experimental treatment strongly affected dispersal behavior: sibling voles were much more likely to attempt dispersal than were nonsibling control voles (a mean of 2.4 voles in the sibling plots versus 1.4 in the control plots). Furthermore, sibling voles attempted to disperse several days before nonsibling voles did (4.6 days in the sibling plots versus 7.1 days in the control plots). From their results, the research team concluded that inbreeding avoidance influences the dispersal behavior of meadow voles. Individuals dispersed in order to move away from their close kin, as predicted by the inbreeding avoidance hypothesis.

So far, we have focused mainly on natal dispersal by juveniles. However, adults also exhibit dispersal behavior when they move from one breeding site to another. Hypotheses to explain these movements typically do not involve competition or inbreeding avoidance. Instead, they propose that individuals are attempting to move to a higher-quality breeding site, as we see next.

10.2 Reproductive success and public information affect breeding dispersal behavior

For many species, a major factor that affects fitness is the ability of parents to successfully raise young. Breeding habitats differ in quality with respect to reproductive success because they differ in food abundance, predators, diseases, and so forth. Animals that settle and breed in a high-quality habitat will tend to have higher fitness than those that settle in a poor-quality habitat. However, it may be difficult for an individual to assess the quality of a habitat accurately, which could result in low reproductive success. Therefore, many animals exhibit **breeding dispersal**— movement to a new breeding site, presumably in search of a higher-quality habitat that will enhance their fitness.

breeding dispersal The abandonment of one breeding site to move to another.

Numerous studies have found a strong relationship between reproductive success and breeding dispersal behavior: individuals that are unsuccessful in a breeding attempt typically disperse to a new site for subsequent breeding attempts (Switzer 1993). In contrast, those that are successful in breeding typically do not disperse: they rebreed at the same site, a pattern known as **site fidelity**. This pattern holds both for individuals that breed seasonally, such as birds in temperate or boreal habitats, and for those that reproduce each day, such as insects. Let's look at one example that illustrates this pattern.

site fidelity Individuals that remain at or return to a previous location to breed.

Featured Research

Reproductive success and breeding dispersal in dragonflies

Paul Switzer experimentally examined the breeding dispersal behavior of male eastern amberwing dragonflies (*Perithemis tenera*) on a pond in Kansas (Switzer 1997). Male eastern amberwing dragonflies defend small sites for oviposition (egg laying) that consist of vegetation or sticks rising above the surface of a pond. When a male spots a female, he leads her back to the oviposition site and courts her. If the female finds the site acceptable, she uses it to lay a clutch of eggs, which

SCIENTIFIC PROCESS 10.1

Breeding dispersal in dragonflies
RESEARCH QUESTION: *What factors affect breeding dispersal in dragonflies?*

Hypothesis: Dragonfly dispersal behavior is affected by their reproductive success.

Prediction: Dragonflies will remain in the same territory if they have high mating success and will disperse to a new territory if they have low mating success.

Methods: The researcher:

- Created 20 breeding territories on a pond using sticks
- Reduced mating success in ten territories (treatment) and kept it high in others (controls)
- Recorded the number of matings for each male and whether he returned to the same territory or dispersed to a different territory on Day 2

Figure 1. Experimental setup. Experimental setup showing the sticks used to manipulate the oviposition site. (*Source:* Switzer 1997)

Results: Control males had higher mating success than treatment males. All control males returned to their oviposition site, whereas only 50% of treatment males returned to their oviposition site.

Conclusions: Mating success affects breeding dispersal: individuals use a win-stay lose-shift strategy when making dispersal decisions.

---

the male fertilizes. Males defend one site per day and may mate with several females in a day. Males leave the pond at night for the safety of dense vegetation and return the next day to select an oviposition site. They may exhibit site fidelity (by selecting the same site) or breeding dispersal (by moving to a new site) (Scientific Process 10.1).

To examine how a male's reproductive success affects his dispersal behavior, Switzer created artificial oviposition sites consisting of two sticks joined together. The longer stick served as a handle to allow for manipulation of the site.

Males were captured and marked on the forewing for visual identification and then released. On the first day of an experimental trial, two males were selected while defending oviposition sticks on the pond. The treatment male was prevented from mating all day, whereas the control male was allowed to mate with females. To prevent a male from mating, Switzer simply lowered the oviposition stick under the water whenever the treatment male had attracted a potential mate to his territory. With the oviposition site submerged, the female would

EVALUATE

Develop a hypothesis to explain why 50% of the treatment males that obtained no matings returned to the same site on Day 2.

quickly abandon the male. When she left, Switzer would return the stick to its original position above the water, which allowed the male to continue to defend the site. To control for this manipulation, Switzer also submerged the control male's oviposition stick when he was on his territory, but only when females were not present. Thus, both males had their sites submerged a similar number of times. Switzer recorded the number of matings and the dispersal decisions of each male by locating them on the pond the following day.

Switzer found that control males mated an average of 6.5 times on the first day, but treatment males never mated successfully. On the following day, all control males reused the same site—that is, all exhibited site fidelity. In contrast, half of the treatment males dispersed to a new site on the following day. On average, the treatment males moved 22 m between Day 1 and Day 2, while control males did not move at all. These results clearly demonstrate that an individual's reproductive success affects its decision to disperse to a new breeding site.

The behavioral pattern exhibited by Switzer's dragonflies, and by many other species as well, is known as a **win-stay lose-shift dispersal** strategy. Here, *win* means high reproductive success, while *lose* means a reproductive failure; *stay* and *shift* refer to the dispersal decision. We can formulate a hypothesis to explain this behavioral pattern by assuming that breeding sites vary in quality and that the quality of a site is consistent between breeding attempts. That is, we assume that (1) in poor sites, reproductive success is low, while in high-quality sites, reproductive success is high; and (2) a site that is high quality currently is likely to be a high-quality site in the future as well. Under these assumptions, the win-stay lose-shift strategy for breeding dispersal makes sense.

So far, we have assumed that the win-stay lose-shift dispersal decision is based on an individual's assessment of breeding patch or site quality. This assessment, in turn, is based solely on the individual's reproductive success in the patch. Thierry Boulinier and Etienne Danchin reasoned that it might be even better for individuals to gather information about the reproductive success of conspecifics breeding in the same patch (Boulinier & Danchin 1997). Using a model, they demonstrated that patch-wide reproductive success of conspecifics provides more accurate information about the quality of the patch than does individual success. They also showed that using such information for dispersal decisions results in higher fitness for an individual compared to relying only on its own experience (Boulinier & Danchin 1997; Doligez et al. 2003). Information about the quality of a resource based on the performance or behavior of others is known as public information (Valone 1989). Over the past decade, several studies have shown that animals use public information to improve their estimates of the quality of mates, food patches, and breeding patches (e.g., Doligez, Danchin, & Clobert 2002; Aparicio, Bonal, & Muñoz 2007; Valone 2007). Let's examine one study showing how public information affects the breeding dispersal decisions of birds.

win-stay lose-shift dispersal
A common behavioral pattern in which individuals return to a previous breeding location after reproductive success there but disperse to a different location after a reproductive failure.

Featured Research

Public information and breeding dispersal in kittiwakes

Kittiwakes (*Rissa tridactyla*) are gulls that nest in large colonies on sea cliffs in northern Europe (Figure 10.6). The rocky habitat often creates subcolonies, or breeding patches, of just a few dozen individuals, aggregated close together and widely separated from other patches. Kittiwakes breed once a year, and previous work suggested that individuals use the win-stay lose-shift strategy when making breeding dispersal decisions from one year to the next (Danchin, Boulinier, & Massot 1998).

Figure 10.6. Kittiwake research. (a) Kittiwakes nest in rock cliff colonies. Note the colored leg bands that researchers use to identify individuals. (b) A researcher examines an experimental bird.

Boulinier and colleagues hypothesized that patch-wide reproductive success in subcolonies might also affect breeding dispersal decisions (Boulinier et al. 2008). To test this hypothesis, the research team examined predictions regarding individuals that experience a reproductive failure. Based solely on its own experience, a failed breeder should estimate that the patch is of low quality and therefore should abandon it in favor of a new site the following year—that is, it should exhibit breeding dispersal. However, if a failed breeder uses patch-wide reproductive success to assess patch quality, and if it is surrounded by many individuals that breed successfully, then it should estimate that the patch is of high quality and so should not abandon it the next year—that is, the individual should exhibit site fidelity.

To test these predictions, Boulinier and colleagues uniquely marked focal birds with color bands in 18 breeding patches. These patches were arranged in nine matched pairs, based on size and proximity to one another. One patch of each pair was the treatment, and the other served as the control. In the treatment patches, all birds, including the focal individual, had their eggs removed 25 days after the first egg was laid and every sixth day after that; thus, all birds in the treatment patch experienced a reproductive failure. In the control patches, the focal bird had its eggs removed at the same time as the treatment birds, but none of the other birds' eggs were manipulated, and so they experienced high levels of reproductive success. The research team observed the behavior of focal birds over two years to determine whether they exhibited breeding dispersal or site fidelity.

The manipulation created a large difference in the patch-wide reproductive success of the paired patches. No bird in the treatment patches successfully raised offspring, while more than 50% of individuals in the control patches successfully raised offspring.

Focal birds in the paired patches exhibited significantly different breeding dispersal behavior. Even though they experienced a reproductive failure, birds in

Figure 10.7. Kittiwake return rates. Mean (+ SE) return rate of unsuccessful individuals was higher for treatment plots with reproductive success of other birds compared to plots with reproductive failure of other birds. (*Source:* Boulinier et al. 2008)

migration Relatively long-distance two-way movement.

the control patches exhibited a high degree of site fidelity, with over 70% returning to breed in the same patch the next year (Figure 10.7). In contrast, fewer than 50% of the focal birds in the treatment patches returned to breed in the same patch, meaning that more than half exhibited breeding dispersal. Because all focal birds experienced a breeding failure, the research team concluded that birds in the control patches, where patch-wide reproductive success was high, used public information from the reproductive success of others to estimate patch quality and to decide to remain in the same patch the next year. This study provided some of the first experimental evidence that information from conspecifics can affect breeding dispersal decisions.

So far, we have examined dispersal behavior in light of the benefits and costs of relatively short-distant movements. Next, we use the same approach to understand the evolution of longer, round-trip annual movements, which are seen in many species.

10.3   Individuals migrate in response to changes in the environment

Many animals undergo longer, two-way movements known as **migration**. Species in a variety of taxa migrate, including marine invertebrates, insects, amphibians, reptiles, fish, birds, and mammals (Table 10.1). One well-known example is the annual East African migration of more than one million wildebeest (*Connochaetes taurinus*), along with thousands of zebras (*Equus burchelli*) and Thomson's gazelles (*Eudorcas thomsonii*) (Figure 10.8). These animals travel from the southern to northern Serengeti in the spring and return to the southern Serengeti each fall. Over the course of this migration, individuals travel over 400 km (roughly

TABLE 10.1 Examples of migration in different taxa. (*Source:* Alerstram, Hedenström, & Åkesson 2003).

TAXON	COMMON NAME	SPECIES	ROUTE	DISTANCE (KM)
Insect	Dragonfly	*Anax junius*	New England to Central America	2,800
Crustacean	Blue crab	*Callinectes sapidus*	Chesapeake Bay	240
Fish	Bluefin tuna	*Thunnus thynnus*	Atlantic	12,000
Amphibian	Common toad	*Bufo bufo*	Hibernation site to breeding site	2–3
Reptile	Loggerhead sea turtle	*Caretta caretta*	California to Japan	11,500
Bird	Arctic tern	*Sterna paradisaea*	Greenland to Antarctica	19,000
Bird	Blackpoll warbler	*Dendroica striata*	Alaska to Bolivia	12,000
Mammal	Gray whale	*Eschrichtius robustus*	Baja California to Chukchi Sea	6,000
Mammal	Elephant seal	*Mirounga leonina*	Sandwich Islands to Antarctica	3,000

250 mi), moving from the short-grass plains in the south to savannah habitats in the north and back again.

The proximate mechanisms involved with the timing of migration vary across taxa and are often related to a combination of internal factors, such as annual rhythms and associated physiological changes; external factors, including changes in photoperiod; and local environmental conditions. For example, in the Serengeti, mammals appear to track both the available biomass of grass and its quality. When grass growth slows and grasses dry out during the onset of drought conditions, grass nutritional value drops, and this change may serve as a cue for the onset of migration (Boone, Thirgood, & Hopcraft 2006; Holdo, Hart, & Fryxell 2009). For birds, the onset of migration is typically triggered by internal physiological changes, whereas the timing and rate of migration are often affected by changes in day length, temperature, and food availability (e.g., Berthold 1993; Marra et al. 2005).

Large-scale movements, however, come at a cost, as the jaeger at the beginning of this chapter illustrates. Not only do long journeys take a physiological toll as a result of the exertion they require, but migrating animals also often suffer risks such as increased predation. Given the costs, what fitness benefits are obtained?

Figure 10.8. Wildebeest migration. Twice each year, wildebeest migrate hundreds of kilometers to more favorable feeding grounds in the Serengeti.

Migration and changing resources

The benefits of migration include the possibility of moving to an area with more favorable conditions (e.g., more food or water, less extreme temperatures). The decision to stay in one location and forgo migration comes with its own benefits and costs: permanent residents do not pay the costs of long-distance movement but must endure harsher conditions. For migration behavior to evolve in a population, the net benefit of migration must exceed the net benefit of being a permanent resident, and so examining variation in these costs and benefits can help us understand variation in the migratory behavior we see today (Pulido 2007).

Let's go back to the example of wildebeest, zebras, and Thomson's gazelles. These large-bodied herbivores require substantial quantities of grass each day, and grass is dependent on rainfall. Rainfall in the Serengeti is highly seasonal. These animals begin the year in the south, feeding on abundant grasses that grow vigorously after the short fall rainy season. By spring, these grasses are depleted, and the migration begins. The herds move north and west to arrive in the northern Serengeti by July, after the onset of summer rains in this region, and stay until October, when the food supply in the north becomes scarce. At that time, herds begin to move south, back to the southern Serengeti, and arrive in December as the grasses there begin to grow rapidly again (Figure 10.9) (Holdo, Hart, & Fryxell 2009). A similar pattern is seen in many insectivorous birds that breed in temperate regions in summer, when insects are plentiful. These birds migrate south in the autumn as insect availability declines, to overwinter in warmer subtropical regions with higher food abundance, and then return north the following spring to breed. Recent changes in global climate are impacting many migratory species (Applying the Concepts 10.1).

Figure 10.9. Wildebeest migration map. This map shows the circular migration route of wildebeest over the course of the year.

APPLYING THE CONCEPTS 10.1

Bird migration and global climate change

Climate change is a phrase heard often in the news. The term *climate* refers to statistical averages of temperature, humidity, wind, and rainfall that are based on data collected over decades, whereas *weather* refers to these same parameters from day to day. From 1850 to 1950, the global average temperature fluctuated but exhibited no discernible trend. However, since 1950, the global average temperature has increased dramatically, demonstrating global climate change. Climate change can lead to mismatches between animals and their habitats, and it can have severe consequences for species.

Over the last 50 years, as a warmer climate has developed, particularly in the Northern Hemisphere, spring conditions have been occurring earlier than in the past (Stenseth & Mysterud 2002). For organisms that migrate to follow resources, this shift may result in a mismatch between the migration date and the availability of those resources.

David Inouye and colleagues examined the spring migration return of American robins (*Turdus migratorius*) back to high-altitude sites in the Colorado Rocky Mountains (Inouye et al. 2000). Over the past 25 years, spring temperatures have increased, but the average date of snowmelt has not changed (perhaps because of increased snowfall). The research team examined the arrival dates of robins from 1974 to the present and the first date of bare ground. They found that robins now arrive back from their winter range 14 days earlier than they did in 1981.

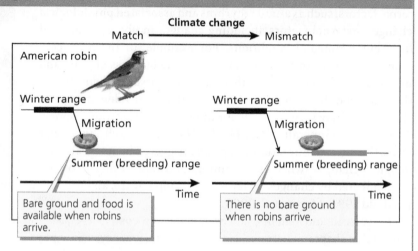

Figure 1. Bird migration and global climate change. When there is a match between the climate and migration times, robins arrive to find bare ground (green) at their summer breeding range (left). When there is a mismatch as a result of climate change (warmer temperatures), robins arrive earlier but must face snow and no bare ground for 18 days (right). (*Source:* Stenseth & Mysterud 2002)

However, snowmelt is now rarely complete when they arrive, and bare ground typically does not appear until 18 days later (Figure 1). This is a problem, because robins need bare ground to search for buried insects. Mismatches resulting from differential responses to climate change can have negative consequences for robins in the Rocky Mountains, as well as other species. ■

Featured Research

Resource variation and migration in neotropical birds

We can hypothesize that migration has evolved in many animal species as an important strategy that allows individuals to (1) take advantage of spatial variation in conditions and (2) avoid seasonal resource depression at different locations. To evaluate this hypothesis, we can test the following prediction: species exposed to a high degree of fluctuation in environmental conditions and resources can benefit by migrating when conditions and resources decline. In contrast, species that live in more stable resource environments have less to gain by migrating and so should be sedentary (Levey & Stiles 1992).

W. Alice Boyle and Courtney Conway tested this hypothesis using neotropical birds in the clade Tyranni (Boyle & Conway 2007). This group includes over 300 species of tyrant flycatchers, manikins, cotingas, and becards. These birds vary greatly in migratory behavior, habitat use, and diet, making them ideal for testing such a hypothesis. The researchers collected previously published data from field guides and reference books on migratory behavior (whether each bird was migrant or sedentary), diet (insectivore or frugivore), and habitat use. They assumed that habitat types varied in degree of environmental fluctuations:

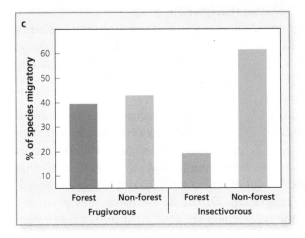

Figure 10.10. **Diet and migratory behavior of neotropical birds.** (a) Frugivorous ochre-bellied flycatcher (*Mionectes oleagineus*); (b) insectivorous royal flycatcher (*Onychorhynchus coronatus*). (c) Insectivorous species are more likely to migrate if they live in nonforested habitats. (*Source:* Boyle & Conway 2007)

tropical forest habitats, for example, tend to have relatively stable temperature and humidity compared with nonforest habitats (which include open and arid habitats, disturbed habitats, and thickets). Less was known about the relative stability of food resources, but the researchers focused on insect and fruit abundance, both of which vary seasonally.

Controlling for the effects of phylogeny, Boyle and Conway found that habitat type did correlate with migration, at least for insectivores. Insectivorous species that inhabited forest habitats were less likely to be migratory than were species inhabiting nonforest habitats (Figure 10.10). Because forest interiors have more stable temperatures and humidity levels, they may provide a more uniform environment that helps stabilize insect abundances. For frugivores, however, habitat did not affect the propensity to migrate, perhaps because forest and nonforest areas do not differ as much in stability of fruit production.

These data suggest that fluctuations in resource levels and environmental conditions can affect the evolution of migratory behavior. Next, we examine how migration can evolve.

Heritability of migration behavior in Eurasian blackcaps

Featured Research

For migratory behavior to evolve, it must be a heritable trait. Francisco Pulido, Peter Berthold, Gabriele Mohr, and Ulrich Querner estimated the heritability of the timing of autumn migration in Eurasian blackcaps (*Sylvia atricapilla*) (Figure 10.11),

Figure 10.11. Eurasian blackcap. A migratory species common throughout Europe.

a small songbird common throughout Europe and Asia (Pulido et al. 2001). Blackcap populations that breed in Eastern Europe and Asia migrate to Africa each autumn.

Many birds, including Eurasian blackcaps, migrate at night and exhibit migratory restlessness, or *Zugunruhe*, prior to the initiation of migration. Previous work had demonstrated that migratory restlessness in captive-reared birds correlates well with the migratory behavior of wild birds (Berthold 1996). Over 11 years, Pulido and his colleagues collected nestlings from a population in Germany, raised them in captivity, and allowed them to breed. All birds were housed in identical conditions and exposed to the natural declining autumnal photoperiod, which stimulates migratory behavior. The researchers defined the onset of each juvenile bird's migratory behavior as occurring on the first night in which it was active during at least five 30-minute observation periods. This behavior was recorded for both parents and their offspring, which allowed the research team to calculate parent-offspring regressions for onset of autumn migratory behavior, and thus the heritability of the behavior (Chapter 3).

Onset of migratory behavior was quite variable in these birds, with an average date of September 17. Parent-offspring regression analysis revealed a significant positive correlation and a heritability estimate of 0.35 (Figure 10.12). This result indicates that the onset of autumn migration is a trait that can evolve.

To confirm this conclusion, the researchers conducted an artificial selection experiment in which they selected for later migratory dates. They used birds collected in one particular year as the initial generation (Generation 0) and selected ten birds that exhibited the latest onset of migratory behavior that fall as breeders for the next generation. Of the 26 offspring they produced (Generation 1), the researchers selected eight birds (four breeding pairs) with the latest migratory behavior dates as breeders to produce the final set of birds (Generation 2). All birds in each generation were scored to calculate the mean onset of autumn migration date for each generation.

Migratory behavior shifted by over seven days from Generation 0 to Generation 2. This result, combined with the heritability estimate from the parent-offspring regression analysis, demonstrates that onset of autumn migration is a heritable trait in this species. As such, they demonstrate that migratory behavior can evolve.

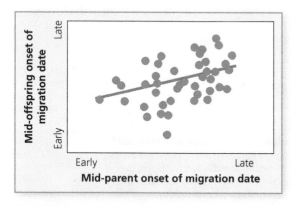

Figure 10.12. Migration date heritability in Eurasian blackcaps. Regression showing positive relationship between mid-parent and mid-offspring onset of migration date. (*Source:* Pulido et al. 2001)

The evolution of migration: a model

Featured Research

Now that we understand that migratory behavior is a heritable trait, how does a population evolve from being sedentary to being migratory? Volker Salewski and Bruno Bruderer proposed a model that involves a sequence of events to explain the evolution of migration within a population (Salewski & Bruderer 2007). The model begins with a resident population of individuals that do not migrate but do disperse when they mature. Dispersal distance varies based on how far an individual must travel to find an unoccupied breeding habitat. It is envisioned that some, but not all, individuals will need to travel to a distant location to breed, and if resource abundance declines strongly in that location in the nonbreeding season, they and their offspring will die unless they move back toward their natal site. That is, selection will favor individuals that return to the resident population at the end of the breeding season. These migrating individuals can again benefit by returning to breed at the same distant area when conditions improve in the following breeding season.

These movement patterns create a population that exhibits partial migration, or a situation in which individuals differ in their migratory behavior: some individuals are year-round residents, while others exhibit seasonal migratory behavior. Over time, if the fitness of migrants is higher than that of residents, eventually all individuals will migrate seasonally. The end result is a migratory population, in which all individuals exhibit migratory behavior. But in this model, the relative fitness benefits and costs of each behavior may vary over time and in different locations, so evolution could also proceed in the opposite direction (Figure 10.13) (e.g., Berthold 1999; Pulido 2007; Salewski and Bruderer 2007). We often see species that exhibit variation in migratory behavior, and indeed, we can find many partial migrant populations (Swingland & Greenwood 1983). The model postulates that variation in the fitness of each behavior can result in rapid changes in the proportion of individuals in a population exhibiting migratory behavior.

Evaluation of this model requires the examination of factors that might favor migration over nonmigration. Partial migrant populations have been particularly useful to understanding the relative fitness costs and benefits of migratory behavior (Chapman et al. 2011). Competition is one general factor thought to influence migratory behavior. If individuals differ in their competitive abilities, perhaps only dominant individuals will be able to secure sufficient food when it is scarce (say, in winter), and so selection can favor migratory behavior by subordinates as a means to find sufficient food. This hypothesis suggests that the density of individuals should affect propensity to migrate because as density increases, competition for food intensifies. Let's examine a recent test of this hypothesis in a migratory amphibian population.

Figure 10.13. Evolution of migration. When some individuals have higher fitness as a result of migrating, a population of residents will exhibit partial migration. If all individuals have higher fitness as a result of migrating, the population will evolve to become migratory. If some individuals in a migratory population have higher fitness by remaining resident, a migratory population will exhibit partial migration. If all individuals have higher fitness by forgoing migration, the population will evolve to become resident. (*Source:* Salewski & Bruderer 2007)

Competition and migratory behavior of newts

Featured Research

Many pond-breeding amphibians like the red-spotted newt (*Notophthalmus viridescens*) breed in spring, migrate to a terrestrial habitat in fall to overwinter, and then return to the same pond to breed the next spring. Some individuals remain in the pond as year-round residents and do not migrate. Residents and migrants exhibit alternate phenotypes. Residents display the aquatic phenotype, with a large dorsal fin and mucus-covered skin, while migrants change to a more terrestrial phenotype, with a reduced tail fin and a skin that is more resistant to desiccation, which allows them to migrate over land. These phenotypes are reversible, and migrants switch between them each year. Residents face the risk of the pond drying up and low oxygen levels, while migrants must pay the physiological costs associated with a change in morphology and face increased costs of travel and associated predation risk.

Kristine Grayson and Henry Wilbur designed an experiment to examine how environmental conditions affect migratory behavior in this species (Grayson & Wilbur 2009). They manipulated the population density and sex ratio of individuals in small enclosures placed around the edge of a large pond such that each enclosure included aquatic and terrestrial substrate. In spring, the enclosures were stocked with different densities of individuals (low, medium, or high) that were either male- or female-biased in their sex ratio (2:1). Enclosures were checked daily for newts out of the water, and the researchers identified these newts as migrants by their morphology. The migrant newts were then removed from the enclosure and released in the terrestrial habitat.

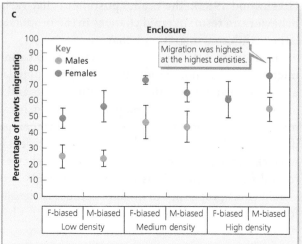

Figure 10.14. Migrating newts. (a) Experimental enclosures on the edge of the pond. (b) Migrating newts. (c) As the density of newts in enclosures increased, a greater mean (± SE) fraction of individuals migrated. (*Source:* Grayson & Wilbur 2009)

Many more newts migrated out of the high-density enclosures (63%) than the low-density enclosures (39%) (Figure 10.14). On average, newts from the high-density enclosures migrated eight days earlier than those from low-density enclosures. However, the sex ratio did not significantly affect the migratory behavior of males or females. These data demonstrate that population density influences the migratory behavior of newts, and that this behavior is plastic. As density increases, competition for food and space becomes more intense and favors individuals that leave.

One interesting trend was that females migrated more often (60%) than did males (37%) in both sex-ratio treatments. Why did a smaller fraction of males migrate? One hypothesis is that competition among males to obtain a territory in the next breeding season favors males that do not migrate, because individuals who arrive earlier on the breeding grounds are more likely to win in a territorial contest (Kokko, López-Sepulcre, & Morrell 2006). A territory can be acquired by simply staying all year—that is, by becoming a resident rather than a migrant (Kokko 1999). Of course, there are risks involved. Residents must survive the harsh winter pond conditions in order to breed the next season, but if they do, they have a higher chance of securing a breeding territory. This may explain the sex-biased migration observed in newts and likely also explains the sex-biased migration often seen in birds: males tend to migrate shorter distances than females (Cristol, Baker, & Carbone 1999).

Maintenance of polymorphism in migratory behavior

In any partial migrant population, you may wonder: How is the migratory behavior polymorphism maintained? Why do partial migrant populations exist—and are they stable? There are two possible answers to these questions. One is that migratory behavior is a fixed genetic trait of an individual, and the two genotypes (and phenotypes) are maintained by frequency-dependent selection (see Chapter 3): the fitness of each strategy declines as it becomes more common. To maintain the polymorphism, both migratory and resident behaviors must result in equal fitness at equilibrium. We have already seen evidence that genes influence migratory behavior in Eurasian blackcaps (Pulido & Berthold 2010). Recent work has also revealed that some variation in migratory tendency among individuals is associated with microsatellite variation in a single gene (Mueller, Pulido, & Kempenaers 2011), but little is known about the genetics of migration in most species.

There is a second way that a migratory polymorphism can be maintained. All individuals in a partial migrant population might possess a genetic disposition to migrate, but the behavior may be expressed only in some, depending on conditions. In this case, the environment or an individual's physical condition, sex, age, or dominance status influences its migratory behavior. In this scenario, each behavior need not have equal fitness; instead, individuals adopt the strategy that is best given their condition. Those in poor condition need to "make the best of a bad job" until their condition improves. Let's examine one recent test of these alternative hypotheses.

Alternative migratory behaviors in dippers: test of hypotheses

Elizabeth Gillis and colleagues examined migratory behavior in American dippers (*Cinclus mexicanus*) (Gillis et al. 2008). Dippers are aquatic birds that live in and around fast-flowing mountain streams. Each year, many wintering populations comprise both year-round residents and migrants, which move upstream to higher elevations in spring to breed and then return to low elevations to overwinter (Price & Bock 1983). Thus, residents and migrants share wintering grounds but breed in different locations.

The research team studied uniquely banded individuals at several sites on the Chilliwack River in British Columbia over a period of seven years. They conducted five censuses each year to locate marked individuals, along with their nests and offspring. Residents were classified as individuals that occupied a single territory year-round. Migrants were identified either as those individuals that wintered on the river and then bred elsewhere or as those that bred on tributaries upstream in spring but wintered elsewhere.

Gillis and colleagues observed hundreds of residents and migrants. On average, migrants traveled almost 6 km to reach their breeding territory. Resident females began nesting almost two weeks earlier than migrants, although clutch size and brood sizes were similar for both resident and migrant females. However, resident females were more likely than migrants to initiate a second brood if their first nesting attempt was successful and so had higher annual reproductive success (Figure 10.15). In contrast, and perhaps unexpectedly, the annual survival of migrants was slightly higher than that of residents. Still, overall, the researchers estimated that a lifelong resident is expected to fledge about 2.6 more offspring than a lifelong migrant; therefore, the fitness of nonmigratory behavior is higher than that of migratory behavior.

Figure 10.15. Dipper reproductive success. (a) Residents had higher mean (+ SE) reproductive success than did migrants in each year of the study. (b) Dipper. (*Source:* Gillis et al. 2008)

The researchers concluded that the persistence of alternative migratory behaviors in dippers cannot be explained by a fixed genetic dimorphism: the behaviors have different fitness, and offspring frequently exhibit a migration strategy that differs from that of their parents. Over 50 banded nestlings eventually became breeders in the population, but these birds adopted the migratory behavior of their parents less than 50% of the time. Instead, Gillis and colleagues suggest that the migratory behavior of dippers is condition dependent: individuals in the best condition (perhaps dominants) do not migrate and experience high fitness. The researchers did observe four birds that changed their migratory behavior over the course of the study, indicating that migratory behavior is flexible. The results of this study indeed suggest that less competitive individuals are "making the best of a bad job." These individuals cannot compete with residents for the limited high-quality breeding territories, but they can reproduce—albeit at lower levels than residents—by migrating to higher elevations to breed.

Dippers and many other animals tend to move in a particular direction during migration. For instance, birds in North America tend to migrate south in autumn and north in spring. How do they determine and maintain their direction? We answer that question next.

10.4   Environmental cues and compass systems are used for orientation when migrating

orientation The process of determining and maintaining a proper direction.

navigation The process of determining a particular location and moving toward it.

Animals that migrate long distances face two issues: (1) **orientation**, or maintaining the proper direction of travel; and (2) **navigation**, or determining how to reach a particular destination. In the 1950s, Gustav Kramer suggested that birds migrate long distances by use of a "compass and map" system (Kramer 1952). He meant that birds can orient in the proper direction and find particular locations using mental maps, much as people use physical maps (Schmidt-Koenig 1979). In the decades since, research has indicated that many animals, not just birds, use a variety of compass systems to orient to a particular direction, and some appear to also possess a mental map that is used to find particular locations. In this section, we examine orientation. In the last section of the chapter, we turn our attention to navigation and mental maps.

Compass systems

A variety of environmental cues can provide directional information for orientation. One type of cue used by a variety of migratory animals, including insects, fish, amphibians, birds, and mammals, is a physical landmark, such as a coastline, island, river system, or distant mountain (e.g., Collett & Graham 2004). It has long been known that some marine mammals and shorebirds migrate along continental coastlines. For instance, gray whales (*Eschrichtius robustus*) in the eastern North Pacific migrate up to 20,000 km from their summer feeding grounds in the Chukchi Sea off the coast of Alaska to their calving grounds off the coast of Baja, California (Figure 10.16). During migrations north and south, whales generally remain within 10 km of the coastline (Poole 1984).

Many migratory animals orient using systems such as a **sun compass**—the use of the location of the sun in the sky (Åkesson & Hedenström 2007; Wehner 1998). Each day, the sun rises in the east and sets in the west, so its position in the sky relative to the time of day provides accurate directional information. Experiments reveal that invertebrates, fish, amphibians, reptiles, birds, and mammals use a solar compass (Åkesson & Hedenström 2007; Wehner 1998). Even on cloudy days, many of these animals can use the pattern of polarized skylight to determine the position of the sun, although the mechanisms of this process are not fully understood (Wehner 2001).

Nocturnal animals or those that move at night can use a **star compass**, because the constellations rotate around the celestial pole (Figure 10.17). A series of experiments by Steve Emlen demonstrated that hand-reared birds can learn to orient based on the rotation of stars in the night sky (Emlen 1967a; Emlen 1967b) (Toolbox 10.1). Subsequent work has revealed that many other taxa, including invertebrates, fish, amphibians, reptiles, and mammals, also use a star compass for orientation.

The earth's magnetic field provides another source of directional information for animals (Figure 10.18). These fields flow from the south magnetic pole to the north magnetic pole. Animals can therefore use a **geomagnetic compass**, or the ability to orient using the earth's magnetic field. Recall from Chapter 5 that the magnetic field provides directional information in three ways. First, it flows in one direction—from south to north. Second, the angle of inclination with respect to the horizontal varies with latitude. Third, the intensity of the field tends to be strongest at the poles and weakest at the magnetic equator. Use of a geomagnetic compass has been documented in dozens of species in many taxa, including snails, crustaceans, insects, fish, amphibians, reptiles, birds, and mammals (Wiltschko & Wiltschko 2005).

Next, we examine two examples illustrating how researchers study the use of compass systems for migratory orientation.

Antennae and the sun compass system in monarchs

One intensely studied animal that uses the sun compass in its annual migration is the monarch butterfly. Monarch butterflies (*Danaus plexippus*) in central and eastern North America migrate to a few mountaintop sites in central Mexico, where millions of individuals gather in massive clusters to spend the winter (Davis & Rendón-Salinas 2010). Depending on the starting

Key

— Northbound migration

— Southbound migration

Figure 10.16. Whale migration. (a) Gray whale. (b) Whales use the coastline for orientation during their migration. Blue indicates southbound migration, and orange indicates northbound migration.

sun compass The use of the sun for orientation.

star compass The use of stars or constellations for orientation.

geomagnetic compass The ability to orient using the earth's magnetic field.

Featured Research

Figure 10.17. Star compass. Time-lapse photography shows how the stars appear to rotate around Polaris in the northern sky in the Northern Hemisphere. Each streak is a star.

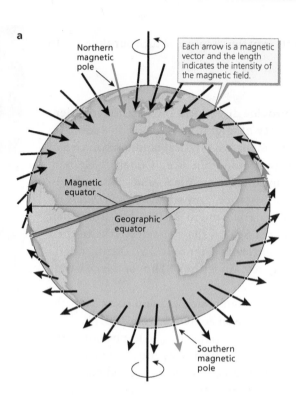

a

Northern magnetic pole

Each arrow is a magnetic vector and the length indicates the intensity of the magnetic field.

Magnetic equator

Geographic equator

Southern magnetic pole

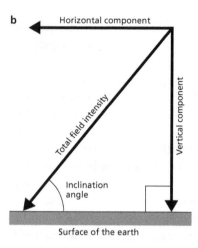

b

Horizontal component

Total field intensity

Vertical component

Inclination angle

Surface of the earth

Figure 10.18. Magnetic field of the earth. (a) The magnetic vectors around the earth. Arrows indicate the local vector and angle of inclination; their length indicates the intensity (strength) of the field. (*Source:* Wiltschko & Wiltschko 2005) (b) The total field intensity is composed of a horizontal component and a vertical component. (*Source:* Lohmann, Putnam, & Lohmann 2008)

a

b

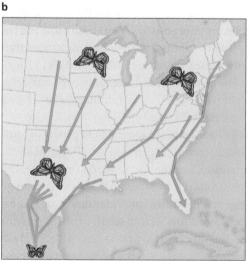

Figure 10.19. Monarch migration. (a) Monarch cluster. (b) Autumn migration routes of eastern monarch butterflies to wintering grounds in Mexico.

location, this journey can take several weeks and cover over 1,000 km (Figure 10.19).

The monarch generation that migrates to Mexico in the fall will not return to its natal area in the spring. Instead, many generations are produced during the spring migration—a cycle known as **intergenerational migration**. The first generation stops in northern Mexico and the southern United States, where the

■ **intergenerational migration**

Migration that occurs over more than one generation.

butterflies mate, lay eggs, and die. The emerging generation heads farther north and again mates, lays eggs, and dies. This pattern is repeated several times as each generation moves farther north (Holland, Wikelski, & Wilcove 2006). This raises a critical question: How do butterflies that have never been to their wintering site get there?

Animals, including monarchs, have internal time-keeping mechanisms that regulate such daily activities as their active and resting cycles. The internal clock is itself regulated by daily light and dark cycles. After a 12-hour dark cycle (and at the beginning of a light cycle), the position of the sun provides directional information, because the sun rises in the eastern sky. In the autumn, after six hours of daylight, the sun has moved to the southern sky (in the Northern Hemisphere), and at the end of a 12-hour light cycle, the position of the sun provides information about the direction west.

A simple "clock-shift" experiment demonstrated that monarchs use a sun compass by bringing two groups of these butterflies into the laboratory (Perez, Taylor, & Jander 1997). One group was "clock-shifted" six hours over the course of several weeks (Figure 10.20). That is, these individuals still experienced a light/dark cycle of 12 hours each, but their lights were turned on at a later time each day until eventually the lights came on at noon, when the sun outside was in the southern sky in the autumn. The control group was kept on the naturally occurring light/dark cycle, so their lights were turned on at 6:00 a.m. Both groups were then released outside (just after their lights turned on) in September and were followed. The control group migrated south/southwest, as is typical in Kansas, where the work was conducted. In contrast, the clock-shifted animals moved in a westward direction. Why? Their internal clock had been reset so that their "daybreak" occurred at noon. At noon, the sun was in the southern sky, but they interpreted this position as east—the normal direction of the sun at daybreak—and so migrated in the wrong direction (Figure 10.21).

Christine Merlin, Robert Gegear, and Steven Reppert investigated whether antennae play a role in the monarch's sun compass (Merlin, Gegear, & Reppert 2009) (Scientific Process 10.2). They examined migratory orientation by tethering individuals in an outdoor flight simulator during fall migration. They first compared the flight direction movement behavior of controls, whose antennae were intact, and that of individuals that had had their antennae removed. Individuals without antennae could fly, but they failed to orient to any particular direction—just as many tried to move north as tried to move

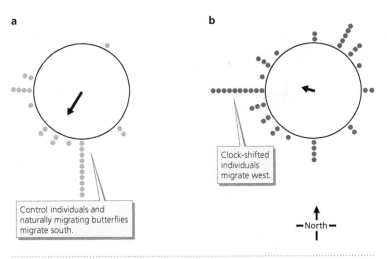

Figure 10.20. Clock-shift prediction. (a) Control butterflies, reared on a natural light cycle, fly 90° clockwise to the rising sun when released at sunrise (6:00 a.m.)—the correct southern migration direction. (b) Treatment butterflies, clock-shifted six hours forward, should also fly 90° clockwise to the sun when released at noon. But because the sun is in the southern sky at noon, they are predicted to fly in the incorrect western direction. (*Source:* Perez, Taylor, & Jander 1997)

Figure 10.21. Clock-shifted butterfly experiment. (a) Control individuals attempted to migrate south/southwest, and (b) clock-shifted butterflies attempted to migrate west. Each dot represents the movement direction of one individual. (*Source:* Perez, Taylor, & Jander 1997)

The role of the antennae in the monarch butterfly sun compass

RESEARCH QUESTION: *How do monarch butterflies use a sun compass?*

Hypothesis (1): The sun compass is associated with the antennae.

Prediction (1): Individuals without antennae will not be able to orient properly.

Methods (1): The researchers:

- Removed antennae from ten butterflies (antennae-less treatment) and left the antennae of ten control butterflies intact
- Tethered individuals and recorded flight behavior direction

Results (1): Control butterflies oriented south/southwest, while antennae-less butterflies did not orient in any particular direction.

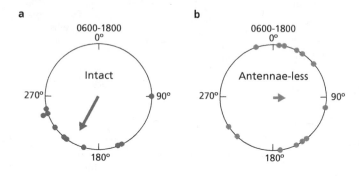

Figure 1. Antennae results. Flight orientation of (a) intact and (b) antennae-less butterflies. (*Source:* Merlin, Gegear, & Reppert 2009)

Hypothesis (2): Antennae contain photoreceptors needed for proper orientation.

Prediction (2): Butterflies with clear paint or no paint on their antennae will be able to orient properly, whereas butterflies with black paint on their antennae will not be able to orient properly.

Methods (2): The researchers:

- Painted butterflies' antennae with opaque black paint, clear paint (Control 1), or no paint (Control 2)
- Tethered individuals and recorded flight behavior direction

Results (2): Butterflies with no paint or clear paint on their antennae oriented south/southeast. Butterflies with black-painted antennae oriented north.

Figure 2. Paint results. Flight orientation of (a) controls, (b) individuals with clear-painted antennae, and (c) individuals with black-painted antennae. (*Source:* Merlin, Gegear, & Reppert 2009)

Conclusions: Antennae and functioning photoreceptors are required for sun compass orientation in monarch butterflies.

EVALUATE

Propose a hypothesis to explain why individuals with black-painted antennae oriented to the north, rather than in a random direction.

south. In contrast, the flight direction of control butterflies was to the south/ southwest, the proper direction for migration in early October.

To determine whether these results were due to differences in light reception in the antennae, the researchers conducted a second set of experiments a few weeks later, from mid-October to mid-November. This time, they painted antennae either an enamel-based black or clear. The black paint blocked perception of light, while the clear paint allowed light to pass through to the antennae. They then again examined the flight direction of tethered individuals in the outdoor flight simulator. The movement behavior of individuals whose antennae were painted with clear paint was to the south/southeast (the proper direction at that time of the year), whereas individuals whose antennae were painted black attempted to move to the north/northwest. The lack of proper movement in individuals with either no antennae or black-painted antennae that blocked light indicates that antennae photoreceptors are crucial to this species' sun compass.

Studies such as this demonstrate how migrating butterflies orient in the proper direction to reach their wintering grounds in Mexico. However, much work still needs to be done to understand how these individuals reach their specific destination. Part of this effort involves the use of citizen scientists all over North America who collect data each year on migrating butterflies (Applying the Concepts 10.2).

APPLYING THE CONCEPTS 10.2

Citizen scientists track fall migration flyways of monarch butterflies

Monarch butterflies are of concern to conservation biologists because of the loss of their required habitat in Mexico, the United States, and Canada. To better monitor monarch butterfly populations, researchers engage citizen scientists to help collect data on the butterflies' life stages. Citizen scientists are individuals, such as yourself, who are trained by biologists to participate in field projects across the country. The monarch butterfly project includes the Monarch Larva Monitoring Project (http://www.mlmp.org/), in which individuals at nature centers throughout the United States learn to conduct weekly surveys of monarch eggs and larvae on plants. Monarch Watch (http://www.monarchwatch.org/), based at the Kansas Biological Survey at the University of Kansas, oversees numerous citizen scientist programs that involve tagging individual butterflies and creating monarch waystations.

One monarch watch program, Journey North (http://www.learner.org/jnorth/) collects citizen scientist reports of monarch sightings during the fall migration. Elizabeth Howard and Andrew Davis used three years of these data to examine the migration flyway of monarchs in eastern North America (Howard & Davis 2009). Their analysis of thousands of sightings indicates that there is one major flyway south through the central portion of the United States (Figure 1).

This is important information, because it allows scientists to know where to concentrate habitat conservation efforts. Such information is the direct result of data collection by

Figure 1. Citizen scientists. The dots represent roost sightings of monarchs reported by citizen scientists in 2005–2007. The star represents the location of the overwinter site in Mexico. The green line separates the central and eastern flyways. (*Source:* Howard & Davis 2009)

citizen scientists. If you want to become a citizen scientist, visit the website of the USDA Forest Service for more information (http://www.fs.fed.us/). ■

Featured Research

The magnetic compass in sea turtles

Many species migrate between breeding and feeding sites that may be thousands of kilometers apart. Loggerhead sea turtles (*Caretta caretta*) (Figure 10.22) lay eggs in underground nests on sandy beaches. Turtle hatchlings from southern Florida emerge and, after traveling to the water, must orient toward their feeding grounds in a large, circular system of ocean currents called the North Atlantic gyre. For loggerhead sea turtle hatchlings, this journey takes several days. Once at the feeding grounds, individuals may spend several years in the ocean, until sexual maturity, when they migrate back to their natal beaches for nesting (Lohmann, Putnam, & Lohmann 2008).

Kenneth and Catherine Lohmann and colleagues have studied sea turtles for over 20 years (e.g., Lohmann et al. 2001). In an early study, Kenneth Lohmann tested the ability of juvenile loggerhead sea turtles to detect and orient using magnetic fields. He used an apparatus to reverse the polarity of the magnetic field experienced by turtles (Lohmann 1991). The apparatus consisted of an open chamber with a walking table, encircled by a large coil that could create a strong magnetic field (Figure 10.23). Individuals were attached to a harness that allowed them to orient and swim in any direction while the apparatus recorded their individual movements and orientation during a trial. One set of turtles was placed in the chamber while the magnetic field was reversed; a control set experienced the normal magnetic field. Turtles in the control treatment oriented and moved to the northeast, the direction of the North Atlantic gyre, and those that experienced a reversed magnetic field oriented and moved in the opposite direction (Figure 10.24). Thus, both groups oriented and moved toward a magnetic north, demonstrating the use of a geomagnetic compass.

Multimodal orientation

The previously described research studies illustrate how researchers have demonstrated the use of different compass systems. But one particular aspect is important to note: most, if not all, animals use multiple cues to orient. For instance, much research has examined the homing behavior of rock doves (*Columba livia*), commonly known as pigeons. Most studies of pigeon homing allow birds to live

Figure 10.22. Loggerhead sea turtle.
(a) An adult turtle migrating in the ocean.
(b) A hatchling.

Figure 10.23. Turtle orientation apparatus. Diagram of the apparatus used to characterize orientation direction of sea turtles. The magnetic field inside the apparatus can be changed. (*Source:* Lohmann 1991)

Figure 10.24. Turtle orientation. (a) Control turtles attempted to move north/northeast. (b) When the magnetic field was reversed, turtles attempted to move toward the new position of magnetic north. Each dot represents data for one turtle, and the mean direction is indicated by the arrow. (*Source:* Lohmann 1991)

in a home loft for several weeks. Birds are then displaced to a new location, and the flight direction or the entire flight of released birds is recorded using radio tags or flight recorders. This domesticated species has a remarkable ability to find its way back to a home roost after being displaced up to 1,800 km. How do they do this?

Pigeons that are temporarily blinded can orient properly and return home, so vision is not required to orient properly (Walcott 2005). However, birds that are clock-shifted six hours fly in the wrong direction—they make a 90° error in their orientation (Schmidt-Koenig 1990), just as we saw for monarch butterflies. This finding demonstrates that pigeons use a sun compass to orient. However, this error only occurs if birds can see the sun; on overcast days, clock-shifted birds orient in the proper direction (Keeton 1974), indicating that pigeons use another cue to orient when the sun is hidden behind clouds. To investigate the possibility that pigeons use a magnetic compass, researchers placed magnets on their back (Figure 10.25) or magnetic coils on their head to alter the magnetic field around each bird. Such birds failed to orient properly, but only when they could not see the sun (Walcott 1996). These experiments reveal that pigeons apparently first rely on the sun compass to orient but use a geomagnetic compass when the sun is not visible. Most birds are thought not to have sophisticated olfactory capabilities. Yet some studies have shown that pigeons that are made anosmic (i.e., lack functioning olfaction) fail to orient properly when released at unfamiliar locations, suggesting a role for olfaction in orientation (Wiltschko & Wiltschko 1989). These studies demonstrate that pigeons can use multiple cues to orient properly. Such redundancy is likely widespread across taxa because it provides backup mechanisms that help animals to travel in the proper direction.

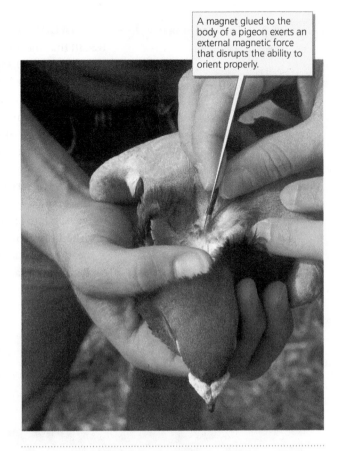

A magnet glued to the body of a pigeon exerts an external magnetic force that disrupts the ability to orient properly.

Figure 10.25. Pigeon magnetoreception. When small magnets are placed on their body, pigeons have trouble orienting properly.

The ability to orient is just one aspect of migration. Animals also need to be able to move to specific locations, like the sea turtles that need to return to their natal beach to nest. Next, we examine how animals accomplish such navigation.

10.5 Bicoordinate navigation allows individuals to identify their location relative to a goal

Some migrating animals are capable of **bicoordinate navigation**: they can identify their geographic location using two varying environmental gradients, much as humans use latitude and longitude. As an animal migrates, its geographic location changes. As long as it knows its present location relative to a target location or goal, such as its starting point or its final destination, it can adjust its direction to arrive at the destination. An important feature of bicoordinate navigation is that it can be used even when an individual is in an unfamiliar location, as might occur when it is blown off course during migration. Because bicoordinate navigation provides positional information, it is often referred to as a mental map.

Here we briefly examine three examples of bicoordinate navigation. We begin by revisiting the work of the Lohmanns and colleagues to see how they determined that sea turtles use bicoordinate navigation and a mental map.

bicoordinate navigation The ability to identify a geographic location using two varying environmental gradients.

Bicoordinate navigation and magnetic maps in sea turtles

Recall that the angle of inclination of the earth's magnetic field changes with latitude (Figure 10.18). In addition, the intensity of the magnetic field also varies across the globe (Figure 10.26). Because these aspects provide two environmental gradients, they can be used for bicoordinate navigation.

To determine whether sea turtles use bicoordinate navigation, the Lohmanns took advantage of an important fact: turtles need to orient themselves in order to remain in the North Atlantic gyre. For example, near Portugal, a northward current springs off the gyre and runs toward the colder northern waters of England. If sea turtles end up in these colder waters, they will die, and so turtles near Portugal need to move south. But turtles in the South Atlantic need to move north to remain in warm water.

The Lohmanns and colleagues (Lohmann et al. 2001) tested the ability of loggerhead sea turtle hatchlings to orient in the gyre. The experiment, conducted in eastern Florida, again used a computerized coil system to control the magnetic field. Turtles were exposed to one of three different magnetic fields that were characteristic of three different geographic locations in the North Atlantic gyre, thanks to variations in the angle of inclination and field intensity. The hatchlings all oriented to stay within the gyre at each of the perceived locations (Figure 10.27). This finding indicates that migrating sea turtles can use the earth's magnetic field to orient in different (and appropriate) directions. In addition, because hatchlings that had no prior experience could navigate properly, this ability must be inherited and not learned.

To further investigate bicoordinate navigation in sea turtles, the Lohmanns and colleagues examined whether green sea turtles (*Chelonia mydas*) use variation in the earth's magnetic field to navigate to specific locations (Lohmann et al. 2004). They captured juveniles near Melbourne, Florida, and transported them to an orientation apparatus at a nearby test site. In this species, juveniles remain in shallow coastal feeding grounds near their natal beach for an extended period. The researchers took advantage of this fact and divided the turtles into two groups. One group was exposed to a magnetic field that simulated the magnetic conditions 337 km north of the capture location. The other group of turtles was exposed to a magnetic field that simulated the magnetic conditions 337 km south of the capture location. Such "displaced" turtles would be expected to try to swim back to their feeding ground. Turtles were allowed to orient and move freely within the apparatus, and the directions in which they swam were recorded.

Figure 10.26. Intensity of the earth's magnetic field. Lines indicate variation in the intensity of the magnetic field, measured in nanoteslas (nT). Intensity ranges from about 25,000 to 65,000 nT.

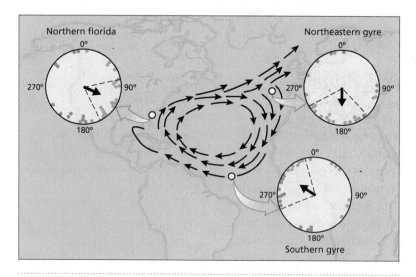

Figure 10.27. Sea turtle gyre results. The North Atlantic gyre is represented by the circular arrows. Each circle represents orientation data from turtles at the geomagnetic fields of three geographic locations. The mean is indicated by each arrow. (*Source:* Lohmann et al. 2001)

The research team found that the turtles in the two test groups behaved in different ways. Those experiencing simulated conditions of an area 337 km north of the capture site oriented and tried to move south, while those experiencing simulated conditions of an area 337 km south of the capture site oriented and tried to move north (Figure 10.28). These results show that sea turtles can use variation in the earth's magnetic field as a map to guide navigation north and south to a specific location—in this case, back toward their capture site. But how do individuals return to a specific location? Do they use landmarks, or do they rely on both the inclination and the intensity of the magnetic field to identify a specific place on earth? Additional work is required to determine exactly how the map sense is used to find a specific location, such as a natal beach.

Turtles are not the only animals that can use bicoordinate navigation. Birds, too, can use bicoordinate navigation, as we see next.

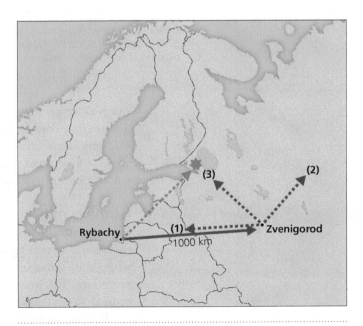

Turtles placed in a simulated magnetic field 337 km north of their capture site attempted to swim south.

Individuals placed in a simulated magnetic field 337 km south of their capture site attempted to move north.

Figure 10.28. Orientation of displaced turtles. Turtles placed in magnetic fields that simulated magnetic conditions north or south of the capture site attempted to return to their site of capture. The mean orientation is indicated by each arrow. (*Source:* Lohmann et al. 2004)

Bicoordinate navigation in birds

Featured Research

Many birds migrate thousands of kilometers and use a geomagnetic compass, a sun compass, or a star compass (for a review, see Wiltschko & Wilschko 2009). But do birds display bicoordinate navigation? To answer this question, Nikita Chernetsov, Dmitry Kishkinev, and Henrik Mouritsen examined migration movements in Eurasian reed warblers (*Acrocephalus scirpaceus*), which spend the winter in sub-Saharan Africa and then migrate to Eurasia to breed.

Chernetsov and colleagues conducted an experiment during the warblers' spring migration (Chernetsov, Kishkinev, & Mouritsen 2008). Over a period of three years, they captured individuals near Rybachy, Russia. Half were flown 1,000 km east to Zvenigorod, while the other half were kept in Rybachy (Figure 10.29). The birds were placed in small Emlen funnels, an apparatus developed by Stephen Emlen to study use of the star compass in birds (Toolbox 10.1). The test birds had a clear view of the sky and so could use a variety of compass cues (Figure 10.30).

The Eurasian reed warblers tested at the capture site oriented northeast, the direction of their destination. The birds displaced 1,000 km east could exhibit one of three behaviors: (1) they might orient due west, trying to return to the capture site, as we saw in green sea turtles; (2) they might orient northeast, the direction they were originally heading; or (3) they might orient northwest, trying to navigate to their final destination. The research team found that displaced birds oriented to the northwest and so were orienting in a new direction to compensate for being displaced, suggesting that they, too, have the capacity for bicoordinate navigation (Figure 10.31). These results suggest that Eurasian reed warblers are able to determine their geographic position and adjust their migration route after being displaced,

Figure 10.29. Displacement predictions. Birds migrating northeast (to the destination indicated by the red dashed line and red star) that were displaced 1,000 km east from Rybachy to Zvenigorod could travel in one of three directions (dashed lines): (1) west, (2) northeast, or (3) northwest. (*Source:* Chernetsov, Kishkinev, & Mouritsen 2008)

Figure 10.30. Reed warbler migration. (a) The Emlen funnels used by Chernetsov and colleagues to study migratory movement in the field. (b) Eurasian reed warbler.

a

Birds at the capture site oriented primarily to the northeast as expected.

b

Displaced birds oriented primarily to the northwest to compensate for their displacement.

Figure 10.31. Orientation results. (a) Orientation of birds at the capture site, predominately to the northeast. (b) Orientation of birds at the displacement site, predominately to the northwest. Arrows indicate mean direction of orientation in each case. (*Source:* Chernetsov, Kishkinev, & Mouritsen 2008)

much as we use maps to find our way home after getting lost—although it is still unclear how they do this. Obviously, this ability can be of adaptive significance if birds are blown off course or need to modify their migratory route.

One final set of well-known migrants include salmon and their relatives. We end by examining how these fish accomplish their navigational feats.

Featured Research

Homing migration in salmon

Salmonids (salmon, char, and trout) are well known for their ability to return to their natal stream to spawn. Many salmon migrate downstream to the ocean, where they spend up to several years feeding prior to their long journey home to reproduce. This journey can be up to thousands of kilometers long and can occur more than ten years after leaving the natal stream. How is this feat accomplished?

Two mechanisms are thought to be used in this migration: olfaction and geomagnetic reception. Both rely on imprinting, which is thought to occur when salmon undergo a physiological change that allows them to adapt to both freshwater and saltwater—a process known as smoltification. Smoltification is associated with increases in several hormones, including thyroxine, which is thought to be associated with olfactory imprinting of the natal stream (Dittman & Quinn 1996). In essence, fish are thought to learn the odor and the magnetic properties of their natal stream in order to return to it.

The olfaction imprinting hypothesis was first proposed by Arthur Haesler and Warren Wisby and assumes that streams possess a characteristic and persistent odor, that fish can detect such odors, and that fish retain a memory of the odor they experience early in life (Haesler & Wisby 1951). Haesler and Wisby demonstrated the importance of olfaction in a simple experiment in which they captured salmon migrating upstream from two forks of a Y-shaped stream near Seattle (Wisby & Haesler 1954). Each fish was uniquely tagged to identify the fork in which it was captured. Half the fish had their olfactory pits blocked to impair their olfactory sense, and the other half (controls) were left untreated. All individuals were then released 1.6 km below the fork so that they could resume their migration upstream. Of the fish recaptured, 89% of controls but only 40% of treatment fish were captured in the same branch in which they were originally captured. The researchers concluded that olfactory-impaired fish randomly selected a fork, whereas those with functioning olfaction used that sense to return to their natal stream. What exactly provides the unique odors of natal streams is less clear. Some work points to amino acids (Yamamoto, Hino, & Ueda 2010), while other work suggests that salmon detect the odor of related, younger fish moving downstream (Nordeng 1977).

For adult salmon in the open ocean, the first need is to find the mouth of their natal river system. It is difficult to do this using odor alone, and so a second mechanism is needed: this may be geomagnetic imprinting. Salmon are hypothesized to imprint on the specific properties of the earth's magnetic field at the mouth of their natal river system (Lohmann, Putnam, & Lohmann 2008). In particular, they are thought to learn the inclination angle and/or intensity of the earth's magnetic field at this location, both of which vary with latitude (Figure 10.32). Studies have shown that salmon can respond to different magnetic field intensities, and juvenile salmon will change their orientation in tanks in response to a change in the external magnetic field (DeBose & Nevitt 2008). We expect further evaluation of this hypothesis in the coming years.

These studies illustrate how researchers study bicoordinate navigation and reveal that a variety of animals possess such a capability. We still have much to learn about the navigational abilities of most species, including humans (Applying the Concepts 10.3).

Figure 10.32. Salmon geomagnetic navigation. Salmon can imprint on the intensity of the earth's magnetic field ($X_{imprint}$) at the mouth of their natal river system (bold line). To return to their natal site from the open ocean (O), they need to travel north (or south) toward this intensity (R arrow) and then turn east and remain along this intensity until they reach the coast. (*Source*: Bracis & Anderson 2012)

APPLYING THE CONCEPTS 10.3

Human magnetic orientation

In this chapter we've seen that many species orient using the earth's magnetic field. Is there any evidence that humans can do so too? In 1980, R. Robin Baker presented evidence that humans, after being displaced, were able to recognize their homeward direction without visual cues (Baker 1980). He conducted a series of experiments in which college students were blindfolded and then transported in a circuitous route to a location 6 to 52 km from where they had lived for at least two years (Manchester University in Great Britain). He asked each student to indicate the direction home while still blindfolded. He found that, on average,

continued...

students tended to indicate the correct direction. In a second experiment, he used the same procedure but placed bar magnets on the heads of half the students and brass bars (with no magnetic properties) on the heads of the other students. Those with magnets on their heads oriented randomly, while those with brass bars tended to correctly point toward home. These experiments stimulated much interest by suggesting that the earth's magnetic field plays a role in human orientation.

In subsequent years, however, other researchers have failed to replicate Baker's findings. Studies on hundreds of subjects in the United States, Australia, and Great Britain have all showed that humans do not correctly orient toward home following a blindfolded displacement (Gould & Abel 1981; Fildes, O'Loughlin, & Bradshaw 1984; Westoby & Partridge 1986), casting much doubt on the ability of humans to orient using a magnetic sense.

Recent work, however, has brought renewed attention to this question. Several studies have revealed that in many species, the ability to detect magnetic fields is dependent on light-sensitive chemical reactions in the retina involving a protein called cryptochrome (CRY) (Thoss et al. 2002). Interestingly, humans have a similar protein in the retina (hCRY2), suggesting that we may also have this ability.

To investigate whether hCRY2 is sensitive to magnetic fields, Lauren Foley, Robert Gegear, and Steven Reppert conducted an experiment with fruit flies (Foley, Gegear, & Reppert 2011). Wild-type flies (with CRY) respond to magnetic fields, but only in the presence of light. Mutant flies that do not express CRY show no consistent response to magnetic fields, which indicates that magnetoreception in flies involves functional CRY. The research team created transgenic fruit flies by inserting hCRY2 into the genome of the mutant flies. They found that the transgenic flies responded to magnetic fields in a light-dependent manner, just like wild-type flies. This work suggests that human hCRY2 may have the molecular capability to be a magnetosensor (i.e., involved in magnetoreception)—a possibility that awaits further testing. ■

TOOLBOX 10.1

Emlen funnels

Early observations about bird migration indicated that caged birds increased their activity levels each spring and fall, a behavior known as migratory restlessness, or *Zugunruhe*. Stephen Emlen studied this behavior in a series of classic experiments using indigo buntings (*Passerina cyanea*). Indigo buntings are beautiful blue birds that breed each spring in the eastern United States and then migrate to spend the winter in southern Mexico, Central America, and the Bahamas. Interestingly, these birds are normally diurnal, except during the migration season, when they become highly active in the evening and throughout the night. These observations suggested that perhaps these birds were using celestial cues such as stars to orient.

Emlen designed a simple yet elegant apparatus to take advantage of the buntings' migratory restlessness behavior (Emlen & Emlen 1966) (Figure 1). An Emlen funnel consists of a small funnel with a wire mesh top that allows a bird to see out. The bottom has an inkpad that transfers ink to a bird's feet. The sides of the funnel are lined with paper. A bird's movements in hopping forward as it exhibits migratory restlessness are recorded by the ink tracks left by its feet on the paper.

Emlen conducted experiments in a planetarium, where he could manipulate the night sky to examine how it affected the birds' orientation and movements (Emlen 1967a; Emlen 1967b). Stars can provide directional information in the Northern Hemisphere, because all stars rotate around the celestial pole. In a series of experiments, he placed birds in Emlen funnels and manipulated the location of stars and their rotation. Emlen convincingly showed that indigo buntings use celestial cues for orientation. In particular, he determined that maturing birds learn to identify the direction around which all stars rotate and use this fixed location as an indicator of north. In the

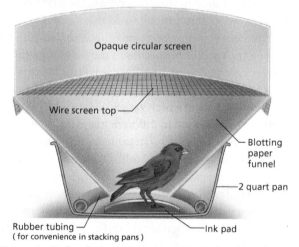

Figure 1. Emlen funnel. Side view of an Emlen funnel, showing a bird standing on an inkpad. Its footprints are then recorded on the paper sides. Inset: Indigo bunting. (*Source:* Emlen & Emlen 1966)

Northern Hemisphere, the celestial pole is centered on the North Star, Polaris. Emlen demonstrated that he could train young birds to use a different star (Betelgeuse, in the constellation Orion) as an incorrect indicator of north after he allowed birds to view all stars in the planetarium rotating around it (Emlen 1970). More recent research indicates that indigo buntings may also use the earth's magnetic field to orient in the proper direction during migration (Sandberg et al. 2000).

Chapter Summary and Beyond

Animals move from one location to another during dispersal and migration. Juveniles embarking on natal dispersal leave their place of birth to find a site for reproduction. Dispersal can benefit individuals by reducing both competition with conspecifics and the likelihood of settling nearby and mating with close kin. Two recent reviews on dispersal behavior suggest that to more completely understand these movements, we need to focus on how the environmental context of an animal (e.g., competition, habitat quality, inbreeding risk) and the internal state of an individual (e.g., body condition, physiology) interact (Handley & Perrin 2007; Clobert et al. 2009). Breeding dispersal by adults is commonly observed after a reproductive failure. Recent work has focused on how individuals combine and use information about their own reproductive success, the success of others nearby, and the density of individuals in the area when deciding whether to disperse to a new site (e.g., Citta and Lindberg 2007). For a comprehensive review on the evolution of dispersal, see Ronce (2007).

Longer, two-way movements are known as migration. Animals use a variety of cues to orient in the proper direction, including the position of the sun and stars and the earth's geomagnetic field. Ongoing work continues to reveal that animals use multiple cues for orientation (Bingman & Chen 2005; Guerra, Gegear, & Reppert 2014; Muheim, Moore, & Phillips 2006; Murray, Estepp, & Cain 2006). Mouritsen and Hore (2012) provide additional details about magnetoreception in birds. The study of migration is being revolutionized by the development of small geolocators that can be attached to individuals to record long-distance movements (Reynolds & Riley 2002; Stutchbury et al. 2009). For a comprehensive review of mechanisms that regulate migration, see Ramenofsky and Wingfield (2007), and for a review of physiological adaptations involved in migration, see Åkesson and Hedenström (2007).

CHAPTER QUESTIONS

1. Identify two ways that dispersal differs from migration behavior. In what way are they alike?
2. Competition for resources and inbreeding avoidance are two explanations for natal dispersal. Discuss each of these hypotheses in the context of proximate and ultimate explanations for natal dispersal.
3. A researcher examined natal dispersal distances in a population of chickadees over the course of two years. The population size and number of juveniles that dispersed was the same in both years. However, in Year 1, the average dispersal distance was 10 km, but in Year 2, the average dispersal distance was only 5 km. Propose an explanation for this difference.
4. What assumption is required to explain the win-stay lose-shift behavioral strategy often seen in breeding dispersal?
5. Flycatchers, like the eastern wood pewee (*Contopus virens*), eat flying insects, including bees, flies, butterflies, and wasps. Woodpeckers, such as the downy woodpecker (*Picoides pubescens*), eat beetle larvae buried under tree bark, as well as berries, seeds, and sap. In temperate climates like midwestern North America, flycatchers migrate each fall to the tropics, whereas woodpeckers do not. In this chapter, we discussed the relationship between resource fluctuations and migration. How would you explain the differences in migratory behavior of flycatchers and woodpeckers? To do this, you need more information. Gather information on the life cycle or life history of these birds' main food items. You can find this information in either the primary or secondary literature. Using the information you collect, describe the relative stability of these birds' food sources over the course of the year.
6. In a northern population of sparrows, you have determined that 80% of individuals migrate each fall to warmer locations, while 20% remain residents year-round and overwinter on the breeding grounds. Climate scientists predict that at this location, winters will become milder over the next 50 years. Explain how this shift might affect the proportion of adults that migrate in the future.
7. How does geomagnetic orientation differ from bicoordinate navigation that relies on the earth's magnetic field?
8. Based on the work of Foley, Gegear, and Reppert (2011) on hCRY2, propose a hypothesis to explain the lack of evidence for magnetosensitivity in most studies of humans.
9. Compare the dispersal studies conducted by Switzer (1993; 1997) and Bollinger, Harper, & Barrett (1993) with the studies on migration behavior conducted by Boyle and Conway (2007) and Merlin, Gegear, and Reppert (2009). Describe which of Tinbergen's four questions each study investigated, and explain your answer.

Chapter 11

Habitat Selection, Territoriality, and Aggression

On a recent outing, we left our vehicle parked near a trail as we hiked. When we returned, a northern mockingbird (*Mimus polyglottos*) was vigorously pecking at the car's side-view mirror (Figure 11.1). We watched for a while, trying to understand this seemingly odd behavior. We realized this bird probably perceived its reflection in the mirror as a rival threat that required active aggression. Once we placed a shirt over the mirror, the behavior stopped. Our observation was not that unusual, and in fact, researchers often use mirror reflections to study aggressive behavior (e.g., Kusayama, Bischof, & Watanabe 2000).

Many animals, like the mockingbird we observed, defend territories, particularly during the breeding season. Before breeding, however, many individuals must select a habitat in which to settle. In this chapter, we'll see that habitat quality and the presence of conspecifics play major roles in habitat selection. The presence of another individual means that resources will be used by multiple individuals, meaning that competition in these areas may be high. Yet the presence of a rival may also indicate that a habitat is of high quality. We'll also see that once an individual has settled in a habitat, territorial defense can provide

Figure 11.1. Mockingbird and mirror. A mockingbird attacks its mirror image, a perceived rival, to defend its territory.

preferential access to food or mates. However, aggressive interactions are costly, and so not all individuals can successfully defend a territory. Next, we examine how hormones influence aggressive behavior. Finally, we discuss how game theory models are used to understand the tremendous variation observed in the intensity and duration of aggressive contests over resources.

11.1 Resource availability and the presence of others can influence habitat selection

Recall that in Chapter 1, we examined a hypothetical situation in which there was variation in the number of robins in yards: some yards had many robins, while others had few. In nature, such variation in the density of individuals in different habitat locations is common. Two factors help explain much of this variation: the amount of resources and the number of other individuals in a habitat. Resource availability is an important aspect of habitat quality, and the number of individuals in a habitat affects the level of competition for those resources. Let's see how researchers have examined the importance of these factors to explain distributions of animals.

The ideal free distribution model

ideal free distribution (IFD) model
A model that explains how animals distribute themselves among habitats or food patches.

Steve Fretwell and Henry Lucas developed the **ideal free distribution (IFD) model** to explain how animals distribute themselves among habitats (Fretwell & Lucas 1969). The model's name comes from its assumption that animals act in an "ideal" manner, in that they select habitats that maximize survival and reproduction and are "free" to enter any habitat.

The model contains five assumptions:

1. Individuals attempt to maximize their fitness when settling in a habitat.
2. Habitat locations differ in the resources they contain.
3. The fitness of individuals in a habitat decreases as more individuals settle there because of increased competition for resources. In other words, fitness is negatively density dependent.
4. Individuals have equal competitive ability and can accurately assess the fitness payoffs of each habitat.
5. Individuals are free to move between habitats at no cost.

The model then makes the following predictions:
Prediction 1: Individuals will settle in habitats based on the relative fitness payoffs of those habitats: the number of individuals in each habitat will be proportional to habitat quality, resulting in more animals settling in higher-quality habitats.
Prediction 2: In the end, all individuals will have the same fitness no matter where they settle.

We can understand the logic of these predictions by examining two hypothetical habitats, or habitat patches, of high and low quality (Figure 11.2). Imagine a situation in which individuals settle sequentially in these habitats. The first few individuals should settle in the high-quality habitat because this will yield the highest fitness. However, as more and more individuals enter the high-quality habitat, the fitness of each individual will decline because of increasing competition for resources, producing a negative relationship between the number of individuals in a habitat and their fitness. At some point, the fitness payoff will be higher for an individual settling in the poor-quality habitat, so a new settler should go there, and additional settlers should then select both of these habitats equally.

This model also allows us to make predictions about the relative number of individuals in each habitat. If habitat quality is determined by resources, the model predicts that the number of individuals in a habitat will be proportionate to the resources available. For just two habitats, if one contains twice the resources of the other, then twice as many individuals should settle there. If the resources differ by a factor of ten, so will the number of individuals. In this way, each individual obtains the same fitness no matter which habitat it settles in. During the settling process, individuals are always free to move to a higher-quality habitat if one is available. However, once all individuals settle, none can do better by moving to another habitat. Let's look at two tests of these predictions.

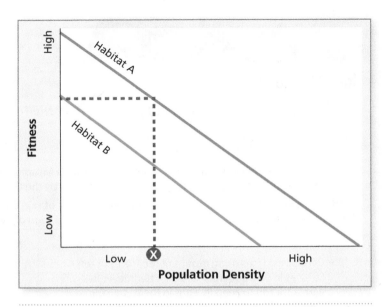

Figure 11.2. Ideal free distribution model. Consider two habitats, A and B. Habitat A is the higher-quality habitat because it yields the highest fitness for any given population density. As individuals first arrive, all individuals should select Habitat A. After more than X individuals arrive, they should settle equally in both habitats.

The ideal free distribution model and guppies

Featured Research

The IFD model is easiest to test at a small spatial scale; therefore, numerous tests have used small food patches. At this smaller scale, patch quality (i.e., habitat quality) can be manipulated by changing food density or provisioning rate. Mark Abrahams did this when he tested the model using guppies *Poecilia reticulata* (Abrahams 1989). Abrahams created two food patches using flasks filled with water that delivered food (fly eggs) to two feeding bars, or patches, on opposite sides of an aquarium. He modified the quality of each patch by changing the number of eggs in each flask. In this way, the relative patch quality varied from equal (1:1 ratio) to highly skewed (1:9 ratio). He then placed ten individuals of the same sex in the aquarium and fed them at three distinct times each day using the same relative patch qualities. In each trial, he recorded the number of fish at each patch (Scientific Process 11.1).

The IFD model predicts that the fish should distribute themselves among the patches in the same ratio as the food delivery ratio. Abrahams found that the model accurately predicted the distribution of fish in the trials held later in the day, but not their distribution in the first trial of the day. Why? Apparently, it took the fish some time each day to learn the relative qualities of the patches. Abrahams also quantified individual feeding rates for a smaller sample of fish by counting the number of eggs eaten by each individual at a patch. He found that all fish had the same feeding intake rate and saw no evidence of unequal competitive ability—a finding that would have violated an assumption of the model. Abrahams's results support the IFD model, demonstrating how resources and competition affect the distribution of individuals among patches.

The IFD model has been very effective in predicting and explaining variation in the number of individuals in food patches. Can the model also explain habitat selection over larger geographic scales and longer time scales? Our next study provides the answer.

SCIENTIFIC PROCESS 11.1

Ideal free guppies

RESEARCH QUESTION: *How do guppies distribute themselves among food patches?*

Hypothesis: Food delivery rate and number of competitors will affect the distribution of individuals among food patches, as outlined by the IFD model.

Prediction: Individuals will distribute themselves so that all obtain equal food intake rates. Their relative abundance will match the relative rate of food delivery.

Methods: The researchers:

- Conducted trials using groups of ten males or ten females in 90 L aquaria. Feeders (Erlenmeyer flasks with a stir rod to keep fly eggs in the water solution) were placed on two sides of the aquarium with a piece of tubing creating a horizontal feeding bar. The presence of different amounts of food in each flask created patches of different quality but kept the over-all total food the same across treatments.
- Conducted three trials per day (one in the morning, one at noon, and one in the evening) using the same relative patch qualities
- Counted the position of the guppies (left feeder, right feeder, or not feeding) every 30 seconds for 24 minutes per trial

Results:

- The distribution of male and female fish was strongly affected by the amount of food in a patch.
- In the morning trial, fewer fish than predicted used the patch with more food.
- In the noon and evening trials, the distributions matched the IFD prediction.
- Within a patch, the feeding intake rate of individuals was similar, as the model assumed.

Figure 1. Experimental tank. Feeders on each side create two feeding patches. A trough under the feeding bar collects extra food. (*Source:* Abrahams 1989)

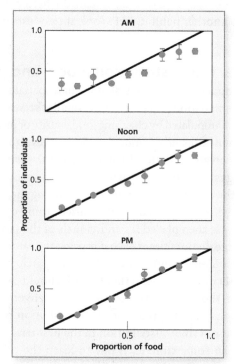

Figure 2. Patch use. The proportion of individuals at the left feeder versus the proportion of food available at that feeder for the morning, noon, and afternoon trials. (*Source:* Abrahams 1989)

Conclusion: Guppies distributed themselves according to the predictions of the IFD model once they had time to learn the distribution of food.

EVALUATE

How would the predictions of the ideal free model change if one individual was a dominant, superior competitor?

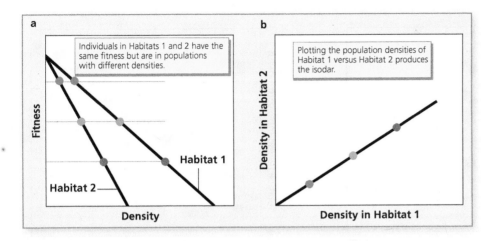

Figure 11.3. Isodar. (a) The ideal free distribution model shows population sizes in two habitats in each of three years (denoted by different colors). Note that in all three years, the fitness for individuals in the two habitats is equal. (b) Data from both densities of each year are plotted against one another to yield a single point. Data from multiple years (different colors) form a linear isodar with a positive slope. (*Source:* Morris 2006)

The ideal free distribution model and pike

Featured Research

Douglas Morris extended the IFD model to predict the distribution of individuals among habitat patches over a number of years as population size changed over time (Morris 1988). Recall that the model predicts that individuals in all patches will have equal fitness. However, if total population size changes each year, so too will the predicted number of individuals in each habitat. Morris showed that if you plot the predicted number of individuals in one habitat against the predicted number in the other habitat over time, the model predicts a straight line with a positive slope, which Morris called an **isodar** (*iso* means same, and *dar* comes from *Darwinian fitness*) (Figure 11.3).

Thrond Haugen and colleagues tested the isodar line prediction of the IFD model for a population of pike (*Esox lucius*) in Lake Windermere in England (Haugen et al. 2006). Lake Windermere is a long, narrow lake with a shallow portion in the middle that creates two deep basins, or habitat patches: one in the north and one in the south. Pike are top predators that often exceed 1 m in length (Figure 11.4). The midpoints of the basins are only 7 km apart, and pike regularly move over 2 km per day, so the cost of moving between basins should be relatively small. Researchers have conducted mark-recapture studies to estimate the density of pike in these basins for 50 years, and repeated captures have revealed that individuals often move back and forth between the basins, as assumed by the model.

Recall that both resource availability and the presence of others affect habitat selection. In this study, the researchers did not examine the effect of differences in habitat quality; rather, they examined the presence of others, taking advantage of year-to-year variations in population size and manipulations of the number of fish in each habitat over time. At Lake Windermere, high-intensity fishing was conducted in the south basin for three years and was then followed by high-intensity fishing in the north basin for three years. These manipulations reduced pike populations by 15 to 28% within each basin in a given year.

isodar The line that results from plotting the number of individuals in one habitat against the number of individuals in another habitat at several points in time.

Figure 11.4. Pike. Pike are common top predators in freshwater systems.

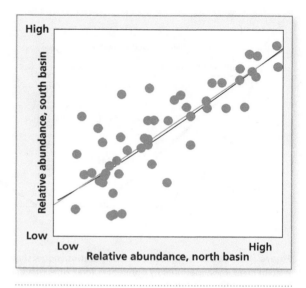

Figure 11.5. Pike fecundity. In both habitat patches, mean fecundity declined as the density of pike increased, and so pike show negative density dependence in fecundity in each basin, as assumed by the IFD model. Circle size reflects estimated body mass of females. (*Source:* Haugen et al. 2006)

Figure 11.6. Pike isodar. The observed isodar (orange line) for the pike data matches the predicted isodar from the IFD model (black line). Each point represents one year of data. (*Source:* Haugen et al. 2006)

Haugen and colleagues first examined the assumption that the fitness of individuals in each basin was negatively density dependent. To determine this, they used the long-term mark-recapture data to plot survivorship and reproductive rates of individuals in years of different density. Within each basin, they found that in years of low pike abundance, both survivorship and fecundity (reproduction) were high, while in years of high abundance, survivorship and reproduction were low, as the IFD model assumes (Figure 11.5).

Over the 50-year period, there was tremendous variation in total population size—and thus the number of pike in each basin. This provided sufficient variation in densities in the basins to plot the isodar, which turned out to be a straight line with a positive slope, just as the IFD model predicts (Figure 11.6).

These examples illustrate how variation in patch quality and the presence of others affect the distribution of individuals in both small-scale food patches and larger-scale habitats in a manner consistent with Fretwell and Lucas's (1969) model. It is important to note that although the model assumes that individuals know the relative quality of their choices, the guppy experiment reveals that it can take time for individuals to assess the relative quality of food patches. How do animals assess the relative quality of larger-scale habitats? Recent research provides insight into this question.

Featured Research

Cuckoos assess habitat quality

One way animals can assess the relative quality of different habitats is by estimating the level of food resources they contain. Nicholas Barber, Robert Marquis, and Wendy Tori used a novel approach to examine how some birds might do this. Black-billed (*Coccyzus erythropthalmus*) and yellow-billed (*C. americanus*) cuckoos migrate from Central and South America to breed in woodland habitats in eastern North America each spring. They eat a variety of insects but consume large quantities of hairy caterpillars, including exotic gypsy moths (*Lymantria dispar*) (Figure 11.7). Gypsy moths, native to Europe and Asia, were accidentally introduced into northeastern North America in the late 1860s and have expanded west and south ever since, making them a major forest pest. Females lay egg masses on tree branches in the fall, and larvae (caterpillars) emerge in spring when trees leaf out. At this stage, larvae can disperse on silken threads to reach new trees. The larvae develop through several molting stages over a period of several weeks and consume leaves as they mature.

Gypsy moth populations fluctuate over space and time. When population densities reach unusually high levels in outbreak years, trees become defoliated as a result of the caterpillars' high leaf consumption (Figure 11.8). Such defoliation is easy to see in aerial photographs. Do cuckoos use defoliation to locate gypsy moth outbreaks and thus habitats with abundant food? To examine this idea, Barber and colleagues obtained digitized maps of gypsy moth outbreaks for each year from 1975 to 2003 (Barber, Marquis, & Tori 2008). They also used data from the Breeding Bird Survey (BBS) to estimate the abundance of cuckoos over a period of 16 years—five years before an outbreak year, the outbreak year, and ten years following the outbreak. BBS data record the abundance of all birds seen or heard each spring along fixed 39 km census routes. The researchers identified BBS routes that came into contact with defoliation areas as focal routes. For each route, they calculated average cuckoo abundance over the 16-year period.

The research team found that cuckoo abundance on focal BBS routes correlated with caterpillar outbreaks. The abundance of cuckoos peaked in the year of defoliation, but in the five years preceding defoliation, abundances were at or below average (Figure 11.9). Black-billed cuckoo densities remained above average in the year following defoliation but returned to average in subsequent years; yellow-billed cuckoo densities were at or below average in all years following an outbreak. This pattern suggests that cuckoos can zero in on local areas in each year that gypsy moths are abundant and defoliating trees.

Were the birds actually tracking the outbreaks? To answer this question, the researchers attempted to characterize patterns of bird movement. Barber and colleagues examined all BBS routes within 500 km of a gypsy moth outbreak. For each route, they determined the distance and compass direction to the area of defoliation. They then compared cuckoo abundance on each route in the outbreak year to its ten-year average. They found that routes within 40 km of an outbreak had substantially higher-than-average cuckoo abundances in outbreak years, whereas routes that were 45 to 140 km away from an outbreak had lower-than-average abundances in outbreak years. This pattern suggests that birds 45 to 140 km away from an outbreak moved toward the outbreak—and did so from all directions.

Barber and his colleagues concluded that gypsy moth outbreaks strongly affect the local abundance of cuckoos. They suggest that cuckoos will move fairly long distances (up to 140 km) to find high-quality habitats, such as those produced by gypsy moth outbreaks. Large defoliated areas may provide the cues for such habitats, but this hypothesis has yet to be directly tested.

Up until now, we have assumed that individuals assess the relative quality of their options and that rivals reduce fitness in a habitat. So if two habitats of equal quality exist and one is already occupied, a new settler should opt for the unoccupied habitat. Sometimes, however, individuals prefer to settle in the same habitat that a competitor already occupies, even when a high-quality, unoccupied habitat is nearby. This requires a new explanation, as we see next.

Conspecific attraction

The IFD model assumes that competition for resources results in a decline in individual fitness as the number of individuals in a habitat increases. Because each

Figure 11.7. Cuckoo and moth.
(a) Yellow-billed cuckoo. (b) Gypsy moth larva.

Figure 11.8. Gypsy moth defoliation. A patch of trees defoliated by gypsy moth larvae.

Figure 11.9. Cuckoo abundance over time. For both species, mean (± SE) cuckoo abundance on a route was highest in the year of gypsy moth defoliation, indicated by the yellow line (zero years since defoliation). The horizontal dashed line is the 16-year average abundance on the route. Abundance was standardized to account for the population declines in both species over the entire study period. (*Source:* Barber, Marquis, & Tori 2008)

additional individual reduces fitness for all, owing to competition, new settlers should avoid rivals if a better option exists. Sometimes, however, individuals exhibit the opposite pattern. Instead of avoiding others, they settle near them, a pattern known as **conspecific attraction**.

Why might individuals exhibit conspecific attraction? Two hypotheses exist. First, for very low population sizes, fitness may increase as more individuals settle, a phenomenon known as an **Allee effect** (Allee 1931). For example, a solitary individual might be at high risk of predation, whereas two or more individuals might experience lower risk. Similarly, a group might be more attractive to potential mates than a single individual.

Second, it may be difficult for individuals to assess the relative quality of some habitats—an ability assumed by the IFD model. Habitat quality is a function of many factors, such as resource levels, the availability of nesting locations, and the number of predators, some of which may be difficult to assess. Instead, a settler can use the presence of another individual as an indicator of habitat quality, a phenomenon called **conspecific cueing** (Stamps 1988). In many species, experienced breeders will resettle in a habitat if they successfully reproduced there the preceding year, as we saw in Chapter 10. If a high-quality habitat last year is likely to be a high-quality habitat this year, then the location of experienced breeders provides a valuable cue. The conspecific cueing hypothesis predicts that inexperienced individuals, whether first-time breeders or new immigrants, are most likely to exhibit conspecific cueing. The next two studies test these hypotheses.

conspecific attraction A phenomenon in which individuals are attracted to others, particularly during habitat selection.

Allee effect A situation in which the fitness of individuals increases with increased population density.

conspecific cueing A hypothesized mechanism to explain conspecific attraction. A settler uses the presence of another individual as a cue to habitat quality.

Conspecific attraction and Allee effects in grasshoppers

The creosote bush grasshopper, or desert clicker (*Ligurotettix coquilletti*), is abundant in the deserts of North America, where it feeds on the leaves of creosote bushes (*Larrea tridentata*). Each spring, males set up mating territories on these bushes, and so a bush represents a breeding habitat. Males tend to be faithful to a site and remain in the territory throughout the spring and summer, when they make a loud clicking sound to attract females (Greenfield & Shelly 1985). Earlier studies indicated that males appear to aggregate on some bushes and not use others at all, even though the bushes seemed to be similar in quality (Greenfield, Shelly, & Downum 1987). Why does such congregation occur? Something other than resource quality appears to affect males' habitat choice.

Katherine Muller wondered whether male desert clickers were using the presence and vocalizations of others as a cue for habitat selection (Muller 1998). At a site in the Mojave Desert, she selected pairs of adjacent creosote bushes. She used playback calls from speakers placed at the base of each bush to standardize the vocalizations. In the first set of experimental trials, she gave males the choice of one bush with a speaker playing conspecific male calls or another bush that was

SCIENTIFIC PROCESS 11.2

Conspecific attraction in grasshoppers

RESEARCH QUESTION: *How do desert clickers select habitat patches?*

Hypothesis: Males use the presence of another male as an indicator of patch quality.

Prediction: Males will prefer creosote bush patches from which conspecifics call.

Methods (1): The researchers:

- Chose pairs of creosote bushes (3.5 m apart) that were similar in size and shape
- Played a conspecific call from one bush and either a heterospecific or no call (silent control) from the other bush
- Placed a caged focal male 2 m from a pair of habitat patches (creosote bushes) and after release, measured which bush the male selected

Focal male

Figure 1. Design for experiments. Males could select one of two habitat bushes.

Results (1): Seventeen out of 20 males flew to the conspecific call bush versus the silent bush. Sixteen out of 20 males flew to the conspecific rather than the heterospecific call bush.

Methods (2): The researchers:

- Used a similar design as in the first set of trials but placed a resident male on one bush and left the other bush empty
- Recorded the presence of the focal males on the bushes at 12 and 36 hours after release

Figure 2. Habitat choice. Males preferred the bush with a conspecific (blue bars) over the empty bush (orange bars). (*Source:* Muller 1998)

Results (2): Most (72 out of 98) focal males selected the bush with a conspecific at 12 hours. More males were also found on the bush that contained a conspecific at 36 hours.

Conclusion: Males prefer habitat patches that contain other males, an example of conspecific attraction.

either unoccupied or had a speaker playing heterospecific male calls. She played the calls at one bush at a time while a focal male was held in a cage equidistant from each bush. She then lifted the lid of the cage and recorded the movement behavior of the focal male (Scientific Process 11.2).

In these trials, at least 80% of males flew to the conspecific call bush. Clearly, males preferred the habitat with an apparent conspecific.

To determine what happens after a male arrives on a bush that also contains a rival, Muller conducted another set of experimental trials using similar protocols. This time, one of the bushes was occupied with a calling conspecific male that had been a resident of the bush for three days, and the other bush was empty. After the focal male was released, Muller counted the number of grasshoppers on both bushes 12 and 36 hours later. Most released males were found on occupied bushes

EVALUATE

Why might females prefer bushes with more than one calling male?

at the end of 12 hours, again displaying conspecific attraction. At the end of 36 hours, some of these males had abandoned the bushes, and a higher proportion had left the occupied bush than had left the empty bush. Muller observed aggressive territorial interactions among males on the bushes and assumed that this behavior explained the decline in males on occupied bushes over time.

Why did focal males prefer the occupied bushes? One explanation is female behavior. In previous work, Muller found that females exhibit a strong preference for bushes with multiple calling males (Muller 1995); Greenfield and Shelly (1985) found that aggregated males have higher mating success than singletons. In other words, female preference may create an Allee effect: males that aggregate on a bush with rivals can obtain more mating partners than can solitary males. Thus, even though a second male may experience increased aggression from an established male, its reproductive success can still be higher than if it settled alone. This example illustrates how an Allee effect can explain conspecific attraction in grasshoppers and their habitat selection behavior. Next, we turn to an example of conspecific cueing to explain conspecific attraction in birds.

Featured Research

Conspecific cueing in American redstarts

American redstarts (*Setophaga ruticilla*) are migratory warblers that breed in deciduous forests in eastern North America (Figure 11.10). Each spring, migrants arrive on their breeding grounds and establish territories. Adult males return first, followed by yearling males that will breed for the first time, and then females. The breeding season is relatively short, and so males need to find a territory quickly in order to breed successfully.

Beth Hahn and Emily Silverman studied conspecific cueing and habitat selection in these birds at a breeding site in Michigan (Hahn & Silverman 2006). Three types of males settle on a territory. The first are returning males, or experienced birds that bred the preceding year at that site. The second are new adult immigrants, or males that bred in a different location the preceding year. The third are first-time breeders, yearling males. These yearlings can be distinguished from older adults by their plumage: they have only scattered black feathers on their head, breast, and back, while adults are solid black (Rohwer, Klein, & Heard 1983).

Both immigrant adults and yearlings lack experience in the habitat. Hahn and Silverman predicted that they would use the presence of a singing male as a cue to the location of a high-quality habitat. In contrast, they predicted that returning males would not be affected by this singing.

The researchers captured and uniquely color-banded all birds on 12 habitat plots measuring 300 m × 300 m. Six plots were designated as treatment plots, and six as controls. In each treatment plot, the researchers played songs of male American redstarts every day, starting before any males arrived and continuing until all males had settled. The next year, the treatment plots were switched: treatment plots became controls, and controls became treatment plots. To play the songs, the researchers placed speakers 2 m up in trees and connected them to a compact disc player. They standardized the redstart playback amplitude so that songs could be heard 100 m away. In order to avoid habituation to the songs, they used a total of 75 individuals to make the discs. Hahn and Silverman surveyed each plot four times per week, recording the arrival of males along with their location, age, and identity. Any new, unmarked adult male was captured and uniquely color-banded.

Figure 11.10. Calling American redstart. A common bird in North American woodlands.

As predicted, new immigrant males strongly preferred to settle on plots that played a conspecific song, while returning adults did not, indicating that conspecific songs do attract new immigrant males to plots through conspecific cueing. However, first-time breeders did not exhibit conspecific attraction after all: they settled equally on treatment and control plots (Figure 11.11). Why the unexpected result? First-time breeders did not arrive at the site until six to 10 days *after* the experienced adults. By then, the adults had already settled and begun to sing, so song production was high on all plots, effectively swamping out the playback treatment. Nonetheless, these results indicate that conspecific attraction is an important mechanism for habitat selection for new males. We now know that many birds respond positively to song production when searching for habitat and breeding territories (Ahlering & Faaborg 2006). Conservation biologists have used this finding to help restore populations of birds threatened with extinction (Applying the Concepts 11.1).

In this section, we have seen that habitat selection is a complex process that is influenced by resource distribution, the number of conspecifics, habitat patch assessment, and Allee effects. Selection of a habitat patch is only the first step for many species: individuals then may need to establish a territory in order to successfully reproduce. We examine this behavior next.

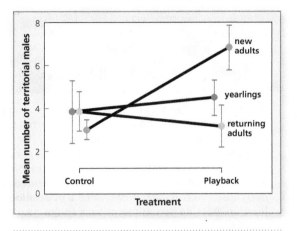

Figure 11.11. Male American redstart settlement on plots. Mean (± SE) number of territorial males on control and treatment plots. New immigrant adults settled preferentially on plots with playback songs of conspecifics, exhibiting conspecific attraction. Yearlings and returning adults were not affected by the playback treatment. (*Source:* Hahn & Silverman 2006)

11.2 Individual condition and environmental factors affect territoriality

When animals settle in an area, they can establish a **territory**, an area defended to obtain the exclusive use of the resources it contains, such as food, nesting sites, or access to potential mates. Territorial defense provides exclusive use of resources but also requires that animals expend effort to defend these resources from rivals. Territories differ from **home ranges**, which are areas of repeated use that are not defended. The home ranges of many individuals can overlap, while territories do not. Territory defense requires physical effort: it takes both time and energy to chase off intruders. Can all individuals defend a territory?

territory An area defended to obtain the exclusive use of the resources it contains.

home range An area of repeated use by an individual that is not defended from others.

Body condition affects territoriality in damselflies

Aggressive behaviors involved in territorial defense require substantial effort, and so only individuals in good body condition may be able to acquire and defend a territory. Body condition can be measured by body mass, body size, or measures of fat, which indicates extra energy storage that could be used in intense behavioral interactions. Odonates (dragonflies and damselflies) defend territories and are easy to observe and manipulate; consequently, they have often been the focus of studies of territoriality. For example, Jorge Contreras-Garduño and colleagues studied how the physiological status of male damselflies affects their ability to acquire and defend a territory.

The researchers studied territorial behavior in the American rubyspot (*Hetaerina americana*), a damselfly found throughout North America near sunny riverbanks. Adults emerge from a pupa and feed on mosquitoes. After several days of feeding and development, individual males reach sexual maturity and then fight for a territory, a perching site. Territory ownership is determined in aerial contests, which can be as short as three seconds or as long as two hours

Featured Research

APPLYING THE CONCEPTS 11.1

Conspecific attraction and conservation

Many bird populations are declining because of human habitat change. Natural habitats are often converted for the purposes of urban development or agriculture. After conversion, they may no longer be suitable for particular species or may contain novel competitors or predators. In response, conservation biologists are working to restore degraded places and even create new habitat areas. Once a population has left a location, how do you attract it back to the new or restored habitat?

One method increasingly used to restore bird populations takes advantage of conspecific attraction. For instance, populations of colonial nesting seabirds such as Atlantic puffins (*Fratercula arctica*) have been restored to former breeding grounds by placing models and dummy nests at former breeding islands to attract individuals (e.g., Kress & Nettleship 1988; Parker et al. 2007).

Song playback has also been used to attract species back to areas. Michael Ward and Scott Schlossberg used conspecific calls to attract endangered black-capped vireos (*Vireo atricapilla*) to restored habitat sites in Texas (Ward & Schlossberg 2004). Black-capped vireos (Figure 1a) are small, territorial songbirds whose populations have declined steeply over the past few decades, largely because of habitat change and an increase in brown-headed cowbirds (*Molothrus ater*), a species that lays eggs in the nests of other birds (hosts) and so decreases hosts' reproductive success.

Ward and Schlossberg used song playbacks to attract birds to settle in restored habitats at Fort Hood, Texas, from which cowbirds had been removed. They selected seven large, restored sites (15–71 ha) from which vireos had been absent for at least two years. Over two breeding seasons, they played black-capped vireo songs in some sites and left others silent as controls, switching treatments and controls between years (Figure 1b). In each year, over 70 vireos were observed at those sites where songs were played, while fewer than five were observed at the control sites. Many of the birds reproduced successfully at treatment sites (Figure 1c). These data indicate that song can be used to attract new settlers to a habitat. This gives resource managers and conservation biologists a new tool for increasing populations of endangered birds. ■

Figure 1. Black-capped vireo experiment. (a) Black-capped vireos are small, endangered songbirds in Texas. (b) The speaker set-up used to play songs on experimental plots. (c) A nesting female.

(Córdoba-Aguilar & Cordero-Rivera 2005). Males that cannot gain a territory simply search for females in their home range, but this tactic results in very limited mating success. What determines whether a male gains a territory? Territory disputes can be energetically expensive, and so the research team examined whether or not territorial males had higher fat reserves than nonterritorial males. Fat serves as fuel for energetically costly behaviors such as fighting, and so larger reserves of fat should allow individuals to fight longer and more aggressively for a territory.

The researchers first compared the fat content of territorial and nonterritorial males of similar age from the Xochitepec River in Mexico (Contreras-Garduño, Canales-Lazcano, & Córdoba-Aguilar 2006). In order to measure body fat

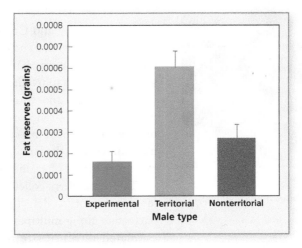

Figure 11.12. Fat reserves in male American rubyspot damselflies. Mean (+ SE) fat reserves in grains (1 grain = 0.06 g). Territorial males had higher fat reserves than nonterritorial males. Inset: Territorial male rubyspot damselfly. (*Source:* Contreras-Garduño, Canales-Lazcano, & Córdoba-Aguilar 2006)

Figure 11.13. Male damselfly fat reserves after experimental challenge. Experimentally challenged males forced to defend their territory had lower mean (+ SE) fat reserves than did nonchallenged territorial males or nonterritorial males (units are grains). (*Source:* Contreras-Garduño, Canales-Lazcano, & Córdoba-Aguilar 2006)

content, they weighed the body both before and after extracting the fat. The difference yielded the fat mass of a male. They found that territorial males had much higher fat reserves than nonterritorial males (Figure 11.12), indicating that territorial males did indeed have higher potential energy reserves. But are these fat reserves used to maintain a territory?

In a second study, Contreras-Garduño and colleagues staged territorial fights. They created "intruders" by capturing sexually mature males and tying them to a wooden stick with a thread attached to their thorax, which still allowed them to fly. These males were used to intrude upon a territorial male in the field. The intruders were brought close to the territorial male repeatedly for 20 minutes to simulate a long territorial contest. After the experiment, the fat reserves of the experimental males were measured and compared to those of nonmanipulated territorial males and those of a set of nonterritorial males.

Territorial males who were forced to interact with intruders for 20 minutes had the lowest fat reserves of all three groups (Figure 11.13). The territorial males that were not manipulated had fat reserves that were almost four times higher than those of the experimental males. The researchers concluded that the long staged contests reduced a territorial male's fat reserves to very low levels, indicating that the aerial displays required for defending a territory are indeed energetically very costly. The study suggests that only males with high energy reserves can hold a territory. As we will continue to see, territory defense is costly. In the next study, we examine how the costs of defense affect territory size in fish.

Environmental factors and territory size in parrotfish

Featured Research

Several factors affect the benefits and costs of territory defense (Brown 1964). One of the most important is the density of competitors. As the number of competitors increases, the cost of defense also increases, because there are more potential intruders. An individual can reduce this cost by defending smaller territories, because they involve less area to defend. This leads to the simple prediction that territory size should be negatively correlated with competitor density.

Figure 11.14. Striped parrotfish. Colorful breeding males defend territories on a reef.

Figure 11.15. Parrotfish territory size. Mean (±SE) territory size declined with an increasing mean (±SE) density of competitors.

Peter Mumby and Colette Wabnitz tested this prediction by examining territory size and competitor density in the striped parrotfish (*Scarus iseri*) (Figure 11.14) (Mumby & Wabnitz 2002). Parrotfish are sequential hermaphrodites, and individuals exhibit distinct color phases: females and young males are drab-colored, while breeding males display a variety of striking colors. A breeding male defends a territory that encompasses one or more female territories, and the male mates with the females each day (Clifton 1989). Such females are known as a male's harem because he defends them from rivals.

The researchers collected data in Belize at multiple sites that differed in parrotfish density. At each site, they recorded the density of parrotfish along multiple transects. Over a two-month period, they estimated territory size for all breeding males by observing each male's repeated pattern of movement as it patrolled its territory on the reef. The research team recorded all interspecific interactions between conspecifics, which were common at territory boundaries. During territory observations, the researchers also noted the presence of all females within a male's territory.

The density of intra- and interspecific competitors varied from about two to ten individuals per 200 m² across sites, and there was a positive correlation between parrotfish density and the number of interspecific interactions at territorial boundaries. Territory size also varied tremendously, ranging from less than 100 m² to over 1,000 m². Male territory size declined as conspecific density increased (Figure 11.15), and male harem size also declined with an increase in the density of rivals. In this species, territory size is strongly affected by the density of competitors, as predicted. As the costs of territory defense rise, males apparently reduce the size of their territory, presumably to better defend their harems from rivals.

These examples demonstrate that individual condition and environmental factors affect territoriality. The acquisition and defense of a territory or other valuable resource often requires aggressive interactions with rivals. Next, we examine how such aggression is regulated.

11.3   Hormones influence aggression

Contests over resources involve a variety of aggressive behaviors, including chasing, pushing, wrestling, and fighting, as well as others that indicate the threat of attack, such as particular postures or displays (Figure 11.16). Some species defend territories year-round and thus may aggressively defend their territories regularly, while others defend a territory only during the breeding season. For example, many birds defend territories during summer (the breeding season) but live in large social groups during winter (the nonbreeding season). Therefore, the aggression levels in these species will vary greatly over the course of a year, with high levels at the start of the breeding season and low levels outside the breeding season.

What regulates aggression? In many vertebrates, variation in aggression is coincident with variation in levels of plasma androgen hormones such as **testosterone**. This steroid hormone is produced in the gonads and is regulated by the hypothalamus and pituitary gland. Androgen hormones have widespread effects on both physiology and behavior. For instance, increased testosterone levels are associated with increases in aggression, sexual behavior, and spermatogenesis. But high testosterone levels are also associated with the suppression of parental

testosterone A steroid hormone produced in the gonads and regulated by the hypothalamus and pituitary gland.

a   **b**

Figure 11.16. Aggression. (a) A swan chases a rival to defend its territory. (b) Wolves bare their teeth during threat displays.

care behavior and fat stores, altered immune system function, and increased mortality (Wingfield, Lynn, & Soma 2001). Thus, elevated testosterone levels enhance aggression and fighting ability, but at a cost to other aspects that affect individual fitness. Such counteracting effects on fitness suggest an important role for tight regulation of androgen levels.

In many species that reproduce seasonally, plasma testosterone levels coincide with changes in photoperiod. There is increased production in spring, as days grow longer, and decreased production in autumn, as days become shorter. John Wingfield and colleagues documented such changes in plasma testosterone levels in male song sparrows (*Melospiza melodia*) (Wingfield et al. 1990; Wingfield et al. 2000; Wingfield, Lynn, & Soma 2001). Free-living males displayed very high spikes in testosterone levels in early spring during territory acquisition, as well as a few weeks later when mating, but testosterone levels declined greatly after that (Figure 11.17). Socially isolated captive birds displayed a similar seasonal pattern, but with much lower spikes (Wingfield, Lynn, & Soma 2001). High levels of testosterone (and associated aggression) may facilitate the acquisition of territories early in spring, with a reduction corresponding to the need for males to provide increased levels of parental care once eggs hatch (Chapter 14).

Why might free-living birds display higher levels of testosterone compared to isolated captives? One possibility is that social interactions also affect testosterone secretion. Wingfield's **challenge hypothesis** states that male–male interactions (over territories, dominance status, or mates) increase plasma testosterone and thus sustain subsequent aggressive behavior (Wingfield, Lynn, & Soma 2001). When males fight over territories and mates, success in a contest can strongly enhance fitness. Enhanced testosterone levels are predicted to increase aggressive behavior and therefore an individual's probability of winning a contest. This can also explain a phenomenon known as the **winner effect**: for many animals, winning an aggressive interaction often enhances the likelihood of winning a subsequent interaction (Hsu, Earley, & Wolf 2006). A combination of the winner effect and the challenge hypothesis, the **winner-challenge effect**, states that winning a challenge increases plasma testosterone levels that enhance aggressive behavior and thus the likelihood of winning subsequent interactions (Oyegbile & Marler 2005).

challenge hypothesis The hypothesis that male–male interactions increase plasma testosterone and thus sustain subsequent aggressive behavior.

winner effect A phenomenon in which winning an aggressive interaction enhances the likelihood of winning a subsequent interaction.

winner-challenge effect A situation in which winning an aggressive interaction increases plasma testosterone levels and so enhances aggressive behavior and the likelihood of winning subsequent interactions.

Figure 11.17. Testosterone variation in male song sparrows. Free-living males (blue circles) had high mean (± SE) plasma testosterone levels during territory formation and reproduction of the first brood. Captive males (orange circles) showed similar patterns of variation over the spring, but with lower levels of testosterone. (*Source:* Wingfield, Lynn, & Soma 2001)

Winner-challenge effect in the California mouse

Temitayo Oyegbile and Catherine Marler examined the winner-challenge effect by exploring the links between testosterone levels, aggressive behavior, and the outcome of contests in the California mouse (*Peromyscus californicus*) (Oyegbile & Marler 2005). California mice aggressively defend their territory year-round, and, like many other species, larger individuals typically win aggressive interactions (Huntingford & Turner 1987).

To test this hypothesis, the research team randomly assigned focal males to one of four training treatments that differed in the number of times (from zero to three) that a male won an interaction with a rival. In a fifth no-interaction control group, males never interacted with a rival. Prior to training, all males were paired with a female for ten days. On Day 11, the pair was placed in a large observation cage that contained a small nest box that allowed the mice to establish residency of the cage. Training interactions were performed on Days 13, 15, and 17 and consisted of either a winning encounter with a rival or simple handling by the experimenters. To create winning encounters, the female was removed and a smaller, mildly sedated male intruder was introduced to the observation cage for ten minutes. On Day 19, the testing phase began. The female was removed, and an unfamiliar and slightly larger male that had previously won an encounter was introduced to the cage. Oyegbile and Marler recorded the resident male's latency to attack, the number of attacks, and the outcome of the contest (winning behaviors included chasing and wrestling; losing behaviors included jumping away, freezing, and retreating). Baseline blood samples were taken from all males before training began and again 45 minutes after the test encounter.

The number of winning encounters a male had in the training phase correlated with an increased likelihood of winning the test encounter. However, the only significant differences were between males with no prior experience and males with three prior winning experiences. Males that won more training-phase encounters also had significantly shorter latency to attack the rival (Figure 11.18). Testosterone levels were higher in mice that had previous winning experiences compared to controls that had none (Figure 11.19). These results support the challenge hypothesis, because variation in plasma testosterone levels correlated with higher levels of aggression and contest outcome. This finding also provides a proximate explanation for the winner effect: winning fights results in elevated plasma testosterone levels, which makes males more aggressive and enhances their likelihood of winning subsequent fights. Together, these results illustrate the winner-challenge effect in mice.

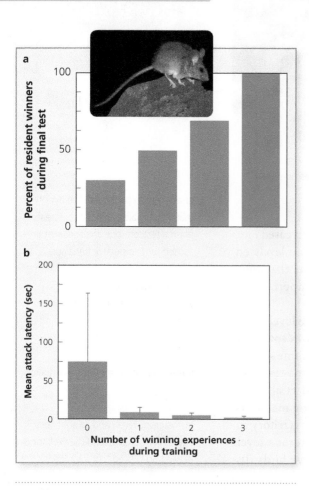

Figure 11.18. Mouse contest results. Mice that won more previous contests in training had (a) a higher winning percentage against a larger rival in the final test and (b) a lower mean (+SE) latency to attack. Inset: California mouse. (*Source:* Oyegbile & Marler 2005)

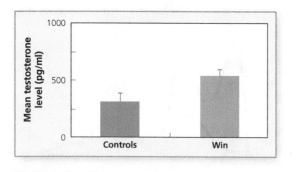

Figure 11.19. Testosterone levels in mice. Mean (+SE) testosterone levels were higher in males that won fights compared to controls. (*Source:* Oyegbile & Marler 2005)

Challenge hypothesis and bystanders in fish

Can the challenge hypothesis be extended to bystanders? Does simply observing aggressive interactions affect hormone levels? Rui Oliveira and colleagues examined these questions using the Mozambique tilapia (*Oreochromis mossambicus*) (Oliveira, Lopos, & Carneiro 2001). These cichlid fish are native to East Africa, and males defend territories during the breeding season.

The researchers first kept all fish in isolation for seven days to minimize any effects of previous interactions. Treatment focal male bystanders were then allowed to view two conspecifics interacting. The conspecifics were initially visually separated by an opaque partition and so could not interact. The partition was then raised for one hour, during which the two rivals engaged in a fight for dominance that the focal bystander could view through a one-way mirror. The research team collected urine from the bystander to assess testosterone levels at different times: two hours before viewing the males fight and then again 30 minutes, two hours, and six hours after the interaction. Control fish were treated similarly, except that the opaque partition was never raised, and so they did not view an aggressive interaction.

Figure 11.20. Testosterone variation in fish. Mean (+ SE) changes in testosterone levels for male bystanders that observed an aggressive interaction (orange) and for controls (blue). Testosterone levels increased in bystanders that had observed a fight 30 minutes earlier compared to controls. Inset: Mozambique tilapia. (*Source:* Oliveira, Lopos, & Carneiro 2001)

Bystanders that observed an aggressive interaction had higher testosterone levels 30 minutes after the interaction than bystanders that did not observe an aggressive interaction (Figure 11.20). This experiment demonstrates that social observations alone can result in increased testosterone levels, thus extending the challenge hypothesis. Individuals do not need to engage in male–male aggressive encounters directly but rather can simply view aggression to experience an increase in their plasma testosterone levels. The researchers suggest that this kind of viewing experience can prepare a bystander for future aggressive interactions: a rapid elevation of testosterone levels can increase aggressive behavior and the likelihood of winning a contest. The challenge hypothesis has even recently been extended to people viewing sporting events and provides a proximate explanation for variation in aggression (Applying the Concepts 11.2).

Juvenile hormone and wasp aggression

Featured Research

The challenge hypothesis was formulated to explain increased testosterone plasma titers (i.e., concentration) in vertebrates in response to a social challenge: upregulating testosterone in response to a rival's challenge maximizes the benefits of short-term increased androgen levels while minimizing the long-term costs of elevated plasma androgen titers. Is there a parallel hormonal response in insects?

Recent work suggests that juvenile hormone (JH) is a likely candidate (Scott 2006). JH influences many aspects of insect physiology and behavior (Nijhout 1994), including metamorphosis, sexual behavior, foraging, and, as we will see in Chapter 14, parental care. High JH titers are associated with fitness benefits, including fertility and dominance, but also with reduced immune system function (Adkins-Regan 2005), similar to the fitness trade-offs found for testosterone in vertebrates.

Elizabeth Tibbetts and Zachary Huang (2010) articulated three predictions of the challenge hypothesis as it relates to JH:

Prediction 1: Aggressive interactions will enhance JH titers.

APPLYING THE CONCEPTS 11.2

Human aggression, testosterone, and sports

Psychologists have long been interested in links between plasma testosterone levels and human aggression (e.g., Persky, Smith, & Basu 1971). Evidence to date suggests a weak association between testosterone levels and aggression, although debate over this relationship continues (Book, Starzyk, & Quinsey 2001; Archer, Graham-Kevan, & Davies 2005). John Archer suggested that the weak association can be explained by the challenge hypothesis: human testosterone levels in males are affected by recent challenges involving competition with men. Archer's examination of existing data provided general support for this hypothesis (Archer 2006). For example, a few studies report that testosterone levels are higher in winners compared to losers after a sporting competition (e.g., Booth et al. 1989).

Justin Carré and Susan Putnam extended the application of the challenge hypothesis to human spectators of a competitive interaction (Carré & Putnam 2010). They collected plasma testosterone samples from the saliva of elite male college ice hockey players before and after these athletes watched videos. In the first experiment, players first watched a video of their team winning a prior contest and then two weeks later watched a video of their team losing a prior contest. In the second experiment, players first watched a video of their team winning a prior contest and then two weeks later watched a neutral video—a documentary film. The researchers collected saliva samples from each player ten minutes before and ten minutes after he watched a video.

Figure 1. Testosterone levels in male hockey players. Mean (+ SE) testosterone levels increased more in male hockey players after they watched a video of their team winning compared to watching either (a) a video of their team losing a prior contest or (b) a neutral video. (*Source:* Carré & Putnam 2010)

In both experiments, there was a significant increase in testosterone concentrations in males' saliva after they had watched a prior winning contest, but not after they had watched a prior losing contest or the documentary (Figure 1). These results suggest that simply viewing an aggressive contest that leads to a victory elevates testosterone levels in men. In fact, the increase in testosterone concentration occurred even in male team members who did not play in that particular game. The researchers suggest that although more study is needed, their results may have ramifications for athletic competition. ■

Figure 11.21. Paper wasps. Workers tending to the nest.

Prediction 2: Individuals with high JH levels will be the most competitive and will have high reproductive success.

Prediction 3: Aggression and JH titers will be positively correlated during periods of social instability (periods of heightened aggression) but will not be correlated during periods of social stability (reduced aggression) because JH levels will be low for all individuals.

They tested these predictions by examining individuals from 12 colonies of the paper wasp (*Polistes dominulus*) (Figure 11.21). This primitive eusocial insect lacks discrete morphological casts, and all workers are capable of reproduction. A colony consists of a single queen and up to about a dozen workers. Workers compete in aggressive contests in which dominant individuals mount subordinates; dominants do not spend time foraging but have enhanced access to food resources brought to the colony by subordinates. In addition, only the most dominant colony member, the queen, reproduces. But if the queen is removed from the colony, remaining workers compete to become the new queen.

Tibbetts and Huang uniquely marked all individuals in each colony and then removed the queen from half the colonies ("queen removal" colonies) to enhance

Figure 11.22. JH titers in wasp workers. Mean (+SE) JH titer of queen removal workers was higher than that of control (constant queen) workers. (*Source:* Tibbetts & Huang 2010)

Figure 11.23. JH titers and aggression. The number of aggressive acts initiated by workers was not correlated with JH titer in (a) control (constant queen) workers but was correlated with this factor in (b) queen-removed workers. (*Source:* Tibbetts & Huang 2010)

the level of aggressive interactions in remaining workers as they competed to become the new queen. In the other half of the colonies, the researchers removed a subordinate worker as a control ("constant queen" colonies). Four days after the removals, each colony was videotaped for two hours to record all aggressive interactions (mounts) and to determine the identity of the replacement queen in the queen removal colonies. The next day, JH levels were measured in all individuals.

Removal of the queen significantly elevated JH titer levels in remaining workers compared to workers in constant queen colonies (Figure 11.22). In addition, individuals that became replacement queens had higher JH titers than other workers in the colony, and JH titers were strongly correlated with the initiation of aggression by workers in these colonies, but not with aggression by workers in constant queen colonies (Figure 11.23).

These results are consistent with the three predictions of the challenge hypothesis as it relates to JH and aggression in paper wasps. The upregulated JH titers of workers in colonies in which the queen was removed were likely due to the increased aggressive interactions that occurred as individuals fought to become the replacement queen. The successful replacement queen had higher JH titer levels than other colony members, indicating a link between JH titer and enhanced competitiveness. And, JH titer was relatively low and uncorrelated with the initiation of aggression in constant queen colonies, while there was a strong positive correlation between JH titer and the initiation of aggression in queen removal colonies.

In sum, aggression can yield fitness benefits such as the acquisition of breeding territories, food, and mates. High titers of androgen hormones in vertebrates and juvenile hormone in insects are associated with elevated aggression and competitiveness, but also with altered immune system function, which can decrease survivorship (Braude, Tang-Martinez, & Taylor 1999). Social stimulation in the form of challenges from a rival often promotes the short-term upregulation of testosterone and JH when aggression and enhanced competitive ability are most beneficial. The short duration of such increases in these hormones minimizes the negative fitness consequences associated with long-term elevated hormone levels.

Examination of hormones provides a proximate explanation for variation in aggressive behavior of individuals. Next, we turn to ultimate explanations for variation in aggressive behavior by examining game theory models.

11.4 Game theory models explain how the decisions of opponents and resource value affect fighting behavior

Figure 11.24. Flies boxing. Fruit flies engage in contests over food.

game theory A cost-benefit modeling approach in which an individual's fitness is affected by the behavior of others.

hawk-dove model A game theory model that examines the behavior of two individuals engaged in a contest over a resource.

Contests over resources involve interactions with others. These contests can be quite complex because individuals often exhibit a variety of aggressive behaviors that differ in intensity. The benefits of aggression are straightforward: the acquisition of a contested resource. But the costs of fighting are more complex because they are affected by the intensity of the aggressive behaviors displayed by both opponents. If an opponent decides to flee, then fighting has a low cost. If an opponent decides to fight intensely and is a superior fighter, then the decision to fight intensely could be costly, resulting in injury.

For example, male fruit flies commonly fight over food. Fights can be as short as a few seconds or longer than four minutes (Chen et al. 2002) and can include chasing, pushing, tussling (in which opponents tumble over one another), and even boxing—a fight in which opponents rear up on their hind legs and strike each other with their forelimbs (Figure 11.24). The longer the contest lasts, the greater the number of behaviors exhibited, and the higher the costs (Chen et al. 2002), but rarely is an individual harmed or killed. Why is there so much variation in contest behavior? Why don't more animals become injured or even die while fighting?

We can answer these questions using **game theory**, a cost-benefit approach used to understand the evolution of behavior when an individual's fitness is affected by how other individuals behave (Chapter 3). Let's see how this works using a very basic game theory model that examines contests between two individuals: the hawk-dove model.

The hawk-dove model

John Maynard Smith and George Price (1973) introduced the **hawk-dove model** to examine the behavior of two individuals engaged in a contest over a resource, such as a territory. The winner of the contest obtains a benefit B that enhances fitness. The loser receives no benefit. In this simple game, both individuals have the same fighting ability and adopt one of two behavioral strategies: hawk or dove. Hawks are always willing to escalate an interaction into an intense, prolonged fight in order to win. In fact, they fight until they are injured (when they give up) or until they win. Losers of these intense fights are injured and pay a cost C. Doves, on the other hand, back down immediately if their opponent escalates its behavior and so never actually fight or pay the cost C. When two doves fight, they engage in a short, no-cost display, and each has a 50% chance of winning.

The model allows us to predict the frequencies of these strategies in a population. When $B < C$, the model predicts that a population will contain both hawks and doves. At the evolutionary stable strategy (ESS; see Chapter 3), the frequency of hawks is predicted to be B/C (Toolbox 11.1), and, by definition, at the ESS, both hawks and doves have the same fitness. Such a population will exhibit tremendous variation in fighting behavior. When two hawks meet, fights will be long and intense. When a hawk interacts with a dove, the contest will be very short, and the individual playing the hawk strategy will always win. Contests between two doves will also be short but will involve only low-cost displays, and so any one individual will win some fights and lose others. This variation is similar to that observed in the real-life contests between animals that Maynard Smith and Price wanted to explain. Interestingly, when $B > C$, the model predicts that a population will consist of individuals exhibiting only the hawk strategy, because that situation yields higher fitness than the dove strategy.

The introduction of the hawk-dove model to animal behavior research was an important conceptual advance in our understanding of fighting behavior for several reasons. First, it showed how individual selection (as opposed to group selection) could explain why individuals did not fight to the death when competing over resources. Second, it predicted variation in fighting behavior in a population depending on the benefits and costs involved in the contest. And third, it focused empirical effort on understanding the benefits and costs of fighting behavior.

While the model has important heuristic value, it was likely never intended to explain in detail the fighting behavior of actual animals (Fawcett & Mowles 2013; Kokko 2013). Recall that one of its assumptions is that animals have identical fighting ability, an unlikely occurrence in nature. Instead, individuals will often differ greatly in their fighting ability or **resource holding potential** (**RHP**) (Parker 1974). Differences in fighting ability will be influenced by an individual's size, condition (health), and endurance (energetic stores), among other factors. Many studies have shown that RHP affects the outcome of contests: the greater the asymmetry in RHP, the more likely that the individual with greater RHP will win a contest (e.g., Kemp & Wiklund 2004; Briffa & Sneddon 2007).

The hawk-dove model also has little to say about contest duration. The duration of actual contests can be quite variable and is often affected by the value of the contested resource. For instance, males often engage in contests with other males for access to a female (or group of females). This is particularly common in amphibians, in which reproduction often involves amplexus, a behavior in which a male grasps a female and holds her for up to several hours before she releases eggs he will fertilize. During amplexus, rival males will attempt to displace an amplexed male by wrestling with him. How much effort should an intruding male invest in these contests?

> **resource holding potential** The ability to win an aggressive encounter.

Wrestling behavior in red-spotted newts

Paul Verrell examined wrestling behavior in red-spotted newts (*Notophthalmus viridescens*) (Verrell 1986). These small salamanders (less than 125 mm in length) are common in eastern North America (Figure 11.25). In amphibians, female fecundity, as measured by number of oocytes, typically increases with body size, indicating that larger females should be more valuable to a male. Verrell first verified that larger females had a higher resource value by observing that larger females laid significantly more eggs than did smaller females.

With this information, Verrell predicted that the fighting intensity of males should increase with the size of the contested female. To test this prediction, he staged encounters between size-matched males in a large aquarium. In each trial, one female, whose body size ranged from 43 mm to 51 mm, was placed in the aquarium along with two males. One of the males quickly initiated amplexus, and in 33 trials, the rival male began a wrestling contest with the amplexed male. The wrestling bout ended either when the rival left the amplexed pair or when he successfully detached the amplexed male from the female.

Featured Research

Figure 11.25. Red-spotted newts. Males wrestle with rivals for access to mates.

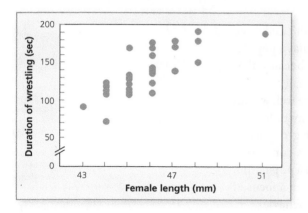

Figure 11.26. Duration of wrestling in male red-spotted newts. The duration of wrestling increased with the size of the female that was amplexed. (*Source:* Verrell 1986)

The total duration of the wrestling contest increased with the size of the female, as predicted (Figure 11.26), and the intruder did displace the amplexed male in four of the trials. These data illustrate how the value of a contested resource affects the intensity of aggressive behavior. Males spent more time wrestling over large females than they did over small females, as one might expect, given that larger females lay more eggs.

Game theory assessment models

Another aspect of contests not fully incorporated into the hawk-dove model is that contests typically begin with opponents exhibiting display behaviors, often with no physical contact. For example, many birds and crickets will begin a contest by singing, fiddler crabs will start a contest by waving their claws, and lizards will begin a contest with head-bob or push-up displays. After some time, contests such as these may escalate as more energetically costly behaviors are used, often leading to physical contact.

One classic example of such a contest sequence of escalated behavior involves fights between male red deer (*Cervus elaphus*). Aggressive interactions during the breeding season typically begin with males engaging in a roaring contest (Clutton-Brock & Albon 1979). As the contest continues, males will subsequently engage in a parallel walk in which they stride next to each other for up to 30 minutes. Finally, if the contest is still not decided, individuals will lock antlers and physically push against one another (Figure 11.27). At any time during this process, one individual may decide to abandon the fight, and so we see tremendous variation in contest duration, ranging from a few minutes to over one hour (Clutton-Brock & Albon 1979).

These observations prompted the development of more sophisticated game theory models that assume that a contest entails time and energy costs, and that these costs increase with the duration of the contest. The models also assume some form of assessment by individuals of the conditions in which the contest takes place. Such conditions can include the perceived value of the resource as well as the relative RHP of the contestants. Unlike the hawk-dove model, these more advanced game theory models predict the amount of effort an individual should allocate to a particular contest and so predict variation in contest duration.

self-assessment game theory models Game theory models that examine the behavior of two individuals engaged in a contest over a resource. Each individual is assumed to assess its resource holding power in deciding how long to fight.

Two basic types of game theory assessment models exist. **Self-assessment game theory models** (e.g., Mesterton-Gibbons, Marden, & Dugatkin 1996; Payne & Pagel 1996) assume each individual assesses the value of the resource and engages in the contest until the individual reaches a cost threshold beyond which it is unwilling to continue, and so gives up. An individual's cost threshold will be based on assessments of its RHP and the value of the resource. A contest will continue until the first individual (the loser) reaches its cost threshold and abandons the fight. These models predict that contests will always be won by the individual with the higher RHP, and that contest duration will increase as the loser's RHP increases and as the value of the resource increases.

Figure 11.27. Red deer contest. Males lock antlers and push against one another during a contest.

Mutual-assessment game theory models, like the sequential assessment model (Enquist & Leimer 1983), assume that each individual assesses the relative cost threshold of itself *and* its opponent and continues the contest until an individual assesses that it is unlikely to win the contest (because it has a lower cost threshold). In essence, these models envision that each individual assesses its RHP and that of its opponent (its relative fighting ability) and decides to give up when it estimates that it has the lower RHP (and so will eventually lose the contest). The models assume that each contestant is unsure of its opponent's RHP and so acquires information about its opponent during the contest. These models also predict that contest duration will increase with the value of the contested resource.

How can we distinguish these models of assessment? Let's examine the predictions they make regarding the duration of a contest and the RHP of winners and losers (Taylor & Elwood 2003). Imagine that in any population, you score the RHP of all individuals from lowest to highest. Now imagine recording the duration of contests between random pairs of individuals and assume that the individual with the higher RHP always wins. Self-assessment models predict that there should be a strong positive correlation between the losers' RHP and contest duration: losers with higher RHP will engage in longer contests before they lose (Figure 11.28). Mutual-assessment models also predict a positive relationship between loser RHP and duration, but for different reasons. In these models, a loser with a low RHP will often lose to a winner with a high RHP, making assessment of the relative difference easy and the contest short. As loser RHP increases, the difference in RHP between the loser and the winner declines, making assessment more difficult and contests last longer.

However, self-assessment and mutual-assessment models make distinctive predictions regarding contest duration and winner RHP. Self-assessment models predict a positive relationship between contest duration and winner RHP because longer contests (i.e., higher loser RHP values) require a higher winner RHP to ensure that the winner wins. Mutual-assessment models, however, predict a negative relationship between winner RHP and contest duration. Winners with low RHP will have beaten losers with similarly low RHP, making assessment difficult and contest duration long. But winners with high RHP will often be matched with low-RHP losers, making assessment easier and contest duration shorter (Taylor & Elwood 2003).

A second prediction can be made for fights between RHP-matched individuals. The self-assessment models predict that for individuals with matched RHP, loser RHP will determine contest duration. Therefore, we will see short contests between two low-RHP individuals and much longer contests between two high-RHP individuals: there will be a positive correlation between mean RHP of rivals and contest duration. In contrast, mutual-assessment models predict no effect of mean RHP: all contests between RHP-matched individuals should be long and similar in duration.

Let's examine one test of these predictions for fights involving male fiddler crabs.

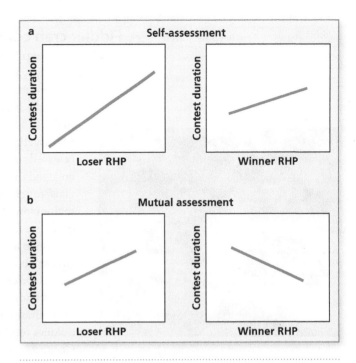

Figure 11.28. Assessment model predictions. (a) Self-assessment models predict that contest duration increases with both loser and winner RHP. (b) Mutual-assessment models predict that contest duration increases with loser RHP but declines with winner RHP. (*Source:* Gammell & Hardy 2003)

mutual-assessment game theory models Game theory models that examine the behavior of two individuals engaged in a contest over a resource. Individuals are assumed to assess the relative fighting ability of their opponent and themselves in deciding how long to fight.

Fiddler crab contests over burrows

Fiddler crabs aggressively defend burrows they create in mudflats. Burrows provide a safe refuge, and females select mates based in part on their burrow characteristics; consequently, possession of a burrow strongly affects male fitness. Leslie Morrell, Patricia Blackwell, and Neil Metcalfe studied fights between males in a population of *Uca mjoebergi* fiddler crabs in Australia (Morrell, Blackwell, & Metcalfe 2005). They captured burrow-holding males and released them far away from their burrow. These "intruder" males quickly engaged in contests with burrow-holding residents. Contests began with claw touching and could escalate to pushing and grappling with the major claw (Figure 11.29). The research team recorded the duration and outcome of 173 contests (the winner occupied the contested burrow), and once a fight was settled, they measured the size of each individual (major claw length) to characterize its RHP.

Figure 11.29. Fiddler crab contest. Fiddler crabs use their major claws during a contest.

Intruders won one-third of the contests. Both claw size (RHP) and status (burrow-holder or intruder) influenced the outcome of fights: the individual with the larger major claw won more contests, and burrow-holders tended to beat intruders when individuals were similar in size. Contests averaged seven seconds in duration, but some lasted over 60 seconds. There was a strong positive correlation between contest duration and loser claw size and a weaker positive correlation between contest duration and winner claw size (Figure 11.30). For size-matched contests, the researchers found a positive correlation between mean claw size of the combatants and contest duration (Figure 11.31). These data are consistent with the predictions of self-assessment models, as contest duration was positively related to both loser RHP (claw size) and winner RHP. Furthermore, for size-matched fights, contest duration increased with size.

Game theory models have been instrumental in providing unique insight into understanding the evolution of, and variation in, fighting across species. These models are also being increasingly used in agricultural settings to manage animal aggression (Applying the Concepts 11.3).

Figure 11.30. Fiddler crab contest duration. Contest duration was positively correlated with (a) loser and (b) winner RHP, as measured by major claw size.

Figure 11.31. Contest duration for size-matched crabs. Contest duration increased with the mean size of the competitors.

APPLYING THE CONCEPTS 11.3

Reducing duration and intensity of piglet fights

In the agricultural production of domestic pigs (*Sus scrofa domesticus*), many litters are born simultaneously. The piglets in a litter are housed together with their mother but are kept physically and visually isolated from other litters. After several weeks, piglets from different litters are mixed together, which results in aggressive fights between piglets. These fights are often intense and can result in injury, making them a major welfare and production issue. Farmers have tried to reduce fighting with drugs and masking odors, but they have had little success.

Per Jensen and Jenny Yngvesson applied game theory models to piglet fights (Jensen & Yngvesson 1998). Fights often begin with a low-intensity nosing phase, in which two individuals walk together and exchange nose contact. After a period of time, fights can escalate to a biting phase, in which individuals push and bite each other until the loser retreats. Jensen and Yngvesson reasoned that if opponents rely on fights to learn each other's relative fighting ability, as assumed by mutual-assessment models, then allowing interactions of

individuals before a contest should produce shorter and lower-intensity fights.

To test this prediction, they staged contests between piglets. Some individuals (the treatment group) were allowed to see each other in adjacent pens separated by a wire mesh. Control animals were housed in adjacent pens separated by a solid wooden wall that prevented visual contact. Contests were conducted between piglets that were either matched for size or differed in weight by about 20%. Jensen and Yngvesson found that fight intensity and duration were affected by both weight difference and prior visual contact. Fights between opponents of similar size lasted longer than fights with a large weight asymmetry, as predicted by the model. In addition, fight duration and intensity were lower in fights between opponents that had had prefight visual contact. The researchers suggested that farmers and animals could benefit from husbandry that allows visual contact between piglets of different litters. ■

TOOLBOX 11.1

The hawk-dove model

John Maynard Smith and George Price introduced the application of game theory to evolutionary biology (Maynard Smith & Price 1973). They wanted to understand why conflicts between individuals of the same species rarely resulted in serious injury, even when individuals possessed effective weapons such as horns or fangs. Their model, the hawk-dove model, examines the outcomes of fights between individuals who can display only two behavioral strategies: hawk or dove. The hawk fights until it wins or is injured. The dove never fights. Maynard Smith and Price assigned benefits and costs to individuals playing these strategies against one another. There are three possible interactions:

Hawk versus hawk:
Each player has a probability of 0.5 of winning or losing.
Winners get a benefit *B*.
Losers receive no benefits, are temporarily injured, and pay a cost of *C*.

When hawks play hawks, the average payoff is $0.5B - 0.5C = 0.5 (B - C)$. In other words, half the time an individual wins and gets a payoff *B*, while half the time the individual loses and pays a cost *C*.

Dove versus dove:
Each player has a probability of 0.5 of winning.
Winners obtain a benefit *B*.
There are no costs, because the players never escalate the conflict and so are never injured.

When doves play doves, the average payoff is 0.5*B*.

Hawk versus dove:
Hawks always win and receive a benefit *B*.
Hawks pay no costs, because the dove always retreats.
Doves always lose and get no benefits, but they also pay no costs.

When a hawk plays a dove, the payoff to the hawk is *B*, and the payoff to the dove is 0.

Let's look at the benefits and costs to a focal hawk or dove in a matrix. The payoffs in the matrix are read left to right and are based on the opponent's behavior:

	Hawk opponent	Dove opponent
Focal hawk payoff	0.5 $(B-C)$	*B*
Focal dove payoff	0	0.5 *B*

Maynard Smith and Price assumed that in this game, the cost of losing a contest was greater than the benefit of winning the resource. This means that $B < C$.

Given this game, what strategy is the ESS? Let's evaluate some possibilities.

A POPULATION OF ALL DOVES IS NOT AN ESS
If all individuals are doves, all interactions are dove versus dove, and so all individuals obtain a fitness of 0.5*B*. However, if a hawk individual enters this population, or if one of the doves changes

continued...

its behavior and adopts the hawk strategy, it will receive a reward B during all encounters with doves and so will have a higher fitness because $B > 0.5B$. *In a population of all doves, hawks can successfully invade, and so pure dove is* not *an ESS.*

A POPULATION OF ALL HAWKS IS NOT AN ESS

In a population of all hawks, each individual receives a payoff of 0.5 ($B − C$). Because we assume that $B < C$, here, individuals actually receive a negative payoff! Remember that they all receive the same negative payoff because all interactions are hawk versus hawk. However, if a dove enters this population, it will lose all contests with hawks but never get injured. The dove's payoff in these encounters will be 0, which is greater than 0.5 ($B − C$) when $B < C$. That means that in a population of all hawks, a dove will have the highest fitness—even though the dove can never win a contest. *Doves will therefore increase in the population, and so pure hawk is also* not *an ESS.*

A POPULATION OF HAWKS AND DOVES IS AN ESS

Given that pure dove and pure hawk are not ESSs, is there an ESS for this game? As you might have guessed, the ESS must be a mixture of the two strategies. In a mixed evolutionary stable strategy, a fraction of individuals play each strategy, or else each individual plays a particular strategy in some fraction of the encounters. To solve for the mixed ESS, we need to find the frequency of hawks and doves in the population such that each strategy receives the same fitness. We can solve this mathematically by letting

p = the frequency of hawks in the population

q = the frequency of doves in the population

By definition, $p + q = 1$, because all individuals are either a hawk or a dove. Next, we calculate the payoffs for a hawk and a dove based on their frequency in the population. Hawks will encounter hawks a fraction p of the time, and doves a fraction q of the time. There are p hawks in the population, and each time a hawk encounters another hawk, it receives a payoff of 0.5 ($B − C$). There are q doves in the population, and each time a hawk encounters a dove, the hawk receives a payoff of B.

The overall payoff to a hawk will then be:

$$p [0.5 (B − C)] + qB$$

By the same reasoning, the overall payoff to a dove will be:

$$p (0) + q (0.5B)$$

Now all we need to do is to set the payoff to a hawk equal to the payoff to a dove:

$$p [0.5 (B − C)] + qB = p (0) + q (0.5B)$$

A little algebra gives us:

$$p = B/C$$

Therefore, in a population at the ESS, the frequency of hawks will be B/C.

The hawk-dove model introduced game theory to animal behavior and has also been used extensively in economics, political science, and even cancer research (e.g., Hanauske et al. 2009; McEvoy 2009).

Chapter Summary and Beyond

Animals select habitats in which to live and reproduce based on the resources those habitats contain and the presence of others. When habitats exhibit negative density-dependent fitness and movement is cost-free, individuals distribute themselves among habitat patches in an ideal free distribution. In this pattern, the number of individuals in a habitat is proportional to habitat quality, and all individuals obtain the same fitness. Additional work has examined how distributions change when individuals are not free to select a habitat because of the behavior of dominant individuals (for recent examples, see Murray, Mane, & Pusey 2007; Purchase & Hutchings 2008). In some cases, fitness does not decline as density increases, and we then observe conspecific attraction, either because of Allee effects or because of conspecific cueing. Much work has been devoted to understanding the relative importance of these mechanisms (Donahue 2006).

Once a habitat is selected, individuals often defend a territory to secure the resources it contains. Territory defense is costly and so only individuals with high energy reserves can defend territories successfully. How territory-holders assess the costs of defense is an active area of research (e.g., Müller & Manser 2007). Territory defense and contests over resources require aggressive interactions that are regulated by androgen hormones such as testosterone. Androgen hormone levels are often elevated by social stimulation, as articulated in the challenge hypothesis. Hirschenhauser and Oliveira (2006) have reviewed empirical studies of the challenge hypothesis in relation to vertebrates, while Scott (2006) has extended the hypothesis to insects. Other work has examined how these hormones mediate winner effects (Gleason et al. 2009; Oliveira, Silva, & Canário 2009).

Animals exhibit much variation in fighting behavior that can be understood by game theory models, which explain how behavior evolves when fitness depends on how others behave. Game theory models have successfully been used to understand the fighting behavior of many taxa (Arnott & Elwood 2008), predicting that fighting behavior will be affected by resource holding potential (fighting ability) and the value of the contested resource. Much recent work has been devoted to understanding how individuals assess their relative fighting ability during a contest (Elias et al. 2008; Hsu et al. 2008; Reddon et al. 2011; Fawcett & Mowles 2013; Elwood & Arnott 2013).

CHAPTER QUESTIONS

1. What information suggests that an Allee effect can explain conspecific attraction in desert clickers?

2. Thick-billed parrots (*Rhynchopsitta pachyrhyncha*) in groups of fewer than ten suffer very high predation by falcons. However, groups with more than ten individuals are relatively safe from predators. Based on these observations, what habitat selection patterns would you expect to find in this species?

3. Imagine that you stage a contest between two male crickets that differ in mass by 20%. The contest lasts for 30 seconds and contains one fighting behavior. Next, you stage a contest between two males that differ in size by 5%. What predictions can you make about fighting behavior in this contest based on mutual-assessment models?

4. Suppose you conduct an experiment with mallards (ducks) at a small pond. You create three food patches 20 m apart from each other. Patch A receives one piece of bread every six seconds, Patch B receives one every 12 seconds, and Patch C receives one every 20 seconds. There are 36 ducks in the pond. What is the predicted number of ducks in each patch according to the IFD model?

5. In the experiment in Question 4, the three patches were placed 20 m apart. How might moving the patches 100 m apart change the results of the experiment? Why?

6. The IFD model is fairly simple to test with a number of species. You can test the predictions of the model by throwing pieces of bread to house sparrows on a lawn or going to a local park with a pond and throwing bread to ducks or koi fish. You can change the quality of each patch by varying the rate of food delivery or food size. You can also watch human behavior. Just as animals choose food patches, humans choose lines at the bank, grocery store, or tollbooths on a highway. In each situation, the object is to get through the line as quickly as possible. Describe how human behavior in these situations would follow the predictions of the IFD model.

7. What prediction can you make about the androgen levels of the winner as compared to the loser in a contest between red-spotted newts, following the interactions observed in Verrell's (1986) experiment?

8. Compare the studies on aggressive behavior conducted by Oyegbile and Marler (2005) on mice with the study by Verrell (1986) on newts. For each study, describe which of Tinbergen's four questions (see Chapter 1) the study investigated, and explain your answer.

Chapter 12

Mating Behavior

Our yard in the Midwest is quiet all winter long—but each spring, this situation changes dramatically. The first sign of spring is the sound of male birds singing each morning. As it gets still warmer, we hear field crickets, and then hundreds of male tree frogs calling in a loud chorus on warm, humid evenings (Figure 12.1). Why the change? Sex. The silence of winter gives way to the sounds of animals attracting mates.

Some years are especially noisy. Every 13 or 17 years, a brood of periodical cicadas emerge from underground, where they have been feeding on the roots of plants, to mate. Cicadas are hemipterans (true bugs), and North America is home to seven periodical species. In 2011, two 13-year species emerged (*Magicicada tredecim* and *M. neotredecim*), and our region became awash with cicadas (Figure 12.2). Millions of males were calling females each day. It was so loud that we couldn't spend much time outside, because the sound level frequently exceeded 90 decibels—the equivalent of a jackhammer.

In this chapter, we start by examining how sexual selection can result in dramatic phenotypic differences between males and females by acting on traits that affect reproduction. We then see why females are often more choosy than males when selecting a mate and discuss the numerous benefits associated with selective mating. Female mate choice typically occurs prior to copulation, but we'll see that it can also occur after mating. When females are choosy, selection can favor the evolution of multiple male mating tactics when males cannot successfully compete for mates. Finally, we'll see how mating preferences can be affected by the mating behavior of others.

Figure 12.1. Calling frog. Male frogs vocalize in spring to attract females for mating.

12.1 Sexual selection favors characteristics that enhance reproductive success

primary sexual characteristics The genitalia and organs of reproduction.

secondary sexual characteristics Morphological differences between the sexes that are not directly involved in reproduction.

sexual selection A form of natural selection that acts on heritable traits that affect reproduction.

As we saw in Chapter 3, Charles Darwin proposed natural selection as a mechanism to explain the evolution of adaptive traits in species (Darwin 1859). Darwin was a devoted naturalist and made copious notes on the morphological differences between male and female genitalia (organs of reproduction), or the **primary sexual characteristics**, in many different species (such traits are often used by researchers to distinguish different species of invertebrates). He also noted many other differences between males and females in traits not directly involved in reproduction, called **secondary sexual characteristics**. For example, in many birds, males have colorful plumage, while females are drab (Figure 12.3). Many male mammals and insects possess large antlers, horns, or horn-like projections that are lacking in females. Darwin observed that such elaborate and exaggerated traits, in conjunction with complex behavioral displays and vocalizations, are often involved in conspecific interactions during the mating season.

Secondary sexual traits puzzled Darwin because they did not seem to fit into his theory of natural selection. Exaggerated morphological and behavioral traits should be energetically expensive to produce and maintain, and they can make individuals more obvious to predators and so reduce their survivorship. How could a male peacock's tail increase its survival? And if it did, why was the exaggerated trait found only in males? In *On the Origin of Species*, Darwin first proposed that exaggerated male traits might be advantageous for reproduction rather than for survival. In a subsequent book, *The Descent of Man and Selection in Relation to Sex*, Darwin in fact hypothesized that these traits might arise from a different form of selection (Darwin 1871). **Sexual selection** represents "the advantage certain individuals have over others of the same sex and species solely in respect of reproduction" (Darwin 1871, p. 210). As such, sexual selection is a subset of natural selection.

How does sexual selection explain the evolution of exaggerated male traits? First, Darwin proposed, in many

Figure 12.2. Cicadas. Periodical cicada species emerge at high densities only in certain years and create a very loud chorus.

Figure 12.3. Exaggerated male traits. (a) Male northern cardinals (*Cardinalis cardinalis*) possess bright red and black plumage that is absent in (b) females. (c) Male stag beetles (*Lucanus cervus*) possess large horn-like projections that females of that species lack.

species there is intense competition within one sex (often males) for mating opportunities with the other sex (often females). For example, males may engage in direct physical combat in which only the winners mate with females. Today, we describe this process as **mate competition** (or **intrasexual selection**)—when members of one sex compete with one another for mating opportunities (Figure 12.4). Such competition can include direct aggressive interactions, as well as contests over resources or breeding territories, as we saw in Chapter 11. While such competition is an important aspect of sexual selection, Darwin recognized another aspect as well: females may be choosy. Darwin thus proposed that males may also compete among themselves to increase their attractiveness to females. In this way, he envisioned that females often play an active role in reproductive decisions through **mate choice** (or **intersexual selection**)—when members of one sex exhibit distinct mating preferences. As we will see, both forms of selection can, and do, occur within a species, often in a sequential manner. Together, mate competition and mate choice create sexual selection, a process that favors characteristics in one sex that allow the trait bearers to be more successful reproductively. Sexual selection is one of the most active areas in animal behavior research today.

Why two sexes?

Obviously, we could not have sexual selection without two sexes, but how did the two sexes evolve? One fundamental difference between males and females is the size of their gametes (Figure 12.5). Males tend to produce many small, motile gametes (sperm), while females tend to produce much larger, nutrient-rich, and nonmotile gametes (eggs), a phenomenon called **anisogamy**. In many algae, fungi, and unicellular protozoans, however, all individuals produce similar-sized gametes; this **isogamy** appears to be the ancestral form (Bell 1978; Bulmer & Parker 2002). How did anisogamy evolve from isogamy?

Geoff Parker and colleagues developed a model to answer this question (Parker, Baker, & Smith 1972; Bulmer & Parker 2002). This model is based on the following assumptions:

1. In the ancestral marine environment, individuals in a population that retain reproductive compatibility, produce different-sized gametes.
2. Each parent has a fixed amount of energy to allocate to gamete production, resulting in a size–number trade-off: as the number of gametes produced increases, their size will decrease.
3. Zygote viability is related to its size. Larger zygotes have higher viability because they contain more resources for survival.

Parker considered the fitness of small, large, and intermediate-sized gametes. Small gametes have a numerical advantage: they will create the most zygotes. On the other hand, large gametes always produce zygotes with the highest survival. Intermediate-sized zygotes have neither advantage and so have the lowest fitness. Thus, there is disruptive selection against intermediate-sized gametes. The result is high fitness for either "proto-males," which produce many small gametes, or "proto-females," which produce fewer large gametes.

We now can understand why one sex (males) produces many small gametes that contain only genetic material, while the other sex (females) produces a smaller number of large, nutrient-rich gametes. The evolution of two different sexes leads to different reproductive strategies and sets the stage for sexual selection, as we see next.

mate competition A situation in which members of the same sex compete for access to the other sex for reproduction. Also called **intrasexual selection**.

mate choice Selection by one sex for members of the other sex for reproduction. Also known as **intersexual selection**.

anisogamy The existence of differently sized gametes (small and large) in the different sexes.

isogamy The production of gametes of the same size by all individuals.

Figure 12.4. Mate competition. Male northern elephant seals (*Mirounga angustirostris*) compete for females.

Figure 12.5. Gametes. A single sea urchin egg cell surrounded by hundreds of sperm cells.

Bateman's hypothesis and parental investment

Angus Bateman was one of the first researchers to examine sexual selection in males and females using fruit flies (*Drosophila melanogaster*). In a series of six experiments, Bateman placed equal numbers of virgin females and males in milk bottles for three or four days and allowed them to mate (Bateman 1948). Adults varied in age from one to six days old. Females can take up to four days to become sexually mature, while males are sexually mature within 24 hours of eclosion (emergence of the adult from the pupal case). Each adult was heterozygous for a different dominant mutation, and so Bateman could determine the parents of three-fourths of the offspring by their phenotype. Across all experiments, 96% of the females produced progeny (as evidenced by the presence of their mutation in some progeny). In contrast, only 79% of the males successfully produced progeny. From these data, Bateman concluded that males had a higher variation in reproductive success than females. Bateman inferred that the intensity of sexual selection, as measured by the variation in reproductive success, was, in general, higher on males than females. He also inferred that this increased intensity of sexual selection was due to male–male competition.

Furthermore, Bateman's experiments suggested that a male's reproductive success increases more strongly with the number of mates obtained than does female reproductive success (Figure 12.6). Bateman concluded that reproductive success for a female is primarily limited by egg production, because a female can obtain sufficient sperm to fertilize all her eggs from a single mate. In contrast, male reproductive success is limited not by sperm depletion, but primarily by the number of matings a male obtains. Because females produce fewer gametes than males, there will be intense competition among males for fertilization of female gametes (Kokko & Jennions 2003). This situation sets the stage for greater variation in male reproductive success—and thus more intense sexual selection on males. In sum, the evolution of male–male competition and female choice follows from competition for access to gametes of the opposite sex (Kokko, Jennions, & Brooks 2006).

In **Bateman's hypothesis**, female reproductive success is most strongly limited by the number and success of eggs that she can produce, whereas male reproductive success is limited by the number of mates he obtains. This difference between the sexes has become a foundation of sexual selection theory. Following Bateman, Robert Trivers expanded on this concept by identifying parental investment as a key difference between the sexes, in addition to gamete size. As defined by Trivers, parental investment includes "any investment by the parent in an individual offspring that increases the offspring's chance of surviving (and hence reproductive success) at the cost of the parent's ability to invest in other offspring" (Trivers 1972, pg. 139). Parental investment can include gestation, incubation, defense, and food provisioning, and these behaviors often differ between males and females. Trivers's **parental investment theory** predicts that the sex that pays the higher cost of parental investment should be choosier when it comes to mates. The other sex will then experience more intense sexual selection. When one sex has greater parental investment, these individuals will not always be sexually receptive, because they are caring for (investing in) current offspring. This "waiting time" in one sex to once again become sexually receptive results in a difference in the number of sexually receptive males and females in a population, known as the **operational sex ratio** (Emlen & Oring 1977). A large skew in the operational sex

Figure 12.6. Bateman's results. (a) Fruit flies mating. (b) Relative reproductive success for males (solid line) and females (dash) in Experiments 1–4 combined. (c) Relative reproductive success in Experiments 5–6 combined. (*Source:* Bateman 1948)

Bateman's hypothesis the hypothesis that female reproductive success is most strongly limited by the number and success of eggs that she can produce, while male reproductive success is limited by the number of mates he has.

parental investment theory The hypothesis that the sex that pays the higher cost of parental investment should be choosier when it comes to mates.

operational sex ratio The ratio of the number of sexually receptive males to sexually receptive females in a population.

ratio toward the noncaring sex with little or no waiting time to mate results in greater intensity of sexual selection on that sex (Emlen & Oring 1977). Because females in most species invest more in offspring, we can expect to see choosy females and a higher variation in male reproductive success.

As a result of sexual selection, we expect that males will often exhibit exaggerated traits used in competition for females. These traits may be morphological or involve behaviors such as vocalizations and courtship rituals. Exaggerated morphological traits used in male–male competition are often called **weapons**, and those used to attract females are known as **ornaments**. Exaggerated male traits are widespread, particularly in mammals and arthropods (Emlen 2008), and many male birds possess ornaments. Next, we examine one example of each.

weapons Exaggerated morphological traits used in competition with individuals of the same sex.

ornaments Exaggerated morphological traits used to attract mates.

Weapon size and mating success in dung beetles

Featured Research

In many beetles, males possess a horn-like projection that females lack (Figure 12.3c). Joanne Pomfret and Robert Knell studied the role of these horns in competition for mates in male dung beetles (*Euoniticellus intermedius*) (Pomfret & Knell 2006). Males fight with one another to defend dung piles, where females dig subterranean tunnels to mate and raise their offspring. Pomfret and Knell observed much variation in body size and horn size among males and conducted an experiment to determine whether large horns provide an advantage in these fights.

They staged fights between males in artificial nesting arenas (similar to ant farms) consisting of two panes of glass placed 5 mm apart and filled with soil and dung. The researchers collected adult males and females from a field and placed a single female in the arena, which quickly dug a tunnel.

Two males, closely matched for body size but not horn length, were then introduced into the arena with the female. Male body length was measured with calipers, and horn length was measured from photographs of the head. The researchers recorded the outcome of subsequent interactions. The winner was deemed to be the male that successfully entered and remained in the tunnel after 24 hours; most winners subsequently mated with the female. The loser was excluded from the tunnel and thus from mating opportunities.

During fights, males tried to hook their horns under their opponent's head. They also used their horns to push an opponent out of the tunnel. In contests between small beetles, the larger male or the male with the longer horn won the most contests. In contests between large males, large horn size alone was the most important predictor of success (Figure 12.7). In all contests, the greater the difference in horn length, the more likely the winner was to have the longer horn. These findings suggest that the maintenance of long horns in male dung beetles can be explained by sexual selection driven by male–male competition.

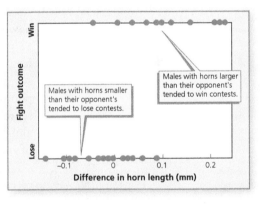

Figure 12.7. Beetle contest outcomes. Focal males with a horn shorter than that of their opponent (negative difference in horn length) won few contests. Focal males with a horn larger than that of their opponent (positive difference in horn length) won most contests. (*Source:* Pomfret & Knell 2006)

Ornaments and mate choice in peafowl

Featured Research

Ever since Darwin, peacocks (*Pavo cristatus*) have been the quintessential example of sexual selection. Males possess extremely large tails whose feathers contain brightly colored ocelli, or eyespots (Figure 12.8). Because males fan out their tail when displaying to females, it seemed likely that female mate choice involves tail morphology. Indeed, early work demonstrated that females prefer to mate with males that have the most ocelli—that is, males with the most ornamented tails (Petrie, Halliday, & Carolyne 1991).

However, before peahens select a peacock mate, males must compete to establish and defend a display site. Many males aggregate on a **lek**, a location that lacks

lek A location where males aggregate and display to females.

Figure 12.8. Peacock tail. Females (peahens) are gray and brown, while males (peacocks) have large, colorful, and elaborate tails.

resources, where they display to females using behaviors such as vocalizations and fanning of the tail. Males fight for status and preferred positions on the lek.

Adeline Loyau, Michel Saint Jalme, and Gabriele Sorci examined mating behavior in a population of peacocks at Parc Zoologique de Clères in France (Loyau, Saint Jalme, & Sorci 2005). The researchers captured and color-banded males and females in early spring. They measured each male's tail length, body size (by measuring the tarsus, a bone in the lower leg that provides an index of body size) and photographed each male's tail when open to count the number of ocelli. The researchers then recorded male–male interactions during the breeding season, along with the number and duration of male tail displays to females, the number of vocalizations, and the number of copulations.

The research team found intense competition for display sites, as only 45 of 61 males successfully defended a site and so over 25% of the males thus had little chance to reproduce. They also found significant positive correlations between the likelihood of defending a display site and both body size and tail length: larger males and those that had longer tails were more successful in acquiring a site (Figure 12.9). Females were very selective in their choice of mates. Only 12 males

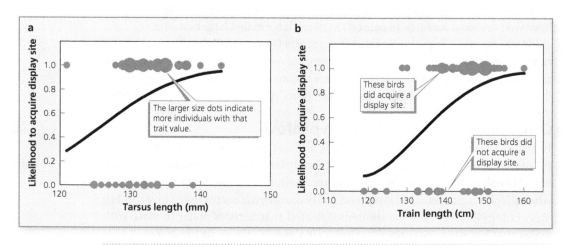

Figure 12.9. Probability of acquiring a display site. (a) Larger males (those with longer tarsus length) and (b) those with longer trains (tails) were more likely to acquire a display site. Dot size indicates number of data points. (*Source:* Loyau, Saint Jalme, & Sorci 2005)

a

Ocelli

b

Males with more eyespots in their tail obtained more copulations.

Number of copulations

Number of eyeposts in the train

Figure 12.10. Number of tail eyespots and copulations. (a) Ocelli. (b) Males that had tails with a large number of ocelli obtained more copulations. Dot size indicates number of data points. (*Source:* Loyau, Saint Jalme, & Sorci 2005)

obtained copulations, and over one-third of all copulations were obtained by a single male. In general, males that had high display rates and the most ocelli obtained the most copulations, although they were not necessarily the largest males (Figure 12.10).

The research team concluded that both mate competition and mate choice have played a role in the evolution of the peacock tail. First, only a subset of males successfully defended display sites, and all of these males were large individuals with long tails. Males frequently used their tail in aggressive interactions to chase nonterritorial males from display sites, a behavior that could also favor longer tails. Second, females strongly preferred males with high display rates and tails with many ocelli. Together, male competition and female choice have favored the evolution of males with long, elaborately ornamented tails and high display rates.

Male mate choice in pipefish

While female mate choice and male exaggerated traits are common, they are by no means universal. Depending on circumstances, males can also be selective, and this choosiness has favored the evolution of female secondary sexual traits in some species (Clutton-Brock 2007). Pipefish are an example of a **sex role–reversed species**, in which females compete for males that invest heavily in parental care. Females transfer their eggs to the males' egg-brooding structure. The male then fertilizes the eggs and cares for them as they develop—that is, males become pregnant. Because males invest more heavily in offspring than do females as a result of their high levels of parental care, sexual selection theory predicts that males should be the choosy sex and that female–female competition for mates may be intense. We might also expect females to display secondary sexual ornaments.

Anders Berglund and Gunilla Rosenqvist have been studying deep-snouted pipefish (*Syngnathus typhle*) for many years. This small (less than 40 cm long), slender species, a close relative of the seahorse, lives in shallow seagrass habitats in coastal Europe (Berglund, Rosenqvist, & Svensson 1986; Berglund & Rosenqvist 1993). Females possess a secondary sexual ornament: a temporary striped pattern on the side of the body that is displayed only during competition between females and when courting males, a behavior known as a courtship dance. Normal pipefish body coloration is quite cryptic, but the ornament makes the female stand out and so is likely to increase predation risk, making it costly (Scientific Process 12.1).

Featured Research

sex role–reversed species A species in which females compete for males that invest heavily in parental care.

SCIENTIFIC PROCESS 12.1

Male mate choice in pipefish
RESEARCH QUESTION: *How do male pipefish select a mate?*

Hypothesis: Males prefer females that have the highest expression of a sexually selected trait.

Prediction: Males should spend more time and mate more often with those females that display their temporary striped pattern (ornament) the most.

Methods: The researchers:

Day 1: (No interactions allowed)

- Conducted a simultaneous-choice test with a single male and two enclosed females that were matched for size
- Recorded the amount of time the male spent in front of each female, how often a female displayed her ornament, and how often the male engaged in a courtship dance with each female

Day 2: (Interactions allowed)

- Conducted the same experiment, but with all dividers removed so that all individuals could interact
- Recorded total time dancing, latency to dance, and copulations with each female
- Removed and counted all eggs from the male's brood pouch

Results:

- Males spent more total time and time dancing, and had a shorter latency to dance with the female that displayed her ornament more.
- The more ornamented female obtained more copulations.

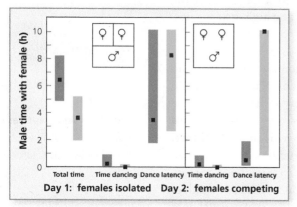

Figure 1. Male behavior. Time spent with the more (blue bars) and less (orange bars) ornamented female. The black square is the median time spent with the female. Each box is the middle 50% of data. (*Source:* Berglund & Rosenqvist 2001)

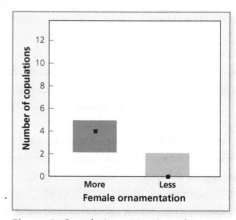

Figure 2. Copulations. Number of copulations with the more (blue bar) and less (orange) ornamented female. The black square is the median time spent with the female. Each box is the middle 50% of data. (*Source:* Berglund & Rosenqvist 2001)

Conclusion: Females with greater expression of the ornament have higher mating success.

EVALUATE
What fitness benefit might males obtain by mating with the more ornamented female?

Berglund and Rosenqvist tested whether this ornament affects male mate choice (Berglund & Rosenqvist 2001). Using a simultaneous-choice experimental design with aquaria divided into three sections, they conducted an experiment that allowed a male to choose between two females on consecutive days. Males spent more time in courtship dancing with the female that displayed her ornament more (the more ornamented female) and copulated with her more often, indicating that males prefer females that are more ornamented. Berglund and Rosenqvist suggest that female ornamentation reflects dominance status and overall vigor such that only females in the best condition can display their ornament at a high rate. It may also make it easier for males to assess female body size. Body size affects egg production, and so males should prefer larger mates.

These studies illustrate how both mate competition and mate choice are important in the evolution of elaborate and extravagant traits used in sexual selection. Each study examined the current function of an exaggerated trait. How did these traits arise? We look at that question next.

The origin of sexually selected traits: the sensory bias hypothesis in guppies

Featured Research

At this point, you may be wondering how a trait—say, in males—becomes a focus of female mate preference. One explanation is the **sensory bias hypothesis**, which states that female mating preferences are a byproduct of preexisting biases in a female's sensory system. These biases are presumed to have evolved in a nonmating context, with males evolving traits to match those biases (West-Eberhard 1984; Basolo 1990; Ryan & Rand 1990). One investigation of this hypothesis involves guppies (*Poecilia reticulata*). Females prefer to mate with males that display the greatest amount of orange coloration, which is derived from carotenoids obtained in the diet (Figure 12.11). Carotenoids are an important nutrient for many species, and they play a key role in immune system function (e.g., Bendich 1989). Colorful guppies are also easier for predators to spot and so suffer higher predation risk (Endler 1980). Prior work revealed that both natural and sexual selection has affected the strength of female preference and male coloration across populations. In stream populations where predation risk is high, female preference for orange is reduced, and males display less orange body color than do males in other populations with low predation risk (Houde & Endler 1990).

sensory bias hypothesis The hypothesis that female mating preferences are a byproduct of preexisting biases in a female's sensory system.

Helen Rodd and her colleagues noticed that guppies in the wild are attracted to and voraciously consume orange-colored fruits that fall into the water (Rodd et al. 2002). These fruits contain high levels of carotenoids, which are rare in other foods. Rodd and colleagues tested the hypothesis that guppies have an innate preference for orange food items because they contain carotenoids, and that this preexisting bias favors males that display orange color on their body. This hypothesis makes a basic prediction: both male and female guppies should be attracted to orange-colored objects because their color is associated with the presence of carotenoids.

The researchers tested the prediction in several natural populations in Trinidad. In one set of studies, they placed small discs on a leaf in the water that was held in place by a small rock. The discs could be red, orange, green, purple, blue, white, yellow, or black (Figure 12.12). For five minutes, the researchers noted all guppy approaches, pecks at the object, and the sex of the individual. To standardize environmental conditions, the researchers also conducted a similar laboratory experiment in which the discs were placed in an aquarium.

Figure 12.11. Male guppies. Males vary in the amount of orange pigment on their body.

Figure 12.12. Color discs. Examples of color discs used to study guppy attraction to different-colored objects. (*Source:* Rodd et al. 2002)

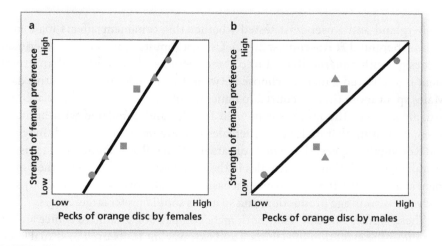

Figure 12.13. Mating and food color preferences. In populations where females most strongly preferred orange males (y-axis), both (a) females and (b) males pecked the orange-colored disc most frequently. Each symbol represents a different population. (*Source:* Rodd et al. 2002)

In both experiments, male and female guppies exhibited a strong attraction to the orange and red discs, and the strength of the preference for orange was correlated with female mating preference for the color orange. In populations with the strongest female preference for orange males, both sexes displayed the strongest attraction to the orange discs (Figure 12.13).

These experiments indicate that both male and female guppies display an innate preference for orange objects. This finding provides support for the hypothesis that female mating preference could be linked to a preexisting preference for orange food objects. In Section 12.3, we'll examine additional hypotheses to explain female trait preferences.

In all these cases, we see that sexual selection can explain the evolution of certain traits because they enhance reproductive success. For traits used in mate competition, the selective advantage is easy to see: larger weapons can provide an advantage in direct competition with rivals. Furthermore, it is straightforward to understand how a trait favored by one sex in their choice of a mate will evolve in the other sex, because those that possess the most extreme form of the trait will have high reproductive success.

Less clear, however, are the benefits of mate choice. At the beginning of the chapter we noted that in most species, females are often the choosy sex. What benefits can females gain by such choosiness? We examine that question next.

12.2   Females select males to obtain direct material benefits

direct material benefits Material resources obtained by a female from mating with a particular male.

indirect genetic benefits Genetic benefits females can obtain for their offspring by mating with males that have high genetic quality.

Parental investment theory predicts that the sex that invests the most time and energy in offspring can benefit by being selective in its choice of mates. As we saw, this usually means that females are choosy. What are the benefits of being selective? Two possibilities exist (Andersson & Simmons 2006). First, females can benefit by choosing males that provide **direct material benefits**, such as food gifts, access to territories with abundant food, or enhanced parental care (Møller & Jennions 2001). Second, females can benefit by mating with males that provide **indirect genetic benefits** that enhance the fitness of offspring. In this section, we examine how females can benefit directly from selective mate choice.

Figure 12.14. Nuptial gift. (a) A male hanging fly (*Bittacus* sp.) (right) presents a nuptial gift (dead fly) to a female (left). (b) The pair copulates. (c) The female takes the gift to eat.

Female choice and nuptial gifts in fireflies

Featured Research

In many arthropods, males provide a **nuptial gift** to a female prior to mating that provides nutrition and so can increase her reproductive success (Figure 12.14). These gifts can be prey, carrion, plant material, glandular secretions, or regurgitated food (Vahed 2007). Sara Lewis, Jennifer Rooney, and Christopher Cratsley studied how nuptial gifts affect female firefly preference for males and the females' resulting fitness (Rooney & Lewis 2002; Cratsley & Lewis 2003; Lewis, Cratsley, & Rooney 2004).

nuptial gift A physical resource such as a food item that an individual provides to a potential mate to enhance mating success.

Male *Photinus ignitus* fireflies produce a spermatophore—sperm packaged within a protein-rich structure produced by male accessory glands—as a nuptial gift (Lewis, Cratsley, & Rooney 2004). Once inside a female, the spermatophore disintegrates, releasing its nutrients, which the female allocates directly to her developing oocytes (Rooney & Lewis 1999). Because adults do not feed, the spermatophore represents an important resource. Does it affect fitness?

Firefly courtship involves species-specific bioluminescent flash patterns; males fly and flash to females sitting on nearby vegetation, and females flash in response to attract a male to her location (see Chapter 6). Cratsley and Lewis (2003) noted variation within a species in the duration of male flashes and examined how this might affect female choice and fitness.

To investigate this question, they first examined whether there was a relationship between male flash duration and spermatophore size. They collected 36 males and brought them into the laboratory, where each male was weighed and video-imaged to analyze its flash pattern. Most had flash durations that ranged from 56 to 89 ms and were highly repeatable among males (Figure 12.15). To quantify

Figure 12.15. Male flash durations. (a) A male producing a bioluminescent flash. (b) Variation in the duration of male flashes in a population. (*Source:* Cratsley & Lewis 2003)

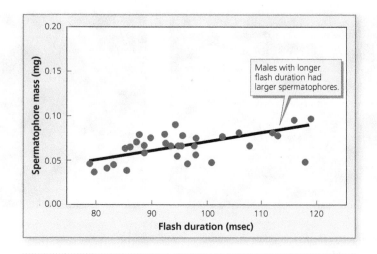

Figure 12.16. Spermatophore mass. There was a significant positive correlation between male flash duration and spermatophore mass. (*Source:* Cratsley & Lewis 2003)

Figure 12.17. Female response to flashes. Mean (+ SE) female response to simulated flashes. Females responded more to longer flash durations. (*Source:* Cratsley & Lewis 2003)

spermatophore mass, the researchers allowed each male to mate and then sacrificed the female to collect and weigh the spermatophore that was transferred. They found a significant positive correlation between male flash duration and spermatophore mass (Figure 12.16). The mechanism behind this correlation is unclear, but the association does mean that females could use variation in flash duration to choose males with the largest spermatophores.

Next, Cratsley and Lewis examined the effect of flash patterns on female mating behavior. Because they wanted to manipulate the duration of the flashes, the research team created simulated male flashes to which females responded by flashing. To examine the effect of the duration of the male flash signal on female behavior, the researchers created flashes of different durations (55, 63, 71, 79, or 87 ms) and presented them to 25 females. They found that females responded most frequently to the longest flashes (Figure 12.17).

Together, these results suggest that females prefer to mate with males that have the longest flash duration and thus likely have large spermatophores. Does this preference affect female fitness? South and Lewis (2011) have recently examined this question from a comparative perspective. Across a broad range of arthropod species, they found a positive relationship between spermataphore nuptial gift size and female fecundity, an important component of fitness. Future experiments will help clarify this relationship in fireflies.

Featured Research

Female choice and territory quality in lizards

Another direct benefit of mate choice for females is access to high-quality resources defended by a territorial male. Ryan Calsbeek and Barry Sinervo examined how territory quality affects female choice and fitness in side-blotched lizards (*Uta stansburiana*) (Figure 12.18) (Calsbeek & Sinervo 2002). Males defend territories where they display to females from rock perches. The quality of a territory is based on its rockiness: rocks provide perches not only to display to females but also to spot predators. Rocks also increase the range of "microclimates" (both hot and cool locations) available for thermoregulation, an important physiological factor for ectotherms like lizards. Previous work demonstrated that offspring growth and survival are greater when females lay eggs within male territories that have many rocks (Calsbeek et al. 2002). In this system, males compete for territories with abundant rocks, and females prefer to mate with males that control the rockiest territories—that is, the best males defend high-quality territories. However, that preference makes it difficult to know whether female mate choice is based on male quality or the quality of his territory.

To find out, the researchers manipulated territory quality after males had established their territories but before females became receptive and selected mates. They first mapped male territories and observed that large, dominant males settled on those territories with the most rocks. Then the researchers moved ten to 40 rocks from the territories of large, dominant males to the territories of small neighboring males. The added rock piles increased territory quality for small males while reducing territory quality for large males, because all males remained on their original territory after the manipulation. This design allowed the researchers to successfully separate high-quality males from high-quality territories. At the end of the breeding season, after females had selected mates, females were captured and brought into the laboratory to lay their eggs, which were incubated under standard conditions. The researchers measured the egg-laying date and egg mass of each female.

They found that females strongly preferred the improved territories, even though they were occupied by small males. Of the 51 females in the population, 37 moved onto the improved plots. Females on improved territories laid eggs sooner and produced larger egg masses (Figure 12.19), demonstrating a direct fitness benefit for females that select high-quality territories. In this system, females appear to select mating partners based on male territory quality and derive a significant fitness benefit from doing so.

We have seen that females can benefit by selecting males that provide access to a material benefit: resources. In many species, however, males do not provide material benefits, and still females are selective in their choice of a mate. Why? This is where indirect benefits enter the equation, as we are about to see.

Figure 12.18. Side-blotched lizard. Males defend territories to attract females.

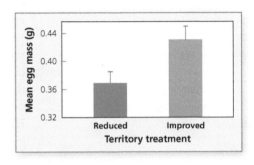

Figure 12.19. Female egg mass. Mean (+ SE) egg mass of females on treatment territories. Females on improved territories produced larger egg masses. (*Source:* Calsbeek & Sinervo 2002)

12.3 Female mate choice can evolve via indirect benefits to offspring

In many species, females benefit not from direct material benefits, but rather from indirect benefits. Females can obtain indirect genetic benefits when potential mates differ in genetic quality that affects their fitness. Females can benefit by mating with high-quality males to obtain those alleles for their offspring. How do females assess the genetic quality of potential mates if they cannot directly observe their genes? One way is by selecting males based on their secondary sexual traits. This idea, that female choice can affect the evolution of male traits, has stimulated much debate and research, starting as early as the nineteenth century.

Fisherian runaway and good genes

Perhaps no area in animal behavior has generated as much controversy as indirect fitness benefits via sexual selection, a debate that started with Darwin and his contemporary Alfred Russell Wallace, who independently proposed a theory of evolution based on natural selection. Recall that Darwin proposed two aspects of sexual selection: male–male (mate) competition and female (mate) choice. Wallace disagreed vehemently and considered *only* male–male competition to be a selective force (Wallace 1878). Their argument about the importance of female choice persisted because there was no clear mechanism to explain how mate choice could drive the evolution of secondary sexual traits.

In 1930, Ronald Fisher wrote a milestone paper proposing a genetic explanation for the evolution of exaggerated secondary sexual traits (Fisher 1930). In his verbal model, Fisher assumed that females select mates based on a particular trait that varies among males. Such a trait could evolve because of (1) its fitness advantage, independent of female choice; and (2) female preference for it. Thus, initially, the trait indicates male quality, because males possessing it have higher fitness and so are more attractive to females. If female preference for a trait and the trait itself have a genetic basis, female offspring will prefer the trait, and male offspring will express it. The intensity of female preference for the trait will increase as long as male offspring have a mating advantage by possessing it.

In sum, the male trait should co-evolve with the female preference and become increasingly exaggerated. Fisher called this a **runaway process**, one that would continue until the benefit it provided through sexual selection was outweighed by the disadvantage it entailed through natural selection (e.g., increased predation risk). Russell Lande formalized Fisher's idea into the notion of runaway selection. He showed mathematically how this process could occur through linkage disequilibrium when the genes for the trait and for female preference for the trait are genetically linked (Lande 1981), even if the trait is unrelated to male fitness. Linkage disequilibrium occurs when the genotype at one locus is not independent of the genotype at another locus.

How common is Fisher's runaway process? The process requires genetic covariation in both the trait and the mating preference. Kasey Fowler-Finn and Rafael Rodríguez (2016) reviewed dozens of empirical studies that characterized the strength of the genetic covariance between sexual traits and the mating preferences for the trait. Of the 43 studies reviewed, 27 detected a significant genetic covariance. These studies included over 20 species of invertebrates and vertebrates, and involved visual, acoustic, and chemical sexual traits. This review suggests that runaway selection may help to explain the evolution of a variety of sexual traits and mating preferences in diverse taxa.

Amotz Zahavi proposed an alternative hypothesis to explain how female choice could affect the evolution of male traits. He suggested that exaggerated secondary sexual characteristics allow the choosing sex to assess mate quality (Zahavi 1975). For Zahavi, the exaggerated trait is "a test imposed on the individual" by natural selection. Only high-quality males should thus be able to display the most exaggerated form of the trait—an idea he called a "handicap." Zahavi's **handicap principle** hypothesis states that well-developed secondary sexual characteristics are costly because they handicap a male's survival. Females should prefer such males because they must have excellent genotypes to overcome the handicap.

Subsequent theory expanded on Zahavi's work to demonstrate that exaggerated secondary sexual traits must be costly to produce in order for them to be a reliable indicator of male genetic quality, or **good genes**. These genes may be associated with an enhanced immune system, greater fighting ability, or increased vigor and viability (Moller & Alatalo 1999). The cost of the trait need not be a handicap to survival; such traits simply need to be costly to produce or maintain (Kodric-Brown & Brown 1984; Grafen 1990). If the trait was not costly, then all males, regardless of their quality, would be able to express it, and it would no longer indicate quality. Thus, a long, physically demanding vocal or behavioral mating display, or physiologically costly morphologies and chemical pheromones

runaway process An evolutionary process in which a male trait co-evolves with a female preference for it and becomes increasingly exaggerated.

handicap principle The hypothesis that well-developed secondary sexual characteristics are costly because they handicap survival.

good genes The alleles of a high-quality individual.

can be reliable indicators of male genetic quality if only the healthiest and most vigorous males (i.e., those with the best genes) can produce them.

At first, researchers debated the merits of Fisher's runaway and Zahavi's handicap principle hypotheses by focusing on their different assumptions. More recent work has shown that these models share important similarities (e.g., Ryan 1998; Kokko et al. 2002; Radwan 2002). In both models, the male trait initially preferred by females indicates a high degree of male vigor and survivorship. In good genes models, a genetic correlation is assumed between the female preference for the male trait and the male "good genes" that provide enhanced physical vigor and are indicated by the trait. In Fisherian runaway selection, there is a presumed genetic correlation between the female preference for the male trait and the male trait itself.

Today, much research on sexual selection focuses on understanding (1) the traits used by females in mate choice, (2) how sexually selected traits indicate male genetic quality, and (3) the fitness benefits of mate choice. Let's see how researchers have examined each of these aspects of sexual selection.

Mate choice for good genes in tree frogs

Featured Research

The good genes hypothesis predicts that females should base their mating decisions on costly male traits to obtain indirect genetic benefits. Many species produce vocalizations to attract mates; for example, among frogs that aggregate in ponds, many males call to attract females in a chorus. Calling is costly, because it is energetically expensive and, as we saw in Chapter 2, can make an individual more susceptible to predators (e.g., Bucher, Ryan, & Bartholomew 1982; Ryan, Tuttle, & Rand 1982). Do females assess male quality on the basis of their vocalizations?

Julie Jaquiéry and her colleagues tested the hypothesis that female European tree frogs (*Hyla arborea*) select males for genetic benefits based on their vocalizations (Figure 12.20) (Jaquiéry et al. 2009). Females initiate all matings, and males do not interrupt a pair that is mating (Friedl & Klump 2005), so male–male competition is assumed to be weak. The species is thus ideal for examining the genetic benefits of female choice.

Each night of the 22-night breeding season, the research team visited four neighboring ponds in western Switzerland to determine which males were calling in the chorus. A total of 15 calling males were captured and identified from a photographic data bank (individuals' black lateral line has unique color patterns), and a buccal (or cheek) sample was taken for genetic analysis (Figure 12.21).

To determine male mating success and the fitness of their offspring, the team visited the ponds every four days in daylight to collect egg masses. All clutches were reared in the laboratory until hatching. Ten offspring from each clutch were reared with unlimited food, and each tadpole was weighed twice to record its growth rate. Tadpole survival was determined at 28 days, and all individuals were then genotyped. The resulting genetic analysis allowed determination of the mating success of adults. From these data, the researchers could determine the attractiveness of each male, defined as

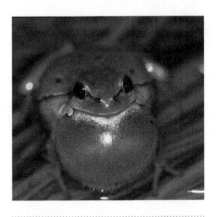

Figure 12.20. European tree frog. A calling male.

Figure 12.21. Buccal swab. Collecting a buccal swab for genetic analysis.

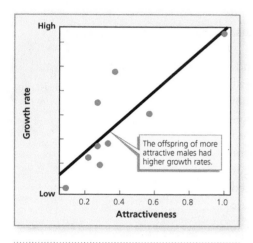

The offspring of more attractive males had higher growth rates.

Figure 12.22. Tadpole growth rate. The offspring of more attractive males had higher growth rates. (*Source:* Jaquiéry et al. 2009)

Featured Research

Hamilton-Zuk hypothesis The hypothesis that parasites and pathogens play an important role in sexual selection when secondary sexual traits are costly and condition dependent.

Figure 12.23. Male jungle fowl. (a) A male jungle fowl with a dull red comb. (b) A male with a brighter red comb that females prefer.

the number of females with which he mated divided by the number of nights he called during the breeding season.

Only ten of the 15 males sired any clutches, indicating that males varied in their attractiveness to females, and that females were selective. Offspring growth rate was positively correlated with the father's attractiveness; males that sired more offspring produced tadpoles with higher growth rates (Figure 12.22). In amphibians, offspring growth rate is positively related to fitness, and so the research team concluded that females do indeed select males for genetic benefits that result in higher fitness for offspring. Additional work is required to determine the features of male vocalizations that attract females. The researchers speculate that these features will be accurate indicators of male quality, because calling is so costly.

This example illustrates how female mate choice for a sexually selected, costly trait can enhance her fitness. Next, we consider a different way that females can benefit by selecting males that have the greatest expression of a costly secondary sexual trait.

Good genes and immune system function in birds

William Hamilton and Marlene Zuk proposed that parasites and pathogens play an important role in sexual selection when secondary sexual traits are costly and condition dependent (Hamilton & Zuk 1982). For example, males may vary in their immune response: some males will have alleles that confer greater resistance to pathogens than will others. If pathogens reduce male vigor, then only males with superior disease resistance will be able to display exaggerated secondary sexual traits. Females that mate with such males will benefit by passing these alleles to their offspring. This is the **Hamilton-Zuk hypothesis**, which makes two predictions:

1. Females should prefer to mate with males that have the greatest expression of secondary sexual traits
2. High parasite loads will reduce that expression in males

Marlene Zuk and colleagues tested these predictions in a captive population of red jungle fowl (*Gallus gallus*) (Zuk et al. 1990).

Jungle fowl are native to Southeast Asia. Like domestic chickens, males possess a variety of secondary sexual traits that involve the plumage and a red comb on their head (Figure 12.23). Zuk's team designed a simple experiment. Half the males were infected with intestinal nematodes at one week of age and kept isolated from control males that were not infected. The researchers measured the size of the comb each week as the birds matured, and at ten months of age (when males were sexually mature), they measured the color intensity of the comb by comparing it to a reference set of Munsell color chips. Next, they conducted mate choice tests in which an adult female could select either a parasitized or a control male. Males were visually isolated from one another and restrained to prevent them from interacting. The researchers recorded the male that the female solicited or with which she copulated.

Parasites affected the development of both comb length and color. Parasitized males had duller and smaller combs compared to controls (Figure 12.24). In the choice tests, females exhibited a strong mating preference for control males, suggesting that females prefer healthy males as mates. The results support the predictions of the Hamilton-Zuk hypothesis and illustrate one way in which females

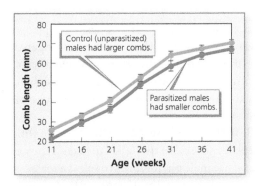

Figure 12.24. Male jungle fowl comb length over time. Mean (± SE) comb length of treatment and control males. As males matured, the combs of parasitized males (blue) were significantly smaller than the combs of unparasitized controls (orange). (*Source:* Zuk et al. 1990)

Figure 12.25. Spitting spider. These spiders expel a sticky substance to aid in prey capture.

can obtain genetic benefits from mate choice if healthy males tend to have alleles that enhance immune system function.

Mate choice fitness benefits in spiders

Featured Research

Another approach to understanding the indirect fitness benefits of mate choice is to manipulate the ability of females to make a choice. In many invertebrates, chemical signals, or pheromones, are commonly used in species recognition, but less is known about their role in mate choice (Johansson & Jones 2007). However, recent work suggests that pheromone production is costly and so may be a sexually selected trait that advertises quality (Johansson, Jones, & Widemo 2005).

Teck Hui Koh and colleagues examined the importance of pheromones as signals of male quality in mate choice in spitting spiders (*Scytodes* sp.) (Koh et al. 2009), which expel a sticky substance when catching prey (Figure 12.25). In this species, males provide no resources to females and so are ideal for investigating the indirect fitness benefits of mate choice.

The research team collected juveniles from two sites in Singapore and raised them individually until they matured. To ensure that females did not have experience with males that might affect their mate choice, the researchers used virgins. They first conducted a two-phase mate choice experiment using a Y-shaped apparatus to test female choice for both live males (Phase 1) and male pheromones alone (Phase 2).

In the first phase, two randomly selected males were placed behind wire mesh screens, one in each long arm, while a female was placed in the end of the stem (Figure 12.26). The screen prevented visual contact but allowed chemical communication. To prevent tactile communication, 3 cm thick foam was placed under the entire apparatus to prohibit any vibrations. Females were given one hour to select a mate; selection was determined by the female making contact with a screen for one minute.

Figure 12.26. Y-shaped test chamber. The top view of the experimental chamber shows males in each end of the Y and the female at the bottom. The side view of the chamber shows the foam that allows for the transmission of chemical but not tactile cues. (*Source:* Koh et al. 2009)

In the second phase, the researchers collected male pheromones. They placed each male from the first phase in a sterile Petri dish with moistened filter paper for 48 hours, during which the male deposited silk draglines and chemicals. The filter paper was then collected from each male and placed in the Y-shaped apparatus with the same female, and she again was allowed to make a choice over a period of one hour.

Of the 44 females that selected a male in the first phase, 42 selected the arm with the same male's filter paper in the second phase. The other two failed to make a choice. These results demonstrate that females can use male pheromones to select a mate.

To examine how mate choice affects female fitness, the 42 females that chose a mate in Phase 2 were divided into two treatments, such that the mean weight of females in the groups was similar. One set was allowed to mate with their preferred male, while the other set was restricted to mating with their nonpreferred male.

Females mated to their preferred male produced significantly larger egg sacs with more and heavier eggs and recorded a higher hatch rate, which should correlate with higher fitness, compared to females mated to their nonpreferred male (Figure 12.27). In this species, males provide no resources, and so females presumably benefit by obtaining indirect fitness benefits that result in enhanced offspring fitness. The researchers suggest that male pheromones may function as if they convey information about male size, nutritional status, immune function, fertility, or other attributes that affect offspring. Additional work is needed to identify the pheromone(s) involved and how it varies with male phenotype. The fitness benefits of mate choice, like those described here, are starting to influence the management of captive breeding programs (Applying the Concepts 12.1).

We've seen that female tree frogs and jungle fowl can obtain indirect fitness benefits by preferring to mate with some males over others. Similarly, in spitting spiders, female egg production and hatch rate were higher when the female mated with a preferred male. One explanation for this outcome is differences in male genetic quality. Another is that the female reduces her reproductive investment based on the quality of her mate. We examine this latter form of sexual selection, which occurs after mating, next.

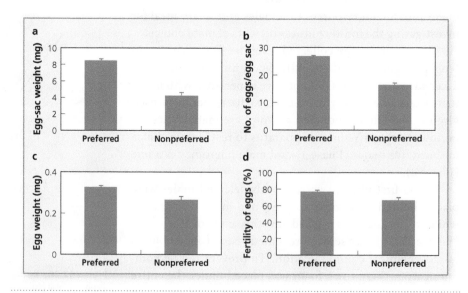

Figure 12.27. Female reproduction. Mean (+ SE) female reproduction. Females mated to their preferred male had (a) larger egg sacs, (b) more eggs per egg sac, (c) higher egg weight, and (d) higher fertility of eggs compared to females mated with their nonpreferred male. (*Source:* Koh et al. 2009)

APPLYING THE CONCEPTS 12.1

Mate choice in conservation breeding programs

Breeding programs can maintain the viability of captive populations and serve as a source for reintroduction efforts. However, it is crucial to maintain genetic variation and minimize inbreeding. Because captive populations are small, two genetic concerns are the primary focus of breeding programs. First, genetic diversity can decline rapidly as a result of genetic drift, and second, inbreeding can lead to reduced viability of offspring. Both issues can create problems for program success, especially given small captive populations.

Historically, managers have selected breeding pairs based on genetic considerations alone and have hoped that viable offspring would result. Many pairings fail, however, and one factor might be the lack of mate choice. Cheryl Asa, Kathy Traylor-Holzer, and Robert Lacy (2011) noted that litter sizes and offspring survival rates are higher when individuals select mates compared to when pairings are forced (see, e.g., Drickamer, Gowaty, & Holmes 2000; Anderson, Kim, & Gowaty 2007).

For instance, the captive breeding program for the Mauritius kestrel (*Falco punctatus*) has attempted to increase the population of birds by hand-rearing eggs. Some eggs are harvested from wild pairs of birds that then re-nest, whereas others are produced by captive birds. All eggs are hand-reared in captivity but differ dramatically in outcome. Over 75% of eggs harvested from the wild produce viable offspring, compared to just 55% of eggs produced in captivity (Cade & Jones 1993). Eggs harvested from the wild presumably result from females that exhibited mate choice, while eggs produced in captivity come from females that cannot exhibit choice because their mate was assigned by researchers.

Asa and colleagues suggest that the success of captive breeding programs can be enhanced by incorporating mate choice while taking into account other genetic concerns (Asa, Traylor-Holzer, & Lacy 2011). For instance, a focal female might be given the option of mating with several genetically acceptable mates. ■

12.4 Sexual selection can also occur after mating

Up to now, we have considered aspects of sexual selection prior to copulation. This is only part of the story. Both mate competition and mate choice can occur after gamete transfer. Postcopulatory male competition can occur when females mate with more than one male and the sperm of different males compete to fertilize the eggs, a phenomenon known as **sperm competition** (Parker 1970). In addition, postcopulatory female choice can occur after gamete transfer when a female influences the fertilization success of sperm from one male over that of others. Both phenomena require that females mate with multiple males, a common behavior that we will address in Chapter 13.

sperm competition A situation that occurs when the sperm of different males compete to fertilize eggs.

Mate guarding in warblers

One behavior that males can adopt to reduce mate competition and enhance their paternity is **mate guarding**: before and after copulation, a male follows his mate to prevent her from mating with rivals. However, this behavior has obvious costs. The time and energy spent guarding one mate can preclude a male from mating with other females or from acquiring food and other resources.

Helen Chuang-Dobbs and her colleagues studied the effectiveness of mate guarding in black-throated blue warblers (*Dendroica caerulescens*) at the Hubbard Brook Experimental Forest in New Hampshire (Chuang-Dobbs, Webster, & Holmes 2001). This small (10 g) passerine bird is common in the forests of eastern North America (Figure 12.28). A single pair will mate and raise offspring together, but females will often mate with other males, and so males risk raising another male's offspring, called **extra-pair young** (Chuang, Webster, & Holmes 1999). The researchers examined the effectiveness of mate guarding through a combination of observations and an experiment.

The researchers monitored the breeding behavior of all adults by capturing and color-banding them for identification. In the observational aspect of their

Featured Research

mate guarding A behavior in which a male remains close to his mate to prevent her from mating with rivals.

extra-pair young Offspring of a pair-bonded female produced outside the pair bond by a third-party male.

Figure 12.28. Mate guarding. Male black-throated blue warblers remain close to their mate.

study, the research team followed focal males and their mates for up to an hour on days just before egg laying, the period of high fertility for females. Every two minutes, they recorded the distance between individuals and noted how often males followed females or vice versa each time the birds moved to new locations. Evidence of effective mate guarding would be short inter-individual distances and a preponderance of male rather than female follows. The researchers then collected a blood sample from all nestlings for genetic analysis to determine parentage.

Observed males varied in their intensity of mate guarding, and this behavior was associated with variation in the paternity of offspring. Males that spent more time mate guarding had fewer extra-pair young than those that spent less time mate guarding (as measured by male follows and time spent within 20 m of the female) (Figure 12.29).

In the experimental part of their study, the team used a different study plot and removed some males for one hour during the female's fertile period. To do this, they attracted males away from their nests by playing conspecific songs of potential rivals from a speaker located near a territory and placed mist nets next to the speakers to capture the responding birds. Five males were captured and held in captivity, while three others spent most of the hour interacting with the speaker and so spent no time with their mate during the song playback. These eight treatment birds were presumed to exhibit a lower level of mate guarding than the eight control males that were not manipulated. The paternity of treatment and control birds was then compared to determine the effect of reduced mate guarding.

The nests of all experimental males that had been away from their mate for one hour contained extra-pair young, while only 50% of control nests contained such young (Figure 12.30). These findings indicate that mate guarding can be an

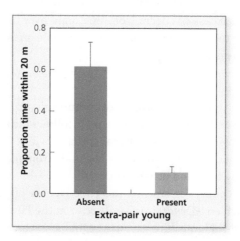

Figure 12.29. Intensity of male mate guarding. Mean (+ SE) proportion of time males spent close to their mate. Males with no ("absent") extra-pair young had spent a greater proportion of their time within 20 m of their mate (mate guarding) than males with extra-pair young present in their nest. (*Source:* Chuang-Dobbs, Webster, & Holmes 2001)

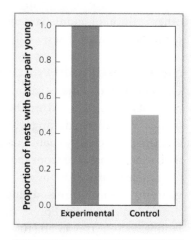

Figure 12.30. Proportion of nests with extra-pair young. Males that were experimentally removed had a higher proportion of nests with extra-pair young than did controls. (*Source:* Chuang-Dobbs, Webster, & Holmes 2001)

effective strategy for increasing paternity assurance and thus fitness for males. However, given that control nests did contain extra-pair young, mate guarding does not prevent females from mating with other males. Furthermore, remember that mate guarding comes at a cost: males that spend more time mate guarding have less time to spend seeking matings with other females.

Sperm competition in tree swallows

Featured Research

Not all males can continuously guard their mates. If males cannot prevent a female from mating with other males, then the sperm of different males may compete to fertilize the eggs. Another way a male can reduce mate competition to enhance his likelihood of success when sperm competition occurs is to swamp a rival male's sperm by mating frequently with a female. Susan Crowe and her colleagues tested the hypothesis that paternity assurance can be enhanced through frequent copulations in tree swallows (*Tachycineta bicolor*) (Crowe et al. 2009). In this species, extra-pair fertilization is common. Males also spend significant time on parental care, so selection should favor behaviors that minimize loss of paternity. Crowe's team predicted that frequent copulation rate would increase paternity for a male.

Figure 12.31. Swallow copulation. Males stand on or hover over the backs of females to copulate.

They tested this prediction on a population of tree swallows at Queen's University Biological Station in Ontario. Adults and offspring were captured and uniquely marked, and blood samples were taken for genetic analysis. In tree swallows, copulations occur mainly at the nest site and so are relatively easy to observe: males hover over females and make cloacal contact (Figure 12.31). Previous work has shown that most fertilizations result from copulations that occur during the three days before egg laying (Lifjeld & Robertson 1992; Lifjeld, Slagsvold, & Ellegren 1997). Consequently, beginning nine days prior to egg laying, each of 43 focal pairs were observed for three hours a day for over two weeks, starting at dawn, when most copulations occur.

The copulation rate varied among pairs and peaked in the three days prior to egg laying, averaging about five copulations per hour. Paternity analysis revealed that 52% of young were sired through extra-pair copulations. However, an increase in the frequency of copulations by a male increased the percentage of young he sired, as predicted (Figure 12.32). Thus, in this species, males can reduce the likelihood of paternity loss from sperm competition by increasing their copulation frequency.

Figure 12.32. Copulation frequency and paternity assurance. There is a significant positive relationship between the frequency of copulations (cloacal contacts per hour) and paternity assurance (proportion of within-pair offspring). (*Source:* Crowe et al. 2009)

High copulation frequency could increase paternity through sperm competition or could act as a type of mate guarding (because frequent copulations prevent other males from mating). Both of these behaviors enhance a male's likelihood of success in mate competition and result in higher rates of fertilization success for a male. Females, however, can also exert mate choice *after* copulation, as we see next.

Cryptic female choice

As we've seen throughout this chapter, female mate choice prior to copulation is usually obvious: females either do or do not mate with a particular male. Research has also found evidence of female choice after copulation. In species with internal fertilization, a female can mate with multiple males and then influence the fertilization success of sperm from one male over that of others, which is known as **cryptic female choice** (Thornhill 1983). The usual result is that one

cryptic female choice A situation that occurs when a female influences the fertilization success of sperm from one male over that from others.

male's sperm fertilizes a disproportionate number of her eggs. Why does cryptic female choice exist?

Cryptic female choice will benefit a female that mates with several males that differ in quality. She can enhance the fitness of her offspring by biasing fertilization success in favor of the higher-quality male. One way to evaluate this idea is to examine the fertilization success of males that differ in degree of relatedness to a female. A close relative is a low-quality mate, because such matings often result in low fitness for offspring, a process known as inbreeding depression. In comparison, an unrelated male is a relatively higher-quality mate. Can females affect the fertilization probability of sperm from males that differ in degree of relatedness?

Featured Research

Inbreeding avoidance via cryptic female choice in spiders

Klaas Welke and Jutta Schneider examined whether female orb spiders (*Argiope lobata*) (Figure 12.33) avoid inbreeding depression by cryptic female choice (Welke & Schneider 2009). Females occupy webs that are visited by roving males and commonly mate with two males. Each female contains two independent sperm storage organs, and these spermathecae, one on each side of the body, are filled as a result of two separate matings, either by the same male or by different males. Males inseminate females using their paired pedipalps, which are filled with sperm, but use only one pedipalp per mating attempt. In this species, the right pedipalp of a male is always inserted into the right spermathecae of a female, and the same is true for the left side.

Welke and Schneider took advantage of this reproductive morphology to conduct double mating trials. Females were mated with either (1) two sibling males (SS), (2) two nonsibling males (NN), or (3) a sibling and a nonsibling male (switching which one was first, SN or NS). Each male was matched for size and age but had one pedipalp removed so that the pair of males had complementary pedipalps (i.e., one only had the right, and the other only had the left).

Subjects were offspring of wild-caught females that had laid egg sacs in the laboratory. Hatchlings from each egg sac were kept in separate groups so that siblings (offspring from the same egg sac) and nonsiblings (offspring from different egg sacs) could be identified. Male and female hatchlings were separated until they reached adulthood. Females were placed in individual containers, in which they built a web, while males were kept in separate containers.

Males were introduced to the female's container one at a time, allowed up to one hour to copulate, and then removed. After both males copulated, the researchers estimated the number of sperm transferred to each spermathecae from the first and second mating by sacrificing a subset of females, dissecting the spermathecae, and counting spermatozoa under a microscope. To determine paternity for each male in each treatment, the researchers conducted a second experiment, in which they sterilized one of the males by irradiating him to make his sperm inviable. They could then calculate the proportion of eggs fertilized by each of the two males, because they knew which one was sterilized, and the eggs his sperm fertilized did not hatch.

Females mated readily with both siblings and nonsiblings, with no signs of inbreeding avoidance prior to copulation. The number of sperm in the spermathecae from the first male mating did not differ based on his relatedness to the female. However, for the second male, females stored more sperm from the nonsibling than they did from the sibling (Figure 12.34). With respect to paternity, the researchers focused on the proportion of eggs fertilized by the second male.

Figure 12.33. Orb spider. Males visit females' webs to mate with them.

Figure 12.34. Sperm storage. Mean (± SE) sperm number stored. Females stored more sperm during the second copulation when it came from nonsibling matings (blue) than when it came from sibling matings (orange). (*Source:* Welke & Schneider 2009)

Figure 12.35. Fertilization success of second males. The mean (± SE) proportion of eggs fertilized by the second male was lower when he was a sibling than when he was a nonsibling. S = sibling; N = nonsibling. (*Source:* Welke & Schneider 2009)

They found that the proportion of eggs fertilized by the nonsibling second male was higher as well (75% versus 48%) (Figure 12.35).

These data demonstrate cryptic female choice by orb spiders. When females mate with two males that differ in relatedness, they store more sperm from the nonsibling male and thus bias paternity toward that male. Both aspects of this choice should reduce inbreeding within a brood.

The above examples illustrate how sexual selection can occur after mating. For males, sperm competition can be intense, and mate guarding and high copulation rates can enhance paternity certainty. Females, too, can affect the outcome of sperm competition via cryptic female choice that biases the fertilization of sperm from one male over that from another. Next we examine how mate choice by females can favor alternative reproductive tactics in males.

12.5 Mate choice by females favors alternative reproductive tactics in males

Throughout the chapter, we have discussed individuals that all use one reproductive strategy to maximize success, such as winning contests over mates, possessing a high-quality territory, or displaying an exaggerated secondary sexual trait. However, as we have seen, not all males are able to win contests, defend high-quality territories, or display exaggerated traits. What do they do? When males cannot effectively compete, selection will favor the evolution of alternative strategies that enhance reproduction, and we then observe multiple behavioral phenotypes in a population, known as **alternative reproductive tactics** (Gross 1996; Taborsky, Oliveira, & Brockmann 2008).

Two different behavioral strategies are commonly observed. In the bourgeois tactic, competitive males defend a territory to monopolize its resources for females or else possess traits attractive to females. Typically, the bourgeois males are large, older, and in the best physiological condition. The parasitic tactic, on the other hand, is used by less competitive males to usurp matings from

alternative reproductive tactics The existence of multiple behavioral mating phenotypes in a population.

bourgeois males. Several types of parasitic behaviors exist. Satellite males, for example, associate with bourgeois males by remaining near them to intercept females that are attracted to the bourgeois individual. In contrast, sneaker males attempt to avoid detection so that they can quickly enter a bourgeois territory to fertilize eggs being deposited in a nest.

The evolution of alternative reproductive tactics

Two hypotheses attempt to explain the evolution of alternative reproductive tactics. One hypothesis views such tactics as a **conditional strategy**. This hypothesis assumes that the bourgeois strategy, on average, leads to higher fitness but is also more costly to adopt. Therefore, only individuals in the best condition (e.g., older, healthier individuals with high body mass) can adopt the bourgeois tactic; those in poorer condition can only adopt the parasitic tactic to obtain some level of reproductive success, even though it will, on average, be lower than that yielded by the bourgeois strategy. Essentially, parasitic males are "making the best of a bad situation" until their condition improves (Kodric-Brown 1986).

> **conditional strategy** A strategy that an individual chooses based on its condition.

The second hypothesis postulates that the tactics coexist in an evolutionary stable strategy (ESS) that is maintained by frequency-dependent selection (see Chapter 3). Here, neither strategy unilaterally leads to higher fitness, but rather the average fitness of each strategy increases as the strategy becomes less common. This idea makes sense when we think about the parasitic tactic. These males do not pay the costs associated with the bourgeois tactic of attracting a mate or defending a resource. They can thus parasitize more individuals and obtain high fitness when there are many more bourgeois males in the population than parasites. However, if too many males in a population exhibit the parasitic tactic, they will, on average, have low fitness, because there will then be too few bourgeois males to parasitize. This hypothesis predicts equal fitness for the tactics when they coexist as an ESS (Gross 1996). Let's see how these hypotheses have been tested.

Featured Research

Conditional satellite males in tree frogs

Male frogs produce advertisement vocalizations near ponds and streams to attract females. Females prefer to mate with those males that produce the most energetic calls (the loudest and at the highest rate) and calls of the lowest frequency. Typically, only large males can produce the lowest-frequency sounds (Gerhardt 1987).

Sarah Humfeld examined male reproductive behavior in green tree frogs (*Hyla cinerea*) in southeastern Missouri (Humfeld 2008). In this population, males exhibit both a bourgeois calling tactic and a parasitic satellite tactic. In the first, they call to attract females. In the second, a satellite male takes up residence near a calling male but remains silent, presumably to intercept females attracted to the caller. Humfeld noticed that some males use both tactics. Could this represent a conditional reproductive strategy? Humfeld's hypothesis made two predictions: (1) males differ in attractiveness to females, and those that produce unattractive calls are more likely to be satellites; and (2) satellite males should exhibit the same call preference as females—that is, both should be attracted to males that exhibit the lowest-frequency calls.

To test these predictions, Humfeld characterized the advertisement calls of males in the wild. Each male in the population was observed and assigned a status of either bourgeois calling male or silent satellite. She recorded 20 consecutive calls and then captured, measured, and marked each male. To obtain calls from satellites, she removed the nearest caller, which induced the satellite to begin calling, and then she recorded his calls. Humfeld found that bourgeois calling males were larger, in better condition, and produced lower-frequency calls than satellites—data that support the first prediction: satellites produce less attractive calls.

Next, Humfeld determined the vocal preference of females and satellite males. To do this, she created pairs of synthesized calls that were identical except for the lowest frequency they reached. The lowest frequencies of one pair were 700 Hz and 840 Hz, and those of the other pair were 840 Hz and 1,120 Hz. Humfeld captured females and placed them individually in a choice arena, covered with a dark cloth and away from all natural populations of calling males. She placed two speakers on either side of the arena and recorded the movement of the female toward one of the speakers, an index of her vocal preference. Then, to determine how different male calls affect satellite males, Humfeld repeated the experiment, but this time with individual satellites in the arena. She found that females and satellites both strongly preferred those male calls with the lowest frequencies, as predicted (Figure 12.36).

Humfeld concluded that male green tree frogs adopt a conditional strategy. Females prefer large males in good condition, the only males that can produce low-frequency calls. Small, perhaps younger, males call when away from an attractive male, but because of their size and condition, they cannot produce attractive calls, and so when an attractive male is nearby, they adopt a satellite strategy and remain silent but stay close to that male. This behavior should increase their encounters with females until they attain a larger size. The experiment did not yield data on male fertilization success, and so the relative fitness of the different strategies is unclear. However, the data suggest that if satellite males attempted to attract females by calling alone, they would be unsuccessful.

Figure 12.36. **Female and satellite male choices.** (a) Green tree frog. (b) Both females (orange) and satellite males (blue) preferred the call with the lowest low-frequency peak in two experiments that differed in the calls presented. Lines connect the two calls of an experiment. (*Source:* Humfeld 2008)

ESS and sunfish sneaker males

Featured Research

As we just saw, testing hypotheses about the evolution of alternative reproductive tactics is challenging because it is often difficult to quantify fitness for different mating tactics. However, the advent of molecular techniques has allowed scientists to establish paternity and thereby understand the fitness consequences of these tactics.

Oscar Rios-Cardenas and Michael Webster examined the mating behavior and reproductive success of male pumpkinseed sunfish (*Lepomis gibbosus*) in a small (4 ha) pond at the Huyck Preserve in upstate New York (Rios-Cardenas & Webster 2008). Pumpkinseed sunfish are medium-sized (15 to 20 cm in length) freshwater fish that are common in ponds and streams throughout North America (Figure 12.37). Some male sunfish, called parentals (i.e., those that adopt a bourgeois tactic), defend nests in breeding territories and provide parental care for the eggs there. Females approach these males and spawn at a nest site within the territory. Parasitic sneaker males do not defend territories but instead intrude while a parental male is spawning with a female in an attempt to fertilize eggs as they are laid.

Figure 12.37. **Pumpkinseed sunfish.** These fish are found in all the Great Lakes and are popular with anglers.

Figure 12.38. Male ages. Relative frequency of sneaker and parental males by age. Sneaker males were younger and smaller than parental males. (*Source:* Rios-Cardenas & Webster 2008)

Figure 12.39. Proportion of young sired by sneakers. Frequency of reproductive success by sneakers. Most sneaker males sired less than 10% of eggs in a nest. (*Source:* Rios-Cardenas & Webster 2008)

Over a period of three years, the researchers mapped the location of all active nests. When a spawning event occurred, the researchers captured nesting males, females, any sneakers involved in the spawning attempt, and all subsequent developing eggs. Tissue samples were taken from the adults and eggs for genetic analysis of paternity. Males were weighed, and their age was determined based on scale morphology.

The researchers observed 435 active nests, 60 spawning events, and 26 sneaker intrusions. They found that sneaker males were younger and smaller than parentals; most sneaker males were about two years old, while parental males were at least four years of age (Figure 12.38). On average, parental males sired about 85% of the offspring in their nest, and sneakers sired 15% (Figure 12.39). At first glance, parental males might seem to have higher fitness than sneakers, but sneakers are much rarer in the population than parentals. In fact, the researchers estimated that sneaker males represent only about 15% of the population. Because the reproductive success of sneakers is in proportion to their abundance in the population, the two strategies actually have equal reproductive fitness.

The researchers concluded that in this population, the two strategies are in an ESS. About 15% of individuals mature rapidly at two years of age and adopt the sneaker tactic. The remainder continue to grow and mature at four years of age, when they begin to compete for territories.

Additional work is required to understand how this reproductive decision occurs, but perhaps physiology plays a role. Rosemary Knapp and Bryan Neff studied closely related bluegill sunfish (*Lepomis macrochirus*) and found that parental and parasitic males exhibited different hormone profiles (Knapp & Neff 2007). Males that had adopted the parasitic tactic had higher testosterone and cortisol and lower 11-ketotestosterone levels than parental males (Figure 12.40). This finding suggests that hormones may regulate the adoption of different reproductive tactics in sunfish.

The above examples on tree frogs and sunfish illustrate that mating behavior is complex and can involve a variety of tactics by males. In the final section, we examine how the decisions of others can also influence mating behavior.

Figure 12.40. Hormone profiles. Mean (+ SE) hormone levels of males. Sneaker and parental males exhibited different levels of (a) testosterone, (b) 11-ketotestosterone, and (c) cortisol. (*Source:* Knapp & Neff 2007)

12.6 Mate choice is affected by the mating decisions of others

Throughout the chapter, we have assumed that the mate choice decisions of individuals are independent. Yet mating for many species takes place in a social environment, where individuals can observe and copy the mating decisions of others, a behavior known as **mate choice copying**, or nonindependent mate choice. We illustrate this behavior with examples from fish and flies.

mate choice copying A situation in which one individual observes and copies the mating decisions of another individual.

Mate copying in guppies

Featured Research

The first experimental evidence of mate choice copying comes from guppies (*Poecilia reticulata*), small freshwater fish native to Trinidad. Recall that males are brightly colored, with spots of orange, black, and iridescence (Figure 12.11). Lee Dugatkin examined whether females copied the mate choice of other females (Dugatkin 1992).

To answer this question, he placed a focal female in a clear Plexiglas container in the center of an aquarium. He then placed males, matched for size and overall coloration, into chambers on either side of the aquarium. Dugatkin first allowed a focal female to choose one of the males. In guppies, the amount of time a female spends near a male is a good indicator of her mate preference. Next, he arranged the aquarium so that a different female (the model) was now near the nonchosen male. This was done to simulate her "choosing" this male as a mate (Figure 12.41). During this time, the model female and male interacted with typical courtship behaviors, which the focal female could view. The model was then removed, and the focal female was released and allowed to swim freely for ten minutes. Dugatkin again recorded the amount of time the focal female spent next to each male.

Of 20 focal females, 17 spent significantly more time next to the male that had been near the model female. These females switched their mate choice preference to copy the mate choice of the model female. In this species, mate choice is not always independent; instead, females will copy the mate choice of others they observe.

Figure 12.41. Choice test design. The test apparatus had separate containers for males and movable doors that were used to adjust what the female viewed. (*Source:* LaFleur, Lozano, & Sclafani 1997)

Featured Research

Mate copying in fruit flies

Mate choice copying has now been documented in many species, but most often in vertebrates (Hoppitt & Laland 2008). Recently, Frédéric Mery and colleagues examined this behavior in fruit flies (*Drosophila melanogaster*) (Mery et al. 2009) (Scientific Process 12.2). The research team gave females a choice of males that were made to differ in quality by raising some on a rich-food medium that contained 100% of standard nutrients (high-quality males) and others on a poor-food medium that contained only 25% of standard nutrients (low-quality males). The researchers then conducted a three-phase experiment in a small Plexiglas box. In the pretest, females were allowed to choose between a high-quality and a low-quality male, and they spent more time with the high-quality male. Next, half the females saw a model female with the high-quality male, and the others saw a model female with the low-quality male. In the posttest, females were again allowed to choose between the high-quality and low-quality males. The behavior of the treatment females differed. Those that saw a model with the low-quality male now spent significantly more time with him—that is, they exhibited mate choice copying. No such change was observed in females that saw a model with the high-quality male.

These examples reveal the effect of the social environment on the mating preferences of different species. The social environment also appears to affect mate preferences in humans (Applying the Concepts 12.2).

The benefit of mate copying

When a female copies the mating behavior of another female, it might be making a poor choice (one that results in low fitness) if the observed female mated with a low-quality mate. Why should an individual copy the mate choice of another individual rather than making the decision on its own; that is, how can mate copying behavior evolve? One answer comes from a simple model (Nordell & Valone 1998). If a female has the option of mating with two males that differ in quality, she will have higher fitness by mating with the higher-quality option. However, it may be difficult for a female to determine which male is of higher quality if the males are similar or if she has little experience discriminating between males. In the absence of other information, she must select one at random and so will have only a 50% chance of making the best decision. Alternatively, she can observe another female and copy *her* mate choice. If the observed female is a better discriminator, the copying female will have a higher chance of selecting the higher-quality male. If not, she is no worse off.

Mate copying in fruit flies

RESEARCH QUESTION: *What factors affect the mating behavior of female fruit flies?*

Hypothesis: Male condition will affect the mate choice copying of female fruit flies.

Prediction: Female fruit flies will spend more time with a nonpreferred male after observing another female associate with him.

Methods: The researchers:

- Created groups of high-quality males (fed nutrient-rich food) and low-quality males (fed nutrient-poor food)
- Conducted a female mate choice trial with one high- and one low-quality male (pretest)
- Allowed half the females to then observe a model female next to the high-quality male, while the other observed a model female next to the low-quality male
- Repeated the mate choice trials (posttest)

Figure 1. Experimental arena. Males were restrained under lids that allowed females to obtain visual and olfactory information. (*Source:* Mery et al. 2009)

Results:

- In the pretest, females spent significantly more time near the high-quality male.
- In the posttest, females that observed a model female with the low-quality male now spent more time near the low-quality male.

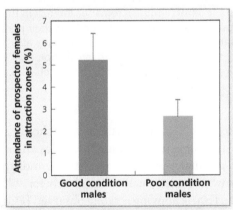

Figure 2. Pretest female behavior. Mean (+ SE) attendance of females. In the pretest, females preferred the high-quality male. (*Source:* Mery et al. 2009)

Figure 3. Posttest female behavior. Mean (+ SE) attendance of females, with blue bars denoting high-quality males and orange bars denoting low-quality males. In the posttest, the attractiveness of the high-quality male did not change. However, the attractiveness of the low-quality male increased significantly. (*Source:* Mery et al. 2009)

Conclusion: Female fruit fly mating behavior is affected by observing the mating decisions of other females.

EVALUATE

Describe how you could use this protocol to test the hypothesis that female experience affects mate choice copying behavior.

APPLYING THE CONCEPTS 12.2

Human mate choice copying

In many animals, mate choice preferences are influenced by observing the mating behavior of others. Is this also true in humans? Skyler Place and his colleagues allowed 40 female college students to observe dating interactions among men and women and examined how such observations affected individual ratings of the perceived attractiveness of a potential mate (Place et al. 2010). Each subject observed photographs of eight males involved in a speed date. Subjects were asked to rank their interest in each of these males as a potential mate for a short-term affair (a one-night stand) and for a long-term committed relationship.

Next, each female subject watched a short video of each of the eight males interacting with a female in a speed date. Video came from the Berlin Speed Dating Study, a carefully controlled study involving young men and women seeking partners. Subjects could hear the daters but could not understand them (because the conversations were in German). They were asked to rate whether the couple was mutually attracted to each other or not. The experimenters had previously selected four interactions in which the speed-dating couple did exhibit mutual interest and four others in which they did not; the subjects, however, were not told this but simply observed the interactions. Following each video, the female subject was again shown an image of each of the eight males and asked to re-rate them based on the same two questions about their short- and long-term interest in them.

Women's perception of a potential mate's attractiveness increased after observing him in a mutual romantic interaction during a speed date. This was true for both short-term and long-term interest (Figure 1). Interestingly, women's level of interest in a potential mate decreased after they observed an interaction that did not involve mutual romantic interest.

These data indicate that observing a mutual romantic interaction changes the degree of attractiveness of a potential

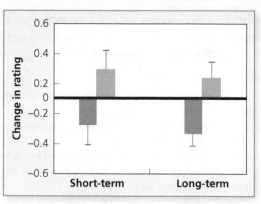

Figure 1. Mean (+SE) change in rating of a potential mate. Changes in rating of a male's attractiveness for short- and long-term relationships by female subjects after they observed a male interacting with a rival female. Blue bars are observations in which the male and rival female were not mutually interested in each other; orange bars are observations in which the male and rival female were mutually interested in each other. On the y-axis, zero represents the initial rating. (*Source:* Place et al. 2010)

mate in human females in a manner consistent with that found in other studies of mate choice copying in animals. Further studies are required to determine how individuals assess mutual romantic interest in the type of interactions observed and how that affects their perception of attractiveness. Interestingly, the same study also tested male subjects by asking them to rate female potential mates, and it found similar findings for mate choice behavior. ■

This model predicts that mate copying should be observed only when discrimination is difficult—that is, when the males are matched for quality. In guppies, for instance, females prefer males with the most orange coloration, and females mate copy when males are matched closely for color. When males are very different in color, female copying is rarely observed (Dugatkin & Godin 1993). In addition, if discrimination ability increases with age and experience, then the model predicts that young females should mate copy more often than older females, a tendency that has been demonstrated in guppies (Dugatkin & Godin 1993).

Featured Research ▶ ## Nonindependent mate choice by male mosquitofish

Now we know that female mate choice is affected by the mating decisions of other females. Is the same true for males? Recall that males compete with one another to fertilize eggs, often by sperm competition inside a female's reproductive tract. We might therefore predict that males should try to avoid mating with females they have recently seen mate.

Bob Wong and Miranda McCarthy asked: Does the risk of sperm competition affect mate choice in male eastern mosquitofish (*Gambusia holbrooki*)? They conducted a three-stage experiment using a standard two-choice test with three aquaria (Figure 12.42) (Wong & McCarthy 2009). In the first stage, Wong and McCarthy examined the preference of focal males for two randomly selected females. In the second stage, the focal male was constrained, and a rival male was placed next to each female. Here, the researchers could control whether the focal male observed these other males. In the final stage, the rivals were removed and the focal male was again allowed to associate with both females.

In the first stage, the focal male could not see either rival. In the second stage, the focal male saw a rival near the female he had preferred in the first phase of the experiment. And in the final third stage, the focal male saw a rival next to only the female he had not preferred in the first phase.

In Stages 1 and 3, males preferred the larger female, and their preference was consistent between the two stages (Figure 12.43). However, in Stage 2, male behavior was affected by observing a rival near the preferred female. When tested again in the final stage, the males now spent less time near their initially preferred female (Figure 12.44). These results show that males have a mating preference for

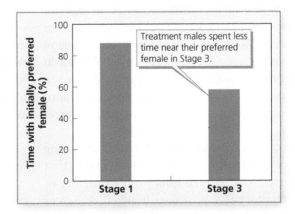

Stage 1:

Stage 2:

One of these partitions was removed to allow the focal male to see the rival.

Stage 3:

Figure 12.42. Male preference experimental setup. Each experiment consisted of three stages. Stage 1 measured male preference, Stage 2 allowed males to see a rival next to a female, and Stage 3 measured male preference again. (*Source:* Wong & McCarthy 2009)

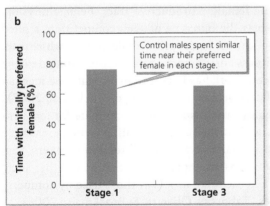

Control males spent similar time near their preferred female in each stage.

Treatment males spent less time near their preferred female in Stage 3.

Figure 12.43. Time in association for control male mosquitofish. (a) Mosquitofish. (b) Median percentage of time males spent in association with an initially preferred female. Males display high repeatability in their initial mate preference. (*Source:* Wong & McCarthy 2009)

Figure 12.44. Time in association for treatment males. Median percentage of time males spent in association with an initially preferred female. Males spent less time with their initially preferred female in Stage 3, after observing her with another male. (*Source:* Wong & McCarthy 2009)

larger females and that their preference of mates is influenced by observing the choice of other males. They will spend less time near a preferred female if they have just seen her associate with another male. This change in behavior is likely to minimize the risk of sperm competition.

These experiments demonstrate that the mate choice of individuals is often affected by the social environment and observations of the mating behavior of others. This phenomenon appears to be widespread, and its implications are just beginning to be explored (e.g., Dubois, Drullion, & Witte 2012).

Chapter Summary and Beyond

The study of mating behavior is organized around the concept of sexual selection—differential reproduction as a result of competition to mate and selective mate choice. Sexual selection theory assumes that, in general, male fitness increases most strongly with the number of sexual partners, whereas female fitness increases more by selecting mates that provide resources or genetic benefits. Therefore, males will often compete for access to females, and females will be choosy in their selection of mates. Sexual selection also explains the evolution of elaborate secondary sexual characteristics and courtship displays, and both are the focus of much research (Jones & Ratterman 2009). Barry and Kokko (2010) summarize why male mate choice is rare in most species. A recent investigation into Bateman's seminal work prompted significant theoretical critiques of his work, but many of his basic conclusions remain valid (Tang-Martínez & Ryder 2005). More proximate studies of mating behavior have focused on how genes regulate the development of traits in one sex but not the other (Williams et al. 2008) and on the neural activation involved in sexual selection (Hoke, Ryan, & Wilczynski 2010). Although sexual selection is typically used to explain traits within a species, recent work has examined how it can promote speciation as well (Gray & Cade 2000).

Females can benefit from mate choice by obtaining direct material benefits when selecting males that provide nutrient-rich nuptial gifts prior to mating or access to a territory that provides abundant resources (Gwynne 2008). Females can also benefit by selecting mates that provide indirect genetic fitness benefits. One way this can occur is through the existence of a genetic covariance between genes that affect female mating preferences and the sexual trait in males they prefer.

Alternatively, females can benefit indirectly by selecting males that possess good genes that will be passed on to her offspring. Much recent work has focused on understanding the indirect benefits of mate choice, including how often genetic covariances or good genes explain mating preferences and sexual traits (e.g., Prokop et al. 2012; Prum 2012; Greenfield et al. 2014). Another focus involves female choice based on immune system function and heterozygosity (e.g., Olsson et al. 2003; Milinski 2006; García-Navas, Ortega, & Sanz 2009). One area of interest is the major histocompatibility complex, genes that encode for proteins involved in the immune system response to pathogens (Tregenza & Wedell 2000). Mate choice based on cognitive traits is also attracting increasing attention (Boogert, Fawcett, & Lefebvre 2011).

Sexual selection also occurs after mating. Sperm from different males often compete to fertilize eggs within the female reproductive tract. Female physiology can bias the outcome of such competition, a phenomenon known as cryptic female choice, which can reduce fitness costs associated with inbreeding depression. Recent work has begun to investigate the mechanisms of cryptic female choice (e.g., Luck et al. 2007). When males cannot successfully compete for females, selection will favor alternative mating tactics. Common examples in many taxa are satellites and sneakers, and recent work has attempted to determine whether such tactics yield the same fitness as those used by competitive males (Taborsky, Oliveira, & Brockmann 2008).

In many species, the choice of a mate is affected by the mating decisions of others. Such mate choice copying occurs in both females and males; Vakirtzis (2011) provides a recent review of the research. While most evidence comes from laboratory studies, recent work with fish has demonstrated mate copying in the wild (Alonso 2008). Mate copying may allow individuals to better discriminate differences in quality among potential mates.

CHAPTER QUESTIONS

1. Under what circumstances would you expect sexual selection to be more intense on females than on males?

2. Bateman's (1948) experiment used flies that varied in age from one to six days old. Given the difference in age of maturation of males and females, how might this design have affected his results?

3. In a polygynous mating system, a few dominant males mate with many females, and many males do not obtain a mate. In a monogamous mating system, individual males and females often form pair bonds, mate, and raise young together. Which mating system would you predict to have the highest intensity of sexual selection on males, and why?

4. Chuang-Dobbs, Webster, and Holmes (2001) used both observation and experiments to examine the hypothesis that mate guarding facilitates paternity assurance for males. Explain why both methods are important.

5. Crowe et al. (2009) studied sperm competition in tree swallows and found a correlation between copulation frequency and paternity. Design an experiment to test the hypothesis that more frequent copulation helps ensure paternity.

6. Imagine that you have discovered a species with alternative reproductive tactics. Large males defend territories to attract females. Small males do not defend territories but instead use a roving tactic: they attempt to copulate with females they encounter while searching for a territory. In the population studied, half of the adult males defend territories and half do not. A genetic analysis of paternity indicates that territorial males sire 80% of all offspring. Is the roving tactic a conditional strategy, or are the two behaviors an ESS in this population?

7. Imagine two populations of the same species that both have identical numbers of adult males and females. In one population, females frequently exhibit mate choice copying, while in the other, this behavior does not occur. How might this difference in female behavior affect the variance in male reproductive success in these populations?

8. Compare the experimental protocol used by Place and colleagues (2010) to study mate choice copying among college students with that used by Dugatkin (1992) to study mate choice copying in guppies. How did each study simulate mate choice by a rival female?

9. Compare the studies conducted by Rodd and colleagues (2002) to those conducted by Cratsley and Lewis (2003). For each study, describe which of Tinbergen's four questions the study investigated and explain your answer.

Chapter 13

Mating Systems

When we are in Arizona, we enjoy taking walks in the desert, especially when animals are busy with the business of reproduction. It is easy to hear and see cactus wrens (*Campylorhynchus brunneicapillus*), the largest wren in North America. A male and female pair up, defend a territory, and raise their offspring together. Hummingbirds are everywhere. Males possess bright iridescent plumage and engage in aerial displays near a female sitting in a tree. Some males may mate with several females, while others fail to find a mate, but males do not help with parental care. Each female builds a nest, incubates the eggs, and feeds the young without aid from a male. And it is always nice to see desert spiny lizards (*Sceloporus magister*) (Figure 13.1). Males engage in aggressive contests as they defend their territories, and males that successfully defend a high-quality territory can often mate with several females.

Wrens, hummingbirds, and lizards differ in the social associations and number of mates they have during a breeding season, and they represent a few of the mating systems that exist. In this chapter, we review mating system diversity and examine theories to explain such variation. We also discuss how social mating systems that are based on behavioral association differ from genetic mating systems, which are based on genetic parentage.

Figure 13.1. Desert spiny lizard. Males defend territories during the mating season.

Sexual conflict and environmental conditions affect the evolution of mating systems

mating system A description of the social associations and number of sexual partners an individual has during one breeding season.

Animals exhibit different **mating systems**—the social associations and number of sexual partners an individual has during a breeding season. Table 13.1 summarizes differences among the five mating systems: monogamy, polygyny, polyandry, polygynandry, and promiscuity. In some mating systems, individuals live in close association with one another during the breeding season and are said to form a pair bond. However, pair bonds are far from universal, and an individual's number of mates and types of social association range widely across taxa. It is important to emphasize that the above breeding seasons are defined within one reproductive season. In many species, individuals mate with a different partner in successive breeding seasons, even if they are, for instance, monogamous within a single season.

Over the past few decades, our understanding of animal mating systems has changed dramatically. In the past, we assumed that each species had a characteristic mating system. Today, we understand that species, populations, and even individuals can exhibit tremendous variation in their mating systems. For instance, four mating systems have been observed in single populations of both dunnocks (*Prunella modularis*), an insectivorous bird found in Europe (Davies & Lundberg 1984), and Milne-Edwards' sifakas (*Propithecus edwardsi*), a social primate that lives in Madagascar (Pochron & Wright 2003). Although these examples are extreme, we do often find more than one type of mating system within a species. To understand why, we need to consider the factors that influence the evolution of mating systems. As we see next, these factors involve the reproductive decisions of males and females. Such decisions are affected by their fitness benefits and costs, which in turn vary with environmental conditions.

The evolution of mating systems

sexual conflict The differential selection on males and females to maximize their fitness.

Stephen Emlen and Lewis Oring argued that mating systems can be understood by examining two factors. One is evolutionary: **sexual conflict**, or the differential selection on males and females to maximize their fitness. The other factor is ecological: the ways in which resource limitation and distribution affect fitness benefits and costs for each sex (Emlen & Oring 1977). We need to examine these factors together to understand the variability in mating systems.

TABLE 13.1 Mating systems. Mating systems differ in the number of mating partners each sex has during one breeding season.

MATING SYSTEM	NUMBER OF FEMALES	NUMBER OF MALES
Monogamy	1	1
Polygyny	More than 1	1
Polyandry	1	More than 1
Polygynandry or plural breeding (mating occurs within social groups)	More than 1	More than 1
Promiscuity (mating is not restricted to specific social associations)	More than 1	More than 1

Emlen and Oring's model assumes that for a female, fitness will often be more strongly limited by the resources she can obtain to invest in offspring than by the number of her sexual partners (Bateman 1948; Emlen & Oring 1977). This idea follows from the assumption that females typically invest more energy in offspring than males do. First, eggs, because of their relatively large size, should require more energy to produce than sperm. For example, female birds and reptiles may invest up to 50% of their body mass in production of a single clutch of eggs (e.g., Vitt & Price 1982; Monaghan & Nager 1997). Second, in many species, females expend more energy on parental care (e.g., incubation, feeding, and defense) than do males, as we will see in Chapter 14. In addition, most females can obtain sufficient sperm from a single male to fertilize all her eggs for one reproductive event. In contrast, for a male, fitness is most strongly affected by the number of sexual partners he obtains. The more partners, the higher his fitness. This difference sets the stage for sexual conflict.

Emlen and Oring suggested that for a female, selection will favor a mating system that provides the greatest access to resources. If the female's sexual partner provides high levels of resources to the young, such as a high-quality territory and a great deal of parental care, **monogamy**—one male with one female—will be favored. If multiple males all provide care to the female's offspring, **polyandry**—one female with multiple males—will be favored. In contrast, for males, selection will favor **polygyny**—one male with multiple females—when males that mate with multiple partners have higher fitness than those that mate with a single female (see Table 13.1). If sexual conflict does not exist because care from both parents is required to successfully raise offspring, as occurs in many birds (e.g., Black 1996), selection will favor monogamy and biparental care. Emlen and Oring suggested that this system might often be favored in resource-poor environments or when resources are difficult to obtain. Biparental care and monogamy might also be favored when predation on unattended young is high.

When biparental care is not required and parental care is female biased, selection will favor the evolution of polygyny. Polygyny is predicted to evolve when environmental conditions lead to the aggregation of females, because such aggregations are more easily defended by a male from rivals. Two factors can promote female aggregation. First, females may aggregate for reasons other than reproduction. For instance, aggregations of females may experience lower predation risk than solitary females, or a group of females may more successfully avoid harassment from males (e.g., Dada, Pilastro, & Bisazza 2005). In such cases, **female defense polygyny** occurs when males monopolize aggregations of females directly (Figure 13.2). The second factor that can lead to aggregation of females is the distribution of resources in the environment. If the resources required by females are clumped, males can monopolize females by defending territories with large amounts of such resources. **Resource defense polygyny** occurs when males defend territories rich in the resources that are used by and attract multiple females. A male then mates with the females that remain near the resources he defends.

Sometimes, however, resources and females may be more uniformly distributed in an environment, rather than clumped. Alternatively, resources may be too unpredictable in space or time, making them costly to defend. If it is too costly for males to successfully defend resources or females, potential mates must seek each other out, which can involve high energetic and predation costs. One way to reduce these costs is for males to settle in fixed locations, called leks, and then display to females there (Figure 13.3). Leks have evolved in many taxa, and males typically form dominance hierarchies within the lek, with dominant males

female defense polygyny A mating system in which a single male monopolizes and mates with two or more females.

resource defense polygyny A mating system in which a male mates with multiple females that are attracted to resources he defends.

Figure 13.2. Female defense polygyny. A male (a) impala (*Aepyceros melampus*) and (b) South American sea lion (*Otaria byronia*) defend their harem of females.

Figure 13.3. **Lek.** Male black grouse (*Tetrao tetrix*) display on a lek.

▮ **male dominance polygyny** A mating system in which a few males on a lek mate with many females.

Figure 13.4. **African lions.** Multiple male lions defend and mate with multiple females.

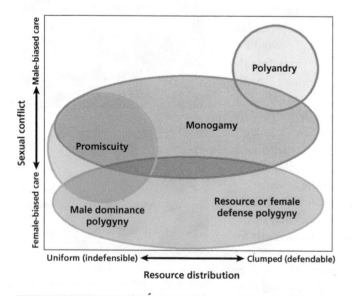

Figure 13.5. **Emlen and Oring's model.** Mating systems are the result of the interaction between sexual conflict and resource distribution.

occupying the most preferred, often central, locations. Such males typically mate with several females, a mating system that Emlen and Oring called **male dominance polygyny**. Males on leks can benefit as a result of reduced predation risk, and such aggregations may be more attractive to females than a single male (Alatalo et al. 1992), thereby increasing encounters with them. Females, too, can benefit from the existence of leks, because their existence reduces the time required to search for a mate: females can quickly assess the relative quality of different males on a lek to find the best option.

In a **polyandrous** mating system, one female associates and mates with multiple males. Emlen and Oring suggest that polyandry can evolve when it is advantageous to both sexes that females be freed from providing parental care, making parental care male biased. One possible situation that might produce a polyandrous system is one in which there is very high predation on offspring. If many offspring are lost to predators, both sexes can benefit if a female can quickly reproduce again. However, recall that egg production is costly. If a male provides more care, a female can provide less care and thus feed more to quickly replenish her energy stores for reproduction in the event that current offspring do not survive. Once males provide high levels of care and females are able to quickly lay multiple clutches, selection can favor females that mate with multiple males and lay a clutch of eggs that is then tended by each one.

Both **polygynandry** and **promiscuity** involve multiple mating partners for each sex but differ in regard to social associations: social associations are found in polygynandry, but not in promiscuity (See Table 13.1). Polygynandry, or plural breeding, is rare but does occur in some social species, and particularly in mammals that defend a territory from other such groups. For example, African lions (*Panthera leo*) live in social groups, which facilitates hunting success (Figure 13.4). A pride consists of two or more adult males, which are often relatives, and up to nine adult females. Males defend their females, and multiple males mate with multiple females (Packer et al. 1991).

Promiscuity, on the other hand, should evolve when the benefits of social living are low. Both females and males are then solitary, and pair-bond formation provides no fitness benefits to either sex. Conditions also favor this system's evolution when the defense of mates or resources is uneconomical, as might occur when population density is high or when it is too costly for males to aggregate on leks, as in high predation risk conditions. Such a system might also occur when individuals cannot afford to forgo feeding to spend time on a lek.

In sum, Emlen and Oring's model suggests that mating system variation can be understood by examining (1) the competing interests of the sexes in attempting to maximize their fitness and (2) how environmental conditions affect the benefits and costs of resource defense and of different mating behaviors for each sex (Figure 13.5).

Clearly, mating systems and parental care behavior have co-evolved. In this chapter, we'll see how the model proposed by Emlen and Oring helps to explain the evolution of, and variation in, mating systems. We cover parental care in the next chapter.

Mating systems in reed warblers

One way to evaluate the Emlen and Oring model is to use the comparative approach to examine a group of related species that exhibit different mating systems. The model allows us to predict, for example, how variation in resource abundance should affect a mating system. In resource-poor environments, biparental care will often be required to successfully raise offspring, and so monogamy should be favored. When resources are more abundant, biparental care will be less essential, and so polygyny or perhaps promiscuity should be favored.

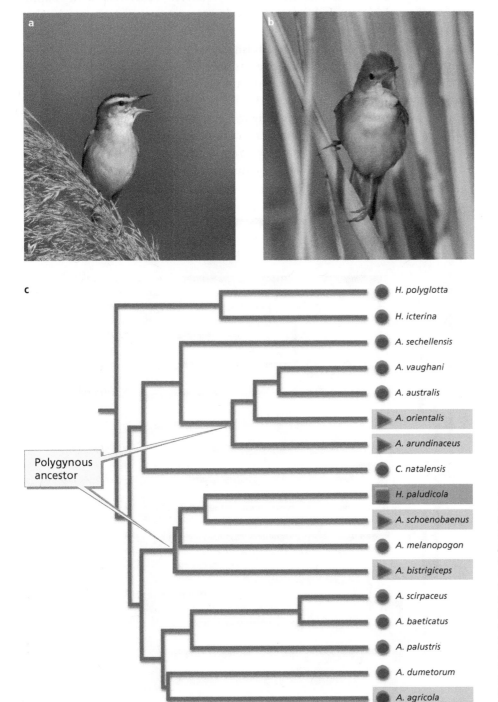

Figure 13.6. Reed warbler phylogeny and mating system. (a) *Acrocephalus schoenobaenus* (polygynous). (b) *A. palustris* (monogamous). (c) Phylogeny and mating system. Only four species display polygyny (triangles), indicating that it is probably not the ancestral condition. Because polygyny is found in two separate groups, it appears to have evolved independently at least twice. The majority of species are monogamous (circles), indicating that this system is likely the ancestral condition. The square indicates a single case of promiscuity. Gray: medium habitat; pink: good habitat. (*Source:* Leisler, Winkler, & Wink 2002)

Bernd Leisler, Hans Winkler, and Michael Wink examined these predictions for the evolution of mating systems in 17 species of Acrocephaline reed warblers (Leisler, Winkler, & Wink 2002). These insectivorous birds are widely distributed throughout the Old World in a variety of marshy and shrubby habitats. All female reed warblers provide more parental care than males through behaviors such as egg incubation and food delivery to young, but the level of male care varies across species. Most species are monogamous, but several are polygynous, and one exhibits promiscuity (Figure 13.6). Leisler and colleagues used the comparative method to address two research questions. First, how does habitat quality—defined as the amount of food resources—correlate with mating system? Second, how are habitat quality and mating system related to the level of male care?

Leisler and his colleagues used published data on food biomass and prey size to estimate habitat quality, categorized as poor (only small prey size), medium (larger prey size), or good (many large prey), for each species. Poor habitats tended to have dense stands of reeds, bushes, or trees that prohibited light penetration. Medium habitats had more light, and good habitats had abundant light and were highly productive, containing many large prey. The researchers then used previously published studies to characterize the relative level of male care as full (equal to that of the female), reduced, or none. Finally, they created a

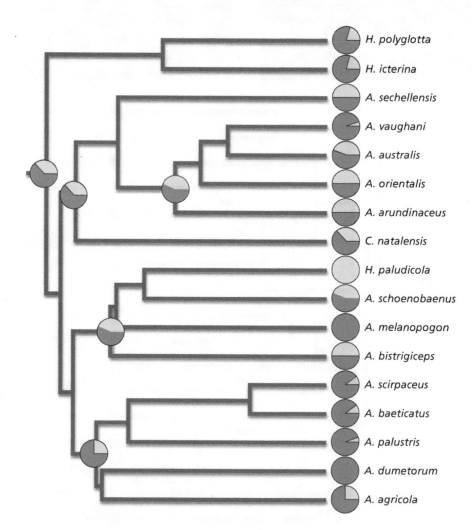

Figure 13.7. Reed warbler phylogeny and paternal care. High levels of paternal care (large amount of blue in circle) appear to be ancestral, with reduced male care (less blue in circle) having evolved several times. (*Source:* Leisler, Winkler, & Wink 2002)

molecular phylogeny using previously published sequences of the cytochrome-b gene.

They found a strong association between habitat quality and mating system across the 17 reed warbler species studied. Monogamy and high levels of male care were predominant in poor-quality habitats, whereas both polygyny and promiscuity and reduced levels of male care were associated with medium- and good-quality habitats. When behavior was mapped onto the phylogeny, two patterns emerged. First, polygyny appears to have evolved independently in this group at least twice. Second, the phylogeny suggests that the ancestral species was monogamous, with a high level of male care (Figure 13.6c; Figure 13.7). Leisler and colleagues' research suggests that in this group of species, male care co-evolved with mating system and was reduced or absent in habitats of higher quality that contained more abundant resources.

This study supports the model of Emlen and Oring, which predicts that habitats with limited resources should favor monogamy and high levels of male care. Changes to higher quality habitats are predicted to be associated with reduced levels of parental care and with polygyny and promiscuity, as observed. Leisler, Winkler, and Wink's work represents one approach to understanding the evolution of mating systems. An alternative is to examine the mating systems of individual species in light of the model proposed by Emlen and Oring. Such work often involves experimental manipulations and tests of different predictions for different mating systems. We now examine examples of such studies for each mating system in turn.

13.2 Biparental care favors the evolution of monogamy

The Emlen and Oring model predicts that monogamy will be observed in situations when biparental care is essential. Monogamy is rare in mammals, an observation consistent with the model, because female lactation—the secretion of highly nutritious milk from mammary glands—alone can provide all the food needed to raise offspring successfully.

Monogamy does occur in some mammals, including many primates. It is rare, however, in rodents but does occur in the California mouse (*Peromyscus californicus*), which is found throughout California and northern Mexico. Adults form exclusive pair bonds, and genetic data indicate that this species does not mate outside the pair bond (Ribble 1991). The pair defends a territory, and males assist with all aspects of parental care except lactation. This assistance includes huddling (maintaining close body contact with pups in order to warm them), grooming the young, and carrying them from one location to another (Gubernick & Alberts 1987). Is male care critical for offspring survival in this species?

California mouse monogamy   Featured Research

David Gubernick and Taye Teferi examined how male parental care affects reproductive success in California mice (Gubernick & Teferi 2000). They conducted an experiment at the Hastings Natural History Reservation in California. All adults were captured regularly with live traps and uniquely marked with numbered ear tags. Females were checked for reproductive condition (pregnant or lactating) and weighed. The researchers created "male-absent" families by

Figure 13.8. Offspring survival. Mean (+ SE) number of young. Families with fathers (blue) had more offspring survive (i.e., emerge from the nest) than did families without fathers (orange). (*Source:* Gubernick & Teferi 2000)

removing a male from a family within three days of the birth of the first litter of the season. In the control "male-present" treatment, males were trapped and handled but then released and allowed to return to their families. The researchers estimated the number of young born to each female by comparing her weight loss between the last days of pregnancy and the time of first capture after giving birth. A standard formula was used to convert body mass loss to offspring number.

There was no difference in the number of offspring born to females that had their male removed and the number of offspring born to those that did not. However, the absence of the father and his parental care significantly reduced offspring survival: significantly more young survived from the male-present families than did from the male-absent families (Figure 13.8). Although males cannot feed offspring, they can assist with thermoregulation and protection. The researchers concluded that the need for biparental care appears to have favored the evolution of monogamy in this species. It is not clear why this rodent is unusual in its requirement for biparental care.

Featured Research

Monogamy and biparental care in poison frogs

A second example of the role biparental care plays in the evolution of monogamy comes from tropical frogs. Jason Brown, Victor Morales, and Kyle Summers examined the mating systems of two closely related species of poison frogs, *Ranitomeya imitator* and *R. variabilis*, in Peru (Brown, Morales, & Summers 2010) (Scientific Process 13.1). Both of these species rear their young in a phytotelma, a small pool of water that collects at the base of leaves or petals in a plant. The small size of these pools reduces predation, but it also limits the quantity of nutrients and other resources available for the developing offspring. *R. imitator* rears its tadpoles in very small pools, averaging 24 mL in volume, while *R. variabilis* prefers larger pools, averaging 112 mL in volume.

Previous work suggested that *R. imitator* was monogamous with biparental care (Brown et al. 2008). Females lay from one to three eggs in dense foliage. Males fertilize the eggs, carry each hatchling to a phytotelma for development, and provide protection. Females deposit trophic (unfertilized) eggs in the phytotelma to nourish the young.

In contrast, *R. variabilis* is promiscuous and raises its young with only male care. Females lay eggs that are fertilized by a male, which then carries the young to a large phytotelma. Males provide parental care, but this care involves only defense, and offspring are not fed by the female (Brown et al. 2008).

Brown and his colleagues examined whether an ecological factor, the size of the breeding pool, is associated with the evolution of parental care. Do larger pools provide enough resources for male-only parental care, whereas smaller pools require biparental care? To examine the importance of parental care for these species, the researchers conducted a transplant experiment. They placed tadpoles of these species into either a large or a small natural phytotelma, with just one species per pool of water. They then placed a mesh screen over the top of the phytotelma to prohibit the addition of resources. None of the experimental tadpoles received parental feeding. In the control, *R. imitator* tadpoles were placed in small pools and did receive parental feeding. The researchers surveyed

SCIENTIFIC PROCESS 13.1

Biparental care and monogamy in poison frogs

RESEARCH QUESTION: *Why does* R. imitator *have a monogamous mating system, while* R. variabilis *is promiscuous?*

Hypothesis: When resources are limited, monogamy will evolve because biparental care is required to successfully raise offspring.

Prediction: Tadpoles in resource-limited small pools will exhibit high survivorship only with biparental care.

Figure 1. Poison frogs with different mating systems. (a) *R. imitator* is monogamous. (b) *R. variabilis* is promiscuous.

Methods: The researchers:

- Manipulated phytotelma pool size. Individual tadpoles of each species were placed in either small (17 mL, on average) or large (39 mL, on average) pools and did not receive female feeding. In the control, *R. imitator* individuals were placed in a small pool and received parental care in the form of female feeding.
- Visited each pool weekly for three weeks to record size (growth rate) and mortality of individuals

Results:

- For both species, individuals in large pools had high growth rates and high survivorship.
- In small pools with no parental care, individuals of both species had low growth rates and low survivorship.
- *R. imitator* controls in small pools with biparental care had high growth rates and high survivorship.

Figure 2. Growth rate. Mean (± SE) percent growth. In the small pool treatments (blue), only offspring that were fed exhibited high growth. All offspring in large pools (orange) exhibited high growth rates, even though they were not fed. (*Source:* Brown, Morales, & Summers 2010)

Conclusion: The need for biparental care in *R. imitator* likely provides strong selection that favors monogamy in this species.

---

the pools each week and measured and weighed each tadpole at the start and end of the experiment.

Tadpole growth rate differed significantly across treatments. With no parental care, individuals of both species in large pools had higher growth rates than those in small pools. However, control *R. imitator* tadpoles in small pools that were fed also had high growth rates. The only mortality observed occurred in small pools where the tadpoles received no feeding: four of the eight *R. variabilis* and three of the eight *R. imitator* young in this treatment died during the experiment.

In this species, offspring growth rate strongly affects survivorship, which demonstrates the benefit of biparental care for *R. imitator* offspring raised in a

EVALUATE

What data would not support the hypothesis that resource limitation favors the evolution of monogamy?

small phytotelma. Because they rear offspring in small pools with limited resources, *R. imitator* adults must provide biparental care and feeding to their offspring in order for these offspring to develop successfully. The researchers concluded that this need has favored the evolution of monogamy, a mating system that is very rare in amphibians. But why don't *R. imitator* lay their eggs in large phytotelmata, where there would be no need for biparental care? One possibility is that *R. imitator* is a poor competitor, and competition for large phytotelmata may be intense. By laying eggs in a small phytotelma, *R. imitator* faces less competition from other species, but doing so requires increased parental investment in the form of biparental care.

Monogamy without biparental care: snapping shrimp

The need for biparental care is thought to explain the evolution of monogamy in many species. Yet monogamy does occur in some species that lack biparental care, including coral reef fishes, some mammals, and several species of crustaceans (Mathews 2002a). Two hypotheses, which are not mutually exclusive, have been proposed to explain monogamy in such species. The **territorial cooperation hypothesis** states that because two individuals can better defend a critical resource, such as a safe refuge, than can a single individual, selection may favor pair formation and shared defense. The **mate guarding hypothesis** states that selection may favor males that guard females by remaining in close association with them over the course of one or more reproductive cycles. This behavior allows a male to monopolize a single female, but it also prevents both partners from seeking other mates. Both hypotheses make assumptions about the environment. In the first, competition for limited territories is assumed to be intense, or predation outside a safe refuge within the territory is assumed to be high. In the second, a male's encounter rate with females is assumed to be low, perhaps because females are rare or difficult to locate.

territorial cooperation hypothesis The hypothesis that two individuals (one of each sex) can better defend a critical resource, such as a safe refuge, than can a single individual, and thus selection may favor pair formation and shared defense.

mate guarding hypothesis The hypothesis that selection favors males that mate with and guard one female over one or more reproductive cycles by remaining in close association with her.

Lauren Mathews tested both hypotheses in the snapping shrimp (*Alpheus angulatus*) in a series of studies (Mathews 2002a; Mathews 2002b; Mathews 2003). Snapping shrimp are small crustaceans (less than 5 cm in length) that live in burrows in intertidal rubble habitats. Most individuals live in male–female pairs within a single burrow (Mathews 2002a). However, the female provides all parental care: she broods the eggs on specialized abdominal appendages and then carries, cleans, and aerates them during their development. Both males and females are territorial and will construct and defend a burrow that provides protection from predators. Females are sexually receptive for only a short period of time after molting. Mate guarding by a male can thus enhance his fitness, particularly if a male can accurately assess time to molting, as a female close to molting should be of high quality.

Mathews tested a prediction for each hypothesis. The territorial defense hypothesis predicts that pairs of shrimp should defend a burrow more successfully than a solitary individual, while the mate guarding hypothesis predicts that males should tend to guard females that are closer to sexual receptivity.

To test the first prediction, she collected shrimp from a site in Florida and housed them in the lab in aquaria (Mathews 2002b). She created four treatments with uniquely marked individuals identified as residents or intruders. Residents (either a single individual or a male–female pair) were placed into aquaria for 24 hours to construct a burrow. Then a single intruder (either female or male) was placed into the tank, and all agonistic interactions were recorded for 30 minutes.

After 24 hours, Mathews determined the location of all shrimp: any shrimp inside the burrow were deemed "winners," while those outside were considered "losers."

Females in pairs were winners in 82% of trials, whereas single females won only 57% of trials. Males in pairs were winners more often than single males, but the difference was not statistically significant. These results indicate that pairs are more effective at territory defense than are single individuals.

Next, Mathews tested the prediction of the second hypothesis, on mate guarding and sexual receptivity. She used the same basic experimental design, introducing a single male intruder to a male–female resident pair in which the female was either close to sexual receptivity (pre-molt, a high-value female) or far from sexual receptivity (post-molt, a low-value female).

Mathews observed that males paired with a high-value female won 72% of their contests, whereas males paired with a low-value female won only 34% of their contests. This finding suggests that males can assess female condition and modify their territorial defense behavior according to female molting state, in line with the mate guarding hypothesis.

To more fully examine this hypothesis, Mathews further examined the ability of males to determine female sexual receptivity (Figure 13.9) (Mathews 2003). In this experiment, a water source was attached to each arm of a Y-shaped maze. One water source (the treatment) contained a shrimp, and the other did not (untreated water). She used seven different treatments, varying the sex of the shrimp and its molting state. Mathews tested water with pre-molt, inter-molt, and post-molt males and females and a control (untreated water) and recorded the behavior of test males as water flowed through the Y-shaped maze. She found that males only moved toward the pre-molt female water, indicating that they are able to determine, and are attracted to, females close to sexual receptivity.

Figure 13.9. Y-shaped maze design. (a) Snapping shrimp. (b) Y-shaped maze used to study snapping shrimp behavior. One source tank contained water that included a shrimp, and the other contained untreated water. A male was placed in the outflow arm and its movement was recorded to determine if it exhibited a preference for one forward arm or the other. Arrows show direction of water flow. (*Source:* Mathews 2003)

In a follow-up experiment, Mathews placed two females in a test aquarium and allowed them to construct a burrow. One was a pre-molt female and the other a post-molt female brooding eggs. She then placed a single male in the aquarium. After 24 hours, over twice as many males had paired with the pre-molt female than had paired with the post-molt female.

Mathews concluded that both territorial cooperation and mate guarding have favored monogamy in snapping shrimp. Paired individuals, especially females, can more successfully defend territories. Males can assess female condition and thus tend to guard females close to sexual receptivity. Mathews does point out that because of the experimental design, she is not able to rule out the possibility that female behavior or the effect of chemical cues could influence male pairing decisions. More work is needed to investigate these possibilities.

The previous three examples reveal conditions that favor the evolution of monogamy, such as the need for biparental care or two individuals defending a breeding site. Next, we turn our attention to the different forms of polygamy.

13.3 Polygyny and polyandry evolve when one sex can defend multiple mates or the resources they seek

Because females tend to invest more in parental care than do males, most examples of polygamy involve polygyny. Much work has centered on determining how males defend and mate with multiple females. According to the Emlen and Oring model, they can do so either directly, by defending these females, or indirectly, by defending resources. Let's see how researchers have distinguished these alternatives.

Featured Research

Female defense polygyny in horses

Feral horses (*Equus caballas*) are found throughout the world in grassland habitats. They live year-round in social units called bands. A typical band consists of one male and up to a dozen adult females. If there are multiple males, a single dominant male obtains most or all of the matings within the band, and so horses exhibit polygyny. Adult males that do not associate with females in bands live in bachelor groups. How can we distinguish between female defense and resource defense polygyny? To do so, we need to determine whether males defend females per se, or whether they simply defend the resources those females use.

Wayne Linklater and his colleagues studied a large population of horses in New Zealand over a three-year period (Linklater et al. 1999; Linklater et al. 2000). Each month, the researchers traveled along regular routes throughout the study site to record the location of bands. Bands were identified by natural markings or by freeze brands previously applied to their coat. The research team plotted the home range of each band using these sightings. In addition, they recorded social behaviors, particularly aggressive interactions between males from different bands.

They found that males aggressively defended their band from other males whenever two bands met or when bachelor males were nearby. During the study, bands overlapped greatly in their home ranges (Figure 13.10). In fact, all bands essentially used the same area, although at different times, and there was no indication of exclusive defense of territories. This finding suggests that males actively defend females and not resources. In other words, they display female defense polygyny.

Figure 13.10. Horse bands. (a) Feral horse aggressive interaction. **(b)** Bands use similar areas. Here, the areas used by 28 bands are plotted as separate polygons, which overlap greatly. (*Source:* Linklater et al. 2000)

Resource defense polygyny in blackbirds

In horses, males defend aggregations of females directly. In other species, males defend an area of resources used by females and so can mate with multiple females. If access to resources affects female reproductive success, the **polygyny threshold model** (Verner 1964; Verner & Wilson 1966; Orians 1969) predicts that females should mate polygynously only when the benefits of doing so—access to greater resources—outweigh the cost of sharing resources with other females. Of course, males differ in their ability to defend resources. Some will defend territories rich enough in resources to meet the requirements of multiple females. Others will only be able to defend territories with lower levels of resources—enough, for example, to fulfill the requirements of just a single female. This variability means that females can mate monogamously on a resource-poor territory or can mate polygynously on a resource-rich territory. Females should select the option that leads to higher fitness (Figure 13.11). The model predicts that only males defending resource-rich territories will mate polygynously.

Stanislav Pribil and William Searcy tested the model using red-winged blackbirds (*Agelaius phoeniceus*), a species known to mate both monogamously and polygynously. Males defend territories in dense vegetation around lakes and ponds. The researchers tested the model by giving females a choice of mating monogamously in a low-quality territory or polygynously in a territory of higher quality (Pribil & Searcy 2001). In this species, territory quality correlates with nest placement: nests in vegetation over water experience less predation and so are of higher quality than those on land.

The experiment was conducted at three marshes in Ontario, Canada. At each marsh, when male blackbirds migrated back to breeding sites and started defending territories, the researchers identified similar adjacent territories, or territory dyads. When the first females arrived to settle on territories, one territory in each dyad was randomly chosen to be high quality, and the other

Featured Research

polygyny threshold model A model that predicts the occurrence of polygyny based on the amount of resources available to females in male territories.

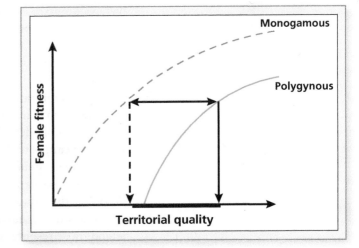

Figure 13.11. Polygyny threshold model. The fitness of a female is plotted as a function of territory quality for two options: mating monogamously or mating polygynously. To benefit by mating polygynously, territory quality must be substantially higher (represented by the thick line on the x-axis).

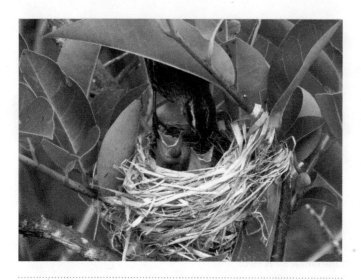

Figure 13.12. Nesting blackbird. Females nest in dense vegetation around ponds.

was chosen to be low quality. Territory quality was manipulated by adding nesting platforms with cattail (*Typha latifolia*) shoots, which created nesting sites for the blackbirds. In the high-quality territory, the platforms were placed over open water, whereas in the low-quality territory, they were placed on land. All females in the low-quality territory were removed, but one female was allowed to remain in the high-quality territory. This set-up created mating options on adjacent territories for newly arriving females. Because the high-quality territory already contained both a resident male and a female, any newly arriving female would mate polygynously if she settled there. In contrast, the low-quality territory contained only a resident male, which ensured that an arriving female would mate monogamously if she chose to settle there. Twice each day, the experimenters recorded the presence of newly settled females (Figure 13.12).

In 12 of the 14 dyads, newly arriving females settled on the high-quality territory. All of these females began building a nest on the platforms, and ten eventually laid a clutch of eggs. The two newly arriving females that settled on the low-quality territory also began nest building. However, one subsequently abandoned her nest, and the other laid a clutch that was lost to predators, demonstrating the territory's low quality. Territory quality clearly affected the mating decisions of females. Females preferred to mate polygynously on high-quality territories over mating monogamously on low-quality territories.

Previous work in this system allowed the research team to calculate the fitness benefits and costs for females settling on the experimental plots (Pribil 2000). Nesting over water rather than land had the benefit of reduced predation—a benefit estimated as approximately one additional offspring raised. Mating polygynously also meant less food provisioning by the male, which reduced the number of offspring that a female could successfully raise by about 0.6 (Pribil 2000). Thus, the net benefit of mating polygynously on the experimental plots was, on average, an extra 0.4 offspring. For females, the benefits of polygyny therefore outweighed the costs, and the newly arriving females were indeed selecting the best option.

These findings offer strong support for the polygyny threshold model because only males defending high-quality territories mated polygynously. This illustrates one important consequence of polygynous mating systems: there will be more variation in male mating success than in female mating success. Males that can successfully defend females or resources will obtain many matings, while males that cannot will obtain few. In contrast, there should be much less variation in female mating success, because most females will mate. Such knowledge about mating systems can provide important information for conservation programs (Applying the Concepts 13.1).

These studies illustrate how resource distribution and the ability of males to defend those resources can affect vertebrate mating systems. Next, we see how the polygyny threshold model also applies to invertebrates.

Featured Research

Resource defense polygyny in carrion beetles

For carrion beetles, the resource defended by a male is very apparent: a vertebrate carcass. Carrion beetles are large insects (2 to 4 cm in length) that live in a variety

APPLYING THE CONCEPTS 13.1

Mating systems and conservation translocation programs

Many species have suffered great population declines and now face an increased probability of extinction. One management strategy to reduce extinction risk is to add individuals to a small population or create new populations by translocating individuals from one place to another. Animals are moved from large captive or wild populations, but translocations are expensive, and many have failed as a result of lack of knowledge about a species' biology or ecology.

A study by Dominique Sigg, Anne Goldizen, and Anthony Pople illustrates the importance of understanding mating systems in carrying out a successful translocation program (Sigg, Goldizen, & Pople 2005). The researchers studied the bridled nailtail wallaby (*Onychogalea fraenata*), a marsupial whose population and geographic range have declined dramatically over the past century. Over a period of three years, a total of 133 uniquely marked animals were translocated to establish a new population. Both captive- and wild-born animals were moved, and an equal number of males and females were released. The researchers collected tissue samples from these adults. Four times each year, animals in the new population were captured, identified, and weighed, and tissue samples were collected from offspring to determine their parents.

Bridled nailtail wallabies are polygynous, and males compete with each other for access to receptive females. Not surprisingly, only about 25% of males bred successfully, whereas almost all of the females did. Male mating success was not affected by origin (captivity or the wild), but body size did strongly affect success. Males larger than 5,800 g were six times more likely than smaller males to sire offspring (Figure 1). The researchers suggest that in polygynous species such as this, it often may be inefficient to introduce equal numbers of males and females into a new environment, as is often done. Instead, they recommend introducing a greater number of females, because most males will not breed successfully. Furthermore, they suggest that only large males should be translocated, because small males will have little chance to mate. These changes could both reduce the costs of translocation and increase the success of such a program. ■

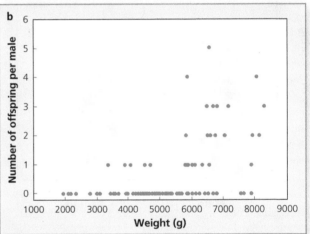

Figure 1. Wallaby reproductive success. (a) Male wallaby. (b) Males over 5,800 g had the highest reproductive success. (*Source*: Sigg, Goldizen, & Pople 2005)

of habitats around the world. They all use small vertebrate carcasses as a food source for their larvae, but these are rare and represent a highly clumped resource. Females lay eggs on or next to a carcass, and after hatching, the larvae feed on the carcass (Figure 13.13). Previous research documented male–male aggression but not female–female aggression over a carcass (Suzuki, Nagano, & Trumbo 2005). Does this variation in aggression among the sexes indicate that these beetles exhibit resource defense polygyny?

Seizi Suzuki and colleagues studied *Ptomaucopus morio*, a carrion beetle found in Japan (Suzuki, Nagano, & Kobayashi 2006). They placed four virgins (two males and two females) in an arena with a piece of meat. Within each sex, individuals differed in size. Large males were dominant over small males, won fights, and actively defended the resource. Females moved throughout the arena, were observed

Figure 13.13. Carrion beetle and carcass. Carrion beetles use vertebrate carcasses for reproduction.

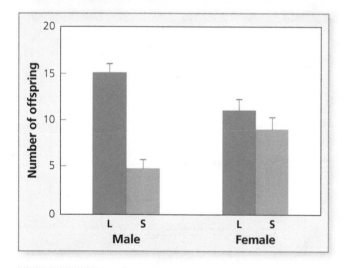

Figure 13.14. Carrion beetle reproductive success. Mean (+ SE) number of offspring. Large males (L) that defended the carcass had three times more reproductive success than did small males (S). In contrast, both the large and small females had similar reproductive success. (*Source:* Suzuki, Nagano, & Kobayashi 2006)

mating with both males, and oviposited eggs on the meat. After a few weeks, the developing larvae crawled away from the meat to pupate. Subsequently, the researchers used DNA fingerprinting (Toolbox 13.1) to determine the parent of each offspring. Large and small females had approximately equal reproductive success, but large males sired three times more offspring than small males (Figure 13.14).

The researchers concluded that *P. morio* exhibits a resource defense polygyny mating system. Large, dominant males defend carcasses to keep smaller males at bay. Females do not exhibit aggression and share access to a carcass. By staying near the carcass, the dominant male obtains multiple matings and sires most of the offspring of visiting females, just as predicted by the polygyny threshold model.

In each of these cases, polygyny involved male defense of a resource. The polygyny threshold model has been used to explain variation in mating systems in many species, even humans (Applying the Concepts 13.2). Next, we turn attention to the evolution of polygyny without resource defense as we examine leks and male dominance polygyny.

Male dominance polygyny: the evolution of leks—hotspots or hotshots?

In horses, blackbirds, and carrion beetles, polygyny results from the defense of either females or resources that are spatially aggregated. However, in many circumstances, resources and females are more uniformly distributed in an environment. It is then too costly for males to defend females or resources, because both are so widely dispersed. In these cases, the Emlen and Oring model predicts the evolution of male dominance polygyny. Males will compete with one another to establish a dominance hierarchy, often on a lek, and females will mate with high-ranking males. A few high-ranking individuals will then obtain most of the matings, while many others will have low reproductive success. So why do low-ranking males decide to aggregate with more attractive males? In other words, how do low-ranking males benefit by lekking?

Two hypotheses explain the evolution of male aggregation into leks. The **hotspot hypothesis** predicts that all males should aggregate where they are likely to encounter many females (Emlen & Oring 1977; Bradbury, Gibson, & Tsai 1986), such as near food resources. Settling in such locations increases encounters with females for all males, even those of low rank. In contrast, the **hotshot hypothesis** states that low-ranking males should aggregate around high-ranking males because females are more likely to visit attractive males (Beehler & Foster 1988). In both cases, low-ranking males can benefit by encountering more females at a lower cost rather than by searching for females on their own.

Distinguishing these hypotheses is challenging because they are not mutually exclusive: a hotshot could be at a hotspot. One approach used to distinguish them is to remove males from a lek, because the two hypotheses then make contrasting predictions about how males and females will respond. If males settle near high-quality (hotshot) males, the removal of hotshots should decrease the lek's

hotspot hypothesis The hypothesis that all males can benefit by aggregating in a location where they are likely to encounter many females.

hotshot hypothesis A proposed explanation for the evolution of leks. It states that males can increase their encounter rate with females by aggregating near an attractive male.

APPLYING THE CONCEPTS 13.2

Human mating systems

Human mating systems vary widely. Across 186 human societies, there is much variation in the number of males and females mated to one another (Marlowe 2000). Monogamy is the sole mating system in about 16% of societies, but polygyny is very common (found in over 80% of societies) and co-occurs in many societies with monogamy. Can the polygyny threshold model explain patterns of polygyny in these societies?

The model predicts that the ability of males to acquire and defend resources will influence the mating system: males that can defend high quantities of resources are predicted to more often mate polygynously; those that cannot are more likely to mate monogamously. Can we find evidence in support of this prediction in a human society?

Thomas Pollet and Daniel Nettle examined variation in mating systems in Uganda (Pollet & Nettle 2009). They examined

2002 census data from 1 million men that allowed the researchers to characterize each man's mating system (his number of wives). Pollet and Nettle examined a variety of factors that might affect the likelihood of a man having multiple wives, such as his age, level of education, locale (urban or rural), and landownership, with landownership serving as a proxy for wealth (resource ownership). In this society, about 10% of men had multiple wives (i.e., were polygynously married). Age, education, and urbanization explained some variation in mating system, but the research team found that the most important factor associated with mating polygynously was landownership: landowners were over two times more likely to have multiple wives than nonlandowners. Given that landownership is a proxy for wealth, this pattern is consistent with the model. ■

attractiveness to both females and remaining males. Removal of lower-quality males near a hotshot should affect only the remaining low-ranking males, which should then move to the vacated territories. In contrast, if males settle in areas of high female activity (a hotspot), removal of some males should not influence the attractiveness of a display location. New males will settle on the vacated sites, which will still be attractive to females. Which hypothesis makes the correct prediction? Two experiments have reached contrasting conclusions.

Lekking behavior in the great snipe

Featured Research

Jacob Höglund and Jeremy Robertson investigated lekking behavior in the great snipe (*Gallinago media*), a small shorebird that breeds in moist terrestrial habitats in Europe (Figure 13.15) (Höglund & Robertson 1990). During the two-week breeding season, males display on leks for several hours each evening. The researchers studied three leks that contained up to 20 males. On each lek, they first observed male interactions to determine dominance relationships. Each lek contained one or two dominant individuals that won most interactions. A few subordinate snipes lost most interactions, and many subdominant males were of intermediate dominance level. Dominant individuals had the highest display rates and mating success on each lek and mated polygynously.

Höglund and Robertson removed individual males for one night and recorded the behavior of the remaining males. When dominant snipes were removed, their territory was left unoccupied, and neighbors tended to move away. In contrast, when subdominant or subordinate males were removed, neighbors quickly added the vacated space into their own territory. These results support the hotshot hypothesis: experimental removal of a dominant male had negative effects on his neighbors, but removal of subdominant and subordinate males

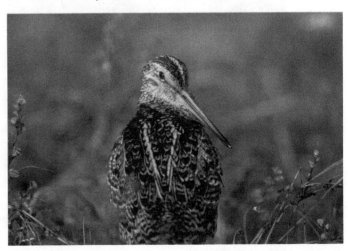

Figure 13.15. Great snipe. During the breeding season, males display to females on leks.

resulted in no major changes on the lek. Other subdominants and subordinates quickly usurped the vacated territories, presumably to move closer to the territory of a dominant male, the hotshot.

Featured Research

Peafowl leks

Höglund and Robertson were among the first experimenters to remove males on leks, and their work was groundbreaking. However, they did not record how the manipulations affected female behavior. Remember that the hotshot hypothesis predicts that if the hotshot is removed, then both males and females should be affected, whereas the hotspot hypothesis predicts that neither should be affected. Adeline Loyau, Michel Saint Jalme, and Gabriel Sorci examined male and female behavior in peafowl (*Pavo cristatus*), a lekking species (Loyau, Saint Jalme, & Sorci 2007).

The research team studied a large lek at the Parc Zoologique de Clères in northern France. The population of the park was around 100 individuals. About 75% of males aggregated and defended sites on the lek, while 25% were "floaters" that did not defend lek territories. The research team recorded male–male interactions, male display rates, female visitation rates, and the mating success of 29 males. The lek was situated near a feeding location, where birds were fed twice each day (Figure 13.16). Males with sites near the feeding location had the highest display rates, attracted the most females, and had a high mating success rate; they also mated polygynously (Figure 13.17). Males farther from the feeding location did obtain some matings, but their success rate was much lower.

To examine the two lek hypotheses, the research team removed each of the 29 males, a few at a time, for about two weeks. The hotshot hypothesis predicts that the removal of the best males should affect female behavior. In contrast, the hotspot hypothesis predicts that such a removal should have no effect on females because of the lek's location near a resource valued by females—the feeding site.

After male removal, floaters quickly settled on the vacated sites and began displaying to females. Display rates of newly settled males were similar to those of the prior site owner. Based on their visitation rates, females found these males just as attractive as the previous settlers (Figure 13.18). In this system, females most

Figure 13.16. Peafowl lek. (a) Peacock display. (b) Diagram of a peafowl lek, showing the location of male display sites (blue) and the feeding place (orange triangle). Size of display dots represents intensity of display by the male. (*Source:* Loyau, Saint Jalme, & Sorci 2007)

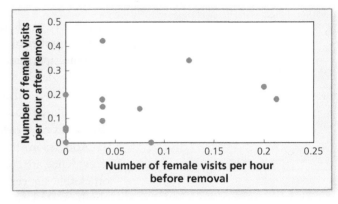

Figure 13.17. Copulations. Males closest to the feeding place obtained the most copulations. (*Source:* Loyau, Saint Jalme, & Sorci 2007)

Figure 13.18. Female visits. Female visitation rates to males on individual display sites (dots) before and after the original male was removed. (*Source:* Loyau, Saint Jalme, & Sorci 2007)

often visited and mated with males whose territory was closest to the feeding location, regardless of male identity. These results support the hotspot hypothesis.

These two studies show that the hotspot and hotshot hypotheses can both explain the evolution of leks in different species. Additional studies are needed to understand the relative importance of these hypotheses in more species.

Up to this point, we have focused on polygyny because it is more common than polyandry. However, polyandry does occur in some birds and fish, as we see next.

Polyandry and sex-role reversal

In polyandry, a single female associates and mates with two or more males. This rare mating system occurs in a few species of pipefish and in some families of birds (Figure 13.19). It is particularly common in shorebirds. Jacanas, for example, are medium-sized birds with very long toes that allow them to walk on floating vegetation. Females in some species aggressively defend large territories from other females, while two or more males settle and defend smaller territories within a female territory. The female mates with each male and lays multiple clutches, one for each male, which then provides most of the parental care; in contrast, females exhibit high levels of territorial defense and aggression. Because these reproductive behaviors usually occur in the opposite sex, the birds are known as a sex role–reversed species.

Polyandry has been a challenge to explain because it is difficult to determine how it benefits both males and females. Malte Andersson proposed a three-step hypothesis (Andersson 2005). First, we start with a species in which parental care is male biased. This behavior might be favored in resource-poor habitats, where females need to spend extensive time feeding to replenish their energy stores for egg production to complete a clutch. Alternatively, as suggested by Emlen and Oring, if predation on clutches is high, selection can favor male-biased care. Again, increased male parental investment gives females time to acquire sufficient energy to lay a replacement clutch.

The second phase in the development of polyandry is the evolution of high female fecundity. In particular, a female may be able to lay more

Figure 13.19. Pipefish and jacana. (a) Pipefish and several shorebirds like (b) this jacana exhibit social polyandry.

eggs than a single male can successfully raise. Female fitness is then constrained by the number of mates she can find, and selection will favor females that mate with multiple males. Here, males become a limiting resource for females, which naturally leads to the final phase—intense competition among females for mates. Increased competition will favor the evolution of increased fighting ability (e.g., large body size) and aggression, both of which are common in polyandrous, sex role–reversed females.

Andersson's hypothesis has yet to be tested experimentally, but multiple lines of evidence are consistent with it. First, in many polyandrous species, predation on clutches is very high. For example, 80% of the clutches of the polyandrous comb-crested jacana (*Irediparra gallinacea*) are lost before hatching, presumably to predation (Mace 2000). Second, polyandry is frequently observed in species in which male care limits female reproduction. For instance, shorebirds produce relatively large eggs, and so a male can properly incubate a clutch of only four or five eggs (e.g., Székely & Cuthill 2000), but females can produce eight. Two clutches of four eggs each are thus typically tended by two males mated to the same female. In contrast, pipefish females deposit eggs on or into a male, which fertilizes and broods them (Wilson et al. 2003). Females can produce more eggs than can be brooded by a single male, and so females typically mate with multiple males, with each male receiving eggs from only one female (Berglund, Rosenqvist, & Svensson 1989; Jones, Walker, & Avise 2001). Third, while many polyandrous birds live in resource-rich environments, they appear to have evolved from species that lived in resource-poor habitats (Andersson 1995). Thus, the ancestral habitat probably favored male-biased care, but the current habitat allows high female fecundity because there is abundant energy for egg production.

The above examples illustrate how sex-biased parental care and resource distribution can favor the evolution of polygyny and polyandry, both of which involve multiple mating by just one sex. Next, we examine mating systems that involve multiple matings by both sexes.

13.4 The presence of social associations distinguishes polygynandry from promiscuity

Polygynandry and promiscuity differ based on the existence of social associations: mating occurs within specific groups in a polygynandrous mating system but is not restricted to occur within specific groups in a promiscuous mating system. Emlen and Oring suggest that polygynandry, or plural breeding, will evolve when group defense of a territory is more effective than defense by a single individual. In these cases, a breeding group will consist of several males and females that together defend a large territory and so associate closely with one another. Plural breeding occurs in social carnivores such as African lions (*Panthero leo*), banded mongooses (*Mungos mungo*), and spotted hyenas (*Crocuta crocuta*) (Figure 13.20), as well as in several species of primates, including chimpanzees (*Pan troglodytes*) and bonobos (*Pan paniscus*).

Such groups are often challenged by rival groups, and large groups are typically more successful than smaller

Figure 13.20. Plural breeding. The spotted hyena (*Crocuta crocuta*) exhibits plural breeding.

ones in a conflict over a territory. For example, encounters and territorial challenges often occur between lion prides, resulting in intense chases. The larger pride wins over 80% of such interactions (Packer, Scheel, & Pusey 1990), demonstrating the advantage of group defense. In many plural breeding species, the number of males in a social group correlates positively with the number of females, because higher male numbers are required to defend all females successfully as group size increases (Clutton-Brock 1989). The males are often relatives, and most sire some proportion of offspring (e.g., Cant 2000; Engh et al. 2002).

Polygynandry in European badgers

Featured Research

One striking example of plural breeding is found in the European badger (*Meles meles*) (Figure 13.21). This large carnivore ranges throughout Europe and Asia and exhibits much variation in sociality. In Great Britain, badgers live in social groups of up to two dozen individuals. Such groups contain multiple reproductively active males and females, suggesting the possibility of polygynandry.

Hannah Dugdale and her colleagues investigated the mating system of a population of badgers in Wytham Woods, Oxford (Dugdale et al. 2007). Over a period of 17 years, adults and cubs from different social groups were captured and uniquely marked with a tattoo. The researchers took blood samples for DNA extraction to determine the parentage of over 300 cubs. By determining the parentage of the cubs, they could also determine which adults were breeding in each group.

On average, social groups contained 12 adults, six of each sex, and approximately 27% of all adult males and 31% of adult females bred successfully each year. Within social groups, an average of 1.6 females per group gave birth each year and 1.5 males sired offspring, but as many as five males and five females reproduced in a social group (Figure 13.22). Because multiple males and females reproduced within social groups, these data provide clear evidence of a plural breeding system. Why do multiple males, rather than just one, mate with many females? Mating occurs year-round in England, which reduces the likelihood of male mate guarding, because such a behavior would likely be too costly for a single male attempting to successfully guard all the females (Yamaguchi, Dugdale, & MacDonald 2006). In addition, females could benefit by mating with multiple males to ensure fertilization, reduce harassment from males, or increase the genetic diversity of their offspring, as we will see later. Males from outside a group do occasionally sire

Figure 13.21. European badgers. A species that lives in social groups in Great Britain and exhibits a plural mating system.

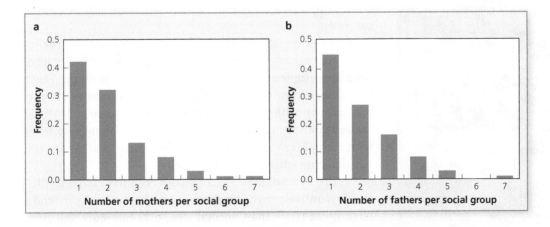

Figure 13.22. Badger reproduction. The frequency of social groups with different numbers of adult (a) females and (b) males that reproduced within the group. In most groups, multiple males and females produced offspring. (*Source:* Dugdale et al. 2007)

offspring in some groups, and perhaps multiple males can better defend a group of females from such outsiders than can a single male. Further work is required to better understand the fitness benefits that such a mating system provides to males.

Promiscuity and scramble competition: seaweed flies and red squirrels

Emlen and Oring suggest that promiscuity will evolve when the costs of the defense of mates or resources exceed the benefits and when there is no need for biparental care. In a promiscuous mating system, individual males and females mate with multiple partners in the absence of social associations. There is either uniparental care, typically provided by the female, or no parental care.

scramble competition The indirect competition of individuals to find and secure copulations with multiple mates.

Promiscuous mating is characterized by **scramble competition**, as individuals compete indirectly with each other to find and secure copulations with multiple mates. This system occurs in many invertebrates. For example, seaweed flies (*Coelopa frigida*) live and breed in decaying seaweed on beaches. They engage in no courtship behavior, and males attempt to copulate with females as soon as they encounter them. A recent study estimated that in one high-density population on St. Mary's Island in England, males mount a female every eight minutes during the day, and a female may mate with hundreds of males during her three-week life span (Blyth & Gilburn 2006).

A less extreme example of promiscuity is found in red squirrels (*Tamiasciurus hudsonicus*). These North American diurnal rodents live in coniferous forests, and males and females defend separate, permanent territories. Jeffrey Lane and colleagues studied the mating system of a population of red squirrels near Kluane National Park in southwest Yukon, Canada (Lane et al. 2008). They captured and marked individuals with numbered ear tags and collected tissue samples for DNA analysis. Each ear tag also contained two uniquely colored wires that facilitated identification from a distance. Females were fitted with radio transmitters and monitored daily to assess reproductive behavior.

In red squirrels, females are receptive for a single day, when they relax territorial defense; on these days, multiple males visit their territory, often simultaneously, and engage in "mating chases." Some chases result in copulations (Figure 13.23).

On these days, the researchers followed the female for an average of 9.5 hours and recorded all chases and copulations. After females gave birth, researchers visited each nest and collected tissue samples from all offspring for DNA analysis to determine paternity.

Females copulated with an average of 5.8 males. For females that raised more than one young, 82.5% of litters were sired by multiple males, with a mean of 2.3 sires per litter. These data demonstrate that the mating system of the red squirrel is promiscuity with scramble competition for mates. Multiple males search for receptive females and then attempt to mate with them. Many males are successful, and so both males and females mate with multiple partners.

In this system, males and females each have large territories. It is unlikely that a single male could successfully defend a territory of twice than normal size so as to secure sufficient food for himself and his mate. In addition, a female's ability to move about freely and her running behavior make economical defense of her unlikely. These two factors therefore favor the evolution of promiscuity in this species. You may be wondering

Figure 13.23. Red squirrels. Each individual red squirrel defends a separate territory from others. Here, two marked individuals mate.

why females mate with multiple males. What benefit do females obtain from this behavior? We address this important question next.

13.5 Social and genetic mating systems differ when extra-pair mating occurs

Mating system classifications were originally developed from observations of individuals and their pair-bond associations, under the assumption that individuals only mated with their social partner(s); these classifications are known as **social mating systems**. With the advent of DNA fingerprinting techniques (Toolbox 13.1), we have learned that social mating systems often differ from **genetic mating systems**, which describe the actual number of sexual partners that contribute to a set of offspring, because both males and females mate outside of existing pair bonds and social associations. For instance, roughly 90% of socially monogamous bird species engage in **extra-pair copulations** (**EPC**s), which result in extra-pair offspring (Griffith, Owens, & Thuman 2002). Such behavior is widespread across taxa (e.g., Sefc et al. 2008; Cohas & Allainé 2009). Together, these observations have led to the definition of extra-pair mating systems, which occur when social and genetic mating systems differ because of extra-pair paternity (e.g., Stutchbury & Morton 1995). Why are these social and genetic mating systems so different?

To understand the evolution of extra-pair mating, consider a socially monogamous, pair-bonded species. From an evolutionary perspective, it is easy to understand why males seek sexual partners outside the pair bond: doing so can increase their reproductive success. The risk to a male of seeking extra-pair partners is that other males are behaving similarly, so while he is away from his partner, he may lose paternity to other males. On the other hand, a female that seeks extra-pair partners may risk losing parental care from her mate if his paternity is not certain. In general, females cannot increase the number of offspring they produce by mating with multiple males. Given the potential cost and apparent lack of benefits, why do socially monogamous females mate with multiple partners?

Several hypotheses explain multi-male mating by socially monogamous females. In some cases, particularly in birds, females can gain additional care for their offspring if each sexual partner provides some parental care (e.g., Blomqvist et al. 2005). Alternatively, in some social mammals with multiple males in the group, dominant males may kill unrelated offspring (infanticide). However, if a female mates with multiple males—especially dominant ones—she may be able to protect her offspring from infanticide because paternity will be uncertain and a male will not kill offspring of a female he has inseminated (Wolff & MacDonald 2004). However, the most general explanation for multi-male mating is that females can increase the genetic quality of their offspring.

How does this work? In social monogamy, females associate with one male. Only one female can form a pair bond with the highest-quality male in a population, and so many females will form a pair bond with lower-ranking males. Mating with a low-quality male can affect not only the survival of a female's offspring but also her own condition: socially monogamous females, in fact, experience higher levels of stress when they mate with low-quality mates (Griffith, Pryke, & Buttemer 2011). From this perspective, it is easier to identify the potential benefits of extra-pair matings for females. First, a female can benefit by mating with a male of higher quality than her social partner, as we saw previously in the good genes hypothesis (see Chapter 12). Second, high levels of heterozygosity commonly increase fitness, because it reduces the likelihood that recessive deleterious alleles will be expressed

social mating system The social associations and presumed mating behavior of individuals based on those associations.

genetic mating system A description of the mating system based on the actual number of sexual partners in a social mating system that contribute to a set of offspring.

extra-pair copulations The copulations of a pair-bonded individual with a third individual outside the pair bond.

genetic quality hypothesis The hypothesis that females that engage in multi-male matings can improve the fitness of their offspring via genetic mechanisms.

(Brown 1997a). A female can benefit by mating with an extra-pair mate that is genetically more dissimilar to herself than her social mate, because this will increase the level of heterozygosity of her offspring. These hypotheses are not mutually exclusive. Taken together, they constitute the **genetic quality hypothesis**, which states that females that engage in multi-male matings can improve the fitness of their offspring via genetic mechanisms. Let's look at two empirical tests of this hypothesis.

Featured Research

Extra-pair mating in juncos

The main prediction of the genetic quality hypothesis is that extra-pair offspring will have higher lifetime fitness than within-pair offspring. This prediction has proven challenging to test, particularly because it requires a direct measure of fitness. Nicole Gerlach and her colleagues tested the hypothesis by analyzing a long-term dataset of socially monogamous dark-eyed juncos (*Junco hyemalis*) (Figure 13.24) from a population at the Mountain Lake Biological Station in Virginia (Gerlach et al. 2012). This population had been studied by Ellen Ketterson and her students since 1983, so detailed behavioral and demographic information, as well as data on mating behavior, were available. DNA analyses were started in 1990, making it possible to determine paternity for thousands of birds. Gerlach and colleagues examined microsatellite data on over 2,200 nestlings.

They found no difference in the fledgling success (F1 generation) between extra-pair and within-pair offspring. However, both male and female F1 individuals from extra-pair matings produced more offspring than those from within-pair matings (Figure 13.25). The researchers found that over a lifetime, females that engaged in extra-pair matings had almost twice as many grand-offspring as females that remained truly monogamous. In this case, the benefit of extra-pair mating is seen in the F2 generation and supports the genetic quality hypothesis. The availability of a long-term dataset allowed the researchers to identify fitness differences in these birds. Now we turn to an example of extra-pair mating in mammals that also used long-term data to assess fitness consequences.

Figure 13.24. Junco. A small sparrow common in North America.

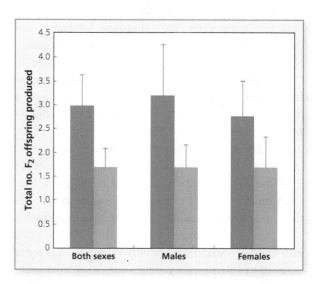

Figure 13.25. Number of junco grand-offspring produced. Mean (+ SE) reproductive success (total F2 offspring produced) of extra-pair (blue) and within-pair (orange) offspring. (*Source:* Gerlach et al. 2012)

Marmot extra-pair mating

Aurélie Cohas and colleagues studied alpine marmots (*Marmota marmota*) for over ten years in La Grande Sassière Nature Reserve in the French Alps (Figure 13.26) (Cohas et al. 2007). These animals live in social groups consisting of a dominant male and female (the breeding pair) and their subordinate offspring. The breeding pair is socially monogamous, but extra-pair matings are common. Cohas and colleagues tested the same prediction of the genetic quality hypothesis investigated by Gerlach and colleagues (2012): extra-pair young should have higher fitness than within-pair young.

Each year, the individuals in this population were trapped, sexed, and marked with ear tags. The researchers collected hair or tissue samples for DNA analysis and fitted the marmots with a transponder to determine their location. The social status of all individuals was assessed by observing dominance interactions.

The research team determined the paternity of 220 offspring. Of these, 45 were extra-pair young sired by an individual other than the dominant male. The remaining 175 were within-pair young sired by the dominant male. The research team characterized the fitness of young by two factors: their survivorship each year and the probability that an individual would attain dominant status during its lifetime. In alpine marmots, most offspring are born to dominant individuals, and so the attainment of dominant status strongly influences fitness.

The survival of extra-pair juveniles, yearlings, and two-year-olds was 15%, 10%, and 30% higher, respectively, than the survival of within-pair young. In addition, extra-pair young were four times more likely to attain dominant status than within-pair young (Figure 13.27). The researchers determined that most extra-pair mates were dispersing males from other populations that were less genetically similar to the female than her pair-bonded mate (DaSilva et al. 2006).

Figure 13.26. Alpine marmot. Marmots are large rodents that live in alpine habitats throughout Europe.

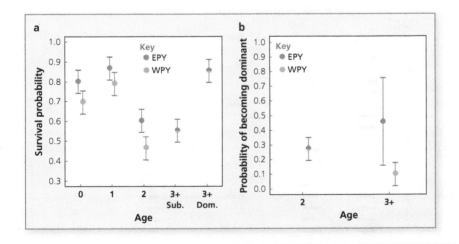

Figure 13.27. Marmot survival. Mean (± SE) probabilities. (a) Young (less than three years) extra-pair young (EPY, denoted by blue) have higher survival probability than within-pair young (WPY, denoted by orange). (b) Older extra-pair young are more likely than within-pair young to attain dominant status. (*Source:* Cohas et al. 2007)

In this species, survivorship is positively related to the level of heterozygosity, and so offspring derived from extra-pair partners have higher fitness than those derived from the pair-bonded mate. Females in this population can improve their fitness by seeking extra-pair partners that will yield more heterozygous offspring than their social mate can.

Molecular techniques have revolutionized our understanding of mating systems. It is becoming clear that, in many species, both males and females mate with multiple partners, despite exhibiting very different types of social associations. We expect to see continued research and the development of new hypotheses to explain why social and genetic mating systems differ so often.

TOOLBOX 13.1

DNA fingerprinting

DNA fingerprinting is the use of molecular techniques to assess relatedness (Queller, Strassmann, & Hughes 1993). One common method for determining paternity uses microsatellites—regions of noncoding DNA that consist of multiple repeats of very short nucleotide sequences. For example, one microsatellite locus might consist of the sequence adenine–cytosine–cytosine repeated multiple times, which we indicate as $(ACC)_n$. Here, n represents the number of times the sequence is repeated. For most microsatellites, individuals are often heterozygous, and there is tremendous variation in the number of repeats in a population. For instance, for a socially monogamous pair, the male's genotype might be $(ACC)_8$ and $(ACC)_{28}$, while his mate might be $(ACC)_{12}$ and $(ACC)_{20}$.

Microsatellite loci often exist adjacent to known DNA sequences, and standard techniques exist to identify both the microsatellite locus and the adjacent DNA. The extracted DNA can then be amplified using the polymerase chain reaction to generate large quantities of the DNA sequence for analysis from a small sample of tissue. The amplified DNA is placed in a polyacrylamide gel that is exposed to an electric current. DNA fragments of different lengths move at varying speeds in the gel, depending on their size, and can be identified by their final location or band pattern. Assessing paternity requires examining the microsatellite fragment lengths from both offspring and parents. In practice, researchers often examine many microsatellite loci for greater certainty about parentage. At each locus, individuals will contain one allele from the mother and one from the father.

In our example, an offspring of the parents noted previously might have a genotype that is $(ACC)_8$ and $(ACC)_{12}$. To determine whether an offspring is the result of an extra-pair partner, we can simply identify whether the social parents could produce its genotype. For instance, an offspring of the previous parents whose genotype is $(ACC)_4$ and $(ACC)_{20}$ can be determined to be the result of an extra-pair mating (Figure 1), because neither social parent possesses the microsatellite locus $(ACC)_4$.

a Paternity not established

b Paternity established

Figure 1. DNA fingerprinting. This technique involves comparing microsatellite bands from offspring and alleged parents. (a) Lack of band sharing between offspring and its social father indicates an extra-pair mating. (b) Band sharing establishes paternity between social father and offspring.

Chapter Summary and Beyond

Males and females mate with different numbers of partners, and their mating systems range from a single mate (monogamy) to multiple mates for one or both sexes (polygamy). Two factors help explain the evolution of mating systems. First, males and females experience different fitness benefits and costs of parental care. Second, environmental conditions affect the ability of individuals to monopolize, through successful defense, either resources or mates.

Recent work on the evolution of mating systems has focused on distinguishing resource defense from female defense polygyny (e.g., Seki et al. 2009) and has examined the role of kin selection in the evolution of leks (Reynolds et al. 2009). In many species, the social mating system differs from the genetic mating system, because individuals mate outside of existing pair bonds. Males can increase their fitness by mating with multiple partners, because the number of sexual partners often limits their reproductive fitness. In contrast, female reproductive success is rarely limited by the number of sexual partners, making it more difficult to explain why females often mate with multiple partners. Leading hypotheses suggest that females can acquire genetic benefits for their offspring through multiple matings, but debate on this question continues (e.g., Westneat & Stewart 2003; Akçay and Roughgarden 2007; Griffith 2007).

CHAPTER QUESTIONS

1. Consider two species of birds that live in the same environment. One catches small flying insects, and the other feeds on seeds on the ground. Imagine that flying insects and seeds on the ground are equally abundant and provide the same energy when eaten. Which species (the insectivore or the granivore) do you predict is more likely to be monogamous, and why?

2. How does an extra-pair mating system differ from promiscuity?

3. Design a study to test the hypothesis that *R. imitator* hatchlings are poor competitors for large phytotelmata.

4. Common crossbills (*Loxia curvirostra*) are socially and genetically monogamous, whereas Gunnison's sage grouse (*Centrocercus minimus*) form leks and are polygynous. What prediction can you make about the degree of variation in male mating success in these two species?

5. The dunnock, or hedge accentor (*Prunella modularis*), displays great variation in mating systems. Females defend exclusive territories from each other, but territory size is negatively correlated with resources: they establish large territories when food is scarce but very small territories when food is abundant. Male territory size is less affected by resource level. When food abundance is moderate, monogamy is typically observed, as female and male territory sizes are similar. What mating system would you expect to observe when food is superabundant? What mating system would you expect to observe when food is scarce?

6. Several species of tropical tent-making bats modify a large leaf into a "tent," which is used as a roost that provides protection during the day when the bats are not active. A tent typically houses several adult females and a single adult male that has mated with them. How would you determine whether this species exhibits female defense polygyny or resource defense polygyny?

7. Leisler, Winkler, and Wink (2002) and Mathews (2002a; 2002b; 2003) examined monogamy in their work. For each study, describe which of Tinbergen's four questions were investigated and explain your answer.

Chapter 14

Parental Care

Each spring we enjoy watching animals care for their young. We have observed both male and female eastern phoebes (*Sayornis phoebe*) feeding their chicks in a nest under our deck and a Virginia opossum (*Didelphis virginiana*) carrying young on her back. Although phoebe offspring receive care from both parents, the opossum young receive care from only one parent, their mother. In many fish, too, care is often provided by just one parent—but which sex provides care varies across species (Figure 14.1). And, while many birds, mammals, and fish provide care for their young, in many other taxa, care is minimal. For example, female spotted salamanders (*Ambystoma maculatum*) lay egg masses in a stream near our house but then depart and provide no direct care for their growing young.

In this chapter, we first examine variation in parental care across taxa and discuss why one sex typically provides more care than the other. We then see how researchers explain variation in parental care among individuals within a species and over their individual lifetimes. Next, we examine brood parasitism, a reproductive strategy that reduces the costs of parental care. Finally, we examine how parental care behaviors are regulated.

Figure 14.1. Parental care. An anemonefish guards eggs in a nest.

14.1 Parental care varies among species and reflects life history trade-offs

parental care Behaviors by a parent to enhance the fitness of offspring, including incubation, feeding, and defense.

life history traits Traits involved with growth, reproduction, and survivorship.

life history theory Hypotheses that assume that natural selection will favor the evolution of behaviors and life history traits that maximize an individual's lifetime reproductive success.

Parental care, the activities of an adult that enhance the survivorship of offspring, includes nourishing and incubating eggs and young, defending them from predators, and sometimes transporting them. If parental care increases offspring survival, why don't all species provide high levels of care to their young? The answer is found in life history theory.

Animals vary greatly in their **life history traits**—traits involved with growth, reproduction, and survivorship that are the result of natural selection. These traits include age at first reproduction, number and size of offspring, level of parental care, and survival rate throughout life. For example, elephants do not become sexually mature until 13 years of age; reproduce only about every five years; have a gestation period of 22 months; give birth to a single, very large calf; and provide parental care for several years. Each calf has a fairly high likelihood of surviving because of the enormous amount of parental care it receives and its large size, which decreases the risk of predation. By contrast, fruit flies mature in a few weeks, and females can produce hundreds of eggs over their brief life span, but parents provide no direct care to their offspring, each of which has a low survivorship. Even more extreme cases are some marine fish, such as cod, that become sexually mature within a few years and can produce 4 to 6 million small eggs at a single spawning. Again, these eggs receive minimal parental care, and only a few may survive to reproduce.

Life history theory proposes that natural selection will favor the evolution of traits that maximize an individual's lifetime reproductive success. For all animals, energy (or effort) is allocated not only to one's self for growth and maintenance but also to offspring in the form of reproduction and parental care. The limited available energy creates an important life history trade-off: effort allocated toward reproduction and parental care reduces the effort that can be allocated toward an individual's own growth and survival (Applying the Concepts 14.1).

This trade-off allows us to understand the evolution of life history traits across species, including parental care behavior (Clutton-Brock 1991). Across species, we often see the evolution of particular sets of life history traits along a continuum. At one end, natural selection can favor individuals that produce many small offspring and provide minimal parental care, while at the other end, selection can favor the evolution of individuals that produce fewer but larger offspring that receive much parental care (Figure 14.2). In general, taxa that have a longer life span produce a relatively small number of large offspring and exhibit higher levels of parental care compared to those with a short life span, which tend to produce many small offspring (e.g., Strathman & Strathman

Life history trait	Species A	Species B	Species C
Fecundity	High	Intermediate	Low
Survivorship	Low	Intermediate	High
Parental care	Low	Intermediate	High

Figure 14.2. Life history continuum. (a) Salamander eggs. (b) Cactus wren (*Campylorhynchus brunneicapillus*) feeding young. (c) High levels of parental care are typically seen in species that produce few large offspring—those with low fecundity and high survivorship, such as Species C. Species A exhibits low levels of care and has high fecundity and low survivorship. Species B exhibits intermediate life history trait values.

APPLYING THE CONCEPTS 14.1

Human life history trade-off

Reproduction and parental care involve high energetic invest-ment for many female animals, particularly mammals. Such costs are part of a common life history trade-off between re-production and survivorship. Do human females experience a life history trade-off between reproduction and longevity?

Rudi Westendorp and Thomas Kirkwood investigated this question using genealogical records from over 2,000 women who were members of the British aristocracy between the years 740 and 1876 (Westendorp & Kirkwood 1998). Detailed records for each woman indicate her age at first childbirth, number of progeny, and age at death.

Among women who lived at least 60 years (and presum-ably had reached menopause), there was a significant nega-tive correlation between number of progeny and life span. In addition, there was a significant positive correlation between longevity and age at which a woman first gave birth.

These data reveal that women of the British aristocracy who had long life spans tended to have fewer offspring and to have begun reproduction at a later age than women who had shorter life spans. This pattern is concordant with a life history trade-off between reproduction and survivorship. ■

1982; Kolm & Ahnesjö 2005; Summers, Sea, & Heying 2006). Let's examine evi-dence of these life history patterns in fish.

Life history variation in fish

Featured Research

Fish display tremendous variation in life history and illustrate how parental care varies across species. For example, in the sunfish family, species vary in the dura-tion of parental care provided to offspring, ranging from one day to a few weeks (Gross & Sargent 1985). And, as one might expect from life history theory, species that provide more care are those that produce fewer and larger eggs (Figure 14.3).

Anna Vila-Gispert and colleagues examined additional variation in life history trait patterns among more than 300 fish species from Europe, North America, and South America (Vila-Gispert, Moreno-Amich, & García-Berthou 2002). For each species, the researchers used published work as well as their own data to estimate life history traits such as age at maturity, maximum body size (stan-dard length), length of breeding season, number of reproductive events per year, average egg size, fecundity (average number of oocytes in a single mature ovary), and level of parental care (amount of protection and nourishment of offspring).

The researchers found that although there was much variation in life history traits among species, species with larger body size tended to have higher fecundity, later maturation, and fewer reproductive events per year. Species with smaller body size showed the opposite trends. Additionally, species with more parental care also tended to have larger eggs. These patterns reflect the basic life history trade-off: species that invest more in individual offspring (indicated by large egg size) tend to also provide greater levels of parental care. The research-ers also found that these patterns held up over different geographic areas and habitats, as well as across different fish taxa.

The great variation in life history traits across taxa helps explain much variation in care provided across species. This care ranges from minimal to care provided only by females (maternal care), care pro-vided only by males (paternal care), or care provided by both sexes (biparental care) (Reynolds, Goodwin, & Freckleton 2002). In verte-brates, certain major trends of parental care are evident across taxa:

Figure 14.3. Parental care and egg diameter. Across species of sunfish, the duration of parental care and egg size co-vary positively. Each point represents one species. (*Source:* Sargent, Taylor, & Gross 1987)

TABLE 14.1 Parental care in selected vertebrate groups. Numbers are the percent of species that exhibit a particular type of care. (*Source:* Reynolds, Goodwin, & Freckleton 2002; Cockburn 2006).

TAXON	TYPE OF PARENTAL CARE (PERCENT OF TAXON)			
	MINIMAL CARE	FEMALE ONLY	MALE ONLY	BIPARENTAL
Fishes (422 families)	80	6	10	4
Cichlid fish (Cichlidae; 182 genera)	0	60	<1	40
Frogs and toads (Anurans; 315 genera)	91	4	4	1
Lizards and snakes (938 genera)	97	3	0	0
Crocodiles and alligators (21 species)	0	62	0	38
Birds (9,450 species)	0	8	1	91
Nonprimate mammals (1,117 genera)	0	91	0	9
Primates (203 species)	0	68	0	32

most amphibians and reptiles provide little parental care, birds provide predominantly biparental care, and mammals often provide extensive and mainly maternal care (Table 14.1). While less common than in vertebrates, parental care has evolved in many invertebrates, including at least 12 orders and 50 families of insects, and 16 lineages of arthropods (Trumbo 2012).

Males and females may both care for offspring in many species, but rarely do both sexes provide the same level of care. We examine that pattern next.

14.2 Sexual conflict is the basis for sex-biased parental care

In species that provide parental care, we see an interesting pattern: one sex typically provides most or all of the parental care. Often, females provide more care than males. Even in birds that typically exhibit biparental care, the female typically provides greater care than her mate. For example, Kirk Stodola and colleagues examined parental care behavior in black-throated blue warblers (*Dendroica caerulescens*). These small songbirds often produce two clutches each summer and exhibit biparental care. Stodola's research team quantified the rate at which males and females brought food to their nestlings. For both broods raised, females consistently brought more food than their mates to offspring (Figure 14.4) (Stodola et al. 2009). Why is parental care so often female biased?

Female-biased parental care

To explain the unequal division of parental care, we must examine the fitness benefits and costs of providing parental care for each sex. Both sexes obtain the same fitness benefit from providing parental care: increased offspring survival. However, as we have seen, such care takes time and energy, and it may increase predation risk for the caring adult. These costs create a conflict of interest between the sexes: each sex will benefit if the other pays the cost of care. The difference

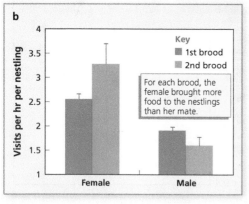

Figure 14.4. Female parental care. (a) Female black-throated blue warbler feeding young. (b) Mean (+ SE) visits to the nest. Female black-throated blue warblers provide more parental care than males in terms of food delivery for their first and second broods. (*Source:* Stodola et al. 2009)

between the sexes in their fitness interests is known as sexual conflict (Trivers 1972; Chapman et al. 2003).

How is this conflict resolved? The sex with the higher cost of care should provide less care, because its net benefit of care will be smaller (Queller 1997). In most species, females have a greater certainty than males about the identity of their offspring, because they lay eggs or give birth. Therefore, the benefit of parental care to females is rather straightforward. What about the benefit for males? Because females can mate with multiple males while fertile, males will always be less certain of whether they sired the offspring of any female they have inseminated. This reduced certainty is an important cost to males, because it results in the possibility that a male will end up providing care to another male's offspring: a very high cost of providing parental care. Thus, natural selection can favor males that reduce their level of parental care. In fact, we can predict that the less certain the paternity, the less care will be provided by males (Trivers 1972; Queller 1997). Let's look at one recent test of this prediction in birds.

Paternity uncertainty and parental care in boobies

Featured Research

The blue-footed booby (*Sula nebouxii*) is a large, colonially nesting seabird (Figure 14.5). In this species, a single male and female form a pair bond and raise offspring together. Parental care is extensive and shared, as they take turns incubating eggs for 40 days and then feed the developing chicks for another 140 days. If a male is uncertain of his paternity, will he provide as much care? Marcela Osorio-Beristain and Hugh Drummond conducted a removal experiment to answer this question (Osorio-Beristain & Drummond 2001).

While fertile, a female copulates frequently with her mate, beginning just before and continuing during egg laying. When her mate is absent, however, she may also copulate with other males. The researchers manipulated certainty of parentage for males by removing pair-bonded males for 11 hours during the fertile period of the females, which occurs a few days before egg laying. Control males were removed for the same period of time, but only when female fertility was low, about 15 days before egg laying. Female fertility was assessed by visual inspection of the cloaca (the urogenital opening). When the cloaca is open and reddish, females are fertile; when it is closed and not reddish, females are not fertile. After the removal period, all males quickly returned to their mate.

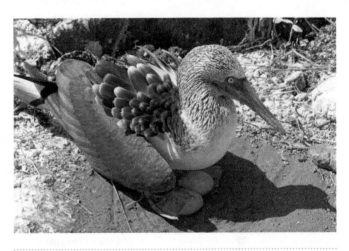

Figure 14.5. Blue-footed booby. A female sits on her clutch of eggs.

Experimental and control males differed in their behavior once eggs were laid. Seven of the 16 experimental males removed the first egg laid by their mate from the nest. (The egg was later destroyed by predators.) No other eggs were removed. None of the 17 control males removed any eggs laid by their mate. The researchers concluded that for experimental males, certainty of paternity was low for the first egg laid, as the male had been away during his mate's fertile period and the female may have mated with another male. Because the pair copulates often during egg laying, the male's certainty of paternity for subsequent eggs should be high. In sum, male boobies reduce parental care, even to the point of egg destruction, when their paternity is uncertain.

The evolution of male-only care

In mammals and birds, females typically provide greater levels of care than males because of a reduced certainty of paternity. Nevertheless, male-only care, while less common, does occur in many species of fish, amphibians, and reptiles. Can sexual conflict and paternity uncertainty explain this difference? Consider the modes of fertilization and growth in these taxa. Mammals and birds have internal fertilization and determinate growth: they stop growing when they reach adult size. In fish, amphibians, and reptiles, many species exhibit external fertilization and indeterminate growth: they never stop growing. External fertilization should increase the level of paternity certainty for males, and so it should also reduce the cost of potentially providing care for unrelated offspring. Indeterminate growth matters because the number of eggs a female can produce increases rapidly with body size: larger females can produce many more eggs than can small ones (Gross & Sargent 1985; van den Berghe & Gross 1989). The large increase in fecundity with an increase in body size establishes a new cost of parental care for females, because providing care to current offspring requires a sacrifice in her own feeding and growth rate, which lessens her future reproductive output. For males, the ability to obtain matings also increases with body size, but at a much lower rate than it does for females (Gross 2005). Thus, the net cost of providing care is higher for females than for males, and so selection can favor male-only parental care.

The mode of fertilization affects paternity certainty, and indeterminate growth affects fecundity. Combining these associations with our understanding of sexual conflict, we can predict that male-only care is most likely to evolve in fish, amphibians, and reptiles with external fertilization and multiple reproduction events (a strategy known as **iteroparity**). Judith Mank and colleagues found exactly this pattern in their survey of parental care in teleost fish (Mank, Promislow, & Avise 2005). For example, smallmouth bass are iteroparous: they have external fertilization, indeterminate growth, and male-only parental care. However, in species such as salmon that reproduce only once and then die (a strategy known as **semelparity**), males do not provide care.

> **iteroparity** A life history strategy in which there are multiple reproductive events throughout the lifetime.

> **semelparity** A life history strategy in which reproduction occurs once in the lifetime.

Featured Research

Paternity uncertainty and male-only care in sunfish

Does certainty of paternity also affect the level of care in fish with male-only care? Our hypothesis predicts that it should. Bryan Neff examined this question in male bluegill sunfish (*Lepomis macrochirus*) (Neff 2003). Sunfish are common in the lakes and rivers of North America. Large breeding males construct nests along the shoreline. Females visit these nests and deposit eggs, which the male then fertilizes and defends. Smaller "sneaker" males adopt a different breeding strategy by hiding near a breeder male's nest (see Chapter 12). When a female begins depositing eggs, a sneaker may quickly dart into the nest to release sperm and frequently succeeds in fertilizing some of the eggs (Fu, Neff, & Gross 2001). Thus, sneaker males represent a threat to a territorial male's paternity.

Neff investigated how males' nest defense behavior was affected by their certainty of paternity (Scientific Process 14.1). On the morning of spawning, territorial males were divided into two groups. For treatment males, he placed four sneaker males at the edge of their nest in clear plastic containers. Empty containers were placed near the nest of control males. The day after spawning, Neff recorded male nest defense behavior in response to a potential egg predator, a pumpkinseed sunfish (*L. gibbosus*), which was placed in a clear plastic container near the nest. Defense behaviors included lateral displays, opercular flares, and bites directed toward the predator.

Paternity certainty and parental care in bluegill sunfish
RESEARCH QUESTION: *What affects the level of male parental care in bluegill sunfish?*

Hypothesis: Males should reduce care as their certainty of paternity declines to avoid caring for unrelated offspring.

Prediction: Males that observe sneakers near their nest during spawning should provide less care to eggs compared to males that do not see sneakers near their nest during this period.

Methods: The researchers:

- Divided territorial bluegill sunfish males into two groups on the day of spawning. For treatment males, four sneaker males were placed in clear plastic containers near the nest. For control males, empty plastic containers were placed near the nest.
- Placed an egg predator (a pumpkinseed sunfish) in a clear plastic container near each nest the next day for two 30-second periods to simulate a threat to the eggs
- Recorded the amount of defense behavior exhibited by each male toward the predator (lateral displays, opercular flares, and bites)

Day 1: Spawning

2 Sneaker fish
or
Empty

Female spawning

Territorial fish

2 Sneaker fish
or
Empty

Day 2: Predator threat

Eggs

Territorial fish

Egg predator

Figure 1. Experimental setup. On Day 1, a sneaker fish or an empty control was placed next to the spawning pair; on Day 2, a predator was placed near the male guarding eggs. (*Source:* Neff 2003)

Results: Treatment males exhibited significantly lower levels of defense behavior than did controls.

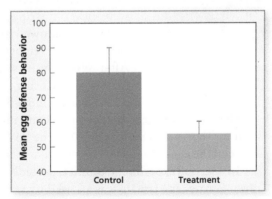

Figure 2. Defense behavior. Treatment males displayed significantly less defense behavior in response to a predator than did control males. (*Source:* Neff 2003)

Conclusion: Parental care by male bluegill sunfish is affected by their certainty of paternity.

EVALUATE

How would the conclusion of the study change if egg defense behavior was the same for control and treatment males?

Neff observed significantly less egg defense by treatment males than by control males. Apparently, the presence of sneakers at the edge of a nest during spawning reduced a male's certainty of paternity and level of parental care for the eggs in his nest, as predicted.

Sunfish display male-only parental care and external fertilization. Not all species with male-only care fertilize eggs externally. Next, we examine male-only care in an insect with internal fertilization.

Paternity assurance and male care in water bugs

Of the insects that provide care, some species exhibit male-only care and have internal rather than external fertilization. Can sexual conflict and uncertainty of paternity still explain such male-biased care? Yes, according to a classic study by Robert Smith on the giant water bug (*Abedus herberti*) (Smith 1979). Females oviposit eggs on the back of their mate, and males protect and aerate the eggs (Figure 14.6). These behaviors enhance offspring survival but are costly, because such care precludes the male from feeding and mating with other females. In addition, females can mate with several males and store sperm for months. How can a male ensure that he is caring only for those eggs he has fertilized? One way is to mate repeatedly with a female. Smith observed a cyclical round of copulation and oviposition in which the pair mated, the female deposited up to three eggs on the male, and then the pair copulated again. This was repeated over and over: in one case he observed, a pair copulated over 100 times in 36 hours, and a total of 144 eggs were deposited on the male!

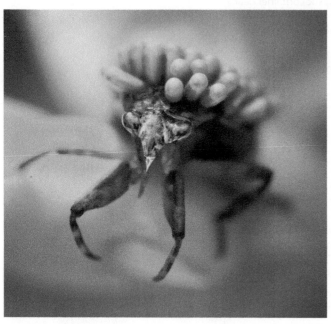

Figure 14.6. Male water bug with eggs. Males care for eggs that females deposit on their back.

To determine the effectiveness of this behavior for ensuring male paternity, Smith conducted several experiments in which he allowed females to mate with two males that differed in a heritable phenotype—males were either unstriped or striped. In one experiment, females were allowed to mate once with one male phenotype and then removed to prevent ovipositing any eggs. After several weeks, the same female was allowed to mate with the other male phenotype and lay all her eggs on his back. Smith recorded the phenotype of nymphs that hatched from the back of the second male. Male paternity was extremely high for the second male. For 20 of the 25 females, the second male fertilized 100% of the eggs.

In a second experiment, Smith repeatedly alternated the two male mates for a given female. The female was allowed to mate and lay three eggs on the back of one male and then was immediately moved to the second male to do the same. Females were moved back and forth until they had oviposited half their eggs on each male. Smith again recorded the phenotype of the nymphs that hatched on each male's back. In this experiment, 245 of 250 nymphs were sired by the male that carried the eggs, demonstrating the very high level of paternity assurance in this species. By and large, males care for their own offspring. In this system, males have high levels of paternity assurance because they frequently copulate with the female whose eggs they accept and then care for. This assurance helps explain the evolution of male-only care.

Up to now, we have seen how differential costs of care between the sexes are linked to sex-biased care among species. We have also examined one life history trade-off involving the certainty of paternity that helps explain variation in care among individuals and species. Next, we examine another life history trade-off that helps to explain additional variation in parental care among individuals within a species.

14.3 Parental care involves fitness trade-offs between current and future reproduction

Above, we saw that sexual conflict offers an explanation for sex-biased parental care that is based on an important cost to males: paternity uncertainty. Another important fitness cost of parental care affects both males and females and involves a second life history trade-off: effort allocated toward current offspring reduces the effort that can be allocated toward future offspring. This trade-off also explains variation in the level of parental care exhibited by individuals in a population, and by an individual to different offspring, as we see next.

Parent-offspring conflict theory

The trade-off that adults face between current and future reproduction was first articulated by Robert Trivers. His **parent-offspring conflict theory** (Trivers 1974) states that parents and their dependent offspring are under different selection pressures: parents should maximize their lifetime reproductive fitness, while offspring should maximize their own fitness by obtaining as much energy and protection as they can from their parents to survive to reproductive age. Recall that Trivers described **parental investment** as costly investments by parents that increase the offspring's chance of survival (Trivers 1974). The benefit of parental investment is increased fitness of the offspring. However, the parent and the offspring obtain different benefits from parental investment, because the parent is related to the offspring by 0.5, whereas the offspring is related to itself by 1.0. The offspring therefore receives twice the fitness benefit from the care provided (Lazarus & Inglis 1986). In terms of reduced future offspring, however, the cost of parental investment is the same for both parent and offspring, since both are related to all future offspring by 0.5 (assuming full sibship) (Figure 14.7). Because the optimal level of parental investment is always higher for the offspring than for the parent, there is a parent-offspring conflict: the fitness of a parent is maximized at a lower level of investment than the fitness of an offspring.

One basic prediction of parent-offspring conflict theory is that if the costs of parental care increase for a parent, the level of care provided to current offspring should decline. In the next section, we examine one test of this prediction in insects.

parent-offspring conflict theory Parents and their dependent offspring are under different selection pressures: parents should maximize their lifetime reproductive fitness, while offspring should maximize the energy and protection they currently receive from their parents in order to survive to reproductive age.

parental investment Any investment by a parent in an offspring that enhances the offspring's fitness at the cost of the parent's ability to invest in other offspring.

A parent only receives half the benefit of care provided to an offspring because the offspring only contains half of the parent's alleles.

Relative benefit to an offspring of receiving care

Relative benefit to a parent for providing care to an offspring.

Figure 14.7. Parent-offspring conflict theory. (a) Eastern bluebird (*Sialia sialis*) chick begging for food. (b) The fitness benefit for offspring (2B) of receiving parental care is twice the fitness benefit for an adult (B) of providing care, because the coefficient of relatedness between parents and offspring is 0.5. C is the cost of parental investment in terms of reduced future offspring. Note that the optimal level of parental investment is higher for the offspring (O) than it is for the parent (P). (*Source:* Trivers 1974)

Predation risk and parental care in golden egg bugs

One common cost of care is predation risk. Adults that provide care are often more obvious to predators and so suffer higher mortality. Parental care to protect offspring from predators can take many forms. In the golden egg bug (*Phyllomorpha laciniata*), females can lay eggs on the backs of adults of both sexes or on vegetation (Figure 14.8). Eggs carried by an adult receive protection, whereas eggs on plants receive no parental care and thus no protection. How does egg carrying affect the adult's predation risk? What is the cost of this behavior?

Piedad Reguera and Montserrat Gomendio examined this cost of parental care in a simple experiment (Reguera & Gomendio 1999). They tested the effect of egg carrying on predation risk using a large, outdoor aviary that housed great tits (*Parus major*), a common avian predator for these insects. The researchers captured golden egg bugs in the field and allowed females to lay eggs on some adults. They then placed four same-sex golden egg bugs in the aviary. Two bugs carried eggs on their back, and the other two did not. The aviary was checked every 20 minutes for the presence of the bugs and to record all predation events. A trial continued either until only one bug was left or for a maximum of two hours.

Figure 14.8. Golden egg bug. Adults carry eggs on their backs as a form of parental care, but this behavior increases their risk of predation.

In ten of 11 trials, the last bug alive was one without eggs on its back, suggesting a high cost of egg carrying. Golden egg bugs are a yellow-brown color, and their morphology makes them cryptic. However, their eggs are shiny, which apparently makes individuals more visible to bird predators. Given this high cost of parental care, why do adults provide it?

Reguera and Gomendio examined this question in both the field and the laboratory. They found a large amount of intraspecific variation in egg-laying behavior. Recall that female golden egg bugs can lay their eggs on either plants or conspecifics. Eggs laid on plants have a much lower probability of survival (only 3%) than eggs laid on conspecifics (25%), and males are the primary egg carriers (Reguera & Gomendio 2002). Females mate with several males, and so a male cannot be certain of paternity for the eggs he carries. Given this uncertainty, why do males provide this costly parental care?

Gomendio and colleagues examined two potential hypotheses to explain egg-carrying behavior in males (Gomendio et al. 2008). The sexual selection hypothesis predicts that males that carry eggs should be more attractive to females (because they are willing to provide care) and so should receive more matings than males that do not carry eggs. However, experimental work revealed that females do not prefer to mate with males that carry eggs compared with those that do not, and so this hypothesis was rejected (Gomendio et al. 2008).

Alternatively, egg carrying can be favored by natural selection when the behavior's benefits exceed its costs for males. This natural selection hypothesis predicts that egg carrying should be more common when it provides a greater net benefit to males. An important cost of this behavior is predation risk, as we saw previously. What are the benefits? Males sire 10% to 40% of the offspring they carry (García-González et al. 2003; Tay, Miettinen, & Kaitala 2003), and so they carry many other males' eggs as well as some of their own. However, the researchers found a large difference in the proportion of males with eggs and levels of egg mortality as a result of egg parasitism across two study sites. In a population in

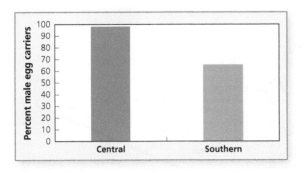

Figure 14.9. Egg carrying in golden egg bugs. More males carried eggs in a population in central Spain compared with a population in southern Spain. (*Source:* Gomendio et al. 2008)

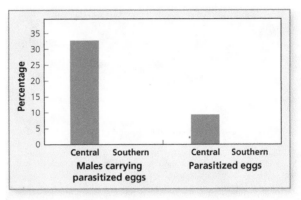

Figure 14.10. Egg parasitism rates. The population in central Spain had more males carrying parasitized eggs and a higher percentage of parasitized eggs. (*Source:* Gomendio et al. 2008)

central Spain, almost all males carried eggs (Figure 14.9), and predation on eggs laid on vegetation as a result of egg parasitism was high (Figure 14.10). In a population in southern Spain, fewer males carried eggs, and there was no incidence of egg parasitism. These results are consistent with the natural selection hypothesis: egg carrying by males is favored when lack of care leads to very low offspring survival. So males provide care to eggs, which increases their predation risk, to enhance the survivorship of their offspring. The low level of paternity assurance may be the reason that egg-carrying behavior varies widely across populations. Males provide parental care only when the egg carrying is required for offspring survival.

This work illustrates one important fitness cost of providing care, increased predation risk, and supports one key prediction of parent-offspring conflict theory. A second important factor that reduces current fitness for parents is missed opportunity cost, or the sacrifice that an individual makes by not engaging in a different behavior that may enhance current fitness. For example, by guarding eggs, a male cannot search for new mates. Let's see one example of opportunity costs in frogs that helps to explain variation in parental care among individuals and provides a second test of parent-offspring conflict theory.

Egg guarding and opportunity costs of parental care in frogs

Featured Research

For many insects, fish, and amphibians, egg attendance is an important aspect of parental care. One parent remains near the eggs to protect them from predators and remove infected or dead eggs. Experimental removal of the guarding parent typically leads to reduced hatch success, showing the benefits of offspring defense as a form of parental care. Just as we saw in egg carrying, the amount or intensity of egg guarding varies among individuals within a population. For example, in big-thumbed frogs (*Kurixalus eiffingeri*), some males attend eggs nearly 100% of the time until they hatch, whereas others rarely stay near eggs to guard them (Figure 14.11). Wei-Chun Cheng and Yeong-Choy Kam investigated this variation (Cheng & Kam 2010), asking the question: If egg attendance increases hatch success, why don't all males attend eggs at a high level?

This species, native to Taiwan, has an extended breeding period that lasts from February through August. During mating, the pair engages in amplexus: a male grasps a female. Males can then fertilize the eggs she deposits in water-filled

Figure 14.11. **Big-thumbed frog.** Males display egg-guarding behavior.

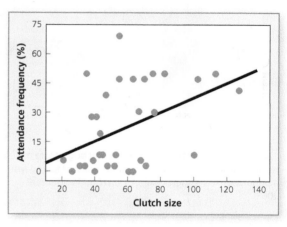

Figure 14.12. **Egg attendance in frogs.** Males with large clutches had higher attendance frequency than males with small clutches. (*Source:* Cheng & Kam 2010)

bamboo stumps or tree holes, making it easy to identify the eggs that a male has fertilized.

Males attending eggs were captured and uniquely marked using a waistband with a small numbered tag. Each night for six days, 38 egg masses were observed six times to determine whether the male was attending the eggs. The research team also noted any egg mortality by counting the number of eggs still in the nest.

Daily egg mortality was negatively correlated with attendance, demonstrating the benefits of male parental care. Male attendance was also positively correlated with egg clutch size: males were more likely to attend larger clutches (Figure 14.12). Why do males with small egg clutches spend less time attending them? The researchers proposed that such males are instead searching for additional mating opportunities to increase their fitness. Time spent attending eggs reduces the ability of a male to search out and mate with additional females. A male with a small clutch of 20 to 30 eggs has less to lose by allocating little time to parental care and more to gain by seeking out other mates. In contrast, males that have already fertilized a large clutch of over 100 eggs can best enhance their fitness by protecting them until they hatch. So here, variation in clutch size and our understanding of life history trade-offs and opportunity costs can explain variation in parental care among males.

In these examples, increased costs to parents led to a reduction in the level of care for current offspring, as predicted by parent-offspring conflict theory. The theory also predicts a fitness trade-off between current and future reproduction, stating that parents that invest more in their current offspring will have higher current reproductive success but also reduced future reproductive success. Let's examine this prediction by first considering parental care and reproduction in insects.

Featured Research

Current versus future reproduction in treehoppers

One way to test parent-offspring conflict theory is to examine how variation in individual parental care affects both current and future offspring. Andrew Zink tested the theory by studying a common species of treehopper, *Publilia concava*, at a site in New York (Figure 14.13) (Zink 2003). In these small, plant-feeding insects, females select a host plant and lay eggs on the underside of mature leaves. Parental care takes the form of egg guarding as females protect their offspring from predators such as mites. In this species, there is tremendous variation among females in guarding behavior: some guard their eggs for three or more weeks until nymphs hatch, whereas others spend no time at all caring for their eggs. There is also variation in the number and size of egg clutches a female can produce in one reproductive season. Are these patterns related?

Zink examined how parental care affected both current and future reproduction of individual females. Based on the existence of a trade-off between current and future reproduction, he made two predictions: (1) egg guarding will increase egg survival and thus enhance current reproduction, and (2) increased parental care (guarding) will decrease future reproductive success.

Zink uniquely marked females with four small dots of colored paint. Because females are generally faithful to individual plants, Zink could follow their behavior and the fate of egg clutches through repeated visits. For each egg mass deposited by a female, Zink measured its size to estimate the number of eggs and checked them every two days until the nymphs hatched. At each check, he also recorded whether the female was present (guarding the eggs) or had abandoned them. Hatching success (an estimate of fitness) was calculated as the number of nymphs observed on the leaf divided by the number of eggs estimated in a clutch.

Zink found that about half the females abandoned their eggs within a few days of their deposition, while the rest remained to guard them for more than three weeks. He also found a positive correlation between the duration of egg guarding and hatching success (Figure 14.14), demonstrating a benefit of care for current reproduction that supports the first prediction. Although the correlation might be due to successful female defense, it might also result from females abandoning clutches that are damaged or otherwise unviable.

To test these alternatives, Zink conducted a removal experiment. He identified 194 plants that contained egg clutches that were similar in size. In half—deemed "removal treatment" clutches—the female was removed to reduce the level of parental care that offspring received. In the other half, clutches were not manipulated (female not removed, the control) and so received a higher level of care. Zink observed that hatch success was significantly higher in controls, providing more evidence that parental care enhances current reproduction. Is there a cost of care in terms of future reproduction? Yes. Females that provided more than ten days of care for their first clutch were less likely to initiate a second clutch than were females that abandoned their first clutch in the first ten days. Another way to characterize the trade-off involved in the cost of care is to examine the benefit of providing little care: females that abandoned their first clutch early laid larger second clutches than females that provided greater care for their first clutch (Figure 14.15). As such, these data support both predictions of the parent-offspring conflict theory.

In this example, Zink relied in part on natural variation in parental care among females to test parent-offspring conflict theory. An alternative approach is to experimentally manipulate the costs of providing care to current offspring and examine the effect on future fitness, as we see in the next example.

Figure 14.13. Treehopper. Treehoppers feed on host plants and lay eggs on the underside of leaves.

Figure 14.14. Effect of egg guarding in treehoppers. There is a positive correlation between the number of days a female guards eggs and the eggs' hatching success. Each point represents one female. (*Source:* Zink 2003)

Figure 14.15. Egg guarding and future reproduction in treehoppers. (a) Mean (+ SE) number of future broods initiated and (b) size of future egg masses. Females that abandoned a clutch (blue bars) as opposed to remaining with their eggs (orange bars) in the first ten days after egg laying initiated more future broods of larger size. Females that abandoned their eggs during Days 11 to 20 after egg laying did not differ from females that remained with their eggs during this period in future egg mass. (*Source:* Zink 2003)

Featured Research

Incubation of eider eggs as a trade-off

In birds, egg incubation is an important component of chick development. Adults sit on eggs and transfer heat to them through their brood patch. This behavior can be critical when the ambient temperature is low.

Sveinn Hanssen and colleagues were interested in how current incubation costs affect future reproduction in the common eider (*Somateria mollissima*) (Hanssen et al. 2005). Eiders are large sea ducks that live primarily along the cold northern coasts of Europe and North America. Females normally lay a clutch of three to six eggs with four to five eggs on average. They incubate the clutch alone without the help of their mate. During the two-week incubation period, females do not leave the nest to eat and so typically lose up to 40% of their body mass—a large parental investment indeed! Assuming that six eggs take more energy to warm than three, Hanssen and colleagues predicted that incubating larger clutches should lead to a greater loss in body mass. Because the eider is a long-lived bird (living up to 12 years), the research team could also examine how mass loss affected future reproduction (Scientific Process 14.2).

The researchers manipulated clutch size to alter incubation costs. For one group of birds, clutch size was reduced to three eggs (a low incubation demand), whereas in the other group of birds, clutch size was increased to six eggs (a high incubation demand).

Females incubating six eggs lost relatively more mass than females incubating only three, confirming the assumption that incubation cost increases with clutch size. In addition, in the next breeding season, females that experienced high incubation costs laid eggs later and had smaller clutch sizes. The researchers concluded that eiders experience a trade-off between current and future reproductive effort. Females that lay a large clutch increase current reproduction but pay high incubation costs, which leads to a poorer body condition and reduced future reproduction.

In both treehoppers and eiders, we see that parental care involves a fitness trade-off between current and future reproduction, as predicted by parent-offspring conflict theory. Females that provide high levels of care enhance their

SCIENTIFIC PROCESS 14.2

Parental care costs in eiders

RESEARCH QUESTION: *Does egg incubation affect future reproduction?*

Hypothesis: Incubation costs increase with clutch size and represent a substantial cost of parental care.

Prediction: Individuals with large clutches will pay high incubation costs and experience lower reproductive fitness in the future compared to those with small clutches.

Methods: The researchers:

- Manipulated the clutch size of uniquely marked females (either decreasing the clutch to three eggs or increasing it to six eggs)
- Recorded the body mass of each female at the beginning and end of the 15-day incubation period
- Recorded egg-laying date and clutch size for each female in the following year

Results: Females with high incubation costs lost more mass and had later egg-laying dates and smaller clutches in the next breeding season than those with low incubation costs.

Figure 1. Incubation costs. Mean ($\pm$ SE) effects of high and low incubation costs on (a) body mass loss, (b) laying date, and (c) clutch size the next year in female common eiders. (*Source*: Hanssen et al. 2005)

Conclusion: High incubation costs in one year reduce reproductive fitness in the next breeding season.

EVALUATE
What is another potential fitness cost associated with high incubation costs?

current reproduction, but at a cost of reduced future reproduction. The parental care trade-off between current and future reproduction can explain much of the variation observed in parental care behavior among individuals. This theory can also be used to predict how the fitness of individuals will be affected by the introduction of novel predation threats that require greater levels of egg guarding (Applying the Concepts 14.2). We can go even further and use parent-offspring conflict theory to understand variation in care provided to siblings, as we see next.

Brood reduction and parent-offspring conflict

Parent-offspring conflict theory can also help us understand the evolution of parental care behavior that perhaps seems outright maladaptive. In many birds and some mammals, parents provide different amounts of care to siblings: some

APPLYING THE CONCEPTS 14.2

Smallmouth bass defend their nests from exotic predators

Exotic species, particularly predators, can reduce the population size of native species. Usually they do so through predation, but recent work suggests that they can also do so by increasing the cost of parental care. Round gobies (*Neogobius melanostomus*) are native to central Eurasia but were unintentionally introduced to Lake Erie in 1993 and are now abundant in the Great Lakes. They can consume an entire nest of eggs of the smallmouth bass (*Micropterus dolomieu*) in just a few minutes (Steinhart, Marschall, & Stein 2004).

Smallmouth bass are effective defenders of their nests, but this defense requires a great deal of effort. Geoffrey Steinhart and colleagues looked at two sites: one in Lake Erie, which has an abundance of gobies, and a control site in Lake Opeongo, where gobies do not occur (Steinhart et al. 2005). Bass in Lake Erie engaged in about nine times as many defense chases as bass in Lake Opeongo (Figure 1). Lake Erie males thus spent almost ten times the energy expended by their Lake Opeongo counterparts in offspring defense. Because of the trade-off between current and future reproduction, researchers speculate that the high energy demand will reduce the lifetime fitness of bass and could lead to strong population declines in areas where nest predators are common. The potential decline of this native species will be due not to actual predation, but to changes in parental care behavior as a result of an introduced exotic species. ■

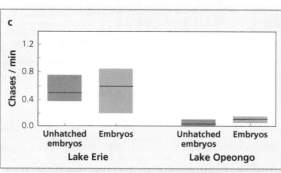

Figure 1. Smallmouth bass nest defense. (a) A male bass defending its nest. (b) Eggs in a nest. (c) Offspring defense (median number of chases per minute) of embryos in different lakes. Each box represents the middle 50% of data. Chases were more frequent in Lake Erie. (*Source:* Steinhart et al. 2005)

---

receive much more care than others. This difference leads to substantial variation in offspring size and, ultimately, survival. The largest, most well-fed siblings tend to survive, while those that receive little food suffer high mortality, which can result in **brood reduction**, or the death of some siblings.

Why don't parents care equally for their offspring or even bias care toward their smallest young, which are at high risk of starvation? At first glance, it might appear that such behavior should increase survivorship of all offspring. However, parent-offspring conflict theory predicts that brood reduction can be adaptive for a parent. If the costs of parental care for a current offspring exceed the benefits of such care, then parents should cease providing care to that offspring. For example, if resources are very scarce, it may require too much effort on the part of the

brood reduction The death of some siblings as a result of reduced parental care to enhance the fitness of surviving siblings.

parents to obtain sufficient food to feed all siblings. A parent can enhance its total lifetime reproduction if it ends care for some current offspring to better provide care to future offspring. While this behavior may seem inconceivable to humans, it does make sense from an evolutionary perspective.

Let's examine two empirical studies of brood reduction that tested predictions derived from parent-offspring conflict theory.

Hatch asynchrony and brood reduction in blackbirds

Featured Research

Many birds lay multiple eggs in a clutch over the course of several days. Often, incubation begins when the first egg is laid, and so offspring begin to develop on different days, resulting in siblings of different ages in the same nest. David Lack suggested that the difference in egg hatch timing, or **hatching asynchrony**, is adaptive because it allows parents to modify their parental care behavior based on current conditions (Lack 1954; Jeon 2008). In years with high levels of resources, all offspring have a high probability of survival, whereas in years of low levels of resources, only the early-hatching, best-developed offspring have a high probability of survival. Brood reduction benefits parents that cannot provide sufficient food for all offspring when resources are scarce. In fact, both parents *and* surviving offspring benefit if some siblings do not survive, because this increases the likelihood that some offspring will survive. Parent-offspring conflict theory predicts that (1) later-hatching offspring should suffer higher mortality as a result of reduced care, and (2) resource availability should affect survivorship.

> **hatching asynchrony** A situation in which chicks in a clutch hatch on different days.

Scott Forbes, Richard Grosshans, and Barb Glassey tested the first prediction in yellow-headed blackbirds (*Xanthocephalus xanthocephalus*) at a site near Winnipeg, Canada (Forbes, Grosshans, & Glassey 2002). Yellow-headed blackbirds typically lay clutches of three to five eggs over a period of two days. Forbes and colleagues classified the first two eggs laid as "core" and any subsequent eggs as "marginal" and examined the growth rate and survivorship of each. They censused nests daily and marked the eggs when laid; each offspring was subsequently marked with nontoxic color markers. Each day, nestlings were weighed and mortality noted until chicks were 12 days old, the age at which they normally leave the nest (Scientific Process 14.3).

Very few of the nests (only 16%) fledged their entire clutch of eggs, and asynchronous hatching did affect offspring growth and survival. The mean mass of core nestlings was higher than that of marginal nestlings in most years, which translated into higher survival. This finding supports the first prediction, that later-hatching siblings are at a competitive disadvantage to and suffer higher mortality (brood reduction) than their earlier-hatching siblings.

What about our second prediction for how resource availability affects survivorship? Evaluating this question requires a long-term study in a variable environment, as we see next.

Brood reduction in fur seals

Featured Research

In some marine mammals as well as some social carnivores, a second offspring is born before the previous one is weaned. This timing creates families with offspring that vary greatly in body size and development, just as in bird clutches with asynchronous hatching. Female Galápagos fur seals (*Arctocephalus galapagoensis*) (Figure 14.16) follow a yearly reproductive cycle. Shortly after giving birth, females copulate and can produce another offspring the following year. However, pups are not weaned until they are two years old. About 35% of females will give birth in two successive years and so have two related offspring of

SCIENTIFIC PROCESS 14.3

Brood reduction in blackbirds

RESEARCH QUESTION: *Does hatch asynchrony affect offspring mortality?*

Hypothesis: Early-hatching offspring will be larger than and competitively dominant to their siblings.

Prediction: Mortality will be highest for later-hatching individuals.

Methods: The researchers:
- Checked nests (*n* = 245) daily. They uniquely marked eggs and resulting chicks with nontoxic colored markers. The first two chicks hatched in a nest were identified as "core" and the others as "marginal."
- Weighed nestlings each day and noted mortality until chicks were 12 days old

Results:
- Core chicks were heavier than marginal ones in each year of the study.
- Nestling survival was highest for core nestlings and significantly lower for marginal offspring.

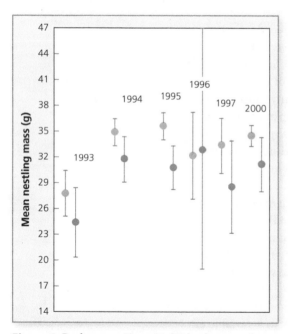

Figure 1. Body mass. Mean (± SE) body mass of core nestlings (orange) was higher than that of marginal nestlings (blue) in most years. (*Source:* Forbes, Grosshans, & Glassey 2002)

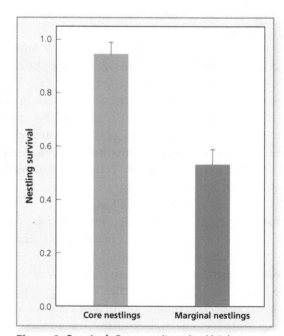

Figure 2. Survival. Core nestlings had higher mean (+ SE) survival than marginal nestlings. (*Source:* Forbes, Grosshans, & Glassey 2002)

Conclusion: Hatching asynchrony facilitates brood reduction by creating offspring of different size and competitive ability.

> **EVALUATE**
> What additional data are needed to test the hypothesis that core nestlings outcompete marginal nestlings for food?

different ages for whom they must care. Lactation for two pups is very costly for the female and leads to high offspring mortality. Why, then, do females give birth in successive years?

These seals live in the eastern Pacific Ocean, where there can be dramatic annual changes in the environment. Fur seals feed on fish and cephalopods whose population sizes are affected by cold-water upwelling, which has a very high nutrient content. In years of high upwelling and cold water, fish are common and food is plentiful for seals. When water temperatures are warmer (no upwelling), food for seals becomes scarce. Upwelling is affected by the Pacific tropical climate system El Niño and is unpredictable from one year to the next. Fritz Trillmich and Jochen Wolf hypothesized that the survivorship of pups with an older sibling is related to resource availability (Trillmich & Wolf 2008). They predicted that second offspring should suffer high mortality in years of scarce resources and have high survivorship in years of abundant resources.

Trillmich and Wolf studied 188 tagged females on Fernandina Island in the western Galápagos archipelago for ten years. Unweaned older pups were marked with ear tags, and their younger siblings were marked with dye. The survivorship of each pup was noted for its first 30 days of life. To estimate food availability, the researchers observed sea surface temperature (SST) in the eastern Pacific. Years of low SST correspond to high marine productivity and abundant food for fur seals.

Findings revealed that the early survivorship of pups with older siblings was related to SST. In cold-water years (with low SST), pup survival was relatively high, approaching the level of survivorship for pups without an older sibling in one year (Figure 14.17). In years of higher SST and lower food availability, the survivorship of pups with an older sibling was dramatically lower. These results support the prediction that brood reduction is a function of food availability. In years of high resources, brood reduction is minimal, and parents can take advantage of the rich food supply. The production of "extra" offspring allows parents to increase their reproductive output. In contrast, in years of low food availability, mortality of the youngest pup is high. Conservation biologists are taking advantage of this relationship to help boost populations of endangered species (Applying the Concepts 14.3).

Why do parents produce more offspring than they can care for? If they didn't, there would be no need for brood reduction. First, "extra" offspring provide some "insurance": if the first-born offspring dies, a second will still be alive. Second, parents do not have information on how much food will be available in the future, when offspring demand for food is high. If the environment contains abundant food, a parent may be able to successfully increase its reproductive fitness for the year. If the benefit of producing an extra offspring is greater than the cost, we can predict that selection can act such that parents will produce "extra" siblings and brood reduction will occur in years of low resources. In this light, brood reduction allows parents to adjust their reproductive effort to environmental conditions.

Figure 14.16. Fur seal with pups. Galápagos fur seal female caring for her pups of different ages.

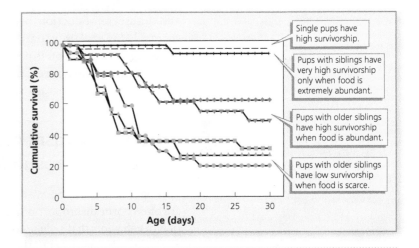

Single pups have high survivorship.

Pups with siblings have very high survivorship only when food is extremely abundant.

Pups with older siblings have high survivorship when food is abundant.

Pups with older siblings have low survivorship when food is scarce.

Figure 14.17. Fur seal pup survival. Cumulative survival of single pups was high in all years (horizontal dashed line). The survival of pups with older siblings was related to water temperature, which was an index of food availability. In cold-water years with high food availability (blue), survival was high. In warm-water years with low food availability (orange), survival was low. (*Source:* Trillmich & Wolf 2008)

APPLYING THE CONCEPTS 14.3

Food supplementation reduces brood reduction in endangered eagles

Brood reduction is common in many birds of prey, such as hawks and eagles. One species that exhibits brood reduction is the endangered Spanish imperial eagle (*Aquila adalberti*). Females lay clutches of one to four eggs over a one-week period, but mortality of chicks is high after the first offspring, so typically only about one chick fledges. The Spanish imperial eagle is one of the rarest birds of prey in the world, with a population size of around 200 breeding pairs. Given brood reduction in the species, conservation biologists have considered two strategies to increase population size. First, biologists can remove second-hatched chicks and place them in other nests or hand-raise them and release them into the wild as juveniles (Meyburg 1978). However, these procedures disturb the nest, are costly, and often are ineffective. An alternative is to provide supplemental food to adults with two chicks.

Luis González and colleagues tested the prediction that supplemental feeding would reduce brood reduction in Spanish imperial eagles in western Spain (González et al. 2006). For 13 years, they provided European rabbit carcasses to 22 nests containing more than one chick. Another 37 nests with multiple chicks served as controls.

Supplemental feeding increased chick survival almost every year (Figure 1). In most years, almost twice as many chicks fledged in nests that received supplemental food than did in control nests. The research team concluded that Spanish eagle reproduction is food-limited, and that population growth can be enhanced by supplemental feeding. This management tool is cheap, nonintrusive, and effective. ◼

a

b

Figure 1. Effect of food supplementation on fledgling success. (a) Spanish imperial eagle. (b) Fledgling rate (number of fledglings per pair) for food-supplemented (blue) and control (orange) nests in different years. (*Source*: González et al. 2006)

In the above examples, we have seen how the fitness trade-off between current and future reproduction, which forms the basis of parent-offspring conflict theory, helps explain variation in parental care both among individuals and to offspring. This trade-off is driven by the substantial cost to parents of caring for their offspring. Next, we examine how selection can favor individuals that reduce such costs via brood parasitism.

14.4   Brood parasitism reduces the cost of parental care and can result in a co-evolutionary arms race

Throughout this section, we have seen that parental care is costly and that the costs are often borne disproportionately by females—a situation that can favor the evolution of behaviors that reduce costs. One way to reduce the cost of parental care is for a female to lay her eggs in the nest of another female, called a "host," a behavior known as **brood parasitism**. The benefit of this behavior is reduced parental care, which can then result in enhanced reproductive success for the brood parasite. Although the benefit of parasitism is clear, it does entail some risk: a parasitized female may reject the added eggs or chicks, an outcome that will not increase fitness for the female parasite. How can a brood parasite reduce the likelihood of egg rejection?

brood parasitism A behavior in which a female (brood parasite) lays an egg in the nest of another female.

The simplest way to avoid egg rejection is to parasitize conspecifics by laying eggs in the nests of other females of the same species. Rejection by the host will be difficult because the eggs of conspecific females look similar. Conspecific brood parasitism is common in birds, with examples from over 200 species (Yom-Tov 2001), and also occurs in fish and insects (Wisenden 1999; Tallamy 2005).

To examine the benefits of conspecific brood parasitism, we might envision three reproductive tactics that a female could use (Sorenson 1991):

1. Lay all eggs in her own nest (nonparasite)
2. Lay all eggs in the nests of other females (pure parasite)
3. Lay some eggs in her own nest and some in the nests of others (a mixed strategy of nesting and parasitizing)

Let's examine one study that attempted to quantify the fitness of these three reproductive strategies.

Conspecific brood parasitism in ducks

Featured Research

Matti Åhlund and Malte Andersson examined the prevalence of conspecific brood parasitism in a population of marked female goldeneye ducks (*Bucephala clangula*) nesting in Sweden (Åhlund & Andersson 2001). They used egg albumin protein fingerprinting to identify the maternity of eggs and hatchlings within a nest. Albumin is deposited by the female into the egg and contains over a dozen proteins. Many of these proteins exhibit genetic polymorphisms, so there is much genetic variation among females. Therefore, each female exhibits a unique protein-banding pattern, allowing researchers to identify the maternity of different eggs (Andersson & Åhlund 2000). Nests were monitored regularly to determine egg laying, and there was no evidence of egg removal by any female. All chicks were marked with numbered tags, and their survival was recorded over four weeks.

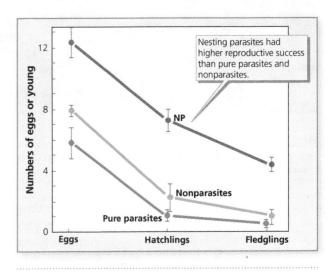

The researchers found that all three strategies existed in the population studied, but females that used the mixed nesting and parasitizing tactic had the highest reproductive success. These females laid about 50% more eggs than nonparasites or pure parasites (Figure 14.18), and the eggs they laid in host nests survived equally as well as nonparasite eggs, indicating that they were not rejected.

The researchers observed variation in the reproductive tactics used by a female across years and found no evidence that nesting parasites suffered lower survival or reproduction in subsequent years, despite laying more eggs. Why this occurred is unclear, and so further work is required to understand the reproductive tactics of conspecific brood parasites in this population. Nonetheless, one factor that likely favors the evolution of brood parasitism in this and other species is its minimization of the costs associated with parental care.

Figure 14.18. Brood parasitism in ducks. Mean (± SE) number of eggs or young for females classified as pure parasites (blue), nonparasites (orange), or nesting parasites (NP, gray). The latter laid the most eggs and had the highest number of hatchlings and fledglings survive. (*Source:* Åhlund & Andersson 2001)

Interspecific brood parasitism and co-evolution

Featured Research

While many species, like the ducks described above, lay eggs in the nests of conspecifics, some brood parasites lay their eggs in the nests of other species and are

Interspecific brood parasitism A behavior in which a female (brood parasite) lays an egg in the nest of another species.

thus known as **interspecific brood parasites**. About 1% of bird species, including many cuckoos and cowbirds, are obligate brood parasites: they *only* lay eggs in the nests of other species, thus avoiding all costs of parental care. This behavior can greatly enhance the fitness of the brood parasite. Female brown-headed cowbirds (*Molothrus ater*) can lay as many as 40 eggs in the nests of hosts during a single reproductive season, all without the cost of parental care. Brood parasitism has a correspondingly negative effect on host fitness; for example, common cuckoo chicks (*Cuculus canorus*) kill host offspring. Similarly, cowbird chicks are competitively dominant to host chicks, and so the parasite receives a disproportionate share of the care (food) provided by the host parent.

Given the fitness costs of brood parasitism, selection should favor hosts that can discriminate their own eggs from those of a parasite so that the latter can be rejected (i.e., removed from the nest). Indeed, egg rejection behavior has evolved in many species that co-occur with brood parasites (Davies 1999).

When hosts evolve this ability, selection can favor brood parasites that produce eggs similar in appearance to host eggs, making discrimination more difficult. Many brood parasites, including the common cuckoo, have evolved traits that reduce the likelihood of egg rejection by a host, including the production of eggs that match the size, color, and shape of host eggs so that it is difficult for the host to distinguish its eggs from that of the parasite (Figure 14.19) (Stokke, Moksnes, & Røskaft 2005). Such a co-evolutionary arms race, driven by the costs of parental care, should continue to affect the evolution of both egg rejection behavior and hosts' discrimination ability.

Not all species that live with brood parasites reject their eggs. For example, brown-headed cowbirds commonly parasitize dozens of species. About 24 of these regularly reject cowbird eggs, but over 30 rarely do so. Why not?

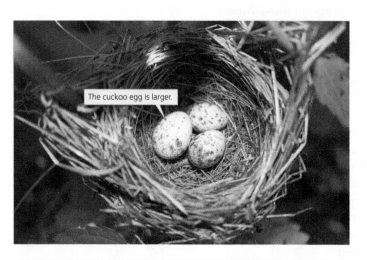

The cuckoo egg is larger.

Figure 14.19. Parasitic egg in nest. A cuckoo egg (the largest) in a redstart (*Phoenicurus phoenicurus*) nest. Note the similarity in the egg coloration.

One explanation, the evolutionary lag hypothesis, notes that the geographic range of brown-headed cowbirds has expanded greatly over time. This species was historically restricted to open habitats in the Great Plains of North America but now occurs throughout the continent as a result of logging that has cleared forest habitat. This range expansion has increased the availability of new host species that have no evolutionary history with brown-headed cowbirds. Such species are more likely to be egg acceptors than are those that have a longer evolutionary history with brown-headed cowbirds.

Another hypothesis focuses on bill size, noting that most egg rejection requires a host to grasp and lift the cowbird egg with the bill. Therefore, only species with sufficiently large bills can successfully remove cowbird eggs. This hypothesis predicts that species with large bills are more likely to be rejectors than are those with small bills. Next we examine a study designed to test these hypotheses.

Featured Research

Acceptance or rejection of brown-headed cowbird eggs by hosts

Brian Peer and Spencer Sealy tested the evolutionary lag and bill size hypotheses by conducting a comparative analysis of brown-headed cowbird hosts (Peer &

Sealy 2004). They identified 12 pairs of closely related species that are known to be parasitized by brown-headed cowbirds and that differ in egg rejection rate. For each pair, they used data from published literature to determine that one species had a high cowbird egg rejection rate (greater than 80%) while the other had a low rejection rate (less than 20%). The research team deemed a species to have a long evolutionary history with brown-headed cowbirds if its historic geographic range overlapped with that of cowbirds and if it nested in open habitats. Bill sizes (length) were obtained for each species from the literature, which was supplemented with measurements of museum specimens.

Analysis of the data supported both hypotheses. For the five species pairs that differed in evolutionary history, the species with the shorter history was always an egg acceptor, while the species with the longer history was always a rejector. In the seven species pairs that differed significantly in bill size, the species with the longer bill was always a rejector, while the species with the shorter bill was always an acceptor. Thus, both length of time in contact with a brood parasite and bill size explain variation in egg rejection behavior in North American host species.

Throughout this chapter we have seen that parental care is costly but can enhance fitness by increasing the survivorship of young. How are parental care behaviors regulated? We examine this proximate question next.

14.5 Hormones regulate parental care

Many adults exhibit a radical change in behavior over the course of the reproductive season. Before mating, individuals concentrate on finding and attracting a mate. As the reproductive season progresses and more individuals mate successfully, mating behavior declines and parental care increases. For example, birds switch from mating displays to nest building, egg incubation, and then chick feeding. Female mammals undergo a change in physiology to begin lactation. What initiates and regulates these changes?

In mammals, steroid hormones such as estradiol and progesterone vary with pregnancy and can strongly affect courtship and reproductive behavior. The peptide hormone **prolactin**, which is involved in milk synthesis, rises during pregnancy and then falls when offspring are weaned (Zarrow, Gandelman, & Denenberg 1971). It is synthesized in and secreted from specialized cells in the anterior pituitary gland. Does prolactin affect parental care?

prolactin A hormone that influences parental care in vertebrates.

Prolactin and maternal care in rats

Featured Research

Robert Bridges and colleagues conducted a series of classic experiments to examine the effect of prolactin, estradiol, and progesterone on maternal behavior in rats (Bridges et al. 1990). They manipulated prolactin levels in female rats that had never given birth. To set up the experiment, they first treated all individuals with the drug bromocriptine to suppress endogenous prolactin secretion so that the rats could not produce their own prolactin. The research team also removed each rat's ovaries to control the levels of other hormones and inserted a cannula—a small tube—into the brain to administer hormones directly into the central nervous system. All females were "primed" with progesterone and estradiol to mimic the changes in these steroid concentrations that occur in pregnant rats. The researchers then examined the behavior of females treated with different levels of either prolactin or a sodium chloride–sodium bicarbonate control. Next, they placed three unrelated pups into the home cage of each female. Over a period of six days, Bridges and colleagues recorded three aspects of parental care: latency to contact the pups,

Figure 14.20. Female rat with young. Females huddle over pups to keep them warm.

whether pups were retrieved and carried to the nest, and instances of crouching behavior (huddling over pups to keep them warm) (Figure 14.20).

There was a rapid increase in maternal behavior in treatment females by Day 3. Females treated with prolactin displayed significantly higher levels of parental care behavior than did controls (Figure 14.21). In follow-up experiments, the research team examined whether prolactin could stimulate maternal behavior without the steroid regimen. The experimental procedures followed were identical to those of the prior study, except that females were not primed with progesterone and estradiol. Without the steroid hormones, females did not display maternal behavior (Figure 14.21). To determine the action site of prolactin, the researchers injected prolactin subcutaneously but found no difference in maternal behavior compared to that of controls. Subsequent work identified prolactin receptors in the brain and revealed that females with nonfunctional prolactin receptors will exhibit very low levels of maternal care behavior (Lucas et al. 1998). Together, these studies demonstrate that prolactin strongly affects parental care in female rats.

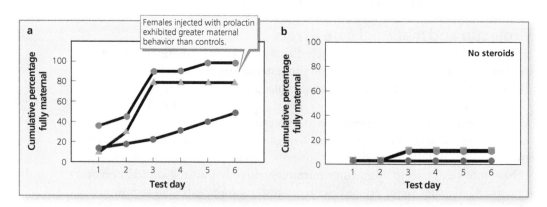

Figure 14.21.   Rat maternal behavior. (a) The cumulative percentage of fully maternal behavior was higher in female rats experimentally injected with prolactin (50 μg, orange; 10 μg, blue) compared to controls (gray). (b) Without the steroid regimen, maternal behavior in female rats experimentally injected with prolactin (blue and orange) was similar to that of controls (gray). (*Source:* Bridges et al. 1990)

Prolactin and incubation in penguins

Incubation of eggs is an important component of parental care in birds, and several studies have shown the importance of prolactin in affecting incubation behavior (e.g., Lormée et al. 1999). Blood prolactin levels tend to increase in birds during egg laying (Vleck 2001), but what is the proximate mechanism that causes this increase? One possibility is that the simple sight of an egg spurs increased prolactin production. Alternatively, birds might require tactile stimulation that occurs when eggs are incubated by contact with the brood patch, a featherless area on the belly that is well vascularized for heat transfer to eggs.

Melanie Massaro, Alvin Setiawan, and Lloyd Davis examined prolactin levels in female yellow-eyed penguins (*Megadyptes antipodes*) at Boulder Beach in New Zealand (Figure 14.22) (Massaro, Setiawan, & Davis 2007). Just before egg laying, the research team placed an artificial egg either in a female's nest or next to the nest in a wire cage. In the first case, the egg provided both tactile and visual stimulation. In the second case, the egg could not be touched, so it provided visual stimulation only. The team used plaster eggs that matched the size, weight, shape, and color of natural yellow-eyed penguin eggs. They divided 33 females into three groups: (1) those sampled for prolactin prior to treatment, on the day the egg was added; (2) those sampled three to four days after the egg was added (experimental birds); and (3) controls, which were sampled at the same time as the treatment group but had no egg placed in their nest. For the first two groups, the researchers recorded the amount of incubation time that occurred after the artificial egg was placed in the nest.

When only visual stimulation was provided, there were no differences compared to control individuals in prolactin levels, incubation behavior, or brood patch development. However, when the artificial egg was placed in the nest, females had higher levels of prolactin, developed a wider brood patch, and spent more time incubating than did controls (Figure 14.23). In other words, for this species, the visual and tactile stimulation resulted in increased prolactin levels, increased maternal care, and an important physiological

Figure 14.22. Penguin. Female yellow-eyed penguins' care for their young begins with egg incubation.

Figure 14.23. Prolactin and brood patch development. Mean (+ SE) (a) prolactin level, and (b) brood patch width in treatment groups. Females with artificial eggs added to their nest (Exp 1) had higher prolactin levels and wider brood patches than females assessed prior to the treatment or controls. (*Source:* Massaro, Setiawan, & Davis 2007)

change: development of a larger brood patch. Additional studies have found that prolactin production is involved in stimulating parental care behavior in many bird species, but our understanding of the mechanism of action in the brain is still limited (Angelier & Chastel 2009).

Juvenile hormones and parental care in earwigs

Featured Research

juvenile hormones (JH) In insects, a class of hormones produced by specialized endocrine glands that vary in a consistent manner with levels of parental care by females.

Although the effects of many hormones on development, the brain, and behavior are rather well studied in vertebrates, much less is known about their relationships in invertebrates. In insects, **juvenile hormones (JH)** are a class of hormones produced by specialized endocrine glands that affect development and molting and vary in a consistent manner with levels of parental care by females (Trumbo 2002; Mas & Kölliker 2008). For example, in earwigs, females care for offspring by constructing a burrow and then grooming and guarding the developing eggs kept there. In these species, JH increases as a female's oocytes develop and peaks a few days prior to oviposition. JH levels then decline and remain low while she cares for her offspring (Figure 14.24). As offspring reach maturity, JH again rises, parental care ceases, and females reproduce again (Rankin et al. 1995; Trumbo 2002). So in this case, parental care is high when JH levels are low, and vice versa.

Susan Rankin, Kelly Fox, and Christopher Stotsky experimentally tested the link between JH level and maternal care in the ring-legged earwig (*Euborellia annulipes*). They topically added JH in acetone to females on the day of oviposition to keep levels artificially high and either did not manipulate or just topically added acetone to controls. Females with enhanced JH exhibited reduced parental care, providing only three days of egg care compared with the eight days provided by untreated controls and acetone-only individuals (Figure 14.25) (Rankin, Fox, & Stotsky 1995). These results illustrate how JH regulates parental care in this species: care is low when JH levels are high, and vice versa.

In this section, we have seen how different hormones regulate parental care in several taxa. But we have only just scratched the surface of this complex field. Ongoing research will continue to provide new insights into interactions among environmental cues, hormonal changes, and parental care behavior (e.g., Dulac, O'Connell, & Wu 2014).

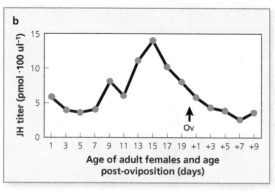

Figure 14.24. Juvenile hormone variation in earwigs. (a) Ring-legged earwig. (b) Juvenile hormone level over time. In earwigs, juvenile hormone increases and peaks a few days prior to oviposition. Juvenile hormone levels remain low while females care for their young (Days +1 through +9). OV represents the day of oviposition. (*Source:* Rankin et al. 1995)

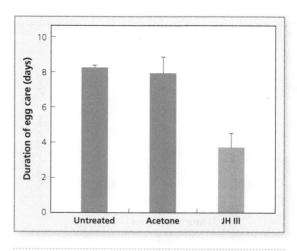

Figure 14.25. Duration of earwig egg care. The mean (+ SE) number of days of egg care by earwigs was five days less for females treated with juvenile hormone and acetone (JH III) than for untreated controls or those treated only with acetone. (*Source:* Rankin et al. 1995)

Chapter Summary and Beyond

Species differ in life history traits, which can explain the variation in parental care across species: those with a long life span or well-developed offspring at birth tend to provide high levels of parental care. Research has focused on understanding the evolution of parental care in different groups (Royle, Smiseth, & Kölliker 2012). For instance, parental care in birds appears to have evolved first in their extinct ancestors, the dinosaurs (Varricchio et al. 2009). Other work has examined the evolution of trade-offs involving parental care and other life history traits in insects (Gilbert & Manica 2010) and fish (Kolm & Ahnesjö 2005). In addition, males and females differ in the fitness benefits they obtain from providing parental care because of the differing costs of care, particularly in terms of the certainty of relatedness to offspring. For most species, and especially for species with internal fertilization, females are more certain of their offspring than are males. This difference favors greater levels of parental care by females. Reduced parental care allows individuals to seek additional sexual partners or allocate more energy to future reproduction.

Adults provide care to increase survivorship of their offspring, but doing so reduces their future reproductive fitness, resulting in a trade-off between allocating care effort to current offspring and allocating such effort to future offspring. This trade-off explains variation in care among individuals, and recent work has begun to examine the mediating effect of hormones in this relationship (McGlothlin, Jawor, & Ketterson 2007). This trade-off also explains why parents often provide unequal care to siblings and how unequal care can result in brood reduction. The high costs of parental care help explain the evolution of intra- and interspecific brood parasitism. Feeney, Welbergen, & Langmore (2012) provide an extensive review of this subject.

Prolactin is an important hormone that regulates parental care in many vertebrates. It initiates lactation and parental care behavior in mammals and also is involved in stimulating parental care in birds, as well as the production of nutrient-rich crop "milk" (a secretion from the alimentary tract) in mourning doves (*Zenaida macroura*) (Miller, Vleck, & Otis 2009). Much research today investigates how stress affects prolactin production. Increases in stress hormones typically suppress prolactin production. This suppression may be an adaptive response: decreases in parental care behavior allow individuals to concentrate on their own survival, but more work is needed to fully understand this pattern (Angelier & Chastel 2009). Prolactin may also be an important factor in the development of paternal care in mammals and fish, but the evidence is conflicting (Wynne-Edwards & Timonin 2007; Bender, Taborsky, & Power 2008). In insects, juvenile hormone production is often negatively associated with parental care behaviors. Mas and Kölliker (2008) have reviewed work on the physiological aspects of parental care in insects and highlight recent work on chemical communication between mother and offspring that affects the level of care provided.

CHAPTER QUESTIONS

1. Consider three female chickadees of the same age and condition. Each successfully raises a clutch of eggs with no help from her mate, but the clutches differ in size. Female A raises four offspring, B raises eight, and C raises 12. What can you predict about the overwinter survivorship and reproductive output of these females the next year?

2. Imagine a bird species with biparental feeding of young and equal biparental care. Suppose you manipulate one individual to increase its cost of flight by removing a few of its flight feathers or adding small weights to the bird. How will this manipulation affect its level of parental care in terms of providing food to offspring?

3. In their study of the effects of prolactin on parental care behavior, why did Bridges et al. (1990) first treat all individuals with bromocriptine to suppress endogenous prolactin secretion?

4. Hatching asynchrony is hypothesized to be an adaptation that allows parents to adjust their parental care and number of offspring to varying environmental conditions. Why would asynchronous clutches lead to higher adult fitness than synchronous clutches of the same size when environmental variation is large?

5. Imagine that you have discovered two new species of fish. One has internal fertilization, and the other fertilizes eggs externally. In both species, only one adult provides parental care for eggs in a nest. In which species is paternal care more likely to have evolved, and why?

6. Develop an alternative hypothesis to explain why some big-thumbed male frogs provide little parental care when they fertilize small clutches.

7. Compare the parental care studies by Bridges and colleagues (1990) on mice with that of Osorio-Beristain and Drummond (2001) on boobies. For each study, describe which of Tinbergen's four questions the study investigated, and explain your answer.

8. In Westendorp and Kirkwood's (1998) study of British women's longevity, why was it important for the researchers to limit the sample to women who had reached age 60?

Chapter 15

Sociality

Around your home or campus, you have probably noted that some animals seem to live in large groups while others appear to be more solitary. For example, in the winter, large flocks of crows often roost in trees. In summer, we often see ants following long foraging trails as they carry food back to their colony. Wasps and bees, too, are highly social, living in colonies of hundreds or even thousands—and remarkably, most of them are sterile (Figure 15.1). Most of these insects spend their lives working to ensure the reproductive success of just one or a few queens.

In this chapter, we examine many aspects of group living. First, we examine the benefits and costs of life in groups to better understand variation in the size of groups. We then focus on a behavior that reduces one cost of group living—the formation of dominance hierarchies. Next we turn our attention to a behavior unique to social groups: cooperation. We examine how cooperation among related individuals can evolve, with a focus on cooperative breeding in vertebrates and the evolution of sterile workers in social insects. Finally, we explore different hypotheses to explain the evolution of cooperation among unrelated individuals.

Figure 15.1. Sociality. Wasps caring for young in a colony.

15.1 Sociality evolves when the net benefits of close associations exceed the costs

a

b

Figure 15.2. Social groups. (a) A large school of fish. (b) A single-sex group of male caribou.

Figure 15.3. Physiological benefits of sociality. Bats huddle to minimize heat loss.

Flocks of birds, schools of fish, swarms of insects, and herds of mammals are examples of social groups—sets of individuals that remain in close proximity and interact with one another (Figure 15.2). Across species, there is tremendous variation in the size, composition, and persistence of social groups—their degree of **sociality**. In some species, such as coyotes (*Canis latrans*), groups contain just a few individuals. In others, such as Atlantic herring (*Clupea harengus*) or the Australian plague locust (*Chortoicetes terminifera*), a school or swarm might contain many thousands or even millions of individuals.

Group composition can differ with respect to sex and age. During the nonreproductive season, for instance, many ungulates (hoofed animals) such as caribou (*Rangifer tarandus*) form single-sex groups, and flocks of common ravens (*Corvus corax*) are often composed of a single age class, juveniles. The time that individuals spend in groups can vary seasonally or daily. For example, European starlings (*Sturnus vulgaris*) tend to join large flocks only in the fall and winter and then live as pairs defending territories during the breeding season. In contrast, many primates, parrots, ants, wasps, and coral reef fish spend their entire life in a social group.

Finally, groups can differ in the relatedness of their members. Many groups, especially large ones, are composed of unrelated individuals; in contrast, smaller family groups are composed of closely related individuals, often containing siblings of different ages or individuals spanning more than two generations. How can we explain all this variation?

Sociality can evolve only if it results in an individual in a group obtaining higher fitness than it would by living alone. In other words, the net benefits of sociality for an individual must be greater than the net costs. Numerous potential benefits and costs of sociality have been identified (e.g., Hoogland & Sherman 1976; Krause & Ruxton 2002). Benefits can include reduced search time to find resources (food or mates), increased feeding success as a result of cooperative hunting, enhanced resource defense, reduced predation risk, lower physiological costs of thermoregulation and movement (Figure 15.3), division of labor, and communal care of young. Costs associated with sociality can involve increased aggression and competition for resources, predator attraction, disease transmission, and, in some species, infanticide and cannibalism. We have already discussed social behavior in regard to predation risk (Chapter 9) and aggression (Chapter 11). Here, we examine additional benefits and costs associated with life in groups.

Featured Research

Foraging benefits: reduced search times for food in minnows

sociality The tendency to live and associate with others.

Food is often not conspicuous in the environment, and so animals spend considerable time searching for it. How might sociality increase foraging success?

Some birds have higher encounter rates with insects when the movements of group members can flush prey out of hiding (Harsha et al. 2007). In other cases, individuals in groups benefit from the food discoveries of others. In Chapter 7, we saw how individuals can use the presence of others to learn about the location of food. This same mechanism produces one benefit of sociality: individuals in groups can find food patches faster because more individuals engage in searching. Once hidden food is found, all group members can move to the patch to feed.

Tony Pitcher, Ann Magurran, and Ian Winfield investigated this benefit of sociality in the common minnow (*Phoxinus phoxinus*), a small fish found throughout Europe in freshwater streams and rivers. Minnows feed on a variety of food items buried in the gravel or sand, including mollusks, insects, and crustaceans.

Pitcher and colleagues tested the prediction that as group size increases, the search time for an individual to find hidden food will decrease (Pitcher, Magurran, & Winfield 1982). They conducted experiments in a large aquarium (2 m × 70 cm × 70 cm) that held schools of two to 20 fish. In order to examine the effect of group size on an individual's time to find food, the researchers uniquely marked one fish, the test fish, and recorded its behavior. Before the start of a trial, the researchers placed a food patch, consisting of 84 ice cube tray wells filled with gravel, in the aquarium. Dried fish protein flakes were randomly added to one of the cubes. The researchers then placed the group in the tank and recorded the amount of time until the test fish found the food. They conducted four replicates of this experiment, using a different test fish, for each school size.

As predicted, the time before the test fish found the food declined as group size increased (Figure 15.4). This finding demonstrates one advantage of group living: the larger the group, the less time it takes an individual to find food, because others are searching for it as well.

Figure 15.4. Time to find food. Mean (± SE) time to find food. Individual minnows find food faster when they feed in shoals of larger sizes. (*Source:* Pitcher, Magurran, & Winfield 1982). *Inset:* minnow.

Antipredator benefits: vigilance in mixed-species flocks

A second benefit of group living is reduced predation risk. Individuals in groups can spot approaching predators sooner than can solitary individuals and so can have more time to take evasive action. In addition, being surrounded by others allows individuals to spend less time being vigilant, which frees up more time for other activities such as feeding. In Chapter 9, we examined these benefits for individuals living with conspecifics, but these same benefits should also extend to individuals living with other species in mixed-species groups.

Patrick Hart and Leonard Freed examined the benefits of sociality in mixed-species flocks of Hawaiian birds. They focused on the Hawaii akepa (*Loxops coccineus*), an endangered Hawaiian honeycreeper that commonly flocks with other species (Figure 15.5). Akepas feed on arthropods in the leaf buds of terminal

Figure 15.5. Akepa. (a) Female and (b) male akepa.

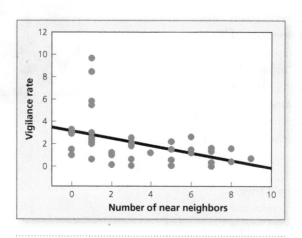

Figure 15.6. Vigilance and group size. Vigilance rate (scans per 20 seconds) declined as group size increased. (*Source:* Hart & Freed 2005)

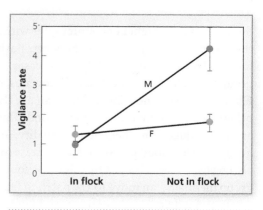

Figure 15.7. Vigilance levels in and out of flocks. Mean (± SE) vigilance rate (scans per 20 seconds) for males (M) was significantly lower when in a flock. In contrast, the vigilance level of females (F) was unaffected by their level of sociality. (*Source:* Hart & Freed 2005)

branches in the tree canopy, a behavior that exposes feeding birds to hawk predators. Akepas also exhibit variation in sociality: they feed both alone and in mixed-species flocks of up to dozens of birds.

Hart and Freed examined how antipredator vigilance behavior varied as a function of flocking (Hart & Freed 2005). They predicted that the vigilance rate of individuals in mixed-species flocks should decline with increasing flock size. At sites in the Hakalau Forest National Wildlife Refuge, the research team conducted focal sampling to record a bird's vigilance behavior while feeding (the number of scans in a 20-second period), its social status (in a flock or not in a flock), and flock size (the number of individuals within a 10 m radius). They found that as overall group size increased, the rate of individual vigilance decreased (Figure 15.6), just as predicted. This result illustrates that individuals in mixed-species flocks can obtain benefits from sociality similar to those derived in single-species flocks.

However, Hart and Freed found that the effect of flocking on vigilance rate differed for male and female akepas: males in flocks had lower rates of vigilance than males not in flocks, but the same relationship did not hold for females. Instead, females exhibited the same level of vigilance whether or not they were in a flock (Figure 15.7). Why? This species exhibits a striking sexual dimorphism: males are bright orange, whereas females are dull green. The researchers hypothesized that males might suffer higher predation risk because predators can spot them more readily. Thus, males not in a flock spend more time engaging in vigilance behavior than females. This study emphasizes the importance of examining the behaviors of males and females separately, because sex-specific selection pressures may have an effect on such behaviors, as we saw in Chapters 12 through 14.

Featured Research

Movement benefits: efficient aerodynamics in flight

One additional benefit of sociality is the aerodynamic advantage of moving within a group. Many large birds such as geese and cranes fly in a V-formation, with fairly uniform spacing between individuals. Why do they adopt this specific formation? One possibility is that it provides an aerodynamic advantage that reduces the cost of flying. Henri Weimerskirch and colleagues examined this hypothesis using trained white pelicans (*Pelecanus onocrotalus*) (Weimerskirch et al. 2001). These researchers placed heart rate monitors on the birds and

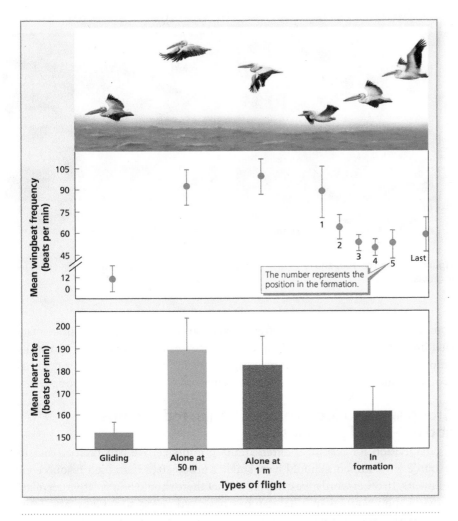

Figure 15.8. Flying effort in pelicans. Pelicans flying in a V-formation had lower mean (± SE) wing-beat frequency (top) and heart rate (bottom) than birds flying alone. (*Source:* Weimerskirch et al. 2001)

videotaped their flight formations as they followed a motorboat or ultralight airplane. Birds flying in a V-formation had a lower wing-beat frequency and heart rate than birds flying alone, probably because they were flying in the wake of another bird (Figure 15.8). These data demonstrate a clear decrease in energy expenditure for birds flying in formation. Interestingly, smaller species like sparrows do not fly in formation, perhaps because their body size does not allow the same aerodynamic advantage, although this idea has not been tested (Hummel 1983).

The costs of sociality

Given the many benefits of sociality, why don't all individuals live in social groups? The answer is that sociality also entails costs: individuals in groups are surrounded by close competitors. **Competition**, when individuals negatively affect one another, can be intense in groups of conspecifics because individuals require the same resources for survival and reproduction. For example, when a group of minnows finds a food patch, all of them exploit the food. If the food is limited, each individual in the group obtains less food than it would have if it fed alone, and its fitness, in terms of energy consumed, is reduced. As group size increases, so does the level of competition and the frequency of aggressive interactions. Therefore,

competition An interaction that results in a reduction in fitness for one or both individuals.

a   b

Figure 15.9. Red colobus and red-tailed guenon. (a) Red colobus and (b) red-tailed guenons feed in groups.

competition for resources will often be keen and is considered a primary factor that limits the size of groups. In addition, diseases are more readily transmitted among individuals that live in groups, and so individual fitness can decline as group size increases. Let's see how researchers study these costs of group living.

Featured Research ▷

Group size and food competition in red colobus and red-tailed guenons

If competition for resources often limits group size, we can predict that low-resource environments should have smaller group sizes than high-resource environments. In other words, we can predict that there should be a positive correlation between the resources in a habitat and observed group size. Colin and Lauren Chapman tested this prediction in two species of primates in Uganda (Chapman & Chapman 2000).

The Chapmans studied red colobus (*Procolobus pennantii*) and red-tailed guenons (*Cercopithecus ascanius*) at four sites in Kibale National Park over a period of two years (Figure 15.9). These primates live in tropical forests in equatorial Africa. Group size ranges from fewer than ten to several dozen individuals. Both species feed on a variety of fruits and vegetation.

In order to examine the influence of competition for resources on sociality in this system, the research team first characterized the specific diets of the species at each site. Both species fed primarily on young leaves, leaf buds, fruits, and flowers from dozens of tree species. Because all food items came from trees, the researchers could estimate food abundance at each site from the density of food trees, the presence of fruits or flowers, and the stage of leaf development. The Chapmans established four to six 200 m paths, or transects, across each site. Each month, the researchers walked along the transects to examine each potential food tree to determine whether it offered food for the monkeys. They also observed the size of all groups at each site for each species.

The Chapmans found a positive relationship between the density of food trees and average group size across the four sites, as predicted (Figure 15.10). This finding supports the

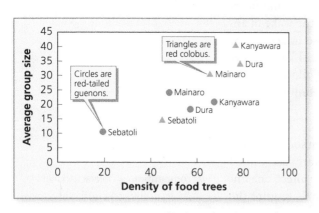

Figure 15.10. Primate group sizes. The average group size in two species of primates correlated positively with the density of food-producing trees. Circles are red-tailed guenons; triangles are red colobus. Names denote populations. (*Source:* Chapman & Chapman 2000)

hypothesis that group size is limited by competition for food. Because the monkeys' sites were relatively close to each other, the researchers assumed that other ecological variables (such as predators) that might affect group size were similar across sites and so had little effect on their results.

Sociality and disease transmission in guppies

Another important cost of group living involves the transmission of infectious diseases. Many parasites and pathogens reduce fitness (Daszak, Cunningham, & Hyatt 2000) and are transmitted by close contact. Contact between individuals increases with degree of sociality, and so it is not surprising that there is often a positive correlation between group size and the proportion of infected individuals in a group. After all, individuals will frequently be in close proximity when group size is large (Brown & Brown 1986; Côte & Poulin 1995).

E. Loys Richards, Cock van Oosterhout, and Joanne Cable examined how the degree of sociality affects parasite transmission in guppies (*Poecilia reticulata*) (Richards, van Oosterhout, & Cable 2010). Guppies are small tropical fish that live in small shoals (or schools). They are frequently infected with external parasitic worms, such as *Gyrodactylus turnbulli*, that attach to their fins, gills, and scales. Female guppies tend to spend more time in schools than males, and the researchers took advantage of this variation in behavior to examine differences in disease spread in schools of either females or males.

They first removed all external parasites from a population of guppies. Groups of six males or six females were housed in separate tanks. The researchers focused on one randomly selected focal guppy in each tank and identified it by its unique color markings. To quantify the degree of sociality, each group was observed every 30 seconds during a 15-minute period for three days, with the research team noting the distance of the nearest neighbor from the focal fish and the maximum school size, defined as the number of guppies within four body lengths of one another. At the end of the third day, the focal fish was infected with approximately 100 *G. turnbulli* worms and then placed back with its school to potentially infect others. Three days later, all fish were inspected for parasite infection.

As expected, female groups spent more time schooling and had tighter groups than males—that is, females exhibited a higher degree of sociality (Figure 15.11). After the parasite was introduced, the researchers found that almost 80% of the females in each tank became infected after three days, while only 40% of the males became infected (Figure 15.12). These results suggest that parasite transmission in guppies is affected by social behavior, with greater time spent near others, as in the female schools, leading to higher rates of transmission.

We have examined several of the many benefits and costs associated with life in groups. When the benefits of sociality are large relative to the costs, we expect to see individuals living in groups. When the benefits are low and the costs are high, we expect to see small group sizes or a lack of sociality. This

Figure 15.11. Guppy shoaling. (a) School of guppies. (b) Mean (± SE) nearest neighbor distance and (c) mean (+ SE) time spent shoaling for males and females. Female guppies were more tightly spaced and spent more time shoaling than males. (*Source:* Richards, van Oosterhout, & Cable 2010)

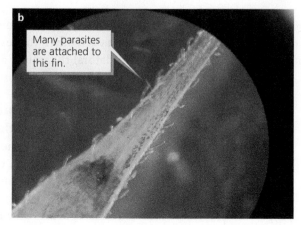

Figure 15.12. Infection rates. (a) Females had higher mean ($\pm$ SE) infection rates than males in the experiment. (b) *G. turnbulli* parasitic worms attached to the fin of a guppy. (*Source:* Richards, van Oosterhout, & Cable 2010)

understanding is being used to benefit social groups housed in captivity (Applying the Concepts 15.1).

As we have seen, competitive interactions are an important cost of living with others, and they can be particularly intense in social groups. In the next section, we examine how the frequency of aggressive interactions, and their associated costs, can be reduced in groups.

15.2 Dominance hierarchies reduce the social costs of aggression

In many groups, individuals will vary in their competitive abilities based on age, size, or condition, and so some individuals will be better competitors than others.

APPLYING THE CONCEPTS 15.1

Group size of social species in captivity

Zoos house many social species, including primates and ungulates. What is the best group size for species in captivity? Group size in the wild is often strongly affected by antipredator benefits or competition for food. For captive animals, these factors are irrelevant, meaning that the group size typical in the wild may often not be the best group size for animals in captivity.

Elizabeth Price and Tara Stoinski have shown that many social species suffer high levels of stress when forced to live in captivity with an inappropriate number of conspecifics (Price & Stoinski 2007). Social species that are housed with too few individuals often exhibit signs of chronic stress and reduced reproduction. For example, gorillas housed as breeding pairs have lower reproductive success than gorillas housed in groups of three or more (Maple & Hoff 1982). Stress, increased levels of aggression, and reduced reproductive success also occur when densities in captivity are too high. Duikers are an instructive example. These African ungulates live in very small social groups or are solitary in the wild. The Los Angeles Zoo initially housed duikers in groups of five, but individuals often developed stress-related jaw abscesses and lived less than five years. When group size was reduced to no more than three, individual health and life span both increased dramatically (Barnes et al. 2002).

Price and Stoinski suggest that naturally observed group size in the wild may not be the best group size in captivity. To minimize stress and aggression from overcrowding, institutions should create captive environments with greater structural complexity. For example, zoos can provide animals with both visual barriers and escape paths so that individuals can modify their proximity to others. ■

In general, more competitive individuals will have greater access to limited resources than will weaker rivals, resulting in greater net benefits of group membership (Vehrencamp 1983). Engaging in aggressive interactions over resources runs the risk of injury for both parties involved, a substantial cost of sociality. In most groups, repeated encounters between two individuals usually end in the same result: the better competitor wins the interaction. Given this predictability of outcomes, one way to minimize aggression within groups is through the formation of a **dominance hierarchy**, an organized social system with dominant and subordinate members. In a **linear dominance hierarchy**, an individual is dominant to each individual below it and subordinate to all above it in rank. Let's look at studies on crayfish and baboons to see how dominance hierarchies reduce the social costs of aggression.

> **dominance hierarchy** An organized social system with dominant and subordinate members.

> **linear dominance hierarchy** An organized social system in which an individual is dominant to each individual below and subordinate to all above it in rank.

Dominance hierarchies and crayfish

Crayfish are small freshwater crustaceans that feed on a wide variety of foods, including living and dead animals and plants (Figure 15.13). They are relatively sedentary and live and feed on stream bottoms. When aggregated in high densities, they often fight over resources and form dominance hierarchies. In contests over resources, crayfish exhibit both aggressive and submissive behaviors. Aggressive behaviors include approach, in which crayfish move toward an opponent; threat display, in which they spread their claws; and attack, in which they move quickly toward an opponent with claws open. Submissive behaviors include retreat and escape. A retreating individual moves slowly away, whereas an escaping individual makes a rapid movement away from an opponent.

Figure 15.13. Crayfish. Crayfish live and feed on stream bottoms.

Jens Herberholz, Catherine McCurdy, and Donald Edwards studied the benefits of a social dominance hierarchy in juvenile crayfish (*Procambarus clarkii*) (Herberholz, McCurdy, & Edwards 2007). Because crayfish do not live in permanent social groups, the research team first investigated whether crayfish form dominance hierarchies in the absence of a resource. Second, they asked whether the formation of a hierarchy reduces aggression and determines access to food.

The research team conducted a simple but elegant experiment. They placed three individuals, uniquely marked with dots of color on their carapace, into an aquarium (15 cm × 30 cm × 20 cm) that was divided into two compartments by an opaque partition. Individuals were placed in the larger compartment, and all agonistic interactions were recorded for 30 minutes in the absence of any resource. Next, a large food item (a piece of chicken liver) was placed in the other compartment, and the opaque partition was lifted. The researchers recorded the number of aggressive and submissive behaviors exhibited by each individual, both in the absence and in the presence of the food, to calculate a dominance index (defined as the number of aggressive acts divided by the total number of aggressive plus submissive acts). They also recorded the proximity of each individual to the food to determine the individual's access to it. This experiment was repeated for ten sets of three crayfish.

In each trio, crayfish displayed high levels of aggression. In the first part of the experiment and in the absence of a resource, aggression levels were high and a

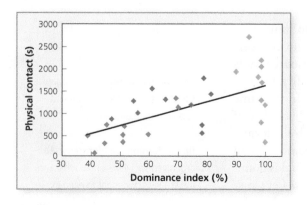

Figure 15.14. Dominance status. The most dominant alpha males (orange) tended to spend the most time in contact with food. Beta individuals (gray) spent more time in contact with food than the lowest-ranking gamma individuals (blue). (*Source:* Herberholz, McCurdy, & Edwards 2007)

linear dominance hierarchy was established. The researchers identified a top-ranked alpha individual that possessed the highest dominance index, a second-ranked beta individual, and a third-ranked gamma individual that possessed the lowest dominance index. However, once the dominance hierarchy had been established, aggression levels decreased by 80% in the second part of the experiment, even though food was present. As expected, dominance rank strongly affected access to the food. Individuals with higher ranks spent more time in contact with the food than did lower-ranked individuals (Figure 15.14). The researchers concluded that dominant crayfish have greater access to resources than subordinate crayfish, and that the formation of a dominance hierarchy significantly reduces aggressive interactions among individuals. Once relative dominance ranks are formed, there is little to gain from continued aggression.

Featured Research

Stable dominance hierarchies in baboons

Aggressive interactions are a source of stress, and **glucocorticoids** are hormones that are secreted by the adrenal gland in vertebrates in response to stressful situations. Chronic stress is associated with altered immune system function and reduced reproduction and thus negatively impacts fitness (Sapolsky 2005). If dominance hierarchies function to reduce aggression within groups, they should also lessen stress. Conversely, any disruption in the dominance hierarchy could result in increased stress levels.

glucocorticoids Hormones secreted by the adrenal gland in vertebrates in response to stressful situations.

Thore Bergman and his colleagues investigated whether an unstable dominance hierarchy could result in increased stress in chacma baboons (*Papio hamadryas ursinus*) (Figure 15.15) (Bergman et al. 2005). For over 14 months, the team studied a population in the Moremi Game Reserve in Botswana. This population had been intensively studied for decades (e.g., Bulger & Hamilton 1987), and so the researchers knew that its males form a strict linear dominance hierarchy in which alpha males have sole access to fertile females. However, this population also lives near other groups (which allows males to easily move between groups), and so changes in rank are common, with the identity of the alpha male changing, on average, about every six months. The research team conducted over 1,000 ten-minute scans to record male aggression and collected 482 fecal samples to measure glucocorticoid levels (using a radio-immunoassay procedure) (Beehner & Whitten 2004). Each sample was time-coded, and at least seven samples were collected for each male.

The researchers divided the study into time periods in which the dominance hierarchy was either stable (times with no change in alpha or beta males) or unstable (times during which the alpha or beta male changed). Over the 14-month study, there were five periods of instability. These periods of instability were interspersed with two stable hierarchy periods of two to three months each.

Instability strongly increased aggression among males, whereas aggression remained below average during stable periods (Figure 15.16). Male

Figure 15.15. Chacma baboons. These Old World primates are found throughout southern Africa.

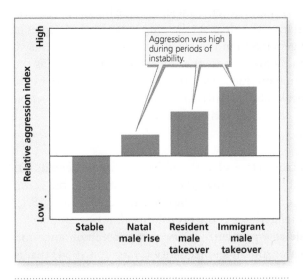

Figure 15.16. Male aggression levels. Male aggression was low when the dominance hierarchy was stable. Aggression levels were higher during periods of instability in the group dominance hierarchy. (*Source:* Bergman et al. 2005)

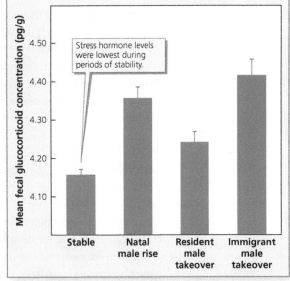

Figure 15.17. Stress hormone levels. Mean (+ SE) male fecal glucocorticoid levels were lowest (indicating low stress) during stable periods. (*Source:* Bergman et al. 2005)

fecal glucocorticoid levels varied with dominance stability, just as predicted. Stress hormone levels were lowest during stable periods but were significantly higher during each period of instability (Figure 15.17). The increase in glucocorticoid levels during periods of instability occurred in both high- and low-ranking males. Therefore, the stable dominance hierarchy has important fitness consequences for all males in the group.

So far, we have examined various benefits and costs of social living and have seen how the formation of dominance hierarchies within groups can reduce one important cost: aggression. Another way that group members can benefit by living with others is by receiving their help, a common behavior in many species. How can natural selection favor cooperative behavior when helpers appear to gain nothing and the act of helping may incur a cost in fitness? We examine how evolutionary biologists explain this behavior in the next section.

15.3 Kin selection favors cooperation among relatives

Cooperative behavior is any behavior by an individual, the actor, that increases the fitness of another individual, the recipient. In many cases, cooperation entails a fitness cost to the individual performing the behavior: this is known as **altruism**. These costs can include the time and energy associated with the behavior, the inability to perform other fitness-enhancing behaviors such as feeding, and possible increased predation risk. Cooperative behavior is only expected to evolve when, on average, the benefits to the actor exceed the costs.

The benefits of cooperation are often indirect; they are not obtained immediately and so can be difficult to identify. The first explanation for altruistic behavior via indirect benefits traces its roots back to Darwin (1859) but was formalized by William Hamilton (1964). Imagine that an individual possesses an allele that enhances the likelihood of being altruistic. If that individual helps another individual reproduce, and if the two individuals are genetically related, there is some

cooperative behavior A mutually beneficial interaction between individuals.

altruism A behavior that results in the increased fitness of another individual and involves a cost to the individual performing the behavior. Also known as helping behavior.

probability that both individuals carry the same allele, because they share genetic relatives. Thus, by helping a relative, the altruist is helping pass identical copies of its own alleles on to the next generation indirectly. It has increased its inclusive fitness, the sum of an individual's own fitness (i.e., offspring produced) plus the fitness derived from the offspring produced by relatives that exist because of the cooperative behavior of the focal individual. Recall from Chapter 3 that selection for behaviors that increase the fitness of relatives is called kin selection.

Hamilton's rule

Hamilton proposed that for altruism to evolve, the additional fitness benefits (B) obtained by the recipient must be greater than the fitness cost (C) to the actor, adjusted for the degree of relatedness (r) between the two individuals (Table 15.1). The degree (or coefficient) of relatedness is the probability that two individuals have inherited the same allele at a particular locus from a common ancestor. Hamilton formalized this relationship in an inequality (Hamilton 1964) that is now known as **Hamilton's rule**:

> **Hamilton's rule** An inequality predicting when altruism should evolve based on the fitness benefits and costs of the altruistic behavior and the degree of relatedness among individuals.

$$\text{Altruism can evolve when } B \times r > C$$

This inequality makes a prediction about when altruism can evolve, but the fitness benefits and costs and the degree of relatedness can be difficult to estimate. Many tests of Hamilton's rule therefore examine a more general prediction of kin selection: helping behavior should be most common between close relatives (when r is relatively high), less common between distant relatives (when r is low), and rare between unrelated individuals (when r is 0). Let's examine one classic test of this prediction.

Featured Research

Belding's ground squirrel alarm calls

Paul Sherman conducted one of the first studies of kin selection by studying alarm-calling behavior in Belding's ground squirrels (*Spermophilus beldingi*) (Sherman 1977). These rodents live in high-density populations and interact frequently in subalpine meadows in western North America (Figure 15.18). Adult males and females live in very different social environments. Females tend to be

TABLE 15.1 Degree of relatedness. Coefficient of relatedness (r) for different relationship types among pairs of animals (e.g., humans).

RELATIONSHIP	COEFFICIENT OF RELATEDNESS (r)
Parent–offspring	0.50
Sibling–sibling	0.50
Grandparent–grandchild	0.25
Aunt/uncle–nephew/niece	0.25
First cousins	0.125
Unrelated individuals	0

Figure 15.18. Belding's ground squirrels. A social mammal that lives in large colonies.

surrounded by female relatives, because juvenile females disperse very short distances from their natal burrow. In contrast, males tend to be surrounded by unrelated squirrels, because juvenile males disperse more than 250 m from their natal burrow, and adult males move an additional 150 m away from the burrow of their mate when the offspring are born.

Ground squirrels are attacked by a variety of predators and produce specific vocalizations, or alarm calls, in response to predators such as long-tailed weasels (*Mustela frenata*) and badgers (*Taxidea taxus*). Alarm calls benefit others, which when they hear them will typically run to the nearest burrow or scan the environment, behaviors that reduce their risk of predation. Alarm calls also entail a risk: producing a vocalization can attract a predator to the caller. Can kin selection explain the evolution of alarm-calling behavior in ground squirrels?

Sherman studied one population over three years. The population contained hundreds of squirrels that had been individually marked. Sherman recorded 102 occasions when a predator was present while squirrels were active: 22 marked squirrels were subsequently stalked or chased. He also noted the first squirrel to produce an alarm call.

He found that adult females were significantly more likely to give the first alarm call than were adult males or juveniles (Figure 15.19). In fact, adult males rarely gave the first alarm call when a predator was present. Sherman also noted that individuals paid a fitness cost of calling: significantly more calling squirrels were hunted by predators than were noncallers, and more callers were killed. Three that were killed gave an alarm call just before being attacked.

Sherman concluded that kin selection can explain the observed variation in alarm-calling behavior because adult females, and not males, are surrounded by close relatives. If alarm calls help increase the survival of close relatives, this behavior could provide genetic benefits to the female callers. Alarm calls are costly to produce because individuals who give them suffer higher predation risk than noncallers, and so adult males, who are not surrounded by close relatives, rarely produce these calls.

These results demonstrate that kin selection can explain variation in helping behavior when animals face predation risk. Next we examine helping behavior in the context of reproduction.

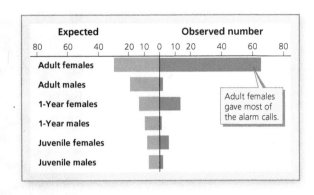

Figure 15.19. Ground squirrel alarm calls. Observed (blue) and expected (orange) number of first alarm calls given based on the numbers of different individuals in each age class in the population. Adult females gave significantly more calls than expected, whereas adult males gave significantly fewer calls than expected. (*Source:* Sherman 1977)

Altruism in turkeys

Featured Research

In the preceding example, helping behavior was most common among relatives and rare between unrelated individuals, as we should expect based on kin selection. It seems a daunting task to measure all the parameters of Hamilton's rule to more precisely test its prediction. However, Alan Krakauer attempted just that in turkeys (*Meleagris gallopavo*).

These large birds (over 1 m in length) live in open fields and woodlands throughout much of North America. Turkeys often form male display partnerships in which a pair of males, called a coalition, displays together to court and defend females against other males (Figure 15.20). However, only one of the two males in a coalition, the dominant one, mates. Why does the subordinate male display with another male when this leads to no direct reproductive fitness benefit? One possibility is that males are closely related and the subordinate individual obtains indirect fitness benefits via kin selection (Watts & Stokes 1971).

Figure 15.20. Male turkeys. Males form coalitions to attract and defend mates.

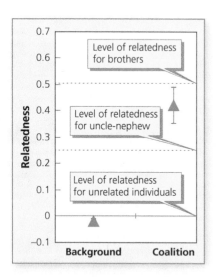

Figure 15.21. Relatedness in turkey coalitions. Males in coalitions are closely related ($r = 0.42$), while random (background) pairs of males in the population are unrelated. Data are mean ± SE. (*Source:* Krakauer 2005)

Krakauer tested this idea by studying a population at Hastings Natural History Reservation near Carmel, California (Krakauer 2005). He captured and uniquely marked 126 turkeys with colored wing tags and collected blood samples to conduct DNA analysis. DNA was genotyped to determine both relatedness among individuals and paternity of offspring, with the latter allowing him to quantify the reproductive success of males. In addition, dozens of birds were fitted with radio transmitters to facilitate rapid location of birds and their nests.

Turkeys were located regularly to observe courtship behavior and associations among males. Coalitions were defined as pairs of males within 2 m of each other that displayed at least twice to a female. The dominant male was identified as the turkey that performed the most complete mating display, strutting. Solitary males, on the other hand, were those adults that did not meet these criteria and were observed displaying alone to a female at least twice.

The combination of behavioral observations and genetic data allowed Krakauer to calculate all three parameters in Hamilton's rule: r, B, and C. He determined that coalitions were composed of very close relatives. The mean degree of relatedness, r, between individuals was 0.42, slightly below the value for full brothers (0.50) but significantly higher than the value for any two random males in the population (Figure 15.21).

Krakauer found that in coalitions, subordinate males fathered no offspring, but a dominant male fathered, on average, 7.0 offspring. In contrast, solo males sired, on average, only 0.9 offspring (Figure 15.22). We can now plug these results into Hamilton's rule. First, the additional fitness benefit, B, that the dominant obtained from the cooperation it received from a subordinate was (7.0 − 0.9), or 6.1 offspring per male. The fitness cost of helping, C, was 0.9, the number of offspring that the subordinate could achieve as a solo male. Finally, the average relatedness, r, between coalition members was 0.42. In sum,

$$B \times r = 6.1 \times 0.42 = 2.56 > 0.9 = C$$

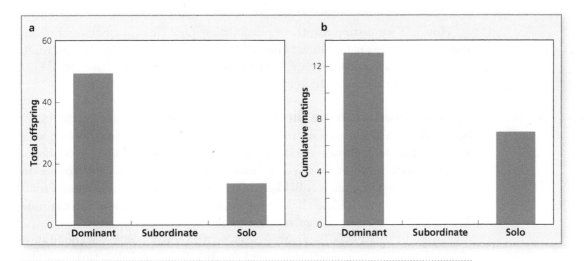

Figure 15.22. Reproductive success in male turkeys. (a) Dominant males sired more offspring and (b) obtained more matings than subordinates and solo males. Note that subordinate males had no direct reproductive success. (*Source:* Krakauer 2005)

Hamilton's rule is satisfied for subordinate males: $B \times r$ is in fact greater than C, and the fitness of altruistic subordinates is higher (thanks to kin selection) than the fitness of nonaltruistic solo males. In this case, subordinate males indeed obtain higher fitness by forming a coalition with and helping a close relative reproduce than they would if they attempted to reproduce on their own.

As these examples illustrate, kin selection is an important concept for understanding the evolution of cooperative behavior among close relatives in a variety of species. For kin selection to explain cooperation, however, actors must bias their cooperative behavior toward their kin. How do individuals discriminate kin from non-kin? We address that question next.

Kin discrimination

Hamilton's rule tacitly assumes that individuals can discriminate kin from non-kin so that altruistic behaviors can benefit relatives. Numerous studies show that individuals, when given a choice, prefer to associate with kin over non-kin in a nonmating context but display the opposite preference when seeking mates (e.g., Ward & Hart 2003; Widdig 2007; Gherardi, Aquiloni, & Tricarico 2012). These studies demonstrate that animals have the ability to discriminate kin from non-kin. But how?

Two mechanisms are thought to provide important explanations for kin discrimination ability, and both involve learning. First, an individual can learn to discriminate kin from non-kin through previous associations, a process called **direct familiarization** (Tang-Martinez 2001). In many species, siblings are raised together in a nest or remain close to their parents for an extended period and so have the opportunity to learn the phenotypes of their close relatives (both their parents and their siblings). Alternatively, individuals could discriminate kin from non-kin via phenotype matching, or **indirect familiarization**. In this process, individuals learn a reference phenotypic cue from themselves or a known close relative (like a parent). One such cue might be a chemical odor profile. Once learned, an individual then assesses the degree of similarity of the learned cue to that cue in others: more closely related individuals should have a higher degree of similarity in the cue.

direct familiarization The ability of individuals to discriminate kin from non-kin through previous associations.

indirect familiarization The discrimination of kin from non-kin via phenotype matching.

How can we distinguish these possibilities? One clear distinction between these mechanisms is that direct familiarization requires prior association while indirect familiarization does not. Let's examine two studies that have tested how individuals discriminate kin from non-kin.

Kin discrimination via direct familiarization in sticklebacks

Schooling with kin often leads to lower levels of aggression than schooling with non-kin (Brown & Brown 1993). Not surprisingly, several studies have shown that fish such as three-spined sticklebacks (*Gasterosteus aculeatus*) prefer to school with kin (Frommen & Bakker 2004; Ward & Hart 2003). Joachim Frommen, Corinna Luz, and Theo Bakker conducted a simple experiment to examine how sticklebacks might recognize kin (Frommen, Luz, & Bakker 2007).

They reared two sets of juveniles in different conditions: one set was raised for five weeks with kin (full siblings) only, while another set was raised for five weeks with both kin (full siblings) and non-kin. In the first experiment, the researchers placed a single kin-reared test fish in the central compartment of an aquarium. Two stimulus schools were present on either side of the test tank, which allowed both visual and olfactory information to be shared between the central and side compartments. One school contained seven kin, while the other contained seven unrelated individuals, and the research team recorded how much time the test fish spent near each school. The findings showed that individuals spent significantly more time on the side that contained familiar kin, a hallmark of kin discrimination.

In a second experiment, using the same setup, test fish reared with both kin and non-kin were presented with one stimulus school of familiar kin and one stimulus school of familiar non-kin. In this experiment, test fish showed no preference for schooling with kin (Figure 15.23). This finding suggests that sticklebacks use direct familiarization based on previous association when selecting shoalmates. In nature, sticklebacks spend several days in a nest after hatching, a time in which they may learn their siblings' phenotypes.

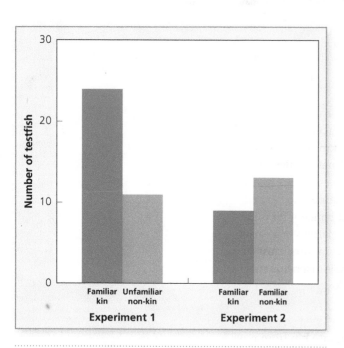

Figure 15.23. Kin discrimination experiment in sticklebacks. In Experiment 1, more test fish preferred to associate with familiar kin over unfamiliar non-kin. In Experiment 2, test fish showed no significant preference between familiar kin and familiar non-kin. (*Source:* Frommen, Luz, & Bakker 2007)

This study suggests that sticklebacks become familiar with other individuals at an early age based on prior association with them, but it does not reveal how they identify those individuals. Test fish had access to both visual and chemical information from the stimulus shoals, so either factor could potentially explain kin recognition. The researchers postulate that sticklebacks use olfactory information to discriminate, given the sophistication of this sensory system in fish (Chapter 5), but this hypothesis awaits testing.

The next study focuses more directly on olfactory information as the basis for kin discrimination in an insect.

Kin discrimination via indirect familiarization in cockroaches

Many insects live in social groups with close relatives and can discriminate kin from non-kin (Gherardi, Aquiloni, & Tricarico 2012). Kin discrimination in insects is thought to involve assessment of chemicals found on an individual's body surface, or cuticle. One diverse class of compounds that may play a role in individual discrimination is cuticular hydrocarbons (CHCs). These hydrocarbons range in chain length from one to several dozen carbon atoms, and the number and relative proportion of different cuticular hydrocarbons present (known as a CHC profile) can vary greatly among individuals in a population (Singer 1998).

Mathieu Lihoreau and Colette Rivault investigated kin discrimination ability in nymphs of the urban cockroach (*Blattela germanica*) (Figure 15.24) (Lihoreau & Rivault 2009). In this nocturnal species, females mate once and produce repeated batches of full siblings that develop through six nymphal instars (distinct developmental stages between molting periods). Offspring often share a shelter to reduce physiological costs related to water loss and usually remain in their hatching area for up to several weeks while they mature. A total of 25 CHCs have been identified in this species, and CHC profiles of siblings are more similar to one another than they are to those of non-kin (Lihoreau & Rivault 2009), suggesting that they potentially play an important role in kin discrimination.

In three experiments, Lihoreau and Rivault investigated individuals' choice of social partners. In the first experiment, they reared nymphs in groups of siblings. To test nymph social preference, they placed individual nymphs in the center of a Petri dish that contained two potential resting sites (small cylinders). One resting site contained 15 sibling nymphs that had been reared with the test individual, and the other contained 15 nonsibling nymphs. In the second experiment, the same conditions were imposed, but each test subject had been reared in isolation. And in the third experiment, one site contained filter paper scented with CHC extracts of siblings, while the other contained filter paper scented with CHC extracts of non-kin. In each experiment, a fine mesh covered the cylinder but allowed the test nymphs to have antennal contact with either the individuals or the filter paper in each potential resting site. The research team recorded the cylinder that each test nymph settled on (i.e., its choice of resting site).

In each experiment, a significantly higher number of nymphs selected the site that contained either kin or the CHC extract of their kin (Figure 15.25). Because individuals without prior association with kin also preferred to associate with kin, and because nymphs preferred to associate with sibling CHC extract over non-kin CHC extract, these experiments demonstrate kin discrimination via indirect familiarization based on CHC profiles. Nymphs apparently use the CHC profiles of individuals to discriminate kin from non-kin and do so even with no prior association. How this is done, however, is still not clear.

Up to this point we have examined cooperative behavior largely outside of reproduction. In the next section, we focus on the evolution of cooperation in the context of reproduction.

Figure 15.24. Urban cockroach. An insect that lives in social groups with close relatives.

Figure 15.25. Kin discrimination in cockroaches. In all experiments, individuals preferred to associate with kin (Experiments 1 and 2) or with their kin's CHC extract (Experiment 3). (*Source:* Lihoreau & Rivault 2009)

15.4 Kinship and ecological constraints favor cooperative reproduction

In many species, individuals in groups cooperatively raise offspring. In such groups, there can be tremendous variation in the relative reproductive success of adults, measured as **reproductive skew**. In groups with low reproductive

■ **reproductive skew** Variation in reproductive success among members of a group.

skew, all adults produce similar numbers of offspring. However, many groups exhibit high reproductive skew because breeding is restricted to a few adults, often just a single pair, and all other adults defend group resources and care for the offspring.

How can we explain the evolution of cooperative behavior in groups with high reproductive skew? We address this question first in vertebrates and then in social insects.

The evolution of cooperative breeding in vertebrates

In hundreds of species of birds, mammals, and fish, individuals spend time and effort caring for the offspring of others, a phenomenon known as **cooperative breeding** (Clutton-Brock 2002). These helpers do not breed themselves but instead provide parental care in the form of incubation, food provision, territorial defense, and antipredator behavior. Helpers are clearly exhibiting altruistic behavior: they pay the energetic costs associated with parental care, sacrifice the chance to breed themselves, and enhance the fitness of others.

Two important factors explain the evolution of nonreproductive helpers. First, in many species, helpers are closely related to the breeding pair; often, they are older offspring of that pair. As such, they are helping relatives and so accrue indirect fitness benefits via kin selection. A second factor that can explain this care behavior is the existence of ecological constraints. Attempting to breed can entail considerable risk based on environmental conditions. For instance, there may be a lack of available breeding sites or very high predation risk when searching for a suitable breeding site. If few breeding sites are available in the environment, or if it is risky to leave a group in order to attempt to breed, it might be more beneficial for an individual to help a breeding pair raise their offspring (Emlen 1982). As we saw at the beginning of the chapter, belonging to a group provides numerous fitness benefits to individuals. Opting to stay in a group to help raise the offspring of others avoids the potential high cost of attempting to find and establish a new breeding territory. In essence, individuals can obtain higher fitness by waiting for better conditions in the future, when new breeding sites or opportunities may become available.

Let's examine cooperative breeding examples in mammals, birds, and fish in which the relative importance of kinship and ecological constraints vary.

Featured Research ▶ ## Cooperative breeding in meerkats

In many social carnivores such as meerkats (*Suricata suricatta*), all adults help protect and feed offspring, even though most of the offspring are produced by a single pair (Figure 15.26). Why? In this species, groups typically contain five to 14 adults and their offspring. The dominant male and female sire about 80% of the pups, and so subordinates do sire some offspring and thus receive direct fitness benefits. In addition, some nonbreeding adults are related to the dominant female and so receive indirect fitness benefits by helping her (Clutton-Brock et al. 1999). Finally, there is a positive relationship between a group's size and the foraging success and survival of adults. Therefore, by helping raise offspring, nonbreeding adults also help increase the size of the social group, which yields fitness benefits to themselves and all group members. This last benefit assumes that helpers contribute to the effort to enhance survivorship of pups so that group size will increase. Is that true?

Tim Clutton-Brock and his colleagues studied 11 groups of meerkats over five years (Clutton-Brock et al. 2001). Across groups and over time, there was much

cooperative breeding In social groups, adults physiologically capable of reproducing forgo breeding and instead help others raise offspring.

Figure 15.26. Meerkat helpers. Adults care for all young in a social group.

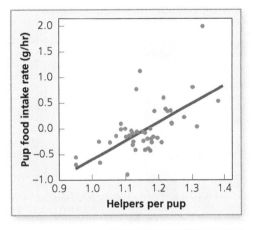

Figure 15.27. Meerkat pup food intake. As the number of helpers per pup increased, pups had higher food intake. (*Source:* Clutton-Brock et al. 2001)

variation in the number of helpers relative to the number of pups in a group. In each group, the research team quantified the food received by each pup from all group members for three months. They found that as the ratio of helpers per pup increased, so did the average feeding rate and daily weight gain of pups: more helpers meant more food delivered to each pup (Figure 15.27). Those that had higher feeding rates also had higher survival rates. This increase in pup survival can ultimately enhance total group size because pups stay with their natal group for up to three years. In this species, a combination of direct and indirect benefits favors the evolution of cooperative reproduction.

Besides mammals, cooperative breeding has been observed in about 100 species of birds. Let's now examine two cases to see how kin selection and ecological constraints have favored cooperative reproduction.

Helpers at the nest: cooperative reproduction in long-tailed tits

Featured Research

Andrew Russell, Ben Hatchwell, and colleagues studied cooperative reproduction and helping behavior in long-tailed tits (*Aegithalos caudatus*) in Great Britain (Figure 15.28) (Russell & Hatchwell 2001; Hatchwell et al. 2004). In spring, monogamous pairs form and begin reproductive behavior (nest building, egg laying, and incubation). However, many nests are destroyed by predators, and when that occurs, the nest builders are unlikely to have sufficient time to successfully raise offspring in the current breeding season. Many failed breeders become helpers by joining with another pair and feeding their offspring. How might this behavior have evolved?

The team captured and uniquely marked individuals in three populations over the course of several years. The researchers used a combination of known pedigrees and genetic data obtained from blood samples to determine the degree of relatedness among individuals in the study populations. They visited all nests regularly and recorded the success or failure of each nest, all instances of helping behavior, and the long-term survivorship of nestlings, as determined by recruitment into a social group the following year.

Figure 15.28. Long-tailed tits. A social bird common in Europe.

Figure 15.29. Helping behavior and relatedness in long-tailed tits. Long-tailed tits helped pairs that contained at least one close relative (kin) more than they helped unrelated pairs (non-kin). (*Source:* Russell & Hatchwell 2001)

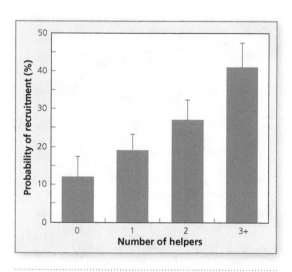

Figure 15.30. Long-tailed tit fledgling success. Mean (+ SE) probability of recruitment into the adult population as a function of the number of helpers at the nest. The probability of fledgling recruitment into a social group the following year increased with the number of helpers at the nest. (*Source:* Hatchwell et al. 2004)

The researchers observed 37 birds they classified as helpers at the nest of a pair. In all cases, these were birds whose own nesting attempt had failed. Nearly all helpers assisted a pair that contained at least one relative: in over 90% of cases, this individual was a close relative, such as a sibling (Figure 15.29). In addition, the number of helpers at a nest positively affected the probability that a nestling would survive and be recruited into the population the following year (Figure 15.30). These data demonstrate that helpers do in fact feed the young of a close relative and do so after their own attempt at reproducing has failed.

Featured Research

Habitat saturation: helping behavior in Seychelles warblers

The Seychelles warbler (*Acrocephalus sechellensis*) is another cooperative breeding species with helpers at the nest. Unlike the long-tailed tit, however, helpers in this species are not failed breeders. Instead, they are offspring from previous clutches that have not yet dispersed. Because the helpers are closely related to the parents and offspring they care for, they gain indirect fitness benefits via kin selection. But why do they stay to help instead of attempting to breed on their own?

One possibility is that there are no available territories for them to disperse into. The **habitat saturation hypothesis** asserts that when few territories are available, a breeding attempt has a low probability of success and likely entails a very high cost in terms of reduced survivorship (Emlen 1982). Individuals can have higher fitness by staying in an established group to gain the associated fitness benefits of sociality and can also obtain indirect fitness benefits by helping to raise siblings. In essence, these individuals are waiting for a better opportunity to breed in the future if territories become available.

Jan Komdeur tested the habitat saturation hypothesis in this species by observing the breeding behavior of 29 translocated Seychelles warblers. These birds were transferred to an island with abundant suitable habitat to initiate a new population (Komdeur et al. 1991; Komdeur 1992). The translocated adults quickly established territories and began to breed. Interestingly, none of the offspring

habitat saturation hypothesis A proposed explanation for cooperative breeding based on limited breeding options for helpers.

born on the new island remained at their natal nest to become helpers, and juveniles left their natal nest at four months of age to begin breeding. In contrast, juveniles on the original island, where population density was high, remained at their natal nest for over 23 months and often became helpers. In this species, the decision by juveniles to stay at their natal nest and help appears to be driven by a lack of breeding opportunities available and kin selection. How juveniles determine the availability of open territories remains to be studied.

In the above examples, both kin selection and ecological constraints appear to have played an important role in the evolution of cooperative breeding. Next, let's examine a species in which only ecological constraints appear to favor cooperative reproduction among groups of unrelated individuals.

Social queuing in clownfish

In many permanent groups, individuals exhibit a dominance hierarchy: subordinates do not breed but instead help in defending the group's territory. Over time, subordinates can move up the social ladder as they age and as higher-ranking individuals suffer mortality, a process called **social queuing** (Kokko & Johnstone 1999). Social queuing provides a delayed direct benefit of helping, because subordinates can eventually inherit a territory and its breeding opportunities (McDonald 1993; Field, Shreeves, & Sumner 1999; Kokko & Johnstone 1999; East & Hofer 2001).

Peter Buston studied social queuing in clownfish (*Amphiprion percula*). These social reef fish live in and around sea anemones, which keep predators at bay (Figure 15.31). Sea anemones are cnidarians: they lack a hard skeleton and possess stinging tentacles that inject a neurotoxin into fish that come into contact with them, an adaptation that is effective in capturing prey. Clownfish have an outer mucus layer that makes them immune to the anemones' sting. These fish live in small groups that consist of a breeding pair and up to four unrelated nonbreeders. Adults are protandrous: on maturation, they develop into males but

social queuing A process in which subordinates eventually move up the social ladder as they age.

Figure 15.31. Clownfish and sea anemones. Clownfish groups live in association with anemones. Note the size difference among group members.

later can develop into females. Within each group, there is a size-based dominance hierarchy in which the breeding female is the largest and most dominant fish, followed by the breeding male and then any nonbreeders. Nonbreeders do not have functioning gonads and so obtain no immediate reproductive benefits from the group, but they often help defend the territory. Why, then, do they stay in a group? Buston predicted that subordinates may obtain future reproductive benefits through social queuing.

To test this idea, he recorded patterns of territory inheritance on three reefs in Papua New Guinea (Buston 2004). Over the course of one year, he observed 97 groups and recorded dominance interactions, the recruitment of new individuals, and changes following the disappearance of one of the breeders. No reversals in rank within the linear dominance hierarchy of groups were observed. Loss of the breeding female always led to a chain reaction: the breeding male changed sex and became the dominant breeding female, and the next-highest-ranking nonbreeder in the group became a sexually mature male. Buston had observed social queuing in action.

Next, he experimentally manipulated 16 groups by removing the breeding male. In each case, nonbreeders moved up the social ladder. The highest-ranking nonbreeder became a sexually mature breeding male, and any new recruit always entered the group as the lowest-ranking member. Buston never observed an outsider usurp a breeding vacancy.

Buston concluded that nonbreeding group members benefit from group membership by eventually inheriting the territory and thus obtaining reproductive success. Clownfish adults are not strong swimmers and are often preyed upon when they leave their anemones. It is risky to venture out in search of breeding vacancies, away from the protection of anemones. This risk may explain why Buston did not observe outsiders taking over a breeding territory. Small larval recruits join existing breeders and always begin at the bottom of the social queue rather than live alone. Biding time within a social queue yields higher fitness for an individual, which helps explain why subordinates remain in clownfish groups even though they are unrelated to the breeders and may receive fewer resources and have fewer mating opportunities than dominant individuals.

To conclude this section, we turn our attention to social insects and examine an extreme form of altruism: the evolution of sterile workers.

Eusocial insects: the evolution of sterile castes

Cooperative breeding also occurs in many invertebrates, often in a nest, hive, or colony. Some cooperatively breeding insects, such as paper wasps, are similar to cooperatively breeding vertebrates because all adults are physiologically capable of reproducing, but some forego reproduction either to help raise close relatives or to wait in a social queue for a breeding opportunity to arise.

But in other species, we see a new phenomenon, the evolution of **castes**—groups of individuals that engage in distinct behaviors. In some species, castes have evolved particular morphologies associated with their different tasks (Figure 15.32): large-bodied soldiers with enlarged forelegs or mandibles, and perhaps even glands that eject toxic chemicals for nest protection (Tian & Zhou 2014); small-bodied workers that tend the nest, feed the young, and search for food; and the very large queen that produces offspring.

In many species, workers and soldiers are sterile: they are physiologically incapable of reproduction. They spend their lives maintaining and defending the nest and helping the queen reproduce. Such **eusocial** species are defined by

castes Morphologically and behaviorally distinct individuals within a social group.

eusocial The social structure of a species that lives in social groups, contains overlapping generations, and has cooperative offspring care by nonparents and a reproductive division of labor.

cooperative care of offspring, reproductive division of labor between breeders and sterile workers, and overlapping generations (Batra 1966; Burda et al. 2000). Eusociality has evolved independently in termites, beetles, aphids, thrips, spiders, and frequently in the insect order Hymenoptera, which includes bees, ants, and wasps. How can such a system of cooperation with sterile workers evolve?

Haplodiploidy hypothesis

Darwin (1859) suggested that relatedness among individuals provided the answer to eusociality. William Hamilton (1963; 1964) focused on this mechanism to explain why eusociality evolved independently multiple times in the order Hymenoptera, in which sterile workers are typically females. This insect order has an unusual mechanism of sex determination known as haplodiploidy: males develop parthenogenetically from unfertilized eggs and so are haploid, while females develop from fertilized eggs and so are diploid. Due to this **haplodiploid genetic system**, females are more closely related to their full sisters ($r = 0.75$) than to their own daughters ($r = 0.5$) (Figure 15.33).

Hamilton suggested that haplodiploidy favors the evolution of sterile daughters because they can achieve higher fitness by helping to raise sisters instead of their own daughters. But this hypothesis cannot be a full explanation for the evolution of sterile workers, because many eusocial species (e.g., termites) have only diploid individuals. In addition, Hamilton ignored the fact that females are related to their full brothers by 0.25 on average, which means they are less closely related to their full brothers than to their own sons ($r = 0.5$). Therefore, for a colony with an equal sex ratio of sons and daughters, the average relatedness among siblings is the same as that between parents and offspring ($r = 0.5$), and so the overall fitness benefits of helping to raise siblings is no higher than that derived from raising offspring (Trivers & Hare 1976; Foster, Wenseleers & Ratnieks 2006).

Figure 15.32. Ant castes. Large soldier and small worker.

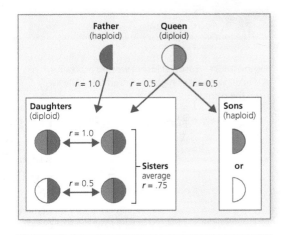

Figure 15.33. Haplodiploidy. Daughters share half of their genome with their mother but inherit all of their father's genome. Sons share half of their genome with their mother and none with their father. Sisters share (on average) three-quarters of their genome.

Kin selection and the ecological constraint hypothesis

The current hypothesis to explain the evolution of sterile workers involves both kin selection and ecological constraints (Alexander 1974; West et al. 2007). With respect to kin selection, all extant eusocial species with sterile castes appear to have descended from ancestor species that exhibited lifetime monogamy in which a single queen mated with a single male, stored his sperm, and used it for the rest of her life to fertilize her eggs (Hughes et al. 2008). As such, all the offspring she produced were full siblings and were related to one another by the same degree, on average, as parents were to their own offspring; the coefficient of relatedness, r, would have been 0.5 in both cases, and this would have been true for both diploid and haplodiploid species (Boomsma 2009). Now consider the behavioral decision faced by potential helpers: leave their natal nest and attempt to produce their own offspring or stay and help to raise siblings. Because the degree of relatedness to their offspring and siblings would have been the same, selection will have favored the latter decision when the net fitness benefit of staying to help parents raise siblings was higher than the net benefit of leaving the natal site in an attempt to produce offspring.

▦ **haplodiploid genetic system**
A genetic system with haploid males and diploid females.

Figure 15.34. Eusocial insect worker mortality. A worker wasp faces high predation risk while away from the colony.

What affects the relative benefits and costs of staying to help versus dispersing to breed? The answer may often depend on the existence of ecological constraints (Emlen 1982; Heg et al. 2004; Queller & Strassman 1998; Ross et al. 2013). Recall that most eusocial species live and reproduce in a nest, hive, or colony. These sites are vigorously defended from threats by competitive rivals (both conspecifics and heterospecific species), and colonies are best defended by many workers and soldiers, which means that small incipient colonies are likely to have low fitness because they will be poorly defended. In addition, venturing outside the nest to forage is risky: many workers are killed by predators (Figure 15.34) (e.g., Schmidt-Hempel & Wolf 1988). And finally, there may be few opportunities in an environment to establish a new colony because of habitat saturation. Together, these three conditions can result in higher fitness for an individual that foregoes any attempt at reproducing and instead helps to raise close relatives. If these conditions existed for ancestral social insects, they could favor the evolution of eusociality.

In the final example of this section, we examine one factor thought to promote the evolutionary success of social insects: the division of labor among caste members.

Featured Research

Division of labor hypothesis in ant castes

Eusocial insects exhibit wide variation in caste structure, ranging from species in which all individuals are morphologically identical to species in which workers vary tremendously in their size and morphology. Why? The division of labor hypothesis asserts that colony fitness is enhanced when individuals specialize on different functions. If these functions are best performed by workers of different size, selection can favor the evolution of workers that occur in distinct size castes. Is there any support for this hypothesis?

Many studies of foraging have found evidence of division of labor among different-sized workers within ant colonies. For instance, in granivorous ants, there is often a positive correlation between worker head width size and the size of the seeds it brings back to the nest (e.g., Kaspari 1996), suggesting that individuals concentrate their foraging on particular seeds. Other studies have indicated that smaller workers are more efficient at finding seeds, while larger workers are more efficient at transporting them back to the colony (Arnan et al. 2011). Such studies suggest that a division of labor occurs, but does it enhance colony fitness?

Stanford Porter and Walter Tschinkel conducted a simple experiment to test the division of labor hypothesis in fire ants (*Solenopsis invicta*) (Porter & Tschinkel 1985). In this species, smaller workers (less than 0.75 mm head width) specialize on brood care, while larger workers (greater than 1.15 mm head width) appear to specialize on nest defense, foraging for insects, and food storage. Medium-sized workers (between 0.75 and 1.15 mm head width) are more generalized in their behavior. Variation in worker size also varies with colony age and size: older, larger colonies tend to have a greater range of worker size classes. Porter and Tschinkel asked whether worker size variation enhances colony fitness.

They created four types of experimental colonies in the laboratory that varied in worker size and variation. Three colony types had monomorphic workers—workers were either all small, all medium, or all large in size. Control colonies were polymorphic: they contained workers of all three sizes. All colonies were standardized for age and total biomass of workers so that colonies of only small workers contained more individuals than colonies of only large workers; polymorphic colonies and colonies of only medium-sized workers contained an intermediate number of workers. A total of seven replicate colonies of each type were established. Each colony was given unlimited food, and the researchers quantified the production of brood eggs by each colony queen over a 40-day period as a measure of colony fitness.

Colonies varied tremendously in brood production. Polymorphic colonies had the highest rate of brood production (similar to that of medium-sized worker colonies), while colonies of only small and only large workers had low rates of brood production (Figure 15.35). In this controlled experiment, colonies with variation in worker size had a fitness advantage over monomorphic colonies of small or large size, providing support for the division of labor hypothesis. The researchers suggest that colonies of only large workers had the lowest brood production because such workers typically spend little time tending broods, and because these colonies contained the fewest number of workers. Additional experiments in the field and under conditions of limited resource abundance can provide a more complete test of the hypothesis.

In this section we have seen that cooperative reproduction is widespread across taxa and that kin selection and ecological constraints alone or in combination can favor its evolution. Eusociality is an extreme form of cooperative reproduction in which some adults forego breeding to help raise the offspring of others. Understanding the origin and evolution of eusociality and social castes will continue to be a vigorous area of research in animal behavior.

Kin selection, facilitated by kin discrimination, helps explain cooperation among relatives. But kin selection is only part of the explanation for the evolution of cooperative behavior, because it cannot explain cooperation that occurs among unrelated individuals. We turn to that subject next.

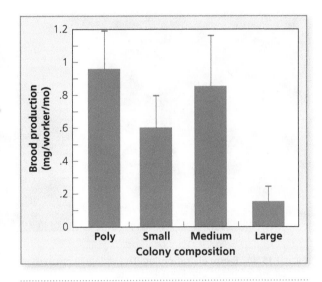

Figure 15.35. Ant colony brood production. Polymorphic colonies with workers of diverse sizes produced broods at higher rates than workers with monomorphic workers. (*Source:* Porter & Tschinkel 1985)

15.5 Cooperative behavior among unrelated individuals involves byproduct mutualisms or reciprocity

Kin selection cannot explain the evolution of cooperation among unrelated individuals. The coefficient of relatedness between such individuals is 0, and so Hamilton's rule predicts that kin selection will not favor the evolution of altruism. How, then, can we explain cooperation among nonrelatives? One possibility is that apparently cooperative behavior is actually selfish. For instance, in many carnivores, nonrelatives often hunt together, and an individual obtains higher fitness (via increased feeding rate) when hunting with others. As a byproduct, all group members also have higher fitness because they can hunt more effectively or on a wider range of prey when they hunt together than they can alone. That is, the decision to hunt with others has a self-serving outcome. In these cases, an individual's behavior can enhance both its own fitness and the fitness of others, a phenomenon known as a **byproduct mutualism**. Natural selection will favor such apparently cooperative behavior.

However, many other forms of cooperation among unrelated individuals involve a fitness cost to the cooperative individual. For example, in many birds, ungulates,

byproduct mutualism When an individual's behavior enhances both its own fitness and the fitness of others.

Figure 15.36. Parrot allogrooming. This harlequin mccaw is removing ectoparasites from an unrelated flockmate.

■ **allogrooming** The grooming by one individual of another individual to remove ectoparasites.

■ **direct reciprocal altruism** An altruistic behavior from which an altruist can benefit if the recipient reciprocates and helps the altruist in the future. Also known as **direct reciprocity**.

■ **tit-for-tat** One solution to the prisoner's dilemma game in which a player always cooperates the first time it interacts with an opponent and then simply matches the behavior of the other player from the previous interaction.

and primates, one individual will often remove ectoparasites from another individual (often a nonrelative), a behavior called **allogrooming** (Figure 15.36). This behavior is altruistic because it benefits the recipient but at the same time entails costs of time and effort for the groomer. How do such behaviors evolve?

Direct reciprocity

Robert Trivers offered one answer: cooperation among unrelated individuals can evolve through **direct reciprocal altruism** or **direct reciprocity** (Trivers 1971), a behavior from which an actor can benefit if the recipient of an altruistic act reciprocates in the future.

Trivers argued that cooperation can evolve between unrelated individuals only when the benefit of helping exceeds the cost. For Trivers, the evolution of direct reciprocal altruism requires two individuals to interact repeatedly so that over many interactions, each individual helps and subsequently is helped. If the benefit received from helping exceeds the cost of helping, individuals that help will obtain higher fitness than nonhelpers through the repeated interaction, and so cooperation can spread in a population.

Trivers also recognized that reciprocal altruism is subject to cheating: an individual that receives help but then does not reciprocate will obtain a fitness benefit but pay no cost. This temptation to cheat can prevent the evolution of helping.

So how can cooperation evolve? Robert Axelrod and William Hamilton proposed one solution called the **tit-for-tat** strategy, a form of reciprocal altruism (Axelrod & Hamilton 1981). Here, a player always cooperates the first time it interacts with a new opponent. In all subsequent interactions with that opponent, it simply matches the behavior of that individual from the previous interaction. When two players use this strategy, they always cooperate on the first encounter (and all subsequent encounters) and so achieve the highest fitness. If a player using tit-for-tat interacts with a betrayer, the player does not cooperate with that individual on all subsequent interactions. If tit-for-tat is common, another strategy cannot take hold in the population (Axelrod & Hamilton 1981).

Given this model, we can outline the conditions required for the evolution of reciprocal altruism: (1) individuals must have the opportunity to interact repeatedly, (2) the fitness benefit of receiving help must exceed the cost of providing help, and (3) individuals must be able to recognize one another so that they can reciprocate the benefit from a past helper and not help a past cheater. Let's examine two tests of the model.

Featured Research

Food sharing in vampire bats

Gerald Wilkinson thought that common vampire bats (*Desmodus rotundus*) (Figure 15.37) might be one species in which the above conditions are satisfied (Wilkinson 1984; Wilkinson 1990). Vampire bats live in relatively stable social groups and have a life span of up to 18 years, providing the opportunity for repeated interactions. A social group, or cluster, living in a tree hollow typically contains a breeding male and eight to 12 breeding females. Many of these females are related, but unrelated females do join clusters. Thus, relatedness between females is high but variable within a cluster.

Vampire bats feed on blood from vertebrates and will share blood meals. Females will often regurgitate blood to their offspring as a form of parental care, but they also share food with other adults. Why? Vampire bats starve if they do not feed on blood for 60 hours, and up to 30% of bats within a cluster fail to find a blood meal each night. Individuals that fail to find a blood meal will often solicit a meal from a successful roostmate through begging behavior. Do unrelated roostmates help each other reciprocally, or is food sharing a form of kin selection?

To answer this question, Wilkinson and Gerald Carter studied a group of 25 captive bats that had been housed together for over two years (Carter & Wilkinson 2013). All bats were uniquely marked and genotyped to determine their degree of relatedness, which varied among different dyads: about 60% were unrelated, while about 20% were closely related ($r > 0.25$).

To quantify food-sharing behavioral interactions, Wilkinson and Carter first withheld food from a randomly selected individual for 24 hours. The researchers then returned the individual to the group of fed bats and recorded its behavioral interactions for two hours, including instances when others shared food with the hungry bat and for how long.

Over 900 food-sharing bouts were observed, and recipients were fed by an average of 3.9 donors. Approximately 65% of food-sharing dyads were distantly ($r < 0.05$) or not related. Individuals were more likely to share food if prior food had been received from the recipient in the past (Figure 15.38), while relatedness was only weakly associated with food donation.

The researchers concluded that the females developed long-term social associations, and that these created a situation for the mutual exchange of food that was largely independent of relatedness. It appears that reciprocity has played a significant role in the evolution of food sharing in this species.

Next, we examine a test of reciprocal altruism that focused on allogrooming behavior in primates.

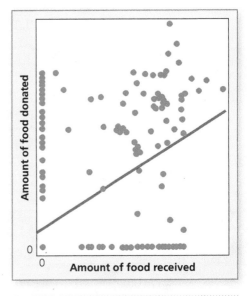

Figure 15.38. Food sharing in vampire bats. Individuals tended to donate food to individuals they had received food from in the past. Values away from the origin represent more food donated and received. Each dot represents one pair of individuals involved in a food-sharing event. (*Source:* Carter & Wilkinson 2013)

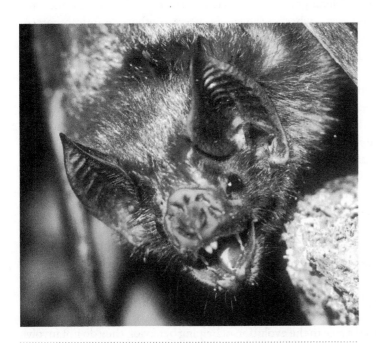

Figure 15.37. Vampire bat. These social species feed on vertebrate blood and sometimes share food with others in their group.

a

b

Figure 15.39. Japanese macaques. Japanese macaques exhibit many social behaviors, including (a) grooming and (b) aggressive interactions.

Allogrooming in Japanese macaques

Gabriele Schino, Eugenia Polizzi de Sorrentino, and Barbara Tiddi studied direct reciprocal altruism in Japanese macaques (*Macaca fuscata*) (Figure 15.39). These macaques, native to Japan, live in large social groups. Two altruistic behaviors occur between adult females. First, macaques frequently exhibit allogrooming. Second, when agonistic interactions occur between group members, adult females will occasionally intervene and provide support for one of the animals. Their support benefits the recipient, which tends not to become injured or lose the encounter. However, such support also takes time, energy, and a risk of injury on the part of the altruistic supporter.

The research team studied a captive population of 23 adult females, all of which were uniquely identifiable, in the Rome Zoo (Schino, Polizzi de Sorrentino, & Tiddi 2007). They used demographic records to determine the macaques' maternal relatedness—whether they were sisters, mothers, grandmothers, or aunts. Over the course of a year, the research team conducted over 1,000 30-minute observations, recording all agonistic interactions, the aggressor and recipient, and any support provided by a third group member. They also conducted one group scan per observation to record allogrooming, noting both the donor and the recipient.

They found that an individual's helping behavior was not correlated with the level of relatedness between individuals, and therefore kin selection could not provide an explanation. Instead, females most often chose to groom others based on the level of grooming they had received from that individual in the past. In addition, females that had received support during an aggressive encounter were more likely to groom the supporting individual (Figure 15.40). Similarly, individual females chose to support others that had supported them in the past and from whom they had received grooming. Reciprocal altruism seems to be the best explanation for allogrooming and agonistic support in this case.

Although these examples suggest that direct reciprocity has played a role in the evolution of helping behavior, they do not involve evidence of the tit-for-tat strategy. Do animals that otherwise cooperate stop helping a cheater? Research on blackbirds suggests the answer is yes.

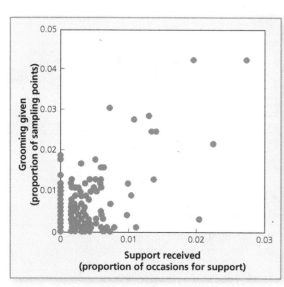

Figure 15.40. Macaque interactions. For females, there was a positive relationship between "grooming given" and "support received" with individuals that had supported them in an aggressive encounter. Each point represents a different dyad. (*Source:* Schino, Polizzi de Sorrentino, & Tiddi 2007)

Tit-for-tat in red-winged blackbirds

Red-winged blackbirds (*Agelaius phoeniceus*) nest in dense aggregations in vegetation that grows at the edge of ponds. Males defend territories and often breed in the same territory for successive years. This behavior allows neighboring males to live side by side for multiple years, which provides the opportunity for repeated interactions. Crows (*Corvus brachyrhynchos*) are common nest predators of blackbirds. When they approach a nesting site, neighboring red-winged blackbirds often engage in cooperative nest defense in the form of mobbing (see Chapter 9) to drive them away (Figure 15.41).

Do unrelated neighboring red-winged blackbirds use a tit-for-tat strategy in cooperative nest defense? Robert Olendorf, Thomas Getty, and Kim Scribner manipulated the cooperative behavior of a neighbor in response to a threatening crow to address this question (Olendorf, Getty, & Scribner 2004). They first identified the territorial boundaries of neighboring males. One male was deemed the focal male and the other the neighbor. Next, they presented a stuffed crow predator at the territorial boundary of pairs of males in three phases of the experiment. In the first phase, the researchers recorded the territory defense behavior of both males in response to the crow. A predator at the territorial boundary represents an equal threat to both territory owners and should encourage cooperative mobbing.

In the second phase, half the focal males were categorized as controls. The others were assigned to the "simulated neighbor defector" treatment. These treatment males were captured and held in a building while the crow was again presented at the territory boundary. During this simulated defection, vocalizations of the defecting neighbor were played from his territory, which provided auditory information that the territory owner was present but not assisting in territorial defense. The researchers again recorded the behavior of focal males during this phase (and for controls, the behavior of both focal and neighboring males). After data collection, the defecting neighbor was released, and he quickly returned to his territory. The next day, during the third phase, the researchers again presented the crow at the territorial boundary of the males and recorded their defense behavior. Territorial defense behaviors were identified as approaches toward the crow, as well as strikes and dives at it. These behaviors were combined into a single value representing overall defense behavior.

During the first phase, focal and neighboring males exhibited similar high levels of defense behavior. In the simulated defection phase, however, focal males whose neighbor had defected exhibited significantly less defense behavior than control males (Figure 15.42). This decrease possibly occurred because the focal males had a higher risk of injury when defending a territory alone against a large predator such as a crow. More important, in the third phase, the reduced level of defense behavior persisted for treatment males whose neighbor had defected compared with the level exhibited by control males. Males whose partner had defected during the simulated defection phase allocated less time and effort toward cooperative mobbing the following day.

These data suggest that male red-winged blackbirds assess the level of cooperation of their neighbor and adjust their level of nest defense accordingly, as described by the tit-for-tat strategy. When both birds engage a predator, cooperative nest defense behavior is high. However, when one partner betrays the other and fails to defend the territory, its neighbor

Figure 15.41. Red-winged blackbird and crow. Crows are common nest predators. Red-winged blackbirds aggressively chase them from their territories during the breeding season.

Figure 15.42. Blackbird tit-for-tat behavior. Mean (± SE) nest defense behavior of experimental (blue and red squares) and control (orange circles) males. In the pre-defection phase, both the (a) focal male and (b) neighbor exhibited similar levels of defense. Both during and after defection, treatment males reduced their level of defense. The neighbor's level of defense was similar in all phases of the experiment. (*Source:* Olendorf, Getty, & Scribner 2004)

immediately reduces its level of defense, and this reduction lasts through a second interaction the following day.

Indirect reciprocity

While the above examples suggest that reciprocal altruism can explain the evolution of cooperation in some species, the temptation to cheat may explain why direct reciprocal altruism has rarely been observed in nature (e.g., Krams et al. 2008). However, Martin Nowak and Karl Sigmund suggested that altruism among unrelated individuals can also evolve through **indirect reciprocity**, in which Individual A helps Individual B because A observed B helping a third individual in the past (Nowak & Sigmund 1998). For indirect reciprocity to emerge, it is crucial that individuals know something about the past helping behavior of other individuals. They must also keep track of others' propensity to help, known as their **image score** or reputation for helping. If A knows that B has helped others frequently in the past, then A will have a high image score for B (Nowak & Sigmund 1998; Nowak & Sigmund 2005). Reciprocity spreads through the population indirectly, because helping individuals are more likely to receive help in the future (Nowak & Sigmund 1998; Nowak & Sigmund 2005; Leimar & Hammerstein 2001).

Is there any evidence that animals use their knowledge of reputations to modify their behavior? Let's examine two studies that suggest the answer is yes.

indirect reciprocity A relationship in which helping individuals are more likely to receive help in the future because they have obtained high image scores by helping others.

image score An assessment of an individual's propensity for helping others based on its observed prior behavior.

Featured Research

Reputations and cleaner fish

Redouan Bshary and Alexandra Grutter examined behavioral interactions between cleaner fish and their clients (Bshary & Grutter 2002; Bshary & Grutter 2006; Grutter & Bshary 2003). Cleaner fish, like the cleaner wrasse *Labroides dimidiatus*, feed on ectoparasites on client fish bodies. However, they actually prefer to feed on client tissue, such as their mucus, because it is high in calories and nutrients (Grutter & Bshary 2003). Client fish approach cleaners to solicit an interaction, and so when cleaners interact with a client, they can "cooperate" by feeding on ectoparasites or "cheat" by feeding on mucus. When a cleaner cheats, clients shake their body as if jolted and then typically chase away the cleaner (Bshary 2002; Bshary & Grutter 2002). Other potential clients can observe these interactions. Bshary and Grutter investigated whether clients form image scores of cleaners based on such observations.

The research team observed interactions between individual cleaners and their clients in the field. For each interaction, they determined whether cleaners cooperated or cheated, based on the client's behavioral response. Did it jolt and chase, or did it swim slowly away? Bshary and Grutter found that when the next client arrived within six seconds, its behavior was affected by the preceding interaction. If the cleaner had previously "cooperated," the next client invited inspection by a cleaner almost all of the time. However, if the cleaner had just "cheated," the next client invited inspection less than 30% of the time (Figure 15.43). These data suggest that client fish observe cleaners and form image scores based on their recent interactions.

Bshary and Grutter further tested this hypothesis in a laboratory experiment that was repeated in two different years (Bshary &

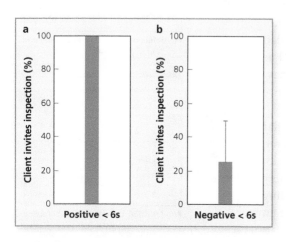

Figure 15.43. Client–cleaner interactions. Mean (+ SE) proportion of interactions in which a client fish invited inspection by a cleaner. Client fish tended to invite interactions with cleaners that had ended (a) a positive interaction within six seconds. They rarely invited interactions with cleaners that had ended (b) a negative interaction within the last six seconds. (*Source:* Bshary 2002)

Figure 15.44. Cleaner–client experimental design. A central focal client observed two cleaners, one at each end of its tank. One cleaner interacted cooperatively with a model (left), whereas the other did not interact with its model (right). (*Source:* Bshary & Grutter 2006)

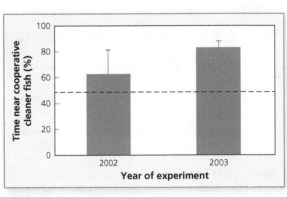

Figure 15.45. Client fish prefer cooperative cleaners. Mean (+ SE) time near the cooperative cleaner. In both years of the experiment, the client fish spent more time near the cooperative cleaner. The dash represents 50%, the null expectation. (*Source:* Bshary & Grutter 2002)

Grutter 2006). A client fish, the two-line monocle beam (*Scolopsis bilineatus*), was allowed to view two cleaner fish, placed at opposite ends of an aquarium, through one-way mirrors (Figure 15.44). One cleaner, the altruist, participated in a cooperative interaction with a model client fish, while the other cleaner did not interact with its model. The researchers found that client fish spent significantly more time near the altruistic cleaner than they did with the unresponsive cleaner (Figure 15.45). Together, these results indicate that client fish can learn about altruistic tendencies by observing cleaners' behavior to form image scores, which then affect the client fish's subsequent behavior.

Chimpanzee image scores

Featured Research

Do other animals form reputations and keep track of image scores? Recent work suggests that some do. Yvan Russell, Josep Call, and Robin Dunbar examined whether captive chimpanzees (*Pan troglodytes*) form image scores of human subjects (Russell, Call, & Dunbar 2008). Chimpanzees witnessed two interactions among human subjects with whom they had had no previous contact. In each interaction, one human subject had food, a bunch of grapes in a transparent container, and the other had no grapes and begged for food. In one interaction, the "nice" person always gave the beggar a grape, but in the "nasty" interaction, the person with grapes never gave the beggar food. Each chimp observed three "nice" interactions followed by three "nasty" interactions, or vice versa. Immediately following these interactions, the chimpanzee subject was moved to a different compartment, where the "nice" and "nasty" people now sat side by side with their box of grapes in front of two Plexiglas windows. The researchers recorded the amount of time the chimpanzee spent at each window (Scientific Process 15.1).

The chimpanzees spent significantly more time in front of the window facing the "nice" person than they did in front of the window facing the "nasty" one.

SCIENTIFIC PROCESS 15.1

Chimpanzee image scoring

RESEARCH QUESTION: *Do chimpanzees learn about the altruistic tendencies of others?*

Hypothesis: Chimpanzees form image scores of others by observing their interactions.

Prediction: Individuals will prefer to associate with individuals that they have observed being altruistic toward others.

Methods: The researchers:

- Allowed a focal individual to observe interactions between humans in a designated human-only area. The "nice" human always gave the beggar grapes, whereas the "nasty" human never gave food to the beggar (Figure 1).
- Recorded the amount of time the focal chimpanzee spent in front of either the "nasty" person or the "nice" person

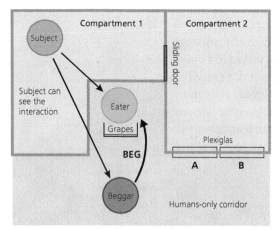

Figure 1. Testing room. Diagram of the experimental testing room. (*Source:* Russell, Call, & Dunbar 2008)

Results: Most chimpanzees (14 of 17) spent more time in front of the "nice" person (Figure 2).

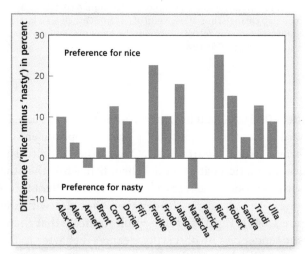

Figure 2. Chimpanzee preference. Most chimpanzees spent more time in front of the "nice" person's window (nice > nasty). (*Source:* Russell, Call, & Dunbar 2008)

Conclusion: Chimpanzees form image scores of humans by observing their interactions.

EVALUATE

Provide an alternative hypothesis to explain why fewer chimps spent time in front of the "nasty" human.

These results suggest that chimpanzees can form image scores for human altruism.

You may not be surprised to learn that humans also care about their reputations for helping others (Applying the Concepts 15.2). This concern may explain the high degree of altruism observed in our species—and not just toward kin.

APPLYING THE CONCEPTS 15.2

Human altruism and reputations

Humans often help unrelated individuals, and indirect reciprocity may explain why. Our reputations depend in part on our altruistic behavior (Nowak & Sigmund 1998; Panchanathan & Boyd 2004). In laboratory experiments, humans show more altruistic behavior when they know they are being watched (Milinski, Semmann, & Krambeck 2002a; Milinski, Semmann, & Krambeck 2002b)—evidence that we care about our reputations.

Melissa Bateson, Daniel Nettle, and Gilbert Roberts examined human altruism in a real-world setting (Bateson, Nettle, & Roberts 2006). They examined the propensity for people to donate money for tea, coffee, and milk in a workplace coffee room by placing money in an "honesty box." Information about costs and donations was posted at eye level on a cupboard above the counter with the box. The notice contained a

15 cm × 3.5 cm banner with an image that alternated each week for ten weeks between a pair of eyes and an image of flowers (Figure 1). Each week, the researchers recorded the amount of money donated and the amount of milk consumed, calculating the money collected per liter of milk. This procedure controlled for weekly variation in consumption. Most, if not all, participants added milk to their tea and coffee in this study site in the United Kingdom.

On average, people donated almost three times more money in weeks when the image was of eyes than they did in weeks with an image of flowers. The researchers concluded that the image of eyes motivated people to be more generous because it created a perception of being watched, and people care about their reputations for altruism. ■

Figure 1. Human altruism. The amount of money donated per week (x-axis) increased when an image of human eyes (blue) was present compared with images of flowers (orange). The weekly image displayed is presented on the y-axis. (*Source:* Bateson, Nettle, & Roberts 2006)

Chapter Summary and Beyond

Living in groups can benefit individuals through reduced search time for food, increased diet breadth, reduced physiological costs, and greater safety from predators. However, it also comes with important costs, such as increased competition for resources, increased aggressive interactions, and enhanced disease transmission. Variation in costs and benefits helps explain observed variation in sociality. Recent work has examined how individuals assess differences in group size (Buckingham, Wong, & Rosenthal 2007) and how groups make decisions (Conradt et al. 2009).

Many social groups are structured by dominance hierarchies. Dominant individuals gain greater access to resources, while subordinates suffer less aggression once the hierarchy is established. Stable dominance hierarchies also reduce stress. However, Gesquiere and colleagues (2011) found that the most dominant individual can experience the highest level of stress within a stable hierarchy, so more work on this subject is needed.

Individuals in social groups often exhibit cooperative behavior. Such altruism is quite common between relatives and is explained by kin selection. In cooperatively breeding species, adults within a group can vary in their reproductive success. In many such groups, some adults forego reproduction and instead help raise the offspring of other breeders. Often these helpers are related to the parents and offspring they care for, and so kin selection explains this cooperative care. In other groups, ecological constraints can result in high current costs of attempting to reproduce. In such conditions, staying to help others and thus waiting to attempt to reproduce in the future, a behavior known as social queuing, yields higher fitness (Field & Cant 2009). Leggett and colleagues (2012) provide a recent review of the evolution of cooperative breeding. In eusocial insects, we see the evolution of sterile workers that defend the nest and help raise offspring produced by another individual. Kin selection and ecological constraints in the form of high costs of attempting to reproduce can explain the evolution of eusociality. The relative importance of these factors is a vigorous area of research (Hunt & Amdam 2005; Pennisi 2009; Nowak, Tarnita, & Wilson 2010; Hunt 2012).

Helping between nonrelated individuals can occur through byproduct mutualisms, reciprocity, or reputations. Understanding their relative importance has been the focus of much recent work, particularly among primatologists (Leinfelder et al. 2001; Payne, Lawes, & Henzi 2003; Manson et al. 2004; Schino & Aureli 2010).

CHAPTER QUESTIONS

1. In the turkey population studied by Krakauer (2005), would you expect to see the formation of coalitions composed of two male first cousins for whom relatedness, r, was 0.125?

2. Imagine that you are studying group size of goldfish feeding in a large pond in two experimental treatments. In one treatment, fish can see the location of food patches, and in the other treatment, food patches are hidden under gravel. What prediction can you make about group size in these treatments?

3. In a group of parrots, you frequently observe allogrooming. What data would allow you to determine whether kin selection or reciprocity was a better explanation for these observations?

4. Imagine that you are studying a group of wolves and notice that one of the individuals loses all of its fights over food. Propose two hypotheses to explain why this most subordinate individual chooses to stay in its group.

5. What features distinguish a cooperatively breeding species from one that is eusocial?

6. Richards, van Oosterhout, and Cable (2010) concluded that disease transmission in guppies appears to be affected by degree of sociality: greater time spent near others, as in female shoals, leads to higher rates of parasite transmission. Propose an alternative hypothesis for these data.

Glossary

adaptation A trait that enhances fitness (survivorship and reproduction). Also, an evolutionary process that results in a population of individuals with traits best suited to the current environment.

aggressive mimicry A situation in which a predator mimics a nonthreatening model.

alarm call A unique vocalization produced by social animals when a predator is nearby.

alarm signal hypothesis The hypothesis that advertisement behavior functions to warn nearby conspecifics of danger.

Allee effect A situation in which the fitness of individuals increases with increased population density.

allogrooming The grooming by one individual of another individual to remove ectoparasites.

alternate hypothesis The statistical hypothesis that the proposed explanation for an observation does significantly affect the behavior of the organism.

alternative reproductive tactics The existence of multiple behavioral mating phenotypes in a population.

altruism A behavior that results in the increased fitness of another individual and involves a cost to the individual performing the behavior. Also known as helping behavior.

ancestral trait A trait found in the common ancestor of two or more species. Also called a **plesiomorphic trait**.

animal behavior Any internally coordinated, externally visible pattern of activity that responds to changing external or internal conditions.

animal culture Differences in multiple traditions among populations.

anisogamy The existence of differently sized gametes (small and large) in the different sexes.

anthropomorphism The attribution of human motivations, characteristics, or emotions to animals.

apomorphic trait A trait found in an organism that was not present in the last common ancestor of a group of two or more species. Also known as a **derived trait**.

aposematic coloration Brightly colored morphology in a species that stands out from the environment and is associated with noxious chemicals or poisons that make individuals unpalatable or dangerous prey.

Bateman's hypothesis Female reproductive success is most strongly limited by the number and success of eggs that she can produce, while male reproductive success is limited by the number of mates he has.

Batesian mimicry The resemblance of a palatable mimic to an unpalatable model that predators have learned to avoid.

behavioral traditions Differences in behavior among populations that are transmitted between generations through social learning.

bicoordinate navigation The ability to identify a geographic location using two varying environmental gradients.

biparental care Both parents provide care for offspring.

breeding dispersal The abandonment of one breeding site to move to another.

brood parasitism A behavior in which a female (brood parasite) lays an egg in the nest of another female.

brood reduction The death of some siblings as a result of reduced parental care to enhance the fitness of surviving siblings.

byproduct mutualism When an individual's behavior enhances both its own fitness and the fitness of others.

bystander A third-party individual that detects a signal transmitted between a signaler and a receiver. Also called an **eavesdropper**.

cache Food stored in a hidden location for later retrieval.

castes Morphologically and behaviorally distinct individuals within a social group.

challenge hypothesis Male–male interactions increase plasma testosterone and thus sustain subsequent aggressive behavior.

chemoreception The process by which an animal detects chemical stimuli.

classical conditioning A type of learning in which a novel stimulus is paired with an existing stimulus and elicits a particular response. Eventually the novel stimulus alone elicits the same response as the existing stimulus. Also called **Pavlovian conditioning**.

cleaner fish Fish that feed on ectoparasites and the dead skin of other fish.

co-evolutionary arms race The back-and-forth process of adaptation in one species favoring counteradaptation in another.

cognition The acquisition, storage, retrieval, and use of information for decision-making.

cognitive buffer hypothesis   Large brains, despite their costs, provide survival benefits through flexible behaviors that are learned.

communication   The process in which a specialized signal produced by one individual affects the behavior of another individual.

communication network   A communication system involving a signaler, a receiver, and at least one bystander.

comparative method   An approach that examines differences and similarities between species to understand the evolution of traits.

compass   An instrument or capability that provides information about direction.

competition   An interaction that results in a reduction in fitness for one or both individuals.

competition hypothesis   Dispersal functions to reduce competition for resources.

conditional strategy   A strategy that an individual chooses based on its condition.

cones   A type of photoreceptor for color vision.

conspecific attraction   A phenomenon in which individuals are attracted to others, particularly during habitat selection.

conspecific cueing   A hypothesized mechanism to explain conspecific attraction. A settler uses the presence of another individual as a cue to habitat quality.

control group   In an experiment, the group that does not experience a manipulation. It provides a comparison to experimentally manipulated groups.

cooperative behavior   A mutually beneficial interaction between individuals.

cooperative breeding   In social groups, adults physiologically capable of reproducing forgo breeding and instead help others raise offspring.

correlation   Two variables that vary together predictably.

cost-benefit approach   A method used to study behavioral adaptations in which the fitness benefits and costs of different traits are examined to determine which has the highest net benefit (calculated as benefit minus cost).

cryptic coloration   Morphological coloration that matches the color of the environment to reduce detection by predators.

cryptic female choice   A situation that occurs when a female influences the fertilization success of sperm from one male over that from others.

dear enemy hypothesis   Territory owners will show reduced aggressive interactions toward territorial neighbors, compared to strangers, as a territory owner becomes more familiar with its neighbor.

dendritic spines   Small protuberances on a dendrite that typically receive synaptic inputs.

derived trait   A trait found in an organism that was not present in the last common ancestor of a group of two or more species. Also known as an **apomorphic trait**.

dilution effect   A reduction in the probability of death as a result of associating with others.

diminishing returns   A decline in instantaneous harvest rate as a food patch is depleted.

direct familiarization   The ability of individuals to discriminate kin from non-kin through previous associations.

direct fitness   Genes contributed to the next generation by an individual due to its own reproduction.

direct material benefits   Material resources obtained by a female from mating with a particular male.

direct reciprocal altruism   An altruistic behavior from which an altruist can benefit if the recipient reciprocates and helps the altruist in the future. Also known as **direct reciprocity**.

direct reciprocity   An altruistic behavior from which an altruist can benefit if the recipient reciprocates and helps the altruist in the future. Also known as **direct reciprocal altruism**.

directional selection   A situation in which individuals with one extreme (highest *or* lowest) trait values in a population possess the highest fitness.

dispersal   A relatively short-distance, one-way movement away from a site.

disruptive selection   A situation in which individuals with either of two extreme (highest *and* lowest) trait values in a population possess the highest fitness.

dominance hierarchy   An organized social system with dominant and subordinate members.

eavesdropper   A third-party individual that detects a signal transmitted between a signaler and a receiver. Also called a **bystander**.

electroreception   The ability to detect weak electric fields.

energetic costs of foraging   The energy used to exploit a patch and the metabolic costs incurred while feeding.

epistasis (I)   Interactions between genes at different loci.

ethogram   A formal description of an animal's behaviors.

eusociality   The social structure of a species that lives in social groups, contains overlapping generations, and has cooperative offspring care by nonparents and a reproductive division of labor.

evolution   Changes in allele frequency or trait values in a population over time.

evolutionary stable strategy (ESS)   A strategy that, if adopted by a population, cannot be invaded by another strategy because it yields the highest fitness.

experimental method   An approach in which scientists manipulate or change a variable to examine how it affects the behavior of an animal.

extended phenotype signal   A signal expressed beyond the body of an individual that often includes modification of the environment.

extra-pair copulations   The copulations of a pair-bonded individual with a third individual outside the pair bond.

extra-pair young   Offspring of a pair-bonded female produced outside the pair bond by a third-party male.

female defense polygyny   A mating system in which a single male monopolizes and mates with two or more females.

fitness   The relative survivorship and reproductive success (ability to produce viable offspring) of an individual.

fixed action pattern   A behavior that is invariant, unlearned, and, once initiated, always completed.

flash disappearance hypothesis   The sudden disappearance of the bright white underside of a deer's tail (a target easily observed by predators) against the dark background of dense woods makes it difficult for the predator to locate its prey and will encourage it to give up its attack.

focal animal sampling   A data collection technique whereby a single focal individual is randomly selected and observed for a specified time period and all pertinent behaviors performed by the individual are recorded.

frequency dependent selection   An evolutionary process in which the fitness of a trait is related to its frequency in a population.

game theory   A cost-benefit modeling approach in which an individual's fitness is affected by the behavior of others.

gene expression   The process in which gene products are produced.

generalist   A forager that consumes a wide variety of items in its diet.

genetic mating system   A description of the mating system based on the actual number of sexual partners in a social mating system that contribute to a set of offspring.

genetic quality hypothesis   Females that engage in multi-male matings can improve the fitness of their offspring via genetic mechanisms.

genotype (G)   All alleles and loci that affect the behavioral phenotype.

geomagnetic compass   The ability to orient using the earth's magnetic field.

giving-up density (GUD)   The density of food items in a food patch after being exploited by a forager.

glucocorticoids   Hormones secreted by the adrenal gland in vertebrates in response to stressful situations.

good genes   The alleles of a high-quality individual.

group selection   A form of selection that favors particular groups of individuals over other such groups.

group size effect   A commonly observed pattern in which the vigilance behavior of an individual declines as group size increases.

gustation   The detection of dissolved chemicals, often within the mouth.

habitat saturation hypothesis   A proposed explanation for cooperative breeding based on limited breeding options for helpers.

habituation   The reduction and then lack of response to a stimulus over time.

Hamilton's rule   An inequality predicting when altruism should evolve based on the fitness benefits and costs of the altruistic behavior and the degree of relatedness among individuals.

Hamilton-Zuk hypothesis   Parasites and pathogens play an important role in sexual selection when secondary sexual traits are costly and condition dependent.

handicap principle   Well-developed secondary sexual characteristics are costly because they handicap survival.

handling time   The amount of time required to manipulate a food item so that it is ready to eat.

haplodiploid genetic system   A genetic system with haploid males and diploid females.

hatching asynchrony   A situation in which chicks in a clutch hatch on different days.

hawk-dove model   A game theory model that examines the behavior of two individuals engaged in a contest over a resource.

heritability   The proportion of phenotypic variation in a population that is due to genetic variation.

heritable   The characteristic of a trait that can be passed from parents to their offspring because it is genetically based.

home range   An area of repeated use by an individual that is not defended from others.

hotshot hypothesis   A proposed explanation for the evolution of leks. It states that males can increase their encounter rate with females by aggregating near an attractive male.

hotspot hypothesis   All males can benefit by aggregating in a location where they are likely to encounter many females.

hypothesis   An explanation based on assumptions that makes a testable prediction.

ideal free distribution (IFD) model   A model that explains how animals distribute themselves among habitats or food patches.

image score   An assessment of an individual's propensity for helping others based on its observed prior behavior.

imprinting   Rapid learning in young animals through observation of parents or other objects. Individuals are typically attracted to objects they imprint on.

inbreeding avoidance hypothesis   Natal dispersal behavior minimizes the likelihood of inbreeding.

inbreeding depression   A reduction in fitness as a result of mating with close relatives.

inclusive fitness   The sum of an individual's direct and indirect fitness.

indirect familiarization   The discrimination of kin from non-kin via phenotype matching.

indirect fitness   The genes contributed to the next generation by an individual as a result of helping non-offspring kin produce additional offspring.

indirect genetic benefits   Genetic benefits females can obtain for their offspring by mating with males that have high genetic quality.

indirect reciprocity   A relationship in which helping individuals are more likely to receive help in the future because they have obtained high image scores by helping others.

individual selection   Natural selection acting on individuals.

insight learning   Spontaneous problem solving without the benefit of trial-and-error learning.

instantaneous sampling   A method of collecting behavioral data at regular time intervals. Also known as **scan sampling**.

intergenerational migration   Migration that occurs over more than one generation.

intersexual selection   Selection by one sex of members of the other sex for reproduction. Also known as **mate choice**.

interspecific brood parasitism   A behavior in which a female (brood parasite) lays an egg in the nest of another species.

intrasexual selection   Selection in which individuals within one sex compete for access to the other sex for reproduction. Also known as **mate competition**.

isodar   The line that results from plotting the number of individuals in one habitat against the number of individuals in another habitat at several points in time.

isogamy   The production of gametes of the same size by all individuals.

iteroparity   A life history strategy in which there are multiple reproductive events throughout the lifetime.

juvenile hormones (JH)   In insects, a class of hormones produced by specialized endocrine glands that vary in a consistent manner with levels of parental care by females.

kin selection   A form of natural selection in which individuals can increase their fitness by helping close relatives, because close relatives share the helper's genes.

knockout technique   A procedure in which a single gene is rendered nonfunctional.

lateral line system   In fish, a network of mechanoreceptors that provide hydrodynamic information.

learning   A relatively permanent change in behavior due to experience.

learning curve   A graphical representation of a change in learning over time.

lek   A location where males aggregate and display to females.

life history theory   Hypotheses that assume that natural selection will favor the evolution of behaviors and life history traits that maximize an individual's lifetime reproductive success.

life history traits   Traits involved with growth, reproduction, and survivorship.

linear dominance hierarchy   An organized social system in which an individual is dominant to each individual below and subordinate to all above it in rank.

local enhancement   A strategy in which an individual uses the presence of another as a cue for focusing on a particular part of the environment.

magnetoreception   The ability to detect and orient using magnetic fields.

male dominance polygyny   A mating system in which a few males on a lek mate with many females.

marginal value theorem   An optimal foraging model that predicts how long an individual should exploit a food patch.

mate choice   Selection by one sex of members of the other sex for reproduction. Also known as **intersexual selection**.

mate choice copying A situation in which one individual observes and copies the mating decisions of another individual.

mate competition A situation in which members of the same sex compete for access to the other sex for reproduction. Also called **intrasexual selection**.

mate guarding A behavior in which a male remains close to his mate to prevent her from mating with rivals.

mate guarding hypothesis Selection favors males that mate with and guard one female over one or more reproductive cycles by remaining in close association with her.

mating system A description of the social associations and number of sexual partners an individual has during one breeding season.

mechanoreceptors Sensory receptors sensitive to changes in pressure.

migration Relatively long-distance two-way movements.

mimicry The adaptive resemblance of one species to another.

missed opportunity costs of foraging The fitness costs of not engaging in other activities while feeding.

mobbing behavior The harassment of a predator by two or more individuals.

monogamy A mating system in which one female mates with one male.

move-on hypothesis Predator harassment functions to cause a predator to abandon an area.

multilevel selection A form of natural selection that may result in circumstances when selection on groups is stronger than selection on individuals.

mutual-assessment game theory models Game theory models that examine the behavior of two individuals engaged in a contest over a resource. Individuals are assumed to assess the relative fighting ability of their opponent and themselves in deciding how long to fight.

natal dispersal A one-time movement away from an individual's place of birth.

natural selection The process of differential reproduction and survivorship among individuals within a population.

navigation The process of determining a particular location and moving toward it.

negative results An outcome in which the null hypothesis is not rejected, and thus the alternate hypothesis is rejected.

null hypothesis The statistical hypothesis that an observation results from chance. Also called the hypothesis of no effect.

numerical competency The ability to recognize numerical quantities.

nuptial gift A physical resource such as a food item that an individual provides to a potential mate to enhance mating success.

observational method An approach in which scientists observe and record the behavior of an organism without manipulating the environment or the animal.

odorant A gaseous compound that is perceived as odorous.

olfaction The detection of airborne chemical stimuli.

operant chamber An enclosure used to study behavioral conditioning. Also known as a **Skinner box**.

operant conditioning A process in which an animal learns to associate a behavior with a particular consequence.

operational sex ratio The ratio of the number of sexually receptive males to sexually receptive females in a population.

optimal behavior The behavior that maximizes fitness in an optimal foraging theory model.

optimal foraging theory An approach to studying feeding behavior that assumes that natural selection has favored feeding behaviors that maximize fitness.

optimal trait value The trait value that confers the highest fitness in a population in a particular environment.

orientation The process of determining and maintaining a proper direction.

ornaments Exaggerated morphological traits used to attract mates.

parental care Behaviors by a parent to enhance the fitness of offspring, including incubation, feeding, and defense.

parental investment Any investment by a parent in an offspring that enhances the offspring's fitness at the cost of the parent's ability to invest in other offspring.

parental investment theory The hypothesis that the sex that pays the higher cost of parental investment should be choosier when it comes to mates.

parent-offspring conflict theory The hypothesis that parents and their dependent offspring are under different selection pressures: parents should maximize their lifetime reproductive fitness, while offspring should maximize the energy and protection they currently receive from their parents in order to survive to reproductive age.

parent-offspring regression A statistical technique used to examine the similarity between parents and their offspring in terms of the traits they possess.

Pavlovian conditioning A type of learning in which a novel stimulus is paired with an existing stimulus and elicits a particular response. Eventually the novel stimulus alone elicits the same response as the existing stimulus. Also called **classical conditioning**.

peer review A process in which editors of scientific journals ask experts to review submitted papers to determine whether to accept or reject a paper for publication.

personality Consistent relative differences in behavior among individuals over time or across different environmental contexts.

phenotype (P) The observable traits of an individual.

pheromones Volatile, organic, species-specific compounds that affect the behavior of conspecifics.

photoreceptor Specialized neurons that are sensitive to light and function as visual receptors.

phylogeny A branching diagram showing hypothesized evolutionary relationships among species. Sometimes called a phylogenetic tree.

pleiotropy A situation in which a single gene affects more than one phenotypic trait.

plesiomorphic trait A trait found in the common ancestor of two or more species. Also called an **ancestral trait**.

plural breeding A mating system of social species in which multiple males and multiple females within one social group produce young during one breeding season.

polyandry A mating system in which a single female associates and mates with multiple males.

polygynandry A mating system of social species in which multiple males and multiple females associate with one another and produce young in one breeding attempt.

polygyny A mating system in which a single male associates and mates with multiple females.

polygyny threshold model A model that predicts the occurrence of polygyny based on the amount of resources available to females in male territories.

predation risk cost The fitness cost associated with being killed by a predator. Often equated with the probability of being killed by a predator.

predator harassment Interactions with a predator to deter attack. Such behavior can involve rapid movements, loud vocalizations, and throwing or kicking objects at the predator.

primary literature The original source of scientific information, typically peer-reviewed scientific journals.

primary sexual characteristics The genitalia and organs of reproduction.

process of science Observing events, organizing knowledge, and providing explanations through the formulation and testing of hypotheses.

producer An individual in a group that searches for food.

profitability For a food item, the energy it contains divided by the handling time.

prolactin A hormone that influences parental care in vertebrates.

promiscuity A mating system in which both males and females mate with multiple partners in the absence of pair bonds.

proximate explanation An explanation that focuses on understanding the immediate causes of a behavior.

public information Information obtained from the activity or performance of others about the quality of an environmental parameter or resource.

pursuit-deterrence hypothesis Advertisement behavior informs a predator that it has been detected and so pursuit is not likely to be successful.

QTL mapping A statistical technique that combines genetic information with trait information to determine which regions of the genome contain the genes that influence the trait QTLs.

quantitative trait loci (QTL) Stretches of DNA that either contain or are linked to genes influencing a trait such as behavior.

quitting harvest rate The forager's instantaneous harvest rate when it leaves a food patch.

reduction An ethical guideline that promotes limiting the number of animals subject to disturbance in research or teaching.

refinement An ethical guideline that involves improving procedures and techniques to minimize pain and stress for animals.

replacement An ethical guideline that encourages the use of computer modeling, videotapes, or other approaches in place of actual animals.

reproductive skew Variation in reproductive success among members of a group.

research question A formal statement of an unknown that begins the scientific method.

resource defense polygyny A mating system in which a male mates with multiple females that are attracted to resources he defends.

resource holding potential The ability to win an aggressive encounter.

rods A type of photoreceptor that is sensitive to low light levels.

rover One type of genetic and behavioral variant in fruit flies. Larval rovers have longer foraging trails than sitters in the presence of food and are more likely to leave a food patch.

runaway process An evolutionary process in which a male trait co-evolves with a female preference for it and becomes increasingly exaggerated.

scan sampling A method of collecting data in which data are collected from all individuals at regular time intervals. Also known as **instantaneous sampling**.

scientific misconduct The violation of ethical behavior standards in science. It includes the falsification or fabrication of data, purposefully inappropriate analysis of data, and plagiarism.

scientific theories Hypotheses that have been tested many times, by many different scientists, and have not been rejected. Scientific theories provide a conceptual framework that explains many phenomena and are well supported by observations and experimental tests.

scramble competition The indirect competition of individuals to find and secure copulations with multiple mates.

scrounger Individuals in a group that exploit the food discoveries of others.

search image The distinctive visual features of a single prey type, which, once learned, can enhance prey detection.

secondary literature A report, which often appears in newspapers, magazines, and books, that summarizes and interprets the primary literature.

secondary sexual characteristics Morphological differences between the sexes that are not directly involved in reproduction.

selection experiment An experiment in which different groups of individuals are subjected to differential selection on a trait.

self-assessment game theory models Game theory models that examine the behavior of two individuals engaged in a contest over a resource. Each individual is assumed to assess its resource holding power in deciding how long to fight.

selfish herd hypothesis A predator is more likely to kill a member on the outside of a group because it will encounter outside individuals first.

semelparity A life history strategy in which reproduction occurs once in the lifetime.

sensory bias hypothesis Female mating preferences are a byproduct of preexisting biases in a female's sensory system.

sex role–reversed species A species in which females compete for males that invest heavily in parental care.

sexual conflict The differential selection on males and females to maximize their fitness.

sexual deception hypothesis Males will produce deceptive signals to females in order to enhance their own reproduction.

sexual dimorphism Morphological differences between the sexes.

sexual selection A form of natural selection that acts on heritable traits that affect reproduction.

signal An evolved trait that is selected for its effect on the behavior of a receiver.

sister species Two species that are more closely related to one another than to any other species; species that share a recent common ancestor.

site fidelity Individuals that remain at or return to a previous location to breed.

sitter One type of genetic and behavioral variant in fruit flies. Larval sitters have shorter foraging trails than rovers in the presence of food and are less likely to leave a food patch.

Skinner box An enclosure used to study behavioral conditioning. Also known as an **operant chamber**.

social group A set of individuals that live near and associate with one another.

social mating system The social associations and presumed mating behavior of individuals based on those associations.

social queuing A process in which subordinates eventually move up the social ladder as they age.

sociality The tendency to live and associate with others.

specialist A forager that has a narrow diet.

sperm competition A situation that occurs when the sperm of different males compete to fertilize eggs.

stabilizing selection A situation in which those individuals in a population with intermediate trait values have the highest fitness in a particular environment.

star compass The use of stars or constellations for orientation.

stotting Jumping high into the air with all four feet off the ground and moving forward.

sun compass The use of the sun for orientation.

teaching The active participation of an experienced individual in facilitating learning by a naïve conspecific.

territorial cooperation hypothesis Two individuals (one of each sex) can better defend a critical resource, such as a safe refuge, than can a single individual, and thus selection may favor pair formation and shared defense.

territory   An area defended to obtain the exclusive use of the resources it contains.

testosterone   A steroid hormone produced in the gonads and regulated by the hypothalamus and pituitary gland.

time budget   A summary of the total time and relative frequency of different behaviors of an individual.

tit-for-tat   A player always cooperates the first time it interacts with an opponent and then simply matches the behavior of the other player from the previous interaction.

trial-and-error learning   Learning to solve a problem through repetition.

ultimate explanation   An explanation of behavior that requires evolutionary reasoning and analysis.

vigilance   A behavior in which an individual scans the environment for predators.

volatile   A substance that has the capability of evaporating, or becoming gaseous.

waggle dance   Behavior performed by a honeybee scout that recruits workers to exploit a food source.

weapons   Exaggerated morphological traits used in competition over resources.

wild type   The typical form of an organism or gene in nature.

winner effect   A phenomenon in which winning an aggressive interaction enhances the likelihood of winning a subsequent interaction.

winner-challenge effect   A situation in which winning an aggressive interaction increases plasma testosterone levels and so enhances aggressive behavior and the likelihood of winning subsequent interactions.

win-stay lose-shift dispersal   A common behavioral pattern in which individuals return to a previous breeding location after reproductive success there but disperse to a different location after a reproductive failure.

Bibliography

Abrahams, M. V. 1989. Foraging guppies and the ideal free distribution: The influence of information on patch choice. *Ethology* 82:116–126.

Abrahams, M. V., Robb, T. L., & Hare, J. F. 2005. Effect of hypoxia on opercular displays: Evidence for an honest signal? *Animal Behaviour* 70:427–432.

Adkins-Regan, E. 2005. *Hormones and animal social behavior.* Princeton, NJ: Princeton University Press.

Adkins-Regan, E., & MacKillop, E. 2003. Japanese quail (*Coturnix japonica*) inseminations are more likely to fertilize eggs in a context predicting mating opportunities. *Proceedings of the Royal Society B* 270:1685–1689.

Ahlering, M. A., & Faaborg, J. 2006. Avian habitat management meets conspecific attraction: If you build it will they come? *Auk* 123:301–312.

Åhlund, M., & Andersson, M. 2001. Female ducks can double their reproduction. *Nature* 414:600–601.

Akçay, E., & Roughgarden, J. 2007. Extra-pair paternity in birds: Review of the genetic benefits. *Evolutionary Ecology Research* 9:855–868.

Åkesson, S., & Hedenström, A. 2007. How migrants get there: Migratory performance and orientation. *Bioscience* 57:123–133.

Akey, J. M., Ruhe, A. L., Akey, D. T., Wong, A. K., Connelly, C. F., Madeoy, J., Nicholas, T. J., & Neff, M. W. 2010. Tracking footprints of artificial selection in the dog genome. *Proceedings of the National Academy of Sciences USA* 107:1160–1165.

Alatalo, R. V., Höglund, J., Lundberg, A., & Sutherland, W. J. 1992. Evolution of black grouse leks: Female preferences benefit males in larger leks. *Behavioral Ecology* 3:53–59.

Alatalo, R. V., Lundberg, A., & Sundberg, J. 1990. Can female preference explain sexual dichromatism in the pied flycatcher, *Ficedula hypoleuca. Animal Behaviour* 39:244–252.

Alerstram, T., Hedenström, A., & Åkesson, S. 2003. Long-distance migration: Evolution and determinants. *Oikos* 103:247–260.

Alexander, R. D. 1974. The evolution of social behavior. *Annual Review of Ecology and Systematics* 5:325–383.

Allee, W. C. 1931. Co-operation among animals. *American Journal of Sociology* 37:386–398.

Allen, C., & Bekoff, M. 2007. Animal minds, cognitive ethology, and ethics. *Journal of Ethics* 11:299–317.

Allman, J., McLaughlin, T., & Hakeem, A. 1993. Brain weight and life-span in primate species. *Proceedings of the National Academy of Sciences U.S.A.* 90:118–122.

Alonso, S. H. 2008. Female mate choice copying affects sexual selection in wild populations of the ocellated wrasse. *Animal Behaviour* 75:1715–1723.

Alquiloni, L., Buřič, M., & Gherardi, F. 2008. Crayfish females eavesdrop on fighting males before choosing the dominant male. *Current Biology* 18:R462–R463.

Altmann, J. 1974. Observational study of behavior: Sampling methods. *Behaviour* 49:227–267.

American Veterinary Medical Association. 2012. https://www.avma.org/KB/Resources/Statistics/Pages/Market-research-statistics-US-pet-ownership.aspx (accessed July 8, 2016).

American Veterinary Society of Animal Behavior (AVSAB). 2008. American Veterinary Society of Animal Behavior position statement—puppy socialization. http://avsabonline.org/uploads/position_statements/puppy_socialization.pdf (accessed July 8, 2016).

Anderson, W. W., Kim, Y. K., & Gowaty, P. A. 2007. Experimental constraints on mate preferences in *Drosophila pseudoobscura* decrease offspring viability and fitness in mated pairs. *Proceedings of the National Academy of Sciences USA* 104:4484–4488.

Andersson, M. 1995. Evolution of reversed sex roles, sexual size dimorphism, and mating system in coucals (Centropodidae: Aves). *Biological Journal of the Linnean Society* 54:173–181.

———. 2005. Evolution of classical polyandry: Three steps to female emancipation. *Ethology* 111:1–23.

Andersson, M., & Åhlund, M. 2000. Host-parasite relatedness shown by protein fingerprinting in a brood parasitic bird. *Proceedings of the National Academy of Sciences USA* 97:13,188–13,193.

Andersson, M., & Simmons, L. W. 2006. Sexual selection and mate choice. *Trends in Ecology and Evolution* 21:296–302.

Angelier, F., & Chastel, O. 2009. Stress, prolactin and parental investment in birds: A review. *General and Comparative Endocrinology* 163:142–148.

Anholt, R. H., & Mackay, T. F. C. 2010. *Principles of behavioral genetics.* London: Elsevier.

Animal Behavior Society. 2006. Guidelines for the treatment of animals in behavioural research and teaching. *Animal Behaviour* 71:245–253.

Aparicio, J. M., Bonal, R., & Muñoz, A. 2007. Experimental tests on public information use in the colonial lesser kestrel. *Evolutionary Ecology* 21:783–800.

Arbelle, S., Benjamin, J., Golin, M., Kremer, I., Belmaker, R. H., & Ebstein, R. P. 2003. Relation of shyness in grade school children to the genotype for the long form of the serotonin transporter promoter region polymorphism. *American Journal of Psychiatry* 160:671–676.

Archer, J. 2006. Testosterone and human aggression: An evaluation of the challenge hypothesis. *Neuroscience and Biobehavioral Reviews* 30:319–345.

Archer, J., Graham-Kevan, N., & Davies, M. 2005. Testosterone and aggression: A re-analysis of Book, Starzyk and Quinsey's (2001) study. *Aggression and Violent Behavior* 10:241–261.

Arnan, X., Ferrandiz-Rovira, M., Pladevall, C., & Rodrigo, A. 2011. Worker size-related task partitioning in the foraging strategy of a seed-harvesting ant species. *Behavioral Ecology and Sociobiology* 65:1881–1890.

Arnott, G., & Elwood, R. W. 2008. Information gathering and decision making about resource value in animal contests. *Animal Behaviour* 76:529–542.

Arnqvist, G. 2006. Sexual exploitation and sexual conflict. *Philosophical Transactions of the Royal Society B* 361:375–386.

Arnqvist, G., & Henriksson, S. 1997. Sexual cannibalism in the fishing spider and a model for the evolution of sexual cannibalism based on genetic constraints. *Evolutionary Ecology* 11:255–273.

Asa, C. S., Traylor-Holzer, K., & Lacy, R. C. 2011. Can conservation-breeding programmes be improved by incorporating mate choice? *International Zoo Yearbook* 45:203–212.

Ascunce, M. S., Yang, C. C., Oakey, J., Calcaterra, L., Wu, W. J., Shih, C. J., Goudet, J., Ross, K. G., & Shoemaker D. 2011. Global invasion history of the fire ant *Solenopsis invicta*. *Science* 331:1066–1068.

Augustsson, H., & Meyerson, B. J. 2004. Exploration and risk assessment: A comparative study of male house mice (*Mus musculus musculus*) and two laboratory strains. *Physiology & Behavior* 81:685–698.

Axelrod, R., & Hamilton, W. D. 1981. The evolution of cooperation. *Science* 211:1390–1396.

Bailey, N.W. 2012. Evolutionary models of extended phenotypes. *Trends in Ecology and Evolution* 27:561–569.

Baker, R. R. 1980. Goal orientation by blindfolded humans after long-distance displacement: Possible involvement of a magnetic sense. *Science* 210:555–557.

Ballantine, B. 2009. The ability to perform physically challenging songs predicts age and size in male swamp sparrows, *Melospiza georgiana*. *Animal Behaviour* 77:973–978.

Baptista, L. F., & Petrinovich, L. 1984. Social interaction, sensitive phases and the song template hypothesis in the white-crowned sparrow. *Animal Behaviour* 32:1359–1371.

Barber, N. A., Marquis, R. J., & Tori, W. P. 2008. Invasive prey impacts the abundance and distribution of native birds. *Ecology* 89:2678–2683.

Bargmann, C. I. 2006. Comparative chemosensation from receptors to ecology. *Nature* 444:295–301.

Barnard, C. J., & Sibley, R. M. 1981. Producers and scroungers: A general model and its application to captive flocks of house sparrows. *Animal Behaviour* 29:543–555.

Barnes, R., Greene, K., Holland, J., & Lamm, M. 2002. Management and husbandry of duikers at the Los Angeles Zoo. *Zoo Biology* 21:107–121.

Barrett, H. C., & Broesch, J. 2012. Prepared social learning about dangerous animals in children. *Evolution and Human Behavior* 33:499–508.

Barry, K. L., & Kokko, H. 2010. Male mate choice: Why sequential assessment can make its evolution difficult. *Animal Behaviour* 80:163–169.

Basolo, A. L. 1990. Female preference predates the evolution of the sword in swordtail fish. *Science* 250:808–881.

Bastock, M. 1956. A gene mutation which changes a behavior pattern. *Evolution* 10:421–439.

Bateman, A. J. 1948. Intra-sexual selection in *Drosophila*. *Heredity* 2:349–368.

Bates, H. W. 1862. Contributions to an insect fauna of the Amazon valley. Lepidoptera: Heliconidae. *Transactions of the Linnaean Society of London* 23:495–566.

Bateson, M., Nettle, D., & Roberts, G. 2006. Cues of being watched enhance cooperation in a real-world setting. *Biology Letters* 2:412–414.

Bateson, P., & Martin, P. 2007. *Measuring behavior: An introductory guide*. Cambridge: Cambridge University Press.

Batra, S. W. T. 1966. Nests and social behavior of halictine bees of India. *Indian Journal of Entomology* 28:375–393.

Beard, C. 1913. *An economic interpretation of the Constitution of the United States*. New York: MacMillan Co.

Beck, W. S., Liem, K. F., & Simpson, G. G. 1991. *Life, an introduction to biology*. 3rd ed. New York: HarperCollins.

Bednekoff, P. A. 2007. Foraging in the face of danger. In *Foraging*, ed. D. W. Stephens, J. S. Brown, & R. C. Ydenberg, 305–329. Chicago: University of Chicago Press.

Bednekoff, P. A., & Kotrschal, T. 2002. Observational learning and the raiding of food caches in ravens, *Corvus corax*: Is it "tactical" deception? *Animal Behaviour* 64:185–195.

Beecher, M. D., & Brenowitz, E. A. 2005. Functional aspects of song learning in songbirds. *Trends in Ecology and Evolution* 20:143–149.

Beecher, M. D., Burt, J. M., O'Loghlen, A. L., Templeton, C. N., & Campbell, S. E. 2007. Bird song learning in an eavesdropping context. *Animal Behaviour* 73:929–935.

Beehler, B. M., & Foster, M. S. 1988. Hotshots, hotspots, and female preference in the organization of lek mating systems. *American Naturalist* 131:203–219.

Beehner, J. C., & Whitten, P. L. 2004. Modifications of a field method for fecal steroid analysis in baboons. *Physiology & Behavior* 82:269–277.

Beier, P. 1991. Cougar attacks on humans in the United States and Canada. *Wildlife Society Bulletin* 19:403–412.

Bekoff, M., Allen, C, & Burghardt, G. M. 2002. *The cognitive animal: Empirical and theoretical perspectives on animal cognition*. Cambridge, MA: MIT Press.

Bekoff, M., & Sherman, P. W. 2004. Reflections on animal selves. *Trends in Ecology and Evolution* 4:176–180.

Bell, G. 1978. The evolution of anisogamy. *Journal of Theoretical Biology* 73:247–270.

Bell, G., Lefebvre, L., Giraldeau, L.-A., & Weary, D. 1984. Partial preferences of insects for male flowers of an annual herb. *Oecologia* 64:287–294.

Belthoff, J. R., & Ritchison, G. 1989. Natal dispersal of eastern screech-owls. *Condor* 91:254–265.

Bender, N., Taborsky, M., & Power, D. M. 2008. The role of prolactin in the regulation of brood care in the cooperatively breeding fish *Neolamprologus pulcher*. *Journal of Experimental Zoology* 309A:515–524.

Bendesky, A., & Bargmann, C. I. 2011. Genetic contributions to behavioural diversity at the gene-environment interface. *Nature Reviews Genetics* 12:809–820.

Bendich, A. 1989. Carotenoids and the immune response. *Journal of Nutrition* 119:112–115.

Bengtsson, B. O. 1978. Inbreeding avoidance—at what cost? *Journal of Theoretical Biology* 73:439–444.

Bengtsson, G., Hedlund, K., & Rundgren, S. 1994. Food- and density-dependent dispersal: Evidence from a soil collembolan. *Journal of Animal Ecology* 63:513–520.

Berec, L., & Křivan, V. 2000. A mechanistic model for partial preferences. *Journal of Theoretical Biology* 58:279–289.

Berglund, A., & Rosenqvist, G. 1993. Selective males and ardent females in pipefishes. *Behavioral Ecology and Sociobiology* 32:331–336.

———. 2001. Male pipefish prefer ornamented females. *Animal Behaviour* 61:345–350.

Berglund, A., Rosenqvist, G., & Svensson, I. 1986. Mate choice, fecundity and sexual dimorphism in two pipefish species (*Syngnathidae*). *Behavioral Ecology and Sociobiology* 19:301–307.

———. 1989. Reproductive success of females limited by males in two pipefish species. *American Naturalist* 133:506–516.

Bergman, T. J., Beehner, J. C., Cheney, D. L., Seyfarth, R. M., & Whittens, P. L. 2005. Correlates of stress in free-ranging male chacma baboons *Papio hamadryas ursinus*. *Animal Behaviour* 70:703–713.

Berthold, P. 1993. *Bird migration: A general survey*. Oxford, UK: Oxford University Press.

———. 1999. A comprehensive theory for the evolution, control and adaptability of avian migration. *Ostrich* 70:1–11.

———. 1996. *Control of bird migration*. London: Chapman & Hall.

Bielsky, I. F., Hu, S-B., Szegda, K. L., Westphal, H., & Young, L. J. 2004. Profound impairment in social recognition and reduction in anxiety-like behavior in vasopressin V1a receptor knockout mice. *Neuropsychopharmacology* 29:483–493.

Biernaskie, J. M., Walker, S. C., & Gegear, R. J. 2009. Bumblebees learn to forage like Bayesians. *American Naturalist* 174:413–423.

Bingman, V. P., & Chen, K. 2005. Mechanisms of animal global navigation: Comparative perspectives and enduring challenges. *Ethology Ecology & Evolution* 17:295–318.

Bird, B. L., Branch, L. C., & Miller, D. L. 2004. Effects of coastal lighting on foraging behavior of beach mice. *Conservation Biology* 18:1435–1439.

Black, J. M., ed. 1996. *Partnerships in birds: The study of monogamy*. Oxford: Oxford University Press.

Blackiston, D., Briscoe, A. D., & Weiss, M. R. 2011. Color vision and learning in the monarch butterfly, *Danaus plexippus* (Nymphalidae). *Journal of Experimental Biology* 214:509–520.

Blackledge, T. A., Scharff, N., Coddington, J. A., Szüts, T., Wenzel, J. W., Hayashi, C. Y., & Agnarsson, I. 2009. Reconstructing web evolution and spider diversification in the molecular era. *Proceedings of the National Academy of Sciences USA* 106:5229–5334.

Blas, J., Bortolotti, G. R., Tella, J. L., Baos, R., & Marchant, T. A. 2007. Stress response during development predicts fitness in a wild, long lived vertebrate. *Proceedings of the National Academy of Sciences USA* 104:8880–8884.

Blomqvist, D., Fessl, B., Hoi, H., & Kleindorfer, S. 2005. High frequency of extra-pair fertilizations in the moustached warbler, a songbird with a variable mating system. *Behaviour* 142:1133–1148.

Blumenrath, S. H., & Dabelsteen, T. 2004. Degradation of great tit (*Parus major*) song before and after foliation: Implications for vocal communication in a deciduous forest. *Behaviour* 141:935–958.

Blumstein, D. T., Lea, A. J., Olson, L. E., & Martin, J. G. A. 2010. Heritability of anti-predator traits: vigilance and locomotor performance in marmots. *Journal of Evolutionary Biology* 23:879–887.

Blyth, J. E., & Gilburn, A. S. 2006. Extreme promiscuity in a mating system dominated by sexual conflict. *Journal of Insect Behavior* 19:447–455.

Boal, J. G., Dunham, A. W., Williams, K. T., & Hanlon, R. T. 2000. Experimental evidence for spatial learning in octopuses (*Octopus bimaculoides*). *Journal of Comparative Psychology* 114:246–252.

Boal, J. G., Prosser, K. N., Holm, J. B., Simmons, T. L., Haas, R. E., & Nagle, G. T. 2010. Sexually mature cuttlefish are attracted to the eggs of conspecifics. *Journal of Chemical Ecology* 36:834–836.

Boggs, C. L. & Dau, B. 2004. Resource specialization in puddling Lepidoptera. *Ecological Entomology* 33:1020–1024.

Bolhuis, J. J., Okanoya, K., & Scharff, C. 2010. Twitter evolution: Converging mechanisms in birdsong and human speech. *Nature Reviews Neuroscience* 11:747–759.

Bolhuis, J. J., & Wynne, C. D. L. 2009. Can evolution explain how minds work? *Nature* 458:832–833.

Bolhuis, J. J., Zijlstra, G. G. O., den Boer-Visser, A. M., & Van der Zee, E. 2000. Localized neuronal activation in the zebra finch brain is related to the strength of song learning. *Proceedings of the National Academy of Sciences USA* 97:2282–2285.

Bollinger, E. K., Harper, S. J., & Barrett, G. W. 1993. Inbreeding avoidance increases dispersal movements of the meadow vole. *Ecology* 74:1153–1156.

Bond, A. 2007. The evolution of color polymorphism: Crypticity, searching images, and apostatic selection. *Annual Review of Ecology, Evolution, and Systematics* 38:489–514.

Bond, A. B., & Kamil, A. C. 1999. Searching image in blue jays: Facilitation and interference in sequential priming. *Animal Learning & Behavior* 27:461– 471.

Bonine, K. E., & Garland, T., Jr. 1999. Sprint performance in phyrnosomatid lizards, measured on a high-speed treadmill, correlates with hindlimb length. *Journal of Zoology* 248:255–265.

Boogert, N. J., Fawcett, T. W., & Lefebvre, L. 2011. Mate choice for cognitive traits: A review of the evidence in nonhuman vertebrates. *Behavioral Ecology* 22:447–459.

Book, A. S., Starzyk, K. B., & Quinsey, V. L. 2001. The relationship between testosterone and aggression: A meta-analysis. *Aggression and Violent Behavior* 6:579–599.

Booker, R. M., Munday, P. L., Chivers, D. P., & Jones, G. P. 2015. You are what you eat: Diet-induced chemical crypsis in a coral-feeding reef fish. *Proceedings of the Royal Society B* 282:20141887. DOI: 10.1098/rspb.2014.1887.

Boomsma, J.J. 2009. Lifetime monogamy and the evolution of eusociality. *Philosophical Transactions of the Royal Society B* 364:3191–3207.

Boone, R. B., Thirgood, S. J., & Hopcraft, J. G. C. 2006. Serengeti wildebeest migratory patterns modeled from rainfall and new vegetation growth. *Ecology* 87:1987–1994.

Boonstra, R., Krebs, C. J., Gaines, M. S., Johnson, M. L., & Craine, I. T. M. 1987. Natal philopatry and breeding systems in voles (*Microtus* spp.). *Journal of Animal Ecology* 56:655–673.

Booth, A., Shelley, G., Mazur, A., Tharp, G., & Kittok, R. 1989. Testosterone, and winning and losing in human competition. *Hormones and Behavior* 23:556–571.

Borgia, G. 1985. Bower quality, number of decorations and mating success of male satin bowerbirds (*Ptilonorhynchus violaceus*): an experimental analysis. *Animal Behaviour* 33:266–271.

Boulinier, T., & Danchin, E. 1997. The use of conspecific reproductive success for breeding patch selection in territorial migratory species. *Evolutionary Ecology* 11:505–517.

Boulinier, T., McCoy, K. D., Yoccoz, N. G., Gasparini, J., & Tveraa, T. 2008. Public information affects breeding dispersal in a colonial bird: Kittiwakes cue on neighbours. *Biology Letters* 4:538–540.

Bouska, M. L., & Beatty, P. A. 1978. Clothing as a symbol of status: Its effect on control of interaction territory. *Bulletin of the Pyschonomic Society* 11:235–238.

Boutin, M. E. 2007. *Learning and behavior: A contemporary synthesis.* Sunderland, MA: Sinauer Associates.

Boyd, R., & Richerson, P. J. 1996. Why culture is common, but cultural evolution is rare. *Proceedings of the British Academy* 88:77–94.

Boyle, W. & Conway, C. 2007. Why migrate? A test of the evolutionary precursor hypothesis. *American Naturalist* 169:344–359.

Boysen, S. T., Berntson, G. G., & Mukobi, K. L. 2001. Size matters: Impact of item size and quantity on array choice by chimpanzees (*Pan troglodytes*). *Journal of Comparative Psychology* 115:106–110.

Bracis, C., & Anderson, J. J. 2012. An investigation of the geomagnetic imprinting hypothesis for salmon. *Fisheries Oceanography* 21:170–181.

Bradbury, J. W., Gibson, R. M., & Tsai, I. M. 1986. Hotspots and the evolution of leks. *Animal Behaviour* 34:1694–1709.

Bradbury, J. W., & Vehrencamp, S. L. 1998. *Principles of animal communication*. Sunderland, MA: Sinauer Associates.

Brain, C. K. 1970. New finds at the Swartkrans australopithecine site. *Nature* 225:1112–1119.

Braude, S., Tang-Martinez, Z., & Taylor, G. T. 1999. Stress, testosterone, and the immunoredistribution hypothesis. *Behavioral Ecology* 10:345–350.

Bridges, R. S., Numan, M., Ronsheim, P. M., Mann, P. E., & Lupini, C. E. 1990. Central prolactin infusions stimulate maternal behavior in steroid-treated, nulliparous female rats. *Proceedings of the National Academy of Sciences USA* 87:8003–8007.

Briffa, M., & Sneddon, L. U. 2007. Physiological constraints on contest behaviour. *Functional Ecology* 21: 627–637.

Bro-Jørgensen, J. 2010. Dynamics of multiple signalling systems: Animal communication in a world of flux. *Trends in Ecology and Evolution* 25:292–300.

Bro-Jørgensen, J., & Pangle, W. M. 2010. Male topi antelopes alarm snort deceptively to retain females for mating. *American Naturalist* 176:E33–E39.

Brönmark, C., & Hansson, L.-A. 2000. Chemical communication in aquatic systems: An introduction. *Oikos* 88:103–109.

Brown, C. R., & Brown, M. B. 1986. Ectoparasitism as a cost of coloniality in cliff swallows (*Hirundo pyrrhonota*). *Ecology* 67:1206–1218.

Brown, G. E., & Brown J. A. 1993. Social dynamics in salmonid fishes: Do kin make better neighbours? *Animal Behaviour* 45:863–871.

Brown, J. L. 1964. The evolution of diversity in avian territorial systems. *Wilson Bulletin* 76:160–169.

———. 1997a. A theory of mate choice based on heterozygosity. *Behavioral Ecology* 8:60–65.

———. 1997b. Long-term memory of an auditory stimulus for food in a natural population of the Mexican jay. *Wilson Bulletin* 109:749–752.

Brown, J. L., Morales, V., & Summers, K. 2010. A key ecological trait drove the evolution of biparental care and monogamy in an amphibian. *American Naturalist* 175:436–446.

Brown, J. L., Twomey, E., Morales, V., & Summers, K. 2008. Phytotelm size in relation to parental care and mating strategies in two species of Peruvian poison frogs. *Behaviour* 145:1139–1165.

Brown, J. S. 1988. Patch use as an indicator of habitat preference, predation risk, and competition. *Behavioral Ecology and Sociobiology* 22:37–47.

———. 1999. Vigilance, patch use, and habitat selection: Foraging under predation risk. *Evolutionary Ecology Research* 1:49–71.

Brown, J. S., & Kotler, B. P. 2004. Hazardous pay duty and the foraging costs of predation. *Ecology Letters* 7:999–1014.

Bshary, R. 2002. Biting cleaner fish use altruism to deceive image-scoring client reef fish. *Proceedings of the Royal Society of London B* 269:2087–2093.

Bshary, R., & Grutter, A. S. 2002. Asymmetric cheating opportunities and partner control in a cleaner fish mutualism. *Animal Behaviour* 63:547–555.

———. 2006. Image scoring and cooperation in a cleaner fish mutualism. *Nature* 441:975–978.

Bucher, T. A., Ryan, M. J., & Bartholomew, G. A. 1982. Oxygen consumption during resting, calling, and nest building in the frog *Physalaemus pustulosus*. *Physiological Zoology* 55:10–22.

Buckingham, J. N., Wong, B. B. M., & Rosenthal, G. G. 2007. Shoaling decisions in female swordtails: How do fish gauge group size? *Behaviour* 144:1333–1346.

Bulger, J. B., & Hamilton, W. J. 1987. Rank and density correlates of inclusive fitness measures in a natural chacma baboon (*Papio ursinus*) population. *International Journal of Primatology* 8:635–650.

Bulmer, M. G., & Parker, G. A. 2002. The evolution of anisogamy: A game-theoretic approach. *Proceedings of the Royal Society B* 269:2381–2388.

Burda, H., Honeycutt, R. L., Begall, S., Locker-Grütjen, O., & Scharff, A. 2000. Are naked mole-rats eusocial and if so, why? *Behavioral Ecology and Sociobiology* 47:293–303.

Burghardt, G. 1985. Animal awareness: Current perception and historical perspective. *American Psychologist* 405:905–919.

Buston, P. M. 2004. Territory inheritance in clownfish. *Proceedings of the Royal Society B* 271 (suppl.): S252–S254.

Cade, T. J., & Jones, C. G. 1993. Progress in restoration of the Mauritius kestrel. *Conservation Biology* 7:169–175.

Caillaud, M. C., & Via, S. 2000. Specialized feeding behavior influences both ecological specialization and assortative mating in sympatric host races of pea aphids. *American Naturalist* 156:606–621.

———. 2012. Quantitative genetics of feeding behavior in two ecological races of the pea aphid, *Acyrthosiphon pisum. Heredity* 108:211–218.

Calsbeek, R., Alonzo, S. H., Zamudio, K., & Sinervo, B. 2002. Sexual selection and alternative mating behaviors generate demographic stochasticity in small populations. *Proceedings of the Royal Society B* 269:157–164.

Calsbeek, R, & Sinervo, B. 2002. Uncoupling direct and indirect components of female choice in the wild. *Proceedings of the National Academy of Sciences USA* 99:14,897–14,902.

Cant, M. A. 2000. Social control of reproduction in banded mongooses. *Animal Behaviour* 59:147–148.

Carazo, P., & Font, E. 2010. Putting information back into biological communication. *Journal of Evolutionary Biology* 23:661–669.

Carmen, R. A., Guitar, A. E., & Dillon, H. M. 2012. Ultimate answers to proximate questions: The evolutionary motivations behind tattoos and body piercings in popular culture. *Review of General Psychology* 16:134–143.

Caro, T. 1995. Pursuit-deterrence revisited. *Trends in Ecology and Evolution* 12:500–503.

———. 2005. *Antipredator defenses in birds and mammals.* Chicago: University of Chicago Press.

Caro, T. M., & Hauser, M. D. 1992. Is there teaching in nonhuman animals? *Quarterly Review of Biology* 67:151–174.

Caro, T. M., Lombardo, L., Goldizen, A. W., & Kelly, M. 1995. Tail-flagging and other antipredator signals in white-tailed deer: New data and synthesis. *Behavioral Ecology* 6:442–450.

Carré, J. M., & Putnam, S. K. 2010. Watching a previous victory produces an increase in testosterone among elite hockey players. *Psychoneuroendocrinology* 35:475–479.

Carter, G. S., & Wilkinson, G. S. 2013. Food sharing in vampire bats: reciprocal help predicts donations more than relatedness or harassment. *Proceedings of the Royal Society B* 280:20122573.

Carter, R. J., Morton, J., & Dunnett, S. B. 2001. Motor coordination and balance in rodents. *Current Protocols in Neuroscience* 15:8–12.

Catania, K. C., Hare, J. F., & Campbell, K. L. 2008. Water shrews detect movement, shape, and smell to find prey underwater. *Proceedings of the National Academy of Sciences USA* 105:571–576.

Chaiken, M., Böhner, J., & Marler, P. 1993. Song acquisition in European starlings *Sturnus vulgaris*. A comparison of the songs of live-tutored, tape-tutored, untutored, and wild males. *Animal Behaviour* 46:1079–1090.

Champion de Crespigny, F. E., & Hoske, D. J. 2007. Sexual selection: Signals to die for. *Current Biology* 17:R853–R855.

Chandrashekar, J., Hoon, M. A., Ryba, N. J. P., & Zuker, S. C. 2006. The receptors and cells for mammalian taste. *Nature* 444:288–294.

Chapman, B. B., Brönmark, C., Nilsson, J-A., & Hansson, L-A. 2011. The ecology and evolution of partial migration. *Oikos* 120:1764–1775.

Chapman, C. A., & Chapman, L. J. 2000. Constraints on group size in red colobus and red-tailed guenons: Examining the generality of the ecological constraints model. *International Journal of Primatology* 21:565–585.

Chapman, T., Arnqvist, G., Bangham, J., & Rowe, L. 2003. Sexual conflict. *Trends in Ecology and Evolution* 18:41–47.

Charnov, E. L. 1976. Optimal foraging: Attack strategy of a mantid. *American Naturalist* 110:141–151.

Chen, S., Lee, A. Y., Bowens, N. M., Huber, R., & Kravitz, E. A. 2002. Fighting fruit flies: A model system for the study of aggression. *Proceedings of the National Academy of Sciences USA* 99:5564–5568.

Cheney, D. L., & Seyfarth, R. M. 1980. Vocal recognition in free-ranging vervet monkeys. *Animal Behaviour* 28:362–367.

———. 1985. Vervet monkey alarm calls: Manipulation through shared information? *Behaviour* 94:150–166.

Cheney, K. L., & Côté, I. M. 2005. Mutualism or parasitism? The variable outcome of cleaning symbioses. *Biology Letters* 1:162–165.

———. 2007. Aggressive mimics profit from a model-signal receiver mutualism. *Proceedings of the Royal Society B* 274:2087–2091.

Cheney, K. L., Grutter, A. S., Blomberg, S. P., & Marshall, N. J. 2009. Blue and yellow signal cleaning behavior in coral reef fishes. *Current Biology* 19:1283–1287.

Cheng, W-C., & Kam, Y-C. 2010. Paternal care and egg survivorship in a low nest-attendance Rhacophorid frog. *Zoological Studies* 49:304–310.

Chernetsov, N., Kishkinev, D., & Mouritsen, H. A. 2008. A long-distance avian migrant compensates for longitudinal displacement during spring migration. *Current Biology* 18:188–190.

Chitty, D. 1996. *Do lemmings commit suicide?* New York: Oxford University Press.

Choe, D.-H., Tsai, K., Lopez, C. M., & Campbell, K. 2014. Pheromone-assisted techniques to improve the efficacy of insecticide sprays against *Linepithema humile* (Hymenoptera: Formicidae). *Journal of Economic Entomology* 107:319–325.

Chuang, H. C., Webster, M. S., & Holmes, R. T. 1999. Extra-pair paternity and local synchrony in the black-throated blue warbler. *Auk* 116:726–736.

Chuang-Dobbs, H. C., Webster, M. S., & Holmes, R. T. 2001. The effectiveness of mate guarding by male black-throated blue warblers. *Behavioral Ecology* 12:541–546.

Citta, J. J., & Lindberg, L. S. 2007. Nest-site selection of passerines: Effects of geographic scale and public and personal information. *Ecology* 88:2034–2046.

Clark, D. L., Roberts, J. A., & Uetz, G. W. 2012. Eavesdropping and signal matching in visual courtship displays of spiders. *Biology Letters* 8:375–378.

Clarke, J. A. 2010. White-tailed ptarmigan food calls enhance chick diet choice: Learning nutritional wisdom? *Animal Behaviour* 79:25–30.

Clayton, D. F., Balakrishnan, C. N., & London, S. E. 2009. Integrating genomes, brain and behavior in the study of songbirds. *Current Biology* 19:R865–R873.

Clifton, K. E. 1989. Territory sharing by the Caribbean striped parrotfish, *Scarus iserti*: Patterns of resource abundance, group size and behaviour. *Animal Behaviour* 37:90–103.

Clobert, J., Le Galliard, J.-F., Cote, J., Meylan, S., & Massot, M. 2009. Informed dispersal, heterogeneity in animal dispersal syndromes and the dynamics of spatially structured populations. *Ecology Letters* 12:197–209.

Clutton-Brock, T. H. 1989. Mammalian mating systems. *Proceedings of the Royal Society B* 236:339–372.

———. 1991. *The evolution of parental care*. Princeton, NJ: Princeton University Press.

———. 2002. Breeding together: Kin selection and mutualism in cooperative vertebrates. *Science* 296:69–72.

———. 2007. Sexual selection in males and females. *Science* 318:1882–1885.

Clutton-Brock, T. H., & Albon, S. D. 1979. The roaring of red deer and the evolution of honest advertisement. *Behaviour* 69:145–170.

Clutton-Brock, T. H., O'Riain, M. J., Brotherton, P. N. M., Gaynor, D., Kansky, R., Griffin, A.S., & Manser, M. 1999. Selfish sentinels in cooperative mammals. *Science* 284:1640–1644.

Clutton-Brock, T. H., Russell, A. F., Sharpe, L. L., Brotherton, P. N. M., McIlrath, G. M., White, S., & Cameron. E. Z. 2001. Effects of helpers on juvenile development and survival in meerkats. *Science* 293:2446–2449.

Cobb, M. 2007. A gene mutation which changed animal behaviour: Margaret Bastock and the yellow fly. *Animal Behaviour* 74:163–169.

Cockburn, A. 2006. Prevalence of different modes of parental care in birds. *Proceedings of the Royal Society B* 273:1375–1383.

Cockrem, J. F., & Silverin, B. 2002. Sight of a predator can stimulate a corticosterone response in the great tit (*Parus major*). *General and Comparative Endocrinology* 125:248–255.

Cohas, A., & Allainé, D. 2009. Social structure influences extra-pair paternity in socially monogamous mammals. *Biology Letters* 5:313–316.

Cohas, A., Bonenfant, C., Gaillard, J.-M., & Allainé, D. 2007. Are extra-pair young better than within-pair young? A comparison of survival and dominance in alpine marmots. *Journal of Animal Ecology* 76:771–781.

Cole, E. F., Morand-Ferron, J., Hinks, A. E., & Quinn, J. L. 2012. Cognitive ability influences reproductive life history variation in the wild. *Current Biology* 22:1808–1812.

Collett, T. S., & Graham, P. 2004. Animal navigation: Path integration, visual landmarks and cognitive maps. *Current Biology* R475–R477.

Conner, J. K., & Hartl, D. L. 2004. *A primer of ecological genetics*. Sunderland, MA: Sinauer Associates, Inc.

Conner, W. E., & Corcoran, A. J. 2012. Sound strategies: The 65-million-year-old battle between bats and insects. *Annual Review of Entomology* 57:21–39.

Conradt, L., Krause, J., Couzin, I. D., & Roper, T. J. 2009. "Leading according to need" in self organizing groups. *American Naturalist* 173:304–312.

Contreras-Garduño, J., Canales-Lazcano, J., & Córdoba-Aguilar, A. 2006. Wing pigmentation, immune ability, fat reserves and territorial status in males of the rubyspot damselfly *Hetaerina americana*. *Journal of Ethology* 24:165–173.

Coolen, I., Ward, A. J. W., Hart, P. J. B., & Laland, K. N. 2005. Foraging nine-spined sticklebacks prefer to rely on public information over simpler social cues. *Behavioral Ecology* 16:865–870.

Coombs, S., & Montgomery, J. C. 1999. The enigmatic lateral line system. In *Comparative hearing: Fish and amphibians*, Springer Handbook of Auditory Research Series, vol. 11, ed. R. R. Fay & A. N. Popper, 319–362. New York: Springer.

Corcoran, A. J., Barber, J. R., & Conner, W. E. 2009. Tiger moth jams bat sonar. *Science* 325:325–327.

Cordero, M. I., Venero, C., Kruyt, N. D., & Sandi, C. 2003. Prior exposure to a single stress session facilitates subsequent contextual fear conditioning in rats: Evidence for a role of corticosterone. *Hormones and Behavior* 44:338–345.

Córdoba-Aguilar, A., & Cordera-Rivera, A. 2005. Evolution and ecology of Calopterygidae (Zygoptera: Odonta): Status of knowledge and research perspectives. *Neotropical Entomology* 34:861–879.

Cork, A., & Park, K. C. 1996. Identification of electrophysiologically-active compounds for the malaria mosquito, *Anopheles gambiae*, in human sweat extracts. *Physiological Entomology* 10:269–276.

Costall, A. 1993. How Lloyd Morgan's canon backfired. *Journal of the History of the Behavioral Sciences* 29:113–122.

Côte, I. M., & Poulin, R. 1995. Parasitism and group size in social animals: A meta-analysis. *Behavioral Ecology* 6:159–163.

Courter, J. R., & Ritchison, G. 2010. Alarm calls of tufted titmice convey information about predator size and threat. *Behavioral Ecology* 5:936–942.

Cox, G. W. 1985. The evolution of avian migration systems between temperate and tropical regions of the New World. *American Naturalist* 126:451–474.

Cratsley, C. K., & Lewis, S. M. 2003. Female preference for male courtship flashes in *Photinus ignitus* fireflies. *Behavioral Ecology* 14:135–140.

Cristol, D. A., Baker, M. C., & Carbone, C. 1999. Differential migration revisited: Latitudinal segregation by age and sex class. *Current Ornithology* 15:33–88.

Crowe, S. A., Kleven, O., Delmore, K. E., Laskemoen, T., Nocera, J. J., Lifjeld, J. T., & Robertson, K. J. 2009. Paternity assurance through frequent copulations in a wild passerine with intense sperm competition. *Animal Behaviour* 77:183–187.

Curio, E., Ernst, U., & Vieth, W. 1978. Cultural transmission of enemy recognition: One function of mobbing. *Science* 202:899–901.

Dada, M., Pilastro, A., & Bisazza, A. 2005. Male sexual harassment and female schooling behavior in the eastern mosquitofish. *Animal Behaviour* 70:463–471.

Daly, J. W., & Myers, C. W. 1967. Toxicity of Panamanian poison frogs (Dendrobates): Some biological and chemical aspects. *Science* 156: 970–973.

Daly, J. W., Spande, T. F., & Garraffo, H. M. 2005. Alkaloids from amphibian skin: A tabulation of over eight hundred compounds. *Journal of Natural Products* 68:1556–1575.

Danchin, E., Boulinier, T. & Massot, M. 1998. Conspecific reproductive success and breeding habitat selection: Implications for the study of coloniality. *Ecology* 79:2415–2428.

Darwin, C. 1859. *On the origin of species by means of natural selection*. London: John Murray.

———. 1871. *The descent of man, and selection in relation to sex*. London: John Murray.

———. 1872. *Expression of the emotions in man and animals*. London: John Murray.

DaSilva, A., Luikart, G., Yoccoz, N., Cohas, A., & Allainé, D. 2006. Genetic diversity-fitness correlation revealed by microsatellite analyses in European alpine marmots (*Marmota marmota*). *Conservation Genetics* 3:371–382.

Daszak, P., Cunningham, A. A., & Hyatt, A. D. 2000. Emerging infectious diseases to wildlife—threats to biodiversity and human health. *Science* 287:443–449.

David, M., & Giraldeau, L.-A. 2011. Zebra finches in poor condition produce more and consume more food in a producer-scrounger game. *Behavioral Ecology* 23:174–180.

David, P., Bjorksten, T., Fowler, K., & Pomiankowski, A. 2000. Condition-dependent signalling of genetic variation in stalk-eyed flies. *Nature* 406:186–188.

Davies, N. B. 1999. Cuckoos and cowbirds versus hosts: Co-evolutionary lag and equilibrium. *Ostrich: Journal of African Ornithology* 70:71–79.

Davies, N. B., Krebs, J. R., & West, S. A. 2012. *An introduction to behavioral ecology*. 4th ed. London: Wiley-Blackwell.

Davies, N. B., & Lundberg, A. 1984. Food distribution and a variable mating system in the dunnock, *Prunella modularis*. *Journal of Animal Ecology* 53:895–912.

Davis, A. K., & Rendón-Salinas, E. 2010. Are female monarch butterflies declining in eastern North America? Evidence of a 30-year change in sex ratios at Mexican overwintering sites. *Biology Letters* 6:45–47.

Dawkins, M. S. 2008. The science of animal suffering. *Ethology* 114:937–945.

Dawkins, M. S., & Guilford, T. 1991. The corruption of honest signaling. *Animal Behaviour* 41:865–873.

Dawkins, R., & Krebs, J. R. 1979. Arms races within and between species. *Proceedings of the Royal Society B* 205:489–511.

DeBose, J. L., & Nevitt, G. A. 2008. The use of odors at different spatial scales: Comparing birds with fish. *Journal of Chemical Ecology* 34:867–881.

De Brito Sanchez, G., & Giurfa, M. 2011. A comparative analysis of neural taste processing in animals. *Philosophical Transactions of the Royal Society B* 366:2171–2180.

De Sousa, V. T. T., Teresa, F. B., & de Cerqueirra Rossa-Feres, D. 2011. Predation risk and jumping behavior in *Pseudopaludicola* aff. *falcipes* tadpoles. *Behavioral Ecology* 22:940–946.

Deutschlander, M. E., Phillips, J. B., & Borland, S. C. 1999. The case for light-dependent magnetic orientation in animals. *Journal of Experimental Biology* 202:891–908.

Dewsbury, D. A. 1985. *Studying animal behavior: Autobiographies of the founders*. Chicago: University of Chicago Press.

Dias, R. I. 2006. Effects of position and flock size on vigilance and foraging behaviour of the scaled dove *Columbina squammata*. *Behavioral Processes* 73:248–252.

Diebel, C. E., Proksch, R., Green, C. R., Neilson, P., & Walker, M. M. 2000. Magnetite defines a vertebrate magnetoreceptor. *Nature* 406:299–302.

Dingemanse, N. J., Both, C., Drent, P. J., Van Oers, K., & Van Noordwijk, A. J. 2002. Repeatability and heritability of exploratory behaviour in great tits from the wild. *Animal Behaviour* 64:929–938.

Dittman, A., & Quinn, T. 1996. Homing in Pacific salmon: mechanisms and ecological basis. *Journal of Experimental Biology* 199:83–91.

Dobzhansky, T. 1973. Nothing in biology makes sense except in the light of evolution. *American Biology Teacher* 35:125–129.

Doligez, B., Cadet, C., Danchin, E., & Boulinier, T. 2003. When to use public information for breeding habitat selection? The role of environmental predictability and density dependence. *Animal Behaviour* 66:973–988.

Doligez, B., Danchin, E., & Clobert, J. 2002. Public information and breeding habitat selection in a wild bird population. *Science* 297:1168–1170.

Donahue, M. J. 2006. Allee effects and conspecific cueing jointly lead to conspecific attraction. *Oecologia* 149:33–43.

Dornhaus, A., Holly, J.-A., Pook, V. G., Worswick, G., & Franks, N. R. 2008. Why do not all workers work? Colony size and workload during emigrations in the ant *Temnothorax albipennis*. *Behavioral Ecology and Sociobiology* 63:43–51.

Doutrelant, C., McGregor, P. K., & Oliveira, R. F. 2001. The effect of an audience on intrasexual communication in male Siamese fighting fish, *Betta splendens*. *Behavioral Ecology* 12:283–286.

Drent, P. J., & Marchetti, C. 1999. Individuality, exploration and foraging in hand raised juvenile great tits. In *Proceedings of the 22nd International Ornithological Congress*, ed. N. J. Adams & R. H. Slotow, 896–914. Johannesburg: Birdlife South Africa.

Drent, P. J., van Oers, K., & van Noordwijk, A. J. 2003. Realized heritability of personalities in the great tit (*Parus major*). *Proceedings of the Royal Society B* 270:45–51.

Drickamer, L. C., Gowaty, P. A., & Holmes, C. M. 2000. Free female mate choice in house mice affects reproductive success and offspring viability and performance. *Animal Behaviour* 59:371–378.

Dubois, F., Drullion, D., & Witte, K. 2012. Social information use may lead to maladaptive decisions: A game theoretic model. *Behavioral Ecology* 23:225–231.

Dücker, G., & Rensch, B. 1977. The solution of patterned string problems by birds. *Behaviour* 62:164–173.

Dugatkin, L. A. 1992. Sexual selection and imitation: Females copy the mate choice of others. *American Naturalist* 139:1384–1389.

Dugatkin, L. A., & Godin, L-G. 1993. Female mate copying in the guppy (*Poecilia reticulata*): Age dependent effects. *Behavioral Ecology* 4:289–292.

Dugdale, H. L., MacDonald, D. W., Pope, L. C., & Burke, T. 2007. Polygynandry, extra-group paternity and multiple-paternity litters in European badger (*Meles meles*) social groups. *Molecular Ecology* 16:5294–5306.

Dukas, R. 2008. Life history of learning: Performance curves of honeybees in the wild. *Ethology* 114:1195–1200.

Dulac, C., O'Donnell, L.A., & Wu, Z. 2014. Neural control of maternal and paternal behaviors. *Science* 345:765–770.

Dunlap, A. S., & Stephens, D. W. 2009. Components of change in the evolution of learning and unlearned preference. *Proceedings of the Royal Society B* 276:3201–3208.

Dunning, D. C. 1968. Warning sounds of moths. *Zeitschrift für Tierpsychologie* 25:129–138.

Dziuk, P. J. 1996. Factors that influence the proportion of offspring sired by a male following heterospermatic insemination. *Animal Reproduction Science* 43:65–88.

East, M. L., & Hofer, H. 2001. Male spotted hyenas (*Crocuta crocuta*) queue for status in social groups dominated by females. *Behavioral Ecology* 12:558–568.

Eaton, M. D. 2005. Human vision fails to distinguish widespread sexual dichromatism among sexually "monochromatic" birds. *Proceedings of the National Academy of Sciences U.S.A.* 102:10942–10946.

Eaton, M. D., & Lanyon, S. M. 2003. The ubiquity of avian ultraviolet plumage reflectance. *Proceedings of the Royal Society B* 270:1721–1726.

Eifler, D. A., Eifler, M. A., & Harris, B. R. 2008. Foraging under risk of predation in desert grassland whiptail lizards (*Aspidoscelis uniparens*). *Journal of Ethology* 26:219–223.

Elgar, M. A., & Schneider, J. M. 2004. The evolutionary significance of sexual cannibalism. *Advances in the Study of Behavior* 34:135–164.

Elias, D. O., Kasumovic, M. M., Punzalon, D., Andrade, M. C. B., & Mason, A. C. 2008. Assessment during aggressive contests between male jumping spiders. *Animal Behaviour* 76:901–910.

Ellis, D. H., Sladen, W. J. L., Lishman, W. A., Clegg, K. R., Duff, J. W., Gee, G. F., & Lewis, J. C. 2003. Motorized migrations: The future or mere fantasy. *Bioscience* 53:260–264.

Ellis, L. L., & Carney, G. E. 2011. Socially-responsive gene expression in male *Drosophila melanogaster* is influenced by the sex of the interacting partner. *Genetics* 187:157–169.

Elwood, R. W., & Arnott, G. 2013. Assessments in contests are frequently assumed to be complex when simple explanations will suffice. *Animal Behaviour* 86:e8–e12.

Emlen, S. 1982. The evolution of cooperative breeding: an ecological constraints model. *American Naturalist* 119:29–39.

Emlen, S. T. 1967a. Migratory orientation in the indigo bunting, *Passerina cyanea*. Part I: Evidence for use of celestial cues. *Auk* 84:309–342.

———. 1967b. Migratory orientation in the indigo bunting, *Passerina cyanea*. Part II: Mechanism of celestial orientation. *Auk* 84:463–489.

———. 1970. Celestial rotation: Its importance in the development of migratory orientation. *Science* 170:1198–1201.

———. 1994. Environmental control of horn length dimorphism in the beetle *Onthophagus acuminatus* (Coleoptera: Scarabaeidae). *Proceedings of the Royal Society B* 256:131–136.

———. 2008. The evolution of animal weapons. *Annual Review of Ecology, Evolution and Systematics* 39:387–413.

Emlen, S. T., & Emlen, J. T. 1966. A technique for recording migratory orientation of captive birds. *Auk* 83:361–367.

Emlen, S. T., & Oring, L. W. 1977. Ecology, sexual selection, and the evolution of mating systems. *Science* 197:215–223.

Endler, J. A. 1980. Natural selection on color patterns in *Poecilia reticulata*. *Evolution* 34:76–91.

———. 1992. Signals, signal conditions, and the direction of evolution. *American Naturalist* 139: S125–S153.

Engh, A. L., Funk, S. M., Van Horn R. C., Scribner, K. T., Bruford, M. W., Libants, S., Szykman, M., Smale, L., & Holekamp, K. E. 2002. Reproductive skew among males in a female-dominated mammalian society. *Behavioral Ecology* 13: 193–200.

Enquist, M., & Leimar, O. 1983. Evolution of fighting behavior: Decision rules and assessment of relative strength. *Journal of Theoretical Biology* 102:387–410.

Ensminger, J. J. 2012. *Police and military dogs*. Boca Raton, FL: CRC Press.

Evans, W. G. 1964. Infrared receptors in *Melanophila acuminata* DeGreer. *Nature* 202:211.

Fausto da Silva, J. J. R., & Williams, R. J. P. 2001. *The biological chemistry of the elements: The inorganic chemistry of life*. Oxford, UK: Oxford University Press.

Fawcett, T. W., & Mowles, S. L. 2013. Assessments of fighting ability need not be cognitively complex. *Animal Behaviour* 86:e1–e7.

Feeney, W. E., Welbergen, J. A., & Langmore, N. E. 2012. The frontline of avian brood parasite-host coevolution. *Animal Behaviour* 84:3–12.

Fernández-Juricic, E., Beauchamp, G., & Bastain, B. 2007. Group-size and distance-to-neighbor effects on feeding and vigilance in brown-headed cowbirds. *Animal Behaviour* 73:771–778.

Ferrari, M. C. O., Sih, A., & Chivers, D. P. 2009. The paradox of risk allocation: A review and prospectus. *Animal Behaviour* 78:579–585.

Fertin, A., & Casas, J. 2007. Orientation towards prey in antlions: Efficient use of wave propagation in sand. *Journal of Experimental Biology* 210:3337–3343.

Field, J., & Cant, M. A. 2009. Social stability and helping in small animal societies. *Philosophical Transactions of the Royal Society B* 364:3181–3189.

Field, J., Shreeves, G., & Sumner, S. 1999. Group size queuing and helping decisions in facultatively eusocial hover wasps. *Behavioral Ecology and Sociobiology* 45:378–385.

Fildes, B. N., O'Loughlin, B. J., Bradshaw, J. L., & Ewens, W. J. 1984. Human orientation with restricted sensory information: No evidence for magnetic sensitivity. *Perception* 13:229–248.

Fiorito, G., & Scotto, P. 1992. Observational learning in *Octopus vulgaris*. *Science* 256:545–547.

Firestein, S. 2001. How the olfactory system makes sense of scents. *Nature* 413:211–218.

Fisher, R. A. 1930. *The genetical theory of natural selection*. Oxford: Clarendon Press.

———. 1966. *The design of experiments*. 8th ed. Edinburgh: Hafner.

Fitzpatrick, B. M., Shook, K., & Izally, R. 2009. Frequency-dependent selection by wild birds promotes polymorphism in model salamanders. *BMC Ecology* 9:12.

Fitzpatrick, M. J., Ben-Shahar, Y., Smid, H. M., Vet, L. E. M., Robinson, G. E., & Sokolowski, M. B. 2005. Candidate genes for behavioural ecology. *Trends in Ecology and Evolution* 20:96–104.

Fleissner, G., Holtkamp-Rötzler, E., Hanzlik, M., Winklhofer, M., Fleissner, G., Petersen, N., & Wiltschko, W. 2003. Ultrastructural analysis of a putative magnetoreceptor in the beak of homing pigeons. *Journal of Comparative Neurology* 458:350–360.

Flint, J. 2003. Analysis of quantitative trait loci that influence animal behavior. *Journal of Neurobiology* 54:46–77.

Foerder, P., Galloway, M., Barthel, T., Moore, D. E., III, & Reiss, D. 2011. Insightful problem solving in an Asian elephant. *PLoS ONE* 6:e23251.

Foley, L. E., Gegear, R. J., & Reppert, S. M. 2011. Human cryptochrome exhibits light-dependent magnetosensitivity. *Nature Communications* 2:356.

Forbes, S., Grosshans, R., & Glassey, B. 2002. Multiple incentives for parental optimism and brood reduction in blackbirds. *Ecology* 83:2529–2541.

Foster, K. R., Wenseleers, T., & Ratnieks, F. L. W., 2006. Kin selection is the key to altruism. *Trends in Ecology and Evolution* 21:57–60.

Foster, W. A., & Treherne, J. E. 1981. Evidence for the dilution effect in the selfish herd from fish predation on a marine insect. *Nature* 293:466–467.

Fowler-Finn, K. D., & Rodríguez, R. L. 2016. The causes of variation in the presence of genetic covariance between sexual traits and preferences. *Biological Reviews*. 91:498–510.

Franks, N. R., & Richardson, T. 2006. Teaching in tandem-running ants. *Nature* 439:153.

Fretwell, S. D., & Lucas, H. L. 1969. On territorial behavior and other factors influencing habitat distributions of birds. *Acta Biotheoretica* 19:16–36.

Friedl, T. W. P., & Klump, G. M. 2005. Sexual selection in the lek-breeding European treefrog: Body size, chorus attendance, random mating and good genes. *Animal Behaviour* 70:1141–1154.

Frommen, J. G. & Bakker, T. C. M. 2004: Adult three-spined sticklebacks prefer to shoal with familiar kin. *Behaviour* 141:1401–1409.

Frommen, J. G., Hiermes, M., & Bakker, T. C. M. 2009. Disentangling the effects of group size and density on shoaling decisions of three-spined sticklebacks (*Gasterosteus aculeatus*). *Behavioral Ecology and Sociobiology* 63:1141–1148.

Frommen, J. G., Luz, C., & Bakker, T. C.M. 2007. Kin discrimination in sticklebacks is mediated by social learning rather than innate recognition. *Ethology* 113:276–282.

Fu, P., Neff, B. D., & Gross, M. R. 2001. Tactic-specific success in sperm competition. *Proceedings of the Royal Society B* 268:1105–1112.

Galef, B. G., Jr. 1998. Edward Thorndike: Revolutionary psychologist, ambiguous biologist. *American Psychologist* 53:1128–1134.

Gammell, M. P., & Hardy, I. C. W. 2003. Contest duration: sizing up the opposition? *Trends in Ecology and Evolution* 18:491–493.

García-González, F., Núñez, Y., Ponz, F., Roldán, E. R. S., & Gomendio, M. 2003. Sperm competition mechanisms, confidence of paternity, and the evolution of paternal care in the golden egg bug (*Phyllomorpha laciniata*). *Evolution* 57:1078–1088.

García-Navas, V., Ortega, J., & Sanz, J. J. 2009. Heterozygosity-based assortative mating in blue tits (*Cyanistes caeruleus*): Implications for the evolution of mate choice. *Proceedings of the Royal Society B* 276:2931–2940.

Garland, T., Jr. 1994. Quantitative genetics of locomotor behavior and physiology in a garter snake. In *Quantitative genetic studies of behavioral evolution*, ed. C. R. B. Boake, 251–277. Chicago: University of Chicago Press.

Gazit, I., Goldblatt, A., & Terkel, J. 2005. Formation of an olfactory search image for explosives in sniffer dogs. *Ethology* 111:669–680.

Gegear, R. J., Foley, L. E., Casselman, A., & Reppert, S. M. 2010. Animal cryptochromes mediate magneto-reception by an unconventional photochemical mechanism. *Nature* 463:804–807.

Gentry, G. L., & Dyer, L. A. 2002. On the conditional nature of neotropical caterpillar defenses against their natural enemies. *Ecology* 83:3108–3119.

Gerhardt, H. C. 1987. Evolutionary and neurobiological implications of selective phonotaxis in the green treeefrog (*Hyla cinerea*). *Animal Behaviour* 35:1479–1489.

Gerlach, N. M., McGlothlin, J. W., Parker, P. G., & Ketterson, E. D. 2012. Promiscuous mating produces offspring with higher lifetime fitness. *Proceedings of the Royal Society B* 279:860–866.

Gesquiere, L. R., Learn, L. H., Simao, M. C. M., Onyango, P. O., Alberts, S. C., & Altmann, J. 2011. Life at the top: Rank and stress in wild male baboons. *Science* 333:357–360.

Gherardi, F., Aquiloni, L., & Tricarico, E. 2012. Revisiting social recognition in invertebrates. *Animal Cognition* 15:745–762.

Gibbons, M. E., Ferguson, A. M., & Lee, D. R. 2005. Both learning and heritability affect foraging behaviour of red-backed salamanders, *Plethodon cinereus*. *Animal Behaviour* 69:721–732.

Gibson, G., & Torr, S. J. 1999. Visual and olfactory responses of haematophagous Diptera to host stimuli. *Medical and Veterinary Entomology* 13:2–23.

Gilbert, J. D. J., & Manica, A. 2010. Parental care trade-offs and life-history relationships in insects. *American Naturalist* 176:212–226.

Gillis, E. A., Green, D. J., Middleton, H. A., & Morrissey, C. A. 2008. Life history correlates of alternative migratory strategies in American dippers. *Ecology* 89:1687–1695.

Giraldeau, L.-A., Soos, C., & Beauchamp, G. 1994. A test of the producer-scrounger foraging game in captive flocks of spice finches, *Lonchura punctulata*. *Behavioral Ecology and Sociobiology* 34:251–256.

Gleason, E. D., Fuxjager, M. J., Oyegbile, T. O., & Marler, C. A. 2009. Testosterone release and social context: When it occurs and why. *Frontiers in Neuroendocrinology* 30:460–469.

Goerlitz, H. R., & Siemers, B. M. 2007. Sensory ecology of prey rustling sounds: Acoustical features and their classification by wild grey mouse lemurs. *Functional Ecology* 21:143–153.

Goerlitz, H. R., ter Hofstede, H. M., Zeale, M.R.K., Jones, G., & Holderied, M. W. 2010. An aerial-hawking bat uses stealth echolocation to counter moth hearing. *Current Biology* 20:1568–1572.

Gomendio, M., García-González, F., Reguera, P., & Rivero, A. 2008. Male egg carrying in *Phyllomorpha laciniata* is favoured by natural not sexual selection. *Animal Behaviour* 75:763–770.

González, L. M., Margalida, A., Sánchez, R., & Oria, J. 2006. Supplmentary feeding as an effective tool for improving breeding success in the Spanish imperial eagle (*Aquila adalberti*). *Biological Conservation* 129:477–486.

Gould, J. L. 2010. Magnetoreception. *Current Biology* 20:R431–R435.

Gould, J. L., & Abel, K. P. 1981. Human homing: An elusive phenomenon. *Science* 212:1061–1063.

Gould, J. L., & Gould, C. G. 1988. *The honey bee.* New York: W. H. Freeman.

Gracheva, E. O., Cordero-Morales, J. F., González-Carachía, J. A., Ingolia, N. T., Manno, C., Aranguren, C. I., Weissman, J. S., & Julius, D. 2011. Ganglion-specific splicing of *TRPV1* underlies infrared sensation in vampire bats. *Nature* 476:88–91.

Graf, S. A., & Sokolowski, M. B. 1989. Rover/sitter *Drosophila melanogaster* larval foraging polymorphism as a function of larval development, food-patch quality, and starvation. *Journal of Insect Behavior* 2:310–313.

Grafen, A. 1990. Biological signals as handicaps. *Journal of Theoretical Biology* 144:517–546.

Gray, D. A., & Cade, W. H. 2000. Sexual selection and speciation in field crickets. *Proceedings of the National Academy of Sciences USA* 97: 14,449–14,454.

Gray, S. M., Dill, L. M., Tantu, F. Y., Loew, E. R., Herder, F., & McKinnon, J.S. 2008. Environment-contingent sexual selection in a colour polymorphic fish. *Proceedings of the Royal Society B* 275: 1785–1791.

Grayson, K. L., & Wilbur, H. M. 2009. Sex- and context-dependent migration in a pond-breeding amphibian. *Ecology* 90:306–312.

Greeney, H. F., Dyer, L. A., & Smilanich, A. M. 2012. Feeding by lepidopteran larvae is dangerous: A review of caterpillars' chemical, physiological, morphological, and behavioral defenses against natural enemies. *Invertebrate Survival Journal* 9:7–34.

Greenfield, M. D., Alem, S., Limousin, D., & Bailey, N. W. 2014. The dilemma of Fisherian sexual selection: Mate choice for indirect benefits despite rarity and overall weakness of trait-preference genetic correlation. *Evolution* 68:3524–3536.

Greenfield, M. D., & Shelly, T. E. 1985. Alternative mating strategies in a desert grasshopper: Evidence of density-dependence. *Animal Behaviour* 33:1192–1210.

Greenfield, M. D., Shelly, T. E., & Downum, K. R. 1987. Variation in host-plant quality: Implications for territoriality in a desert grasshopper. *Ecology* 68:828–838.

Greenwood, P. J. 1980. Mating systems, philopatry and dispersal in birds and mammals. *Animal Behaviour* 28:1140–1162.

Griffin, A. S. 2004. Social learning about predators: A review and prospectus. *Learning & Behavior* 32:131–140.

Griffin, D. R. 1976. *The question of animal awareness: Evolutionary continuity of mental experience.* New York: Rockefeller University Press.

———. 1984. *Animal thinking.* Cambridge, MA: Harvard University Press.

Griffin, D. R., & Speck, G. B. 2004. New evidence of animal consciousness. *Animal Cognition* 7:15–18.

Griffith, S. C. 2007. The evolution of infidelity in socially monogamous passerines: Neglected components of direct and indirect selection. *American Naturalist* 169:274–281.

Griffith, S. C., Owens, I. P. F., & Thuman, K. A. 2002. Extra pair paternity in birds: A review of interspecific variation and adaptive function. *Molecular Ecology* 11:2195–2212.

Griffith, S. C., Pryke, S. R., & Buttemer, W. A. 2011. Constrained mate choice in social monogamy and the stress of having an unattractive partner. *Proceedings of the Royal Society B* 278:2798–2805.

Griffiths, P. E. 2008. Ethology, sociobiology, evolutionary psychology. In *A companion to the philosophy of biology*, ed. S. Sarkar & A. Plutyinski, 393–414. Blackwell Companions to Philosophy. Oxford, UK: Wiley-Blackwell.

Grimm, D. 2011. Are dolphins too smart for captivity? *Science* 332:526–529.

Gross, M. R. 1996. Alternative reproductive strategies and tactics: Diversity within sexes. *Trends in Ecology and Evolution* 11:92–98.

———. 2005. The evolution of parental care. *Quarterly Review of Biology* 80:37–45.

Gross, M. R., & Sargent, R. C. 1985. The evolution of male and female parental care in fishes. *American Zoologist* 25:807–822.

Grutter, A. S. 1994. Spatial and temporal variations of the ectoparasites of 7 reef fish species from Lizard Island and Heron Island, Australia. *Marine Ecology Progress Series* 115:21–30.

Grutter, A. S., & Bshary, R. 2003. Cleaner wrasse prefer client mucus: Support for partner control mechanisms in cleaning interactions. *Proceedings of the Royal Society B* 270 (suppl.): S242–S244.

Grütter, C., Balbuena, M. S., & Farina, W. M. 2008. Informational conflicts created by the waggle dance. *Proceedings of the Royal Society B* 275:1321–1327.

Grütter, C., & Farina, W. M. 2009. The honeybee waggle dance: Can we follow the steps? *Trends in Ecology and Evolution* 24:242–247.

Gubernick, D. J., & Alberts, J. R. 1987. The biparental care system of the California mouse, *Peromyscus californicus*. *Journal of Comparative Psychology* 101:169–177.

Gubernick, D. J., & Teferi, T. 2000. Adaptive significance of male parental care in a monogamous mammal. *Proceedings of the Royal Society B* 267:147–150.

Guerra, P. A., Gegear, R. J., & Reppert, S. M. 2014. A magnetic compass aids monarch butterfly migration. *Nature Communications* 5:4164. DOI: 10.1038/ncomms5164.

Gusset, M., & Dick, G. 2011. The global reach of zoos and aquariums in visitor numbers and conservation expenditures. *Zoo Biology* 30:566–569.

Gwynne, D. T. 2008. Sexual selection over nuptial gifts in insects. *Annual Review of Entomology* 53: 83–101.

Haesler, A. D., & Wisby, W. J. 1951. Discrimination of stream odors by fishes and its relation to parent stream behavior. *American Naturalist* 85:223–238.

Haesler, S., Rochefort, C., Georgi, B., Licznerski, P., Osten, P., & Scharff, C. 2007. Incomplete and inaccurate vocal imitation after knockdown of *FoxP2*

in songbird basal ganglia nucleus area X. *PLoS Biology* 5:e321.

Haesler, S., Wada, K., Nshdejan, A., Morrisey, E. E., Lints, T., Jarvis, E. D., & Scharff, C. 2004. *FoxP2* expression in avian vocal learners and non-learners. *Journal of Neuroscience* 24:3164–3175.

Hahn, B. A., & Silverman, E. D. 2006. Social cues facilitate habitat selection: American redstarts establish breeding territories in response to song. *Biology Letters* 2:337–340.

Hamilton, W. D. 1963. The evolution of altruistic behavior. *American Naturalist* 97:354–356.

———. 1964. The genetical evolution of social behavior. *Journal of Theoretical Biology* 7:1–52.

———. 1971. Geometry for the selfish herd. *Journal of Theoretical Biology* 31:295–311.

Hamilton, W. D., & May, R. M. 1977. Dispersal in stable habitats. *Nature* 269:578–581.

Hamilton, W. D., & Zuk, M. 1982. Heritable true fitness and bright birds: A role for parasites? *Science* 218:384–387.

Hanauske, M., Kunz, J., Bernius, S., & König, W. 2009. Doves and hawks in economics revisited. An evolutionary quantum game theory-based analysis of financial crises. *Physica A: Statistical Mechanics and Its Applications* 389:5084–5102.

Handley, L. J. L., & Perrin, N. 2007. Advances in our understanding of mammalian sex-biased dispersal. *Molecular Ecology* 16:1559–1578.

Hansen, B. T., Erik, L., & Slagsvold, T. 2008. Imprinted species recognition lasts for life in free-living great tits and blue tits. *Animal Behaviour* 75:921–927.

Hanssen, S. A., Hasselquist, D., Folstad, I., & Erikstad, K. E. 2005. Cost of reproduction in a long-lived bird: Incubation effort reduces immune function and future reproduction. *Proceedings of the Royal Society B* 272:1039–1046.

Harsha, S., Satischandra, K., Kudavidanage, E. P., Kotagama, S. W., & Goodale, E. 2007. The benefits of joining mixed-species flocks for greater racket-tailed drongos *Dicrurus paradiseus*. *Forktail* 23:145–148.

Hart, P. J., & Freed, L. A. 2005. Predator avoidance as a function of flocking in the sexually dichromatic Hawaii akepa. *Journal of Ethology* 23:29–33.

Hartl, D. 2000. *A primer of population genetics*. 3rd ed. Sunderland, MA: Sinauer Associates.

Hatchwell, B. J., Russell, A. F., MacColl, A. D. C., Ross, D. J., Fowlie, M. K., & McGowan, A. 2004. Helpers increase long-term but not short-term productivity in cooperatively breeding long-tailed tits. *Behavioral Ecology* 15:1–10.

Haugen, T. O., Winfield, I. J., Asbjörn Vøllestad, L., Fletcher, J. M., James J. B., & Stenseth, N. C. 2006. The ideal free pike: 50 years of fitness-maximizing dispersal in Windermere. *Proceedings of the Royal Society B* 273:2917–2924.

Healy, S. D., & Rowe, C. 2013. Costs and benefits of evolving a larger brain: Doubts over the evidence that large brains lead to better cognition. *Animal Behaviour* 86:e1–e3.

Hedrick, A. V. 1994. The heritability of mate-attractive traits: A case study on field crickets. In *Quantitative genetic studies of behavioral evolution*, ed. C. R. B. Boake, 228–250. Chicago: University of Chicago Press.

Heg, D., Bachar, Z., Brouwer, L., & Taborsky, M. 2004. Predation risk is an ecological constraint for helper dispersal in a cooperatively breeding cichlid. *Proceedings of the Royal Society of London B* 271:2367–2374.

Heinrich, B., & Bugnyar, T. 2005. Testing problem solving in ravens: String-pulling to reach food. *Ethology* 111:962–976.

Herberholz, J., McCurdy, C., & Edwards, D. H. 2007. Direct benefits of social dominance in juvenile crayfish. *Biological Bulletin* 213:21–27.

Hering, H., & Sheng, M. 2001. Dendritic spines: Structure, dynamics and regulation. *Nature Reviews Neuroscience* 2:880–888.

Herman, C. S., & Valone T. J. 2000. The effect of mammalian predator scent on the foraging behavior of *Dipodomys merriami*. *Oikos* 91:139–145.

Hill, G. E. 1990. Female house finches prefer colourful males: Sexual selection for a condition-dependent trait. *Animal Behaviour* 40:563–572.

Hill, P. S. M. 2009. How do animals use substrate-borne vibrations as an information source? *Naturwissenschaften* 96:1355–1371.

Hirschenhauser, K., & Oliveira, R. F. 2006. Social modulation of androgens in male vertebrates: Meta-analyses of the challenge hypothesis. *Animal Behaviour* 71:265–277.

Hoare, D. J., Couzin, I. D., Godin, J.-G. J., & Krause, J. 2004. Context-dependent group size choice in fish. *Animal Behaviour* 67:155–164.

Hoefler, C. D., Carlascio, A. L., Persons, M. H., & Rypstra, A. L. 2009. Male courtship repeatability and potential indirect genetic benefits in a wolf spider. *Animal Behaviour* 78:183–188.

Hoefler, C. D., Persons, M. H., & Rypstra, A. L. 2008. Evolutionarily costly courtship displays in a wolf spider: a test of viability indicator theory. *Behavioral Ecology* 19:974–979.

Höglund, J., & Robertson, J. G. M. 1990. Female preferences, male decision rules and the evolution of leks in the great snipe *Gallinago media*. *Animal Behaviour* 40:15–22.

Hoke, K. L., Ryan, M. J., & Wilczynski, W. 2010. Sexually dimorphic sensory gating drives behavioral differences in túngara frogs. *Journal of Experimental Biology* 213:3463–3472.

Holdo, R. M., Holt, R. D., & Fryxell, J. M. 2009. Opposing rainfall and plant nutritional gradients best explain the wildebeest migration in the Serengeti. *American Naturalist* 173:431–445.

Holland, B., & Rice, W. R. 1998. Chase-away sexual selection: Antagonistic seduction versus resistance. *Evolution* 52:1–7.

Holland, R. A., Wikelski, M., & Wilcove, D. S. 2006. How and why do insects migrate? *Science* 313:794–796.

Hoogland, J. L., & Sherman, P. W. 1976. Advantages and disadvantages of bank swallow (*Riparia riparia*) coloniality. *Ecological Monographs* 46:33–58.

Hoppitt, W., & Laland, K. N. 2008. Social processes influencing learning in animals: A review of the evidence. *Advances in the Study of Behavior* 38:105–165.

Horn, G. 1998. Visual imprinting and the neural mechanisms of recognition memory. *Trends in Neurosciences* 21:300–305.

———. 2004. Pathways of the past: The imprint of memory. *Nature Reviews Neuroscience* 5:108–120.

Horowitz, A. 2009. Disambiguating the "guilty look": Salient prompts to a familiar dog behavior. *Behavioral Processes* 81:447–452.

Horváth, G., & Varjú, D. 2004. *Polarized light in animal vision*. Berlin: Springer Science & Business Media.

Houck, L. D., & Drickamer, L. C. 1996. *Foundations of animal behavior: Classic papers with commentaries*. Chicago: University of Chicago Press.

Houde, A. E., & Endler, J. A. 1990. Correlated evolution of female mating preference and male color patterns in the guppy *Poecilia reticulata*. *Science* 248:1405–1408.

How, M. J., Hemmi, J. M., Zeil, J., & Peters, R. 2008. Claw waving display changes with receiver distance in fiddler crabs, *Uca perplexa*. *Animal Behaviour* 75:1015–1022.

Howard, E., & Davis, A. K. 2009. The fall migration flyways of monarch butterflies in eastern North America revealed by citizen scientists. *Journal of Insect Conservation* 13:279–286.

Howell, T. F., & Bennett, P. C. 2011. Puppy power! Using social cognition research tasks to improve socialization practices for domestic dogs (*Canis familiaris*). *Journal of Veterinary Behavior* 6:195–204.

Hristov, N. I., & Conner, W. E. 2005. Sound strategy: Acoustic aposematism in the bat-tiger moth arms race. *Naturwissenschaften* 92:164–169.

Hsu, Y., Earley, R. L., & Wolf, L. L. 2006. Modulation of aggressive behaviour by fighting experience: Mechanisms and contest outcomes. *Biological Reviews* 81:33–74.

Hsu, Y., Lee, S., Chen, M.-H., Yang, S.-Y., & Cheng, K.-C. 2008. Switching assessment strategy during a contest: Fighting in killifish *Kryptolebias marmoratus*. *Animal Behaviour* 75:1641–1649.

Huang, Y.-C., & Hessler, N. A. 2008. Social modulation during songbird courtship potentiates midbrain dopaminergic neurons. *PLoS One* 3:e3281.

Hubbs, A. H., Millar, J. S., & Wiebe, J. P. 2000. Effect of brief exposure to a potential predator on cortisol concentrations in female Columbian ground squirrels (*Spermophilus columbianus*). *Canadian Journal of Zoology* 78:578–587.

Hughes, W. O. H., Oldroyd, B. P., Beekman, M., & Ratnieks, F. L. W. 2008. Ancestral monogamy shows kin selection is key to the evolution of eusociality. *Science* 320:1213–1216.

Hult, C. A. 2002. *Research and writing across the curriculum*. New York: Longman.

Humfeld, S. C. 2008. Intersexual dynamics mediate the expression of satellite mating tactics: Unattractive males and parallel female preferences. *Animal Behaviour* 75:205–215.

Hummel, D. 1983. Aerodynamic aspects of formation flying in birds. *Journal of Theoretical Biology* 104:321–347.

Hunt, J. H. 2012. A conceptual model for the origin of worker behaviour and adaptation of eusociality. *Journal of Evolutionary Biology* 25:1–19.

Hunt, J. H., & Amdam, G. V. 2005. Bivoltinism as an antecedent to eusociality in the paper wasp genus *Polistes*. *Science* 308:264–267.

Hunt, S., Low, J., & Burns, K. C. 2008. Adaptive numerical competency in a food-hoarding songbird. *Proceedings of the Royal Society B* 275:2373–2379.

Huntingford, F., & Turner, A. K. 1987. *Animal conflict*. London: Chapman & Hall.

Hutchinson, J. M. C., Wilke, A., & Todd, P. M. 2008. Patch leaving in humans: Can a generalist adapt its rules to dispersal of items across patches? *Animal Behaviour* 75:1331–1349.

Inouye, D. W., Barr, B., Armitage, K. B., & Inouye, B. 2000. Climate change is affecting altitudinal migrants and hibernating species. *Proceedings of the National Academy of Sciences USA* 97:1630–1633.

Institute for Laboratory Animal Research. 1996. *Guide for the care and use of laboratory animals*. Washington, D.C.: National Academy Press.

Iyengar, E. V. 2008. Kleptoparasitic interactions throughout the animal kingdom and a re-evaluation, based on participant mobility, of the conditions promoting the evolution of kleptoparasitism. *Biological Journal of the Linnean Society* 93:745–762.

James, P. C., & Verbeek, N. A. M. 1984. Temporal and energetic aspects of food storage in northwestern crows. *Ardea* 72:207–215.

Jaquiéry, J., Broquet, T., Aguilar, C., Evanno, G., & Perrin, N. 2009. Good genes drive female choice for mating partners in the lek-breeding European treefrog. *Evolution* 64:108–115.

Jensen, P., & Yngvesson, J. 1998. Aggression between unacquainted pigs—sequential assessment and effects of familiarity and weight. *Applied Animal Behaviour Science* 58:49–61.

Jeon, J. 2008. Evolution of parental favoritism among different-aged offspring. *Behavioral Ecology* 19:344–352.

Johansson, B. G., & Jones, T. M. 2007. The role of chemical communication in mate choice. *Biological Reviews* 82:265–289.

Johansson, B. G., Jones, T. M., & Widemo, F. 2005. Cost of pheromone production in a lekking *Drosophila*. *Animal Behaviour* 69:851–858.

Johnson, J. C., & Sih, A. 2005. Precopulatory sexual cannibalism in fishing spiders (*Dolomedes triton*): A role for behavioral syndromes. *Behavioral Ecology and Sociobiology* 58:390–396.

Johnsson, J. I., & Kjallman-Eriksson, K. 2008. Cryptic prey colouration increases search time in brown trout (*Salmo trutta*): Effects of learning and body size. *Behavioral Ecology and Sociobiology* 62: 1613–1620.

Johnstone, R. A. 2008. Sexual selection, honest advertisement and the handicap principle: Reviewing the evidence. *Biological Reviews* 70:1–65.

Johnstone, R. A., & Grafen, A. 1993. Dishonesty and the handicap principle. *Animal Behaviour* 46:759–764.

Jones, A. G., & Ratterman, N. L. 2009. Mate choice and sexual selection: What have we learned since Darwin? *Proceedings of the National Academy of Sciences USA* 106:10,001–10,008.

Jones, A. G., Walker, D., & Avise, J. C. 2001. Genetic evidence for extreme polyandry and extraordinary sex-role reversal in a pipefish. *Proceedings of the Royal Society B* 267:677–680.

Jones, J. E., & Kamil, A. C. 2001. The use of relative and absolute bearings by Clark's nutcrackers, *Nucifraga columbiana*. *Animal Learning and Behavior* 29: 120–132.

Kajiura, S. M., & Holland, K. N. 2002. Electroreception in juvenile scalloped hammerhead and sandbar sharks. *Journal of Experimental Biology* 205:3609–3621.

Karolinska Institutet. 1973. Nobel Prize in physiology or medicine—1973 [press release]. http://www.nobelprize.org/nobel_prizes/medicine/laureates/1973/press.html (accessed July 8, 2106).

Kasai, H., Fukuda, M., Watanabe, S., Hayashi-Takagi, A., & Noguchi, J. 2010. Structural dynamics of dendritic spines in memory and cognition. *Trends in Neurosciences* 33:121–129.

Kaspari, M. 1996. Worker size and seed size selection by harvestor ants in a Neotropical forest. *Oecologia* 105:397–404.

Kaspari, M., Chang, C., & Weaver, J. 2010. Salted roads and sodium limitation in a northern forest ant community. *Ecological Entomology* 35:543–548.

Kats, L. B., & Dill, L. M. 1998. The scent of death: Chemosensory assessment of predation risk by prey animals. *Ecoscience* 5:361–394.

Keagy, J., Savard, J.-F., & Borgia, G. 2009. Male satin bowerbird problem-solving ability predicts mating success. *Animal Behaviour* 78:809–817.

Keeton, W. T. 1974. The navigational and orientational basis of homing in birds. *Advances in the Study of Behavior* 5:47–132.

Kelber, A., Vorobyev, M., & Osorio, D. 2003. Animal colour vision—behavioural tests and physiological concepts. *Behavioural Reviews* 78:81–118.

Kemp, D. J., & Wiklund, C. 2004. Residency effects in animal contests. *Proceedings of the Royal Society B* 271:1707–1711.

Kennedy, J. S. 1983. Zigzagging and casting as a programmed response to wind-borne odour: A review. *Physiological Entomology* 8:109–120.

Kennedy, P. L., & Ward, J. M. 2003. Effects of experimental food supplementation on movements of juvenile northern goshawks (*Accipiter gentilis atricapillus*). *Oecologia* 134: 284–291.

Kent, C. F., Daskalchuk, T., Cook, L., Sokolowski, M. B., & Greenspan, R. J. 2009. The *Drosophila* foraging gene mediates adult plasticity and gene-environment interactions in behavior, metabolites, and gene expression in response to food deprivation. *PLoS Genetics* 5:e1000609.

Kilpatrick, A. M. 2003. The impact of thermal regulatory costs on foraging behavior: A test with American crows (*Corvus brachyrhynchos*) and eastern grey squirrels (*Sciurus carolinensis*). *Evolutionary Ecology Research* 5:781–786.

Kirschvink, J. L., Walker, M. M., & Diebel, C. E. 2001. Magnetite-based magnetoreception. *Current Opinion in Neurobiology* 11:462–467.

Knapp, R., & Neff, B. D. 2007. Steroid hormones in bluegill, a species with male alternative reproductive tactics including female mimicry. *Biology Letters* 3:628–631.

Kodric-Brown, A. 1986. Satellites and sneakers: Opportunistic male breeding tactics in pupfish (*Cyprinodon pecosensis*). *Behavioral Ecology and Sociobiology* 19:425–432.

Kodric-Brown, A., & Brown, J. H. 1984. Truth in advertising: The kinds of traits favored by sexual selection. *American Naturalist* 124:309–323.

Koh, T. H., Seah, W. K., Yap, L.-M. Y. L., & Daiqin, L. 2009. Pheromone-based female mate choice and its effect on reproductive investment in a spitting spider. *Behavioral Ecology and Sociobiology* 63:923–930.

Köhler, W. 1925. *The mentality of apes*. New York: Harcourt, Brace and Co.

Kokko, H. 1999. Competition for early arrival in migratory birds. *Journal of Animal Ecology* 68: 940–950.

———. 2013. Conflict and restraint in animal species. In *War, peace, and human nature: The convergence of evolutionary and cultural views*, ed. D. P. Fry, 38–53. New York: Oxford University Press.

Kokko, H., Brooks, R., Jennions, M. D., & Morley, J. 2002. The evolution of mate choice and mating biases. *Proceedings of the Royal Society B* 270:653–664.

Kokko, H., & Heubel, K. U. 2011. Prudent males, group adaptation, and the tragedy of the commons. *Oikos* 120:641–656.

Kokko, H., & Jennions, M. 2003. It takes two to tango. *Trends in Ecology and Evolution* 18:103–104.

Kokko, H., Jennions, M. D., & Brooks, R. 2006. Unifying and testing models of sexual selection. *Annual Review of Ecology and Systematics* 37: 43–66.

Kokko, H., & Johnstone, R. A. 1999. Social queuing in animal societies: A dynamic model of reproductive skew. *Proceedings of the Royal Society B* 266:571–578.

Kokko, H., López-Sepulcre, A., & Morrell, L. J. 2006. From hawks and doves to self-consistent games of territorial behavior. *American Naturalist* 167: 901–912.

Kolm, N., & Ahnesjö, I. 2005. Do egg size and parental care coevolve in fishes? *Journal of Fish Biology* 66:1499–1515.

Komdeur, J. 1992. Importance of habitat saturation and territory quality for evolution of cooperative breeding in the Seychelles warbler. *Nature* 358:493–495.

Komdeur, J., Bullock, I. D., & Rands, M. R. W. 1991. Conserving the Seychelles warbler *Acrocephalus sechellensis* by translocation: a transfer from Cousin Island to Aride Island. *Bird Conservation International* 1:177–185.

Kondo, N., & Watanabe, S. 2009. Contact calls: Information and social function. *Japanese Psychological Research* 51:197–208.

Kotrschal, A., Rogell, B., Bundsen, A., Svensson, B., Zajitschek, S., Brännström, I., Immler, S., Maklakov, A. A., & Kolm, N. 2013. Artificial selection on relative brain size in the guppy reveals costs and benefits of evolving a large brain. *Current Biology* 23:168–171.

Krakauer, A. H. 2005. Kin selection and cooperative courtship in wild turkeys. *Nature* 434:69–72.

Kramer, G. 1952. Experiments on bird orientation. *Ibis* 94:265–285.

Krams, I., Krama, T., Iguane, K., & Mänd, R. 2008. Experimental evidence of reciprocal altruism in the pied flycatcher. *Behavioral Ecology and Sociobiology* 62:599–605.

Krause, J., & Ruxton, G. D. 2002. *Living in groups*. Oxford, UK: Oxford University Press.

Krebs, J. R., & Dawkins, R. 1984. Animal signals: Mind-reading and manipulation. In *Behavioural ecology: An evolutionary approach*, 2nd ed, ed. J. R. Krebs & N. B. Davies, 380–402. Oxford, UK: Blackwell.

Krebs, J. R., Sherry, D. F., Healy, S. D., Perry, H., & Vaccarino, A. L. 1989. Hippocampal specialization of foodstoring birds. *Proceedings of the National Academy of Sciences USA* 86:1388–1392.

Kress, S. W., & Nettleship, D. N. 1988. Re-establishment of Atlantic puffins (*Fratercula arctica*) at a former breeding site in the gulf of Maine. *Journal of Field Ornithology* 59:161–170.

Krieger, M. J. B., & Ross, K. G. 2002. Identification of a major gene regulating complex social behavior. *Science* 295:328–332.

Kuba, M. J., Byme, R. A., Meisel, D. V., & Mather, J. A. 2006. When do octopuses play? Effects of repeated testing, object type, age, and food deprivation on object play in *Octopus vulgaris*. *Journal of Comparative Psychology* 120:184–190.

Kuchta, S. R., Krakauer, A. H., & Sinervo, B. 2008. Why does the yellow-eyed Ensatina have yellow eyes? Batesian mimicry of Pacific newts (genus *Taricha*) by the salamander *Ensatina eschscholtzii xanthoptica*. *Evolution* 62:984–990.

Kulahci, I., Dornhaus, A., & Papaj, D. R. 2008. Multimodal signals enhance decision making in foraging bumblebees. *Proceedings of the Royal Society B* 275:797–802.

Kümmerli, R., & Keller, L. 2007. Reproductive specialization in multiple-queen colonies of the ant *Formica exsecta*. *Behavioral Ecology* 18:375–383.

Kusayama, T., Bischof, H.-J., & Watanabe, S. 2000. Responses to mirror-image stimulation in jungle crows (*Corvus macrorhynchos*). *Animal Cognition* 3:61–64.

Lack, D. 1954. *The natural regulation of animal populations*. London: Oxford University Press.

LaFleur, D. L., Lozano, G. A., & Sclafani, M. 1997. Female mate-choice copying in guppies, *Poecilia reticulata*: A re-evaluation. *Animal Behaviour* 54:579–586.

Laine, C., & Mulrow, C. 2003. Peer review: Integral to science and indispensible to annals. *Annals of Internal Medicine* 139:1038–1040.

Laiolo, P. 2010. The emerging significance of bioacoustics in animal species conservation. *Biological Conservation* 143:1635–1645.

Laland, K. N., & Janik, V. M. 2006. The animal culture debate. *Trends in Ecology and Evolution* 21:542–547.

Laland, K. N., Sterelny, K., Odling-Smee, J., Hoppitt, W., & Uller, T. 2011. Cause and effect in biology revisited: Is Mayr's proximate-ultimate dichotomy still useful? *Science* 334:1512–1516.

Lande, R. 1981. Models of speciation by sexual selection on polygenic traits. *Proceedings of the National Academy of Sciences USA* 78:3721–3725.

Lane, J. E., Gunn, M. R., Slate, J., & Coltman, D. W. 2008. Female multiple mating and paternity in free-ranging North American red squirrels. *Animal Behaviour* 75:1927–1937.

Laufer, G., & Barreneche, J. M. 2008. Re-description of the tadpole of *Pseudopaludicola falcipes* (Anura: Leiuperidae), with comments on larval diversity of the genus. *Zootaxa* 1760:50–58.

Laundré, J. W., Hernández, L., & Altendorf, K. B. 2001. Wolves, elk, and bison: Reestablishing the "landscape of fear" in Yellowstone National Park, U.S.A. *Canadian Journal of Zoology* 79:1401–1409.

Lazarus, J., & Inglis, I. R. 1986. Shared and unshared parental investment, parent-offspring conflict and brood size. *Animal Behaviour* 34:1791–1804.

Leadbeater, E., Raine, N. E., & Chittka, L. 2006. Social learning: Ants and the meaning of teaching. *Current Biology* 16:R323–R325.

Leggett, H. C., Mouden, C. E., Wild, G., & West, S. 2012. Promiscuity and the evolution of cooperative breeding. *Proceedings of the Royal Society B* 279:1405–1411.

Lehner, P. N. 1998. *Handbook of ethological methods*. Cambridge: Cambridge University Press.

Leigh, E. G., Jr. 2010. The group selection controversy. *Journal of Evolutionary Biology* 23:6–19.

Leimar, O., & Hammerstein, P. 2001. Evolution of cooperation through indirect reciprocity. *Proceedings of the Royal Society B* 268:745–753.

Leinfelder, I., de Vries, H., Deleu, R., & Nelissen, M. 2001. Rank and grooming reciprocity among females in a mixed-sex group of captive hamadryas baboons. *American Journal of Primatology* 55:25–42.

Leisler, B., Winkler, H., & Wink, M. 2002. Evolution of breeding systems in Acrocephaline warblers. *Auk* 119:379–390.

Levey, D. J., & Stiles, F. G. 1992. Evolutionary precursors of long-distance migration: Resource availability and movement patterns in neotropical landbirds. *American Naturalist* 140:447–476.

Levitis, D. A., Lidicker, W. Z., Jr., & Freund, G. 2009. Behavioural biologists do not agree on what constitutes behaviour. *Animal Behaviour* 78:103–110.

Lewis, S. M., Cratsley, C. K., & Rooney, J. 2004. Nuptial gifts and sexual selection in *Photinus* fireflies. *Integrative and Comparative Biology* 44:234–237.

Lifjeld, J. T., & Robertson, R. J. 1992. Female control of extra-pair fertilization in tree swallows. *Behavioral Ecology and Sociobiology* 31:89–96.

Lifjeld, J. T., Slagsvold, T., & Ellegren, H. 1997. Experimental mate switching in pied flycatchers: Male copulatory access and fertilization success. *Animal Behaviour* 53:1225–1232.

Lihoreau, M., & Rivault, C. 2009. Kin recognition via cuticular hydrocarbons shapes cockroach social life. *Behavioral Ecology* 20:46–53.

Lima, S. L. 1985. Maximizing feeding efficiency and minimizing time exposed to predators: A trade-off in the black-capped chickadee. *Oecologia* 66:60–67.

———. 1998. Stress and decision-making under the risk of predation: Recent developments from behavioral, reproductive and ecological perspectives. *Advances in the Study of Behavior* 27:215–290.

Lima, S. L., Mitchell, W. A., & Roth, T. C. 2003. Predators feeding on behaviorally responsive prey: Some implications for classical models of optimal diet choice. *Evolutionary Ecology Research* 5:1083–1102.

Lima, S. L., Valone, T. J., & Caraco, T. 1985. Foraging-efficiency-predation-risk trade-off in the grey squirrel. *Animal Behaviour* 33:155–165.

Lindburg, D. G., & Fitch-Snyder, H. 1994. Use of behavior to evaluate reproductive problems in captive mammals. *Zoo Biology* 13:433–445.

Linklater, W. L., Cameron, E. Z., Minot, E. O., & Stafford, K. J. 1999. Stallion harassment and the mating system of horses. *Animal Behaviour* 58:295–306.

Linklater, W. L., Cameron, E. Z., Stafford, K. J., & Veltman, C. J. 2000. Social and spatial structure and range use by Kaimanawa wild horses (*Equus caballus*: Equidae). *New Zealand Journal of Ecology* 24:139–152.

Lishman, W. A., Teets, T. L., Duff, J. W., Sladen, W. J. L., Shire, G. G., Goolsby, K. M., Bezner Kerr, W. A., & Urbanke, R. P. 1997. A reintroduction technique for migratory birds: Leading Canada geese and isolation-reared sand-hill cranes with ultralight aircraft. *Proceedings of the North American Crane Workshop* 7:96–104.

Lloyd, J. E. 1975. Aggressive mimicry in *Photuris* fireflies: Signal repertoires by femmes fatales. *Science* 187:452–453.

Lohmann, K. J. 1991. Magnetic orientation by hatchling loggerhead sea turtles (*Caretta caretta*). *Journal of Experimental Biology* 155:37–49.

Lohmann, K. J., Cain, S. D., Dodge, S. A., & Lohmann, C. M. 2001. Regional magnetic fields as navigational markers for sea turtles. *Science* 294:364–366.

Lohmann, K. J., Lohmann, C. M. F., Ehrhart, L. M., Bagley, D. A., & Swing, T. 2004. Animal behaviour: Geomagnetic map used in sea-turtle navigation. *Nature* 428:909–910.

Lohmann, K. J., Putnam, N. F., & Lohmann, C. M. F. 2008. Geomagnetic imprinting: A unifying hypothesis of long-distance natal homing in salmon and sea turtles. *Proceedings of the National Academy of Sciences USA* 105:19,096–19,101.

Lonsdorf, E. V. 2005. Sex differences in the development of termite-fishing skills in the wild chimpanzee, *Pan troglodytes schweinfurthii*, of Gombe National Park, Tanzania. *Animal Behaviour* 70:673–683.

Lorenz, K. 1935. Companions as factors in the bird's environment. *Studies in Animal and Human Behavior* 1:101–258.

———. 1950. The comparative method in studying innate behavior. *Symposium of the Society for Experimental Biology* 4:221–268.

Lorenz, K. Z., & Tinbergen, N. 1957. Taxis and instinct: Taxis and instinctive action in the egg-retrieving behavior of the graylag goose. In *Instinctive behavior: The development of a modern concept*, ed. C. H. Schiller, 176–208. New York: International Universities Press.

Lormée, H., Jouventin, P., Chastel, O., & Mauget, R. 1999. Endocrine correlates of parental care in an Antarctic winter breeding seabird, the emperor penguin, *Aptenodytes forsteri*. *Hormones and Behavior* 35:9–17.

Loyau, A., Saint Jalme, M., & Sorci, G. 2005. Intra- and intersexual selection for multiple traits in the peacock (*Pavo cristatus*). *Ethology* 111:810–820.

———. 2007. Non-defendable resources affect peafowl lek organization: A male removal experiment. *Behavioural Processes* 74:64–70.

Lozano, G. A. 2009. Multiple cues in mate selection: The sexual interference hypothesis. *Bioscience Hypotheses* 2:37–42.

Lu, T., Qiu, Y. T., Wang, G., Kwon, J. Y., Rutzler, M., Kwon, H.-W., Pitts, R. J., et al. 2007. Odor coding in the maxillary palp of the malaria vector mosquito *Anopheles gambiae*. *Current Biology* 17:1533–1544.

Lucas, B. K., Ormandy, C. J., Binart, N., Bridges, R. S., & Kelly, P. A. 1998. Null mutation of the prolactin receptor gene produces a defect in maternal behavior. *Endocrinology* 139:4102–4107.

Lucas, J. R., Brodin, A., de Kort, S. R., & Clayton, N. S. 2004. Does hippocampal size correlate with degree of caching specialization? *Proceedings of the Royal Society B* 271:2423–2429.

Luck, N., Dejonghe, B., Fruchard, S., Huguenin, S., & Joly, D. 2007. Male and female effects on sperm precedence in the giant sperm species *Drosophila bifurca*. *Genetica* 130:257–265.

Mace, T. R. 2000. Time budget and pair-bond dynamics in the comb-crested jacana *Irediparra gallinacea*: A test of hypothesis. *Emu* 100:31–41.

Mackey, T. K. C. 2001. The genetic architecture of quantitative traits. *Annual Review of Genetics* 35:303–339.

MacLean, E. L., Matthews, L. J., Hare, B. H., Nunn, C. L., Anderson, R. C., Aureli, F., Brannon, E. M., et al. 2012. How does cognition evolve? *Phylogenetic Comparative Psychology* 15:223–238.

MacNab, R. M., & Koshland, D. E. 1972. The gradient-sensing mechanism in bacterial chemotaxis. *Proceedings of the National Academy of Sciences U.S.A.* 69:2509–2512.

Magnhagen, C. 1991. Predation risk as a cost of reproduction. *Trends in Ecology & Evolution* 6:183–186.

Mank, J. E., Promislow, D. E. L., & Avise, J. C. 2005. Phylogenetic perspectives in the evolution of parental care in ray-finned fishes. *Evolution* 59:1570–1578.

Manríquez, K. C., Pardo, L. M., Wells, R. J. D., & Palma, A. T. 2008. Crypsis in *Paraxanthus barbiger* (Decapoda: Brachyura): Mechanisms against visual predators. *Journal of Crustacean Biology* 28:473–479.

Manson, J. H., Navarrete, C., Silk, J. B., & Perry, S. 2004. Time-matched grooming in female primates? New analyses from two species. *Animal Behaviour* 67:493–500.

Maple, T., & Hoff, M. P. 1982. *Gorilla behavior*. New York: Van Nostrand Reinhold Company.

Marchetti, K. 1993. Dark habitats and bright birds illustrate the role of the environment in species diversity. *Nature* 362:149–152.

Marks, C., West, T. N., Bagatto, B., & Moore, F. B.-G. 2005. Developmental environment alters conditional aggression in zebrafish. *Copeia* 2005:901–908.

Marler, P., & Peters, S. 1977. Selective vocal learning in a sparrow. *Science* 198:519–521.

Marlowe, F. 2000. Paternal investment and the human mating system. *Behavioural Processes* 51:45–61.

Marra, P. P., Francis, C. M., Mulvihill, R. S., & Moore, F. R. 2005. The influence of climate on the timing and rate of spring bird migration. *Oecologia* 142:307–315.

Marriner, L. M., & Drickamer, L. C. 1994. Factors influencing stereotyped behaviors of primates in a zoo. *Zoo Biology* 13:267–275.

Martin, G. R. 2012. Through birds' eyes: Insights into avian sensory ecology. *Journal of Ornithology* 153 (Suppl. 1): S23–S48.

Martin, P., & Bateson, P. 1993. *Measuring behaviour: An introductory guide*. 2nd ed. Cambridge: Cambridge University Press.

Martin, R. A., & Pfennig, D. W. 2009. Disruptive selection in natural populations: The roles of ecological specialization and resource competition. *American Naturalist* 174:268–281.

Mas, F., & Kölliker, M. 2008. Maternal care and offspring begging in social insects: Chemical signaling, hormonal regulation and evolution. *Animal Behaviour* 76:1121–1131.

Mason, G. J. 2010. Species differences in response to captivity: Stress, welfare and the comparative method. *Trends in Ecology and Evolution* 25:713–721.

Massaro, M., Setiawan, A. N., & Davis, L. S. 2007. Effects of artificial eggs on prolactin secretion, steroid levels, brood patch development, incubation onset and clutch size in the yellow-eyed penguin (*Megadyptes antipodes*). *General and Comparative Endocrinology* 151:220–229.

Mather, J. 2008. Cephalopod consciousness: Behavioral evidence. *Consciousness and Cognition* 17:37–48.

Mathews, L. M. 2002a. Tests of mate-guarding hypothesis for social monogamy: Does population density, sex ratio, or female synchrony affect behavior of male snapping shrimp (*Alpheus angulatus*). *Behavioral Ecology and Sociobiology* 51:426–432.

———. 2002b. Territorial cooperation and social monogamy: Factors affecting intersexual behaviours in pair-living snapping shrimp. *Animal Behaviour* 63:767–777.

———. 2003. Tests of mate-guarding hypothesis for social monogamy: Male snapping shrimp prefer to associate with high-value females. *Behavioral Ecology* 14:63–67.

Mäthger, L. M., Denton, E. J., Marshall, N. J., & Hanlon, R. T. 2009. Mechanisms and behavioural functions of structural coloration in cephalopods. *Journal of the Royal Society Interface* 6 (suppl. 2): S149–S163.

Matos, R. J., & Schlupp, I. 2005. Performing in front of an audience: Signalers and the social environment. In *Animal communication networks*, ed. P. K. McGregor, 63–83. Cambridge: Cambridge University Press.

Matsumasa, M., Murai, M., & Christy, J.H. 2013. A low-cost sexual ornament reliably signals male condition in the fiddler crab *Uca beebei. Animal Behaviour* 85:1335–1341.

Maynard Smith, J. M. 1964. Group selection and kin selection. *Nature* 201:1145–1147.

———. 1978. Optimization theory in evolution. *Annual Review of Ecology and Systematics* 9:31–56.

———. 1982. *Evolution and the theory of games.* Cambridge: Cambridge University Press.

Maynard Smith, J., & Harper, D. 2003. *Animal signals.* Oxford, UK: Oxford University Press.

Maynard Smith, J., & Price, G. R. 1973. The logic of animal conflict. *Nature* 246:15–18.

Mayr, E. 1961. Cause and effect in biology. *Science* 134: 1501–1506.

———. 1982. *The growth of biological thought.* Cambridge, MA: Harvard University Press.

McComb, K., Reby, D., Baker, L., Moss, C., & Sayialel, S. 2003. Long-distance communication of acoustic cues to social identity in African elephants. *Animal Behaviour* 65:317–329.

McDonald, D. B. 1993. Delayed plumage maturation and orderly queues for status: A manakin mannequin experiment. *Ethology* 94:31–45.

McDonald, F. 1958. *We the people: The economic origins of the Constitution.* Chicago: University of Chicago Press.

McEvoy, J. W. 2009. Evolutionary game theory: Lessons and limitations, a cancer perspective. *British Journal of Cancer* 101:2060–2061.

McGlothlin, J. W., Jawor, J. M., & Ketterson, E. D. 2007. Natural variation in a testosterone-mediated trade-off between mating effort and parental care. *American Naturalist* 170:864–875.

McGregor, P., ed. 2005. *Animal communication networks.* Cambridge: Cambridge University Press.

McNamara, J. M., & Houston, A. I. 1987. Starvation and predation as factors limiting population size. *Ecology* 68:1515–1519.

Mello, C. V., Vicario, D. S., & Clayton, D. F. 1992. Song presentation induces gene expression in the songbird forebrain. *Proceedings of the National Academy of Sciences USA* 89:6818–6822.

Meredith, R. M., McCabe, B. J., Kendrick, K. M., & Horn, G. 2004. Amino acid neurotransmitter release and learning: A study of visual imprinting. *Neuroscience* 126:249–256.

Merlin, C., Gegear, R. J., & Reppert, S. M. 2009. Antennal circadian clocks coordinate sun compass orientation in migratory monarch butterflies. *Science* 325:1700–1704.

Mery, F., & Kawecki, T. J. 2002 Experimental evolution of learning ability in fruit flies. *Proceedings of the National Academy of Sciences USA* 99:14,274–14,279.

Mery, F., Varela, S. A. M., Danchin, E., Blanchet, S., Parejo, D., Coolen, I., & Wagner, R. H. 2009. Public versus private information for mate copying in an invertebrate. *Current Biology* 19:730–734.

Mesterton-Gibbons, M., Marden, J. H., & Dugatkin, L. A. 1996. On wars of attrition without assessment. *Journal of Theoretical Biology* 181:65–83.

Meyburg, B. U. 1978. Sibling aggression and cross-fostering of eagles. In *Endangered birds management techniques for threatened species*, ed. S. A. Tempe, 195–200. Madison: University of Wisconsin Press.

Milá, B., Smith, T. B., & Wayne, R. K. 2006. Postglacial population expansion drives the evolution of long-distance migration in a songbird. *Evolution* 60:2403–2409.

Miles, D. B. 2004. The race goes to the swift: Fitness consequences of variation in sprint performance in juvenile lizards. *Evolutionary Ecology Research* 6:63–75.

Milinski, M. 2006. The major histocompatibility complex, sexual selection, and mate choice. *Annual Review of Ecology Evolution and Systematics* 37:159–186.

Milinski, M., Semmann, D., & Krambeck, H. J. 2002a. Donors to charity gain both indirect reciprocity and political reputation. *Proceedings of the Royal Society B* 269:881–883.

———. 2002b. Reputation helps solve the "tragedy of the commons." *Nature* 393:573–577.

Miller, D. A., Vleck, C. M., & Otis, D. L. 2009. Individual variation in baseline and stress-induced corticosterone and prolactin levels predicts parental effort by nesting mourning doves. *Hormones and Behavior* 56:457–464.

Miller, P. 2001. *The power of positive dog training.* New York: Howell.

Mitchell, M. D., McCormick, M. I., Ferrari, M. C. O., & Chivers, D. P. 2011. Coral reef fish rapidly learn to identify multiple unknown predators upon recruitment to the reef. *PLoS One* 6:e15764.

Mitchell, W. A., & Valone, T. J. 1990. The optimization research program: Studying adaptations by their functions. *Quarterly Review of Biology* 65:43–52.

Mo, J., Glover, M., Munro, S., & Beattie, G. A. 2006. Evaluation of mating disruption for control of lightbrown apple moth (Lepidoptera: Tortricidae) in citrus. *Journal of Economic Entomology* 99:421–426.

Moller, A. P., & Alatalo, R. V. 1999. Good-genes effects in sexual selection. *Proceedings of the Royal Society B* 266:85–91.

Møller, A. P., & Jennions, M. D. 2001. How important are direct fitness benefits of sexual selection? *Naturwissenschaften* 88:401–415.

Monaghan, P., & Nager, R. G. 1997. Why don't birds lay more eggs? *Trends in Ecology and Evolution* 12:270–274.

Mora, C. V., Davison, M., Wild, J. M., & Walker, M. M. 2004. Magnetoreception and its trigeminal mediation in the homing pigeon. *Nature* 432:508–511.

Morand-Ferron, J., Varennes, E., & Giraldeau, L.-A. 2011. Individual differences in plasticity and sampling when playing behavioural games. *Proceedings of the Royal Society B* 278:1223–1230.

Morgan, C. L. 1894. *An introduction to comparative psychology.* London: W. Scott.

———. 1903. *An introduction to comparative psychology.* 2nd ed. London: W. Scott.

Morissette, S., & Himmelman, J. H. 2000. Subtidal food thieves: Interactions of four invertebrate kleptoparasites with a sea star *Leptasteria polaris. Animal Behaviour* 60:531–543.

Morrell, L. J., Blackwell, P. R. Y., & Metcalfe, N. B. 2005. Fighting in fiddler crabs *Uca mjoebergi:* What determines duration? *Animal Behaviour* 70:653–662.

Morris, D. W. 1988. Habitat-dependent population regulation and community structure. *Evolutionary Ecology* 2:253–269.

———. 2006. Moving to the ideal free home. *Nature* 443:645–646.

Morton, E. S. 1975. Ecological sources of selection on avian sounds. *American Naturalist* 109:17–34.

Mottley, K., & Giraldeau, L.-A. 2000. Experimental evidence that group foragers can converge on predicted producer-scrounger equilibria. *Animal Behaviour* 60:341–350.

Mouritsen, H., & Hore, P. J. 2012. The magnetic retina: Light-dependent and trigeminal magnetoreception in migratory birds. *Current Opinion in Neurobiology* 22:343–352.

Mueller, J. C., Pulido, F., & Kempenaers, B. 2011. Identification of a gene associated with avian migratory behavior. *Proceedings of the Royal Society B* 278:2848–2856.

Muheim, R., Moore, F. R., & Phillips, J. B. 2006. Calibration of magnetic and celestial compass cues in migratory birds: A review of cue-conflict experiments. *Journal of Experimental Biology* 209:2–17.

Müller, C. A., & Manser, M. B. 2007. "Nasty neighbors" rather than "dear enemies" in a social carnivore. *Proceedings of the Royal Society B* 274:959–965.

Muller, K. L. 1995. Habitat settlement in territorial species: The effects of habitat quality and conspecifics. Ph.D. diss., University of California, Davis.

———. 1998. The role of conspecifics in habitat settlement in a territorial grasshopper. *Animal Behaviour* 56:479–485.

Mumby, P. J., & Wabnitz, C. C. C. 2002. Spatial patterns of aggression, territory size, and harem size in five sympatric Caribbean parrotfish species. *Environmental Biology of Fishes* 63:265–279.

Mundinger, P. C. 2010. Behaviour genetic analysis of selective song learning in three inbred canary strains. *Behaviour* 147:705–723.

Murray, C. M., Lonsdorf, E. V., Eberly, L. E., & Pusey, A. E. 2009. Reproductive energetics in free-living female chimpanzees (*Pan troglodytes schweinfurthii*). *Behavioral Ecology* 20:1211–1216.

Murray, C. M., Mane, S. V., & Pusey, A. E. 2007. Dominance rank influences female space use in wild chimpanzees, *Pan troglodytes:* Towards an ideal despotic distribution. *Animal Behaviour* 74:1795–1804.

Murray, E. A., Kralik, J. D., Wise, S. P. 2005. Learning to inhibit prepotent responses: Successful performance by rhesus macaques, *Macaca mulatta,* on the reversed-contingency task. *Animal Behaviour* 69:991–998.

Murray, J. A., Estepp, J., & Cain, S. D. 2006. Advances in the neural bases of orientation and navigation. *Integrative and Comparative Biology* 46:871–879.

Nakano, R., Ishikawa, Y., Tatsuki, S., Surlykke, A., Skals, N., & Takanashi, T. 2006. Ultrasonic courtship song in the Asian corn borer moth, *Ostrinia furnacalis*. *Naturwissenschaften* 93:292–296.

National Academy of Sciences. 1998. *Teaching about evolution and the nature of science*. Washington, D.C.: National Academies Press.

Neff, B. D. 2003. Decisions about parental care in response to perceived paternity. *Nature* 422: 716–719.

Nelissen, R. M. A., & Meijers, M. H. C. 2011. Social benefits of luxury brands as costly signals of wealth and status. *Evolution and Human Behavior* 32:343–355.

Newman, J. A., & Elgar, M. A. 1991. Sexual cannibalism in orb-weaving spiders: An economic model. *American Naturalist* 138:1372–1395.

Nicholls, J. A., & Goldizen, A. W. 2006. Habitat type and density influence vocal signal design in satin bowerbirds. *Journal of Animal Ecology* 75:549–558.

Niimura, Y., Matsui, A., & Touhara, K. 2014. Extreme expansion of the olfactory receptor gene repertoire in African elephants and evolutionary dynamics of orthogolous gene groups in 13 placental mammals. *Genome Research* 24:1485–1496.

Nijhout, H. F. 1994. *Insect hormones*. Princeton, NJ: Princeton University Press.

Noad, M. J., Cato, D. H., Bryden, M. M., Jenner, M.-N., & Jenner, K. C. S. 2000. Cultural revolution in whale songs. *Nature* 408:537.

Nonacs, P., & Dill, L. M. 1988. Foraging response of the ant *Lasius pallitarsis* to food sources with associated mortality risk. *Insectes Sociaux* 35: 293–303.

———. 1990. Mortality risk vs. food quality trade-offs in a common currency: Ant patch preferences. *Ecology* 71:1886–1892.

Nordby, J. C., Campbell, S. E., & Beecher, M. D. 2001. Late song learning in song sparrows. *Animal Behaviour* 61:835–846.

Nordell, S. E. 1994. Observations of the mating behavior and dentition of the round stingray, *Urolophus halleri*. *Environmental Biology of Fishes* 39:219–229.

Nordell, S. E., & Valone, T. J. 1998. Mate choice copying as public information. *Ecology Letters* 1:74–76.

Nordeng, H. 1977. A pheromone hypothesis for homeward migration in anadromous salmonids. *Oikos* 28:155–159.

Nowak, M. A., & Sigmund, K. 1998. Evolution of indirect reciprocity by image scoring. *Nature* 393:573–577.

———. 2005. Evolution of indirect reciprocity. *Nature* 437:1291–1298.

Nowak, M. A., Tarnita, C. E., & Wilson, E. O. 2010. The evolution of eusociality. *Nature* 466: 1057–1062.

Ohlsson, T., Smith, H. G., Raberg, L., & Hasselquist, D. 2002. Pheasant sexual ornaments reflect nutritional conditions during early growth. *Proceedings of the Royal Society B* 269:21–27.

Olendorf, R., Getty, T., & Scribner, K. 2004. Cooperative nest defense in red-winged blackbirds: Reciprocal altruism, kinship or by-product mutualism? *Proceedings of the Royal Society B* 271:177–182.

Oliveira, R. F., Lopos, L. A., & Carneiro, A. V. M. 2001. Watching fights raises fish hormone levels. *Nature* 409:784.

Oliveira, R. F., Silva, A., & Canário, A. V. M. 2009. Why do winners keep winning? Androgen mediation of winner but not loser effects in cichlid fish. *Proceedings of the Royal Society B* 276:2249–2256.

Olofsson, M., Eriksson, S., Jakobsson, S., & Wiklund, C. 2012. Deimatic display in the European swallowtail butterfly as a secondary defence against attacks from great tits. *PLoS ONE* 7:e47092.

Olsson, M., Madsen, T., Nordby, J., Wapstra, E. Ujvari, B., & Wittsell, H. 2003. Major histocompatability complex and mate choice in sand lizards. *Proceedings of the Royal Society B* 270:S254–S256.

Orians, G. H. 1969. On the evolution of mating systems in birds and mammals. *American Naturalist* 103:589–603.

Osorio, D., & Vorobyev, M. 2008. A review of the evolution of animal color vision and visual communication signals. *Vision Research* 48:2042–2051.

Osorio-Beristain, M., & Drummond, H. 2001. Male boobies expel eggs when paternity is in doubt. *Behavioral Ecology* 12:16–21.

Östlund-Nilsson, S., & Holmlund, M. 2003. The artistic three-spined stickleback (*Gasterosteous aculeatus*). *Behavioral Ecology and Sociobiology* 53:314–220.

Otte, D. 1974. Effects and function in the evolution of signaling systems. *Annual Review of Ecology and Systematics* 5:385–417.

Outlaw, D. C., Voelker, G., Milá, B., & Girman, D. J. 2003. Evolution of long-distance migration in and historical biogeography of *Catharus* thrushes: A molecular phylogentic approach. *Auk* 120: 299–310.

Owen, M. J., Rendell, D., & Ryan, M. J. 2010. Redefining animal signaling: Influence versus information in communication. *Biology & Philosophy* 25:755–780.

Owen, P. C., & Perrill, S. A. 1998. Habituation in the green frog, *Rana clamitans*. *Behavioral Ecology and Sociobiology* 44:209–213.

Oyegbile, T. O., & Marler, C. A. 2005. Winning fights elevates testosterone levels in California mice and enhances future ability to win fights. *Hormones and Behavior* 48:259–267.

Packer, C., Gilbert, D. A., Pusey, A. E., & O'Brien, S. J. 1991. A molecular genetic analysis of kinship and cooperation in African lions. *Nature* 351:562–565.

Packer, C., Scheel, D., & Pusey, A. E. 1990. Why lions form groups: Food is not enough. *American Naturalist* 136:1–19.

Pak, A. A. 2010. The synergy of laboratory and field studies of dolphin behavior and cognition. *International Journal of Comparative Psychology* 23:538–565.

Panchanathan, K., & Boyd, R. 2004. Indirect reciprocity can stabilize cooperation without the second-order free rider problem. *Nature* 432:499–502.

Parker, G. A. 1974. Assessment strategy and the evolution of fighting behavior. *Journal of Theoretical Biology* 47:223–243.

———. 1970. Sperm competition and its evolutionary consequences in the insects. *Biological Reviews* 45:455–616.

Parker, G. A., Baker, R. R., & Smith, V. G. F. 1972. The origin and evolution of gamete dimorphism and the male-female phenomenon. *Journal of Theoretical Biology* 36:529–553.

Parker, G. A., & Hammerstein, P. 1985. Game theory and animal behavior. In *Evolution: Essays in honour of John Maynard Smith*, ed. J. Maynard Smith, J. Greenwood, P. H. Harvey, & M. Slatkin, 73–94. New York: Cambridge University Press.

Parker, M. W., Kress, S. W., Golightly, R. T., Carter, H. R., Parsons, E. B., Schubel, S. E., Boyce, J. A., McChesney, G. J., & Wisley, S. M. 2007. Assessment of social attraction techniques used to restore a common murre colony in central California. *Waterbirds* 30:17–28.

Parsons, M. H., Lamont, B. B., Kovacs, B. R., & Davies, S. J. J. F. 2007. Effects of novel and historic predator urines on semi-wild western grey kangaroos. *Journal of Wildlife Management* 71:1225–1228.

Patterson, E. M., & Mann, J. 2011. The ecological conditions that favor tool use and innovation in wild bottlenose dolphins (*Tursiops* sp.). *PLoS One* 6:e22243.

Pavey, C. R., & Smyth, A. K. 1998. Effects of avian mobbing on roost use and diet of powerful owls, *Ninox strenua. Animal Behaviour* 55:313–318.

Pavey, C. R., Smyth, A. K., & Mathieson, M. T. 1994. The breeding season diet of the powerful owl *Ninox strenua* at Brisbane, Queensland. *Emu* 94:278–284.

Pavlov, I. P. 1927. *Conditioned reflexes: An investigation of the physiological activity of the cerebral cortex.* Trans. and ed. G. V. Anrep. London: Oxford University Press.

Payne, H. F. P., Lawes, M. J., & Henzi, S. P. 2003. Competition and the exchange of grooming among female samango monkeys (*Cercopithecus mitis erythrarchus*). *Behaviour* 140:453–471.

Payne, R. J. H., & Pagel, M. 1996. Escalation and time costs in displays of endurance. *Journal of Theoretical Biology* 183:185–193.

Pearce, J. M. 2008. *Animal learning and cognition: An introduction*, 3rd ed. East Sussex, UK: Psychology Press.

Peer, B. D., & Sealy, S. G. 2004. Correlates of egg rejection in hosts of the brown-headed cowbird. *Condor* 106:580–599.

Pennisi, E. 2005. A genomic view of animal behavior. *Science* 307:30–32.

———. 2009. On the origin of cooperation. *Science* 325:1196–1199.

Pepperberg, I. M. 2006. Grey parrot numerical competence: A review. *Animal Cognition* 9:377–391.

Perez, S. M., Taylor, O. R., & Jander, R. 1997. A sun compass in monarch butterflies. *Nature* 387:29.

Perry, C. J., Barron, A. B., & Cheng, K. 2013. Invertebrate learning and cognition: Relating phenomena to neural substrate. *Wiley Interdisciplinary Reviews: Cognitive Science* 4:561–582.

Persky, H., Smith, K. D., & Basu, G. K. 1971. Relation of psychologic measures of aggression and hostility to testosterone production in man. *Psychosomatic Medicine* 33:265–277.

Peters, R. A., Hemmi, J. M., & Zeil, J. 2007. Signaling against the wind: Modifying motion-signal structure in response to increased noise. *Current Biology* 17:1231–1234.

Petrie, M., Halliday, T., & Carolyne, S. 1991. Peahens prefer peacocks with elaborate trains. *Animal Behaviour* 41:323–331.

Pfennig, D. W. 1992. Polyphenism in spadefoot toads as a locally adjusted evolutionarily stable strategy. *Evolution* 46:1408–1420.

Pfennig, D. W., Rice, A. M., & Martin, R. A. 2007. Field and experimental evidence for competition's role in phenotypic divergence. *Evolution* 61:257–271.

Pfennig, K. S., & Pfennig, D. W. 2005. Character displacement as the "best of a bad situation": Fitness trade-offs resulting from selection to minimize resource and mate competition. *Evolution* 59:2200–2208.

Phelps, S. M., Rand, A. S., & Ryan, M. J. 2007. The mixed-species chorus as public information: Túngara frogs eavesdrop on a heterospecific. *Behavioral Ecology* 18:108–114.

Piep, M., Radespiel, U., Zimmermann, E., Schmidt, S., & Siemers, B. M. 2008. The sensory bias of prey detection in captive born grey mouse lemurs (*Microcebus murinus*). *Animal Behaviour* 75:871–878.

Pietrewicz, A. T., & Kamil, A. C. 1977. Visual detection of cryptic prey by blue jays (*Cyanocitta cristata*). *Science* 195:580–582.

———. 1979. Search image formation in the blue jay (*Cyanocitta cristata*). *Science* 204:1332–1333.

Pirolli, P. 2005. Rational analyses of information foraging on the web. *Cognitive Science* 29:343–373.

Pitcher, T. J., Magurran, A. E., & Winfield, I. J. 1982. Fish in larger shoals find food faster. *Behavioral Ecology and Sociobiology* 10:149–151.

Place, S. S., Todd, P. M., Penke, L., & Asendorpf, J. B. 2010. Humans show mate copying after observing real mate choices. *Evolution and Human Behavior* 31:320–325.

Platt, J. R. 1964. Strong inference. *Science* 146:347–353.

Plusquellec, P., & Bouissou, M. 2001. Behavioural characteristics of two dairy breeds of cows selected (Hérens) or not (Brune des Alpes) for fighting and dominance ability. *Applied Animal Behaviour Science* 72:1–21.

Pochron, S. T., & Wright, P. C. 2003. Variability in adult group composition of a prosimian primate. *Behavioral Ecology and Sociobiology* 54:285–293.

Pohlmann, K., Atema, J., & Breithaupt, T. 2004. The importance of the lateral line in nocturnal predation of piscivorous catfish. *Journal of Experimental Biology* 207:2971–2978.

Pohlmann, K., Grasso, F. W., & Breithaupt, T. 2001. Tracking wakes: The nocturnal predatory strategy of piscivorous catfish. *Proceedings of the National Academy of Sciences USA* 98:7371–7374.

Pollet, T. V., & Nettle, D. 2009. Market forces affect patterns of polygyny in Uganda. *Proceedings of the National Academy of Sciences U.S.A.* 106:2114–2117.

Pomfret, J. C., & Knell, R. J. 2006. Sexual selection and horn allometry in dung beetle *Euoniticellus intermedius*. *Animal Behaviour* 71:567–576.

Poole, M. M. 1984. Migration corridors of gray whales along the central California coast, 1980–1982. In *The gray whale*, Eschrichotus robustus, ed. M. L. Jones, S. L. Swartz, & S. Leatherwood, 389–407. Orlando, FL: Academic Press.

Popper, K. R. 1959. *The logic of scientific discovery*. New York: Basic Books.

Porter, S. D., & Tschinkel, W. R. 1985. Fire ant polymorphism: the erognomics of brood production. *Behavioral Ecology and Sociobiology* 16:323–336.

Praw, J. C., & Grant, J. W. A. 1999. Optimal territory size in the convict cichlid. *Behaviour* 136:1347–1363.

Pribil, S. 2000. Experimental evidence for the cost of polygyny in the red-winged blackbird *Agelaius phoeniceus*. *Behaviour* 137:1153–1173.

Pribil, S., & Searcy, W. A. 2001. Experimental confirmation of the polygyny threshold model for red-winged blackbirds. *Proceedings of the Royal Society B* 268:1643–1646.

Price, E. E., & Stoinski, T. S. 2007. Group size: Determinants in the wild and implications for the captive housing of wild mammals in zoos. *Applied Animal Behaviour Science* 103:255–264.

Price, F. E., & Bock, C. E. 1983. Population ecology of the dipper (*Cinclus mexicanus*) in the front range of Colorado. *Studies in Avian Biology* 7:1–84.

Prokop, Z. M., Michalczyk, Ł., Drobniak, S. M., Herdegen, M., & Radwan, J. 2012. Meta-analysis suggests choosy females get sexy sons more than "good genes." *Evolution* 66:2665–2673.

Prum, R. O. 2012. Aesthetic evolution by mate choice: Darwin's *really* dangerous idea. *Philosophical Transactions B* 367:2253–2265.

Pruvost, M., Bellone, R., Benecke, N., Sandoval-Castellanos, E., Cieslak, M., Kuznetsova, T., Morales-Muniz, A., et al. 2011. Genotypes of predomestic horses match phenotypes painted in Paleolithic works of cave art. *Proceedings of the National Academy of Science USA* 108:18,626–18,630.

Pryke, S. R., Andersson, S., & Lawes, M. J. 2001. Sexual selection of multiple handicaps in the red-collared widowbird: Female choice of tail length but not carotenoid display. *Evolution* 55:1452–1463.

Pulido, F. 2007. The genetics and evolution of avian migration. *Bioscience* 57:165–174.

Pulido, F., & Berthold, P. 2010. Current selection for lower migratory activity will drive the evolution of residency in a migratory bird population. *Proceedings of the National Academy of Sciences USA* 107:7341–7346.

Pulido, F., Berthold, P., Mohr, G., & Querner, U. 2001. Heritability of the timing of autumn migration in a natural bird population. *Proceedings of the Royal Society B* 268:953–959.

Pulliam, H. R. 1973. On the advantages of flocking. *Journal of Theoretical Biology* 38:419–422.

———. 1974. On the theory of optimal diets. *American Naturalist* 108:59–74.

Purchase, C. F, & Hutchings, J. A. 2008. A temporally stable spatial pattern in the spawner density of a freshwater fish: Evidence for an ideal despotic distribution. *Canadian Journal of Aquatic and Fisheries Sciences* 65:382–388.

Pusey, A. 1987. Sex-biased dispersal and inbreeding avoidance in birds and mammals. *Trends in Ecology and Evolution* 2:295–299.

Pusey, A., & Wolf, M. 1996. Inbreeding avoidance in animals. *Trends in Ecology and Evolution* 11:201–206.

Queller, D. C. 1997. Why do females care more than males? *Proceedings of the Royal Society B* 264:1555–1557.

Queller, D.C., & Strassman, J.E. 1998. Kin selection and social insects. *Bioscience* 48:165–175.

Queller, D. C., Strassmann, J. E., & Hughes, C. R. 1993. Microsatellites and kinship. *Trends in Ecology and Evolution* 8:285–288.

Radford, A. N., & Ridley, A. R. 2007. Individuals in foraging groups may use vocal cues when assessing their need for antipredator vigilance. *Biology Letters* 3:249–252.

Radwan, J. 2002. Good genes go Fisherian. *Trends in Ecology and Evolution* 17:539.

Raine, N. E., & Chittka, L. 2007. Adaptive significance of sensory bias in a foraging context: Floral colour preferences in the bumblebee *Bombus terrestris*. *PLoS ONE* 2:e556.

———. 2008. The correlation of learning speed and natural foraging success in bumble-bees. *Proceedings of the Royal Society B* 275:803–808.

Ramenofsky, M., & Wingfield, J. C. 2007. Regulation of migration. *Bioscience* 57:135–143.

Rankin, S. M., Fox, K. M., & Stotsky, C. E. 1995. Physiological correlates to courtship, mating, ovarian development and maternal behavior in the ring-legged earwig. *Physiological Entomology* 20:257–265.

Rankin, S. M., Palmer, J. O., Yagi, K. J., Scott, G. L., & Tobe, S. S. 1995. Biosynthesis and release of juvenile hormone during the reproductive cycle of the ring-legged earwig. *Comparative Biochemistry and Physiology C: Pharmacology, Toxicology & Endocrinology* 110:241–251.

Réale, D. R., Reader, S. M., Sol, D., McDougall, P. T., & Dingemanse, N. J. 2007. Integrating animal temperament within ecology and evolution. *Biological Reviews* 82:291–318.

Reaney, L. T. 2007. Foraging and mating opportunities influence refuge use in the fiddler crab, *Uca mjoebergi*. *Animal Behaviour* 73:711–716.

Reddon, A. R., Voisin, M. R., Menon, M., Marsh-Rollo, S. E., Wong, M. Y. L., & Balshine, S. 2011. Rules of engagement for resource contests in social fish. *Animal Behaviour* 82:93–99.

Rees, P. A. 2009. Activity budgets and the relationship between feeding and stereotypic behaviors in Asian elephants (*Elephas maximus*) in a zoo. *Zoo Biology* 28:79–97.

Reeve, H. K., & Sherman, P. W. 1993. Adaptation and the goals of evolutionary research. *Quarterly Review of Biology* 68:1–32.

Reguera, P., & Gomendio, M. 1999. Predation costs associated with parental care in the golden egg bug *Phyllomorpha laciniata* (Heteroptera: Coreidea). *Behavioral Ecology* 10:541–544.

———. 2002. Flexible oviposition behavior in the golden egg bug (*Phyllomorpha laciniata*) and its implications for offspring survival. *Behavioral Ecology* 13:70–74.

Rendell, D., Owen, M. J., & Ryan, M. J. 2009. What do animal signals mean? *Animal Behaviour* 78:233–240.

Resasco, J., Haddad, N. M., Orrock, J. L., Shoemaker, D., Brudvig, L. A., Damschen, E. L. Tewksbury, J. J., & Levey, D. J. 2014. Landscape corridors can increase invasion by an exotic species and reduce diversity of native species. *Ecology* 95: 2033–2039.

Reynolds, D. R., & Riley, J. R. 2002. Remote-sensing, telemetric and computer-based technologies for investigating insect movement: A survey of existing and potential techniques. *Computers and Electronics in Agriculture* 35:271–307.

Reynolds, J. D., Goodwin, N. B., & Freckleton, R. P. 2002. Evolutionary transition in parental care and live bearing in vertebrates. *Philosophical Transactions of the Royal Society B* 357:269–281.

Reynolds, S. M., Christman, M. C., Uy, A. C., Patricelli, G. L., Braun, M. J., & Borgia, G. 2009. Lekking satin bowerbird males aggregate with relatives to mitigate aggression. *Behavioral Ecology* 20: 410–415.

Ribble, D. O. 1991. The monogamous mating system of *Peromyscus californicus* as revealed by DNA fingerprinting. *Behavioral Ecology and Sociobiology* 29:161–166.

Rice, W. R., & Holland, B. 1997. The enemies within: Intergenomic conflict, interlocus evolution (ICE) and the intraspecific red queen. *Behavioral Ecology and Sociobiology* 41:1–10.

Richards, E. L., van Oosterhout, C., & Cable, J. 2010. Sex-specific differences in shoaling affect parasite transmission in guppies. *PLoS ONE* 5:e13285.

Richardson, H., & Verbeek, N. A. M. 1986. Diet selection and optimization by northwestern crows feeding on Japanese littleneck clams. *Ecology* 67:1219–1226.

Riechert, S. E., & Hammerstein, P. 1983. Game theory in the ecological context. *Annual Review of Ecology and Systematics* 14:377–409.

Rieucau, G., & Giraldeau, L.-A. 2011. Exploring the costs and benefits of social information use: An appraisal of current experimental evidence. *Philosophical Transactions of the Royal Society B* 366:949–957.

Riley, J. R., Greggers, U., Smith, A. D., Reynolds, D. R., & Menzel, R. 2005. The flight paths of honeybees recruited by the waggle dance. *Nature* 435:205–207.

Rios-Cardenas, O., & Webster, M. S. 2008. A molecular genetic examination of the mating system of pumpkinseed sunfish reveals high pay-offs for specialized sneakers. *Molecular Ecology* 17:2310–2320.

Ritchie, M. L. 1990. Optimal foraging and fitness for Columbian ground squirrels. *Oecologia* 83:56–67.

Ritz, T., Adem, S., & Schulten, K. 2000. A model for vision-based magnetoreception in birds. *Biophysical Journal* 78:707–718.

Robb, G. N., McDonald, R. A., Chamberlain, D. E., Reynolds, S. J., Harrison, T. J. E., & Bearhop, S. 2008. Winter feeding of birds increases productivity in the subsequent breeding season. *Biology Letters* 4:220–223.

Roberts, J. A., Galbraith, E., Milliser, J., Taylor, P. W., & Uetz, G. W. 2006. Absence of social facilitation of courtship in the wolf spider, *Schizocosa ocreata* (Hentz) (Araneae: Lycosidae). *Acta Ethologica* 9:71–77.

Robinson, G. E., Fernald, R. D., & Clayton, D. F. 2008. Genes and social behavior. *Science* 322:896–900.

Rodd, F. H., Hughes, K. A., Grether, G. F., & Baril, C. T. 2002. A possible non-sexual origin of mate preference: Are male guppies mimicking fruit? *Proceedings of the Royal Society B* 269:475–481.

Rohwer, S., Klein, W. P., Jr., & Heard, S. 1983. Delayed plumage maturation and the presumed pre-alternate molt in American redstarts. *Wilson Bulletin* 95:199–208.

Romanes, G. J. 1882. *Animal intelligence*. London: Kegan Paul, Trench & Co.

———. 1888. *Mental evolution in man: Origin of human faculty*. London: Kegan Paul, Trench & Co.

Ronce, O. 2007. How does it feel to be like a rolling stone? Ten questions about dispersal evolution. *Annual Review of Ecology, Evolution and Systematics* 38:231–253.

Rooney, J., & Lewis, S. M. 1999. Differential allocation of male derived nutrients in two lampyrid beetles with contrasting life history characteristics. *Behavioral Ecology* 10:97–104.

———. 2002. Fitness advantage from nuptial gifts in female fireflies. *Ecological Entomology* 27:373–377.

Rosenthal, G. G. 2007. Spatiotemporal dimensions of visual signals in animal communication. *Annual Reviews of Ecology Evolution and Systematics* 38:155–178.

Rosenzweig, M. L. 2001. Optimality—the biologist's tricorder. *Annales Zoologici Fennici* 38:1–3.

Røskaft, E., Bjerke, T., Kaltenborn, B., Linnell, J. D. C., & Andersen, R. 2003. Patterns of self-reported fear towards large carnivores among the Norwegian public. *Evolution and Human Behavior* 24:184–198.

Ross, K. G. 1997. Multilocus evolution in fire ants: Effects of selection, gene flow, and recombination. *Genetics* 145:961–974.

Ross, K. G., & Keller, L. 2002. Experimental conversion of colony social organization by manipulation of worker genotype composition in fire ants (*Solenopsis invicta*). *Behavioral Ecology and Sociobiology* 51:287–295.

Ross, L., Gardner, A., Hardy, N., & West, S.A. 2013. Ecology, not the genetics of sex determination, determines who helps in eusocial populations. *Current Biology* 23:2383–2387.

Roth, T. C., II, Brodin, A., Smulders, T. V., LaDage, L. D., & Pravosudov, V. V. 2010. Is bigger always better? A critical appraisal of the use of volumetric analysis in the study of the hippocampus. *Philosophical Transactions of the Royal Society B* 365:915–931.

Rowe, L., Cameron, E., & Day, T. 2005. Escalation, retreat, and female indifference as alternative outcomes of sexually antagonistic coevolution. *American Naturalist* 165:S5–S18.

Rowell, J. T., Ellner, S. P., & Reeve, H. K. 2006. Why animals lie: How dishonesty and belief can coexist in a signaling system. *American Naturalist* 168:E180–E204.

Royle, N. J., Smiseth, P. T., & Kölliker, M. (eds.) 2012. *The evolution of parental care*. Oxford University Press. Oxford, UK.

Ruefenacht, S., Gebhardt-Henrich, S., Miyake, T., & Gaillard, C. 2002. A behavior test on German shepherd dogs: Heritability of seven different traits. *Applied Animal Behavior Science* 79:113–132.

Rundus, A. S., Owings, D. H., Joshi, S. S., Chinn, E., & Giannini, N. 2007. Ground squirrels use an infrared signal to deter rattlesnake predation. *Proceedings of the National Academy of Sciences USA* 104:14,372–14,376.

Russell, A. F., & Hatchwell, B. J. 2001. Experimental evidence for kin-based helping in a cooperatively breeding vertebrate. *Proceedings of the Royal Society B* 268:2169–2174.

Russell, Y. I., Call, J., & Dunbar, R. I. M. 2008. Image scoring in great apes. *Behavioral Processes* 78:108–111.

Russo, N. F., & Denmark, F. L. 1987. Contributions of women to psychology. *Annual Reviews of Psychology* 38:279–298.

Ruxton, G. D. 2009. Non-visual crypsis: A review of the empirical evidence for camouflage to senses other than vision. *Philosophical Transactions of the Royal Society B* 364:549–557.

Ruxton, G. D., Sherratt, T. N., & Speed, M. P. 2005. *Avoiding attack: The evolutionary ecology of crypsis, warning signals and mimicry*. New York: Oxford University Press.

Ryan, M. J. 1980. Female mate choice in a neotropical frog. *Science* 290:523–525.

———. 1998. Principle with a handicap. *Quarterly Review of Biology* 73:477–479.

———. 2011. Replication in field biology: the case of the frog-eating bat. *Science* 334:1229–1230.

Ryan, M. J., & Rand, A. S. 1990. The sensory basis of sexual selection for complex calls in the túngara frog, *Physalaemus pustulosus* (sexual selection for sensory exploitation). *Evolution* 44:305–314.

Ryan, M. J., Tuttle, M. D., & Rand, A. S. 1982. Bat predation and sexual advertisement in a neotropical anuran. *American Naturalist* 119:136–139.

Rypstra, A. L., Weig, C., Walker, S. E., & Persons, M. H. 2003. Mutual mate assessment in wolf spiders: Differences in the cues used by males and females. *Ethology* 109:315–325.

Sætre, G.-P., Dale, S., & Slagsvold, T. 1994. Female pied flycatchers prefer brightly coloured males. *Animal Behaviour* 48:1407–1416.

Sætre, G-P., Moum, T., Stanislav, B., Miroslav, K., Adamjan, M. & Moreno, J. 1997. A sexually selected character displacement in flycatchers reinforces premating isolation. *Nature* 387:589–592.

Safer, A. B., & Grace, M. S. 2004. Infrared imaging in vipers: Differential responses of crotaline and viperine snakes to paired thermal targets. *Behavioural Brain Research* 154:55–61.

Salewski, V., & Bruderer, B. 2007. The evolution of bird migration—a synthesis. *Naturwissenschaften* 94:268–279.

Sánchez, F. 2006. Harvest rates and patch use strategy of Egyptian fruit bats in artificial food patches. *Journal of Mammalogy* 87:1140–1144.

Sandberg, R., Bäckman, J., Moore, F. R., & Lõhmus, M. 2000. Magnetic information calibrates celestial cues during migration. *Animal Behaviour* 60:453–462.

Sapolsky, R. M. 2005. The influence of social hierarchy on primate health. *Science* 308:648–652.

Sapolsky, R. M., Romero, L. M., & Munck, A. U. 2000. How do glucocorticoids influence stress responses? Integrating permissive, suppressive, stimulatory and preparative actions. *Endocrine Reviews* 21:55–89.

Saporito, R. A., Zuercher, R., Roberts, M., Gerow, K. G., & Donnelly, M. A. 2007. Experimental evidence for aposematism in the Dendrobatid poison frog *Oophaga pumlio*. *Copeia* 2007:1006–1011.

Sargent, B. L., & Mann, J. 2009. Developmental evidence for foraging traditions in wild bottlenose dolphins. *Animal Behaviour* 78:715–721.

Sargent, R. C., Taylor, P. D., & Gross, M. R. 1987. Parental care and the evolution of egg size in fishes. *American Naturalist* 129:32–46

Schaedlin, F. C., & Taborsky, M. 2006. Mating craters of *Cyathopharynx furcifer* (Cichlidae) are individually characteristic extended phenotypes. *Animal Behaviour* 72: 753–761.

———. 2009. Extended phenotypes as signals. *Biological Reviews* 84:293–313.

Scheuber, H., Jacot, A., & Brinkof, M. W. G. 2003. The effect of past condition on a multicomponent sexual signal. *Proceedings of the Royal Society B* 270:1779–1784.

Schino, G., & Aureli, F. 2010. The relative roles of kinship and reciprocity in explaining primate altruism. *Ecology Letters* 13:45–50.

Schino, G., Polizzi de Sorrentino, E., & Tiddi, B. 2007. Grooming and coalitions in Japanese macaques (*Macaca fuscata*): Partner choice and the time frame of reciprocation. *Journal of Comparative Psychology* 121:181–188.

Schlenoff, D. H. 1985. The startle responses of blue jays to *Catocala* (Lepidoptera: Noctuidae) prey models. *Animal Behaviour* 33:1057–1067.

Schmidt, K. A., & Ostfeld, R. S. 2008. Eavesdropping squirrels reduce their future value of food under the perceived presence of cache robbers. *American Naturalist* 171:386–393.

Schmidt-Hempel, P., & Wolf, T. 1988. Foraging effort and lifespan of workers in a social insect. *Journal of Animal Ecology* 57:500–521.

Schmidt-Koenig, K. 1979. *Avian orientation and navigation*. London: Academic Press.

———. 1990. The sun compass. *Experimentia* 16:336–342.

Schmitz, H., & Bleckmann, H. 1998. The photomechanic infrared receptor for the detection of forest fires in the beetle *Melanophila acuminata* (Coleoptera: Buprestidae). *Journal of Comparative Physiology A* 182:647–657.

Scott, J. P. 1943. Effects of single genes on the behavior of *Drosophila*. *American Naturalist* 77:184–190.

Scott, M. L., Holm, E. R., & Reynolds, R. E. 1963. Studies on the protein and methionine requirements of young bobwhite quail and young ring-necked pheasants. *Poultry Science* 42:676–680.

Scott, M. P. 2006. Resource defense and juvenile hormone: The "challenge hypothesis" extended to insects. *Hormones and Behavior* 49:276–281.

Scott-Phillips, T. C. 2008. Defining biological communication. *Journal of Evolutionary Biology* 21:387–395.

Seed, A., & Byrne, R. 2010. Animal tool-use. *Current Biology* 23:R1032–R1039.

Seed, A., & Tomasello, M. 2010. Primate cognition. *Topics in Cognitive Science* 2:407–419.

Seeley, T. D. 1985. *Honey bee ecology: A study of adaptation in social life*. Princeton, NJ: Princeton University Press.

Sefc, K. M., Mattersdorfer, K., Sturmbauer, C., & Koblmüller, S. 2008. High frequency of multiple paternity in broods of a socially monogamous cichlid fish with biparental nest defense. *Molecular Ecology* 17:2531–2543.

Seki, S., Kohda, M., Takamoto, G., Karino, K., Nakashima, Y., & Kuwamura, T. 2009. Female defense polygyny in the territorial triggerfish *Sufflamen chrysopterum*. *Journal of Ethology* 27:215–220.

Semple, S., Gerald, M. S., & Suggs, D. N. 2009. Bystanders affect the outcome of mother-infant interactions in rhesus macaques. *Proceedings of the Royal Society B* 276:2257–2262.

Seyfarth, R. M., Cheney, D. L., Bergman, T., Fischer, J., Züberbühler, K., & Hammerschmidt, K. 2010. The central importance of information in studies of animal communication. *Animal Behaviour* 80:3–8.

Seyfarth, R. M., Cheney, D. L., & Marler, P. 1980. Monkey responses to three different alarm calls: Evidence of predator classification and semantic communication. *Science* 210:801–803.

Sherman, P. W. 1977. Nepotism and the evolution of alarm calls. *Science* 197:1246–1253.

Shettleworth, S. J. 2001. Animal cognition and animal behaviour. *Animal Behaviour* 61:277–286.

Shier, D. M., & Owings, D. H. 2007. Effects of social learning on predator training and postrelease survival in juvenile black-tailed prairie dogs, *Cynomys ludovicianus*. *Animal Behaviour* 73:567–577.

Shipman, P. 2010. The animal connection and human evolution. *Current Anthropology* 51:519–538.

Siemers, B. M., Goerlitz, H. R., Robsomanitrandrasana, E., Piep, M., Ramanamamjato, J.-B., Rakotondravony, D., Ramilijaona, O., & Ganzhorn, J. U. 2007. Sensory basis of food detection in wild *Microcebus murinus*. *International Journal of Primatology* 28:291–304.

Sigg, D. P., Goldizen, A. W., & Pople, A. R. 2005. The importance of mating system in translocation programs: Reproductive success of released male bridled nailtail wallabies. *Biological Conservation* 123:289–300.

Sih, A., & Christensen, B. 2001. Optimal diet theory: When does it work and when and why does it fail? *Animal Behaviour* 61:379–390.

Sih, A., Bell, A., & Johnson J. C. 2004. Behavioral syndromes: An ecological and evolutionary overview. *Trends in Ecology and Evolution* 19:372–378.

Sih, A., Kats, L. B., & Maurer, E. F. 2003. Behavioural correlations across situations and the evolution of antipredator behavior in a sunfish-salamander system. *Animal Behaviour* 65:29–44.

Sih, A., Krupa, J., & Travers, S. 1990. An experimental study on the effects of predation risk and feeding regime on the mating behavior of the water strider. *American Naturalist* 135:284–290.

Siitari, H., Honkavaara, J., Huhta, E., & Viitala, J. 2002. Ultraviolet reflection and female mate choice in the pied flycatcher, *Ficedula hypoleuca*. *Animal Behaviour* 63:97–102.

Singer, T. L. 1998. Roles of hydrocarbons in the recognition systems of insects. *American Zoologist* 38:394–405.

Skinner, B. F. 1938. *The behavior of organisms: An experimental analysis*. New York: Appleton-Century.

Slabbekoorn, H., & den Boer-Visser, A. 2006. Cities change the songs of birds. *Current Biology* 16:2326–2331.

Sladen, W. J. L., Lishman, W. A., Ellis, D. H., Shire, G. G., & Rininger, D. L. 2002. Teaching migration routes to Canada geese and trumpeter swans using ultralight aircraft, 1990–2001. *Waterbirds* 25:132–137.

Smith, B. R., & Blumstein, D. T. 2008. Fitness consequences of personality: A meta-anlaysis. *Behavioral Ecology* 19:448–455.

Smith, E. A., & Wishnie, M. 2000. Conservation and subsistence in small-scale societies. *Annual Review of Anthropology* 29:493–524.

Smith, R. L. 1979. Repeated copulation and sperm precedence: Paternity assurance for a male brooding water bug. *Science* 205:1029–1031.

Smrt, R. D., & Zhao, X. 2010. Epigenetic regulation of neuronal dendrite and dendritic spine development. *Frontiers in Biology* 5:304–323.

Sober, E., & Wilson, D. S. 2011. Adaptation and natural selection revisited. *Journal of Evolutionary Biology* 24:462–468.

Sokolowski, M. B. 2001. *Drosophila*: Genetics meets behavior. *Nature Reviews Genetics* 2:879–890.

Sokolowski, M. B., Pereira, H. S., & Hughes, K. 1997. Evolution of foraging behavior in *Drosophila* by density-dependent selection. *Proceedings of the National Academy of Sciences USA* 94:7373–7377.

Sol, D., Székely, T., Liker, A., & Lefebvre, L. 2007. Big-brained birds survive better in nature. *Proceedings of the Royal Society B* 274:763–769.

Soler, M., Soler, J. J., Møller, A. P., Moreno, J., & Linden, M. 1996. The functional significance of sexual display: Stone carrying in the black wheatear. *Animal Behaviour* 51:247–254.

Sorenson, M. D. 1991. The functional-significance of parasitic egg-laying and typical nesting in redhead ducks—an analysis of individual behavior. *Animal Behaviour* 42:771–796.

South, A., & Lewis, S. M. 2011. The influence of male ejaculate quantity on female fitness: A meta-analysis. *Biological Reviews* 86:299–309.

Spady, T. C., & Ostrander, E. A. 2008. Canine behavioral genetics: Pointing out the phenotypes and herding up the genes. *American Journal of Human Genetics* 82:10–18.

Springer, M. S., Teeling, E. C., Madsen, O., Stanhope, M. J., & de Jong, W. W. 2001. Integrated fossil and molecular data reconstruct bat echolocation. *Proceedings of the National Academy of Sciences U.S.A.* 98:6241–6246.

Staaterman, E. R., Claverie, T., & Patek, S. N. 2010. Disentangling defense: The function of spiny lobster sounds. *Behaviour* 147:235–258.

Stalleicken, J., Labhart, T., & Mouritsen, H. 2006. Physiological characterization of the compound eye in monarch butterflies with focus on the dorsal rim area. *Journal of Comparative Physiology A* 192:321–331.

Stam, H. J., & Kalmanovitch, T. 1998. E. L. Thorndike and the origins of animal psychology. *American Psychologist* 53:1135–1144.

Stamps, J. A. 1988. Conspecific attraction and aggregation in territorial species. *American Naturalist* 131:329–347.

Stamps, J. A., & Groothuis, T. G. G. 2010. Developmental perspectives on personality: Implications for ecological and evolutionary studies of individual differences. *Philosophical Transactions of the Royal Society B* 365:4029–4041.

Stankowich, T., & Coss, R. G. 2007. Effects of risk assessment, predator behavior and habitat on escape behavior in Columbian black-tailed deer. *Behavioral Ecology* 18:358–367.

Stapput, K., Thalau, P., Wiltschko, R., & Wiltschko, W. 2008. Orientation of birds in total darkness. *Current Biology* 18:602–606.

Stebbins, R. C. 1949. *Speciation in salamanders of the plethodontid genus* Ensatina. University of California Publication in Zoology, vol. 48. Berkeley: University of California Press.

Steinhart, G. A., Marschall, E. A., & Stein, R. A. 2004. Round goby predation on smallmouth bass offspring in nests during simulated catch-and-release angling. *Transactions of the American Fisheries Society* 133:121–131.

Steinhart, G. A., Sandrene, M. E., Weaver, S., Stein, R. A., & Marschall, E. A. 2005. Increased parental care cost for nest-guarding fish in a lake with hyperabundant nest predators. *Behavioral Ecology* 16:427–434.

Stenseth, N. C., & Mysterud, A. 2002. Climate, changing phenology, and other life history traits: Nonlinearity and match-mismatch to the environment. *Proceedings of the National Academy of Sciences USA* 99:13,379–13,381.

Stevens, M. 2013. *Sensory ecology, behavior, and evolution.* Oxford: Oxford University Press.

Stirling, D. G., Réale, D., & Roff, D. A. 2002. Selection, structure and the heritability of behaviour. *Journal of Evolutionary Biology* 15:277–289.

Stodola, K. W., Linder, E. T., Buehler, D. A., Franzreb, K. E., & Cooper, R. J. 2009. Parental care in the multi-brooded black-throated blue warbler. *Condor* 111:497–502.

Stokke, B. G., Moksnes, A., & Røskaft, E. 2005. The enigma of imperfect adaptations in hosts of avian brood parasites. *Ornithological Science* 4:17–29.

Strathman, R. R., & Strathman, M. F. 1982. The relationship between adult size and brooding in marine invertebrates. *American Naturalist* 119:91–101.

Stutchbury, B. J., & Morton, E. S. 1995. The effect of breeding synchrony on extra-pair mating system in songbirds. *Behaviour* 132:675–690.

Stutchbury, B. J. M., Tarof, S. A., Done, T., Gow, E., Kramer, P. M., Tautin, J., Fox, J. W., & Afanasyev, V. 2009. Tracking long-distance songbird migration by using geolocators. *Science* 323:896.

Summers, K., Sea, C. S., & Heying, H. 2006. The evolution of parental care and egg size: A comparative analysis in frogs. *Proceedings of the Royal Society B* 273:687–692.

Suzuki, S., Nagano, M., & Kobayashi, N. 2006. Mating competition and parentage assessment in *Ptomascopus morio* (Coleoptera: Silphidae): A case for resource defense polygyny. *European Journal of Entomology* 103:751–755.

Suzuki, S., Nagano, M., & Trumbo, S. T. 2005. Intrasexual competition and mating behavior in *Ptomascopus morio* (Coleoptera: Silphidae Nicrophorinae). *Journal of Insect Behavior* 18:233–242.

Swingland, I. R., & Greenwood, P. J. 1983. *The ecology of animal movement.* Oxford, UK: Clarendon Press.

Switzer, P. V. 1993. Site fidelity in predictable and unpredictable habitats. *Evolutionary Ecology* 7:533–555.

———. 1997. Past reproductive success affects future habitat selection. *Behavioral Ecology and Sociobiology* 40:307–312.

Székely, T., & Cuthill, I. 2000. Trade-off between mating opportunities and parental care: Brood desertion by female Kentish plovers. *Proceedings of the Royal Society B* 267:2087–2092.

Taborsky, M., Oliveira, R. F., & Brockmann, H. J. 2008. The evolution of alternative reproductive tactics: Concepts and questions. In *Alternative reproductive tactics*, ed. R. F. Oliveira, M. Taborsky, & H. J. Brockmann, 1–21. Cambridge: Cambridge University Press.

Tack, E. J., Robson, T. E., Putland, D. A., & Goldizen, A. W. 2005. Geographic variation in vocalisations of satin bowerbirds, *Ptilonorhynchus violaceus*, in southeastern Queensland. *Emu* 105:27–31.

Tallamy, D. W. 2005. Egg dumping in insects. *Annual Review of Entomology* 50:347–370.

Tanaka, Y. 2000. Realized heritability of behavioral responsiveness to an oviposition pheromone in the azukibean weevil *Callosobruchus chinensis*. *Entomologia Experimentalis et Applicata* 96:239–243.

Tang-Martinez, Z. 2001. The mechanisms of kin discrimination and the evolution of kin recognition in vertebrates: a critical re-evaluation. *Behavioural Processes* 53:21–40.

Tang-Martínez, Z., & Ryder, T. B. 2005. The problem with paradigms: Bateman's worldview as a case study. *Integrative and Comparative Biology* 45:821–830.

Tarsitano, M. S., & Jackson, J. J. 1992. Influence of prey movement on the performance of simple detours by jumping spiders. *Behaviour* 123:106–120.

Tay, W. T., Miettinen, M., & Kaitala, A. 2003. Do male golden egg bugs carry eggs they have fertilized? A microsatellite analysis. *Behavioral Ecology* 14:481–485.

Taylor, M. L., Wigmore, C., Dodgson, D. J., Wedell, N., & Hosken, D. J. 2008. Multiple mating increases female fitness in *Drosophila simulans*. *Animal Behaviour* 76:963–970.

Taylor, P. W., & Elwood, R. W. 2003. The mismeasure of animal contests. *Animal Behaviour* 65:1195–1202.

Temeles, E. J. 1994. The role of neighbours in territorial systems: When are they "dear enemies"? *Animal Behaviour* 47:339–350.

Temussi, P. A. 2009. Sweet, bitter and umami receptors: A complex relationship. *Trends in Biochemical Sciences* 34:296–302.

Thaker, M., Lima, S. L., & Hews, D. L. 2009. Alternative antipredator tactics in tree lizard morphs: Hormonal and behavioural responses to a predator encounter. *Animal Behaviour* 77:395–401.

Thorndike, E. L. 1911. *Animal intelligence: Experimental studies*. New York: MacMillan.

Thornhill, R. 1983. Cryptic female choice and its implications in the scorpionfly *Harpobittacus nigricepts*. *American Naturalist* 122:765–788.

Thornton, A., Isden, J., & Madden, J. R. 2014. Toward wild psychometrics: Linking individual cognitive differences to fitness. *Behavioral Ecology* 25:1299–1301.

Thornton, A., & Lukas, D. 2012. Individual variation in cognitive performance: Developmental and evolutionary perspectives. *Philosophical Transactions of the Royal Society of London B* 367:2773–2783.

Thoss, F., Bartsch, B., Tellschaft, D., & Thoss, D. 2002. The light sensitivity of human visual system depends on the direction of view. *Journal of Comparative Physiology A* 188:235–237.

Tian, L. & Zhou, X. 2014. The soldiers in societies: defense, regulation, and evolution. *International Journal of Biological Sciences* 10:296–308.

Tibbetts, E. A., & Dale, J. 2007. Individual recognition: It is good to be different. *Trends in Ecology and Evolutionary Biology* 22:529–537.

Tibbetts, E. A., & Huang, Z. Y. 2010. The challenge hypothesis in an insect: Juvenile hormone increases during reproductive conflict following queen loss in *Polistes* wasps. *American Naturalist* 176:123–130.

Tieleman, B. I., Dijkstrab, T. H., Klasing, K. C., Visser, G. H., & Williams, J. B. 2008. Effects of experimentally increased costs of activity during reproduction on parental investment and self-maintenance in tropical house wrens. *Behavioral Ecology* 19:949–959.

Tinbergen, N. 1951. *The study of instinct*. New York: Oxford University Press.

———. 1960. The natural control of insects in pinewoods. I. Factors influencing the intensity of predation by songbirds. *Archives Néerlandaises de Zoologie* 13:265–343.

———. 1963. On aims and methods in ethology. *Zeitschrift für Tierpsychologie* 20:410–433.

Tome, M. W. 1988. Optimal foraging: Food patch depletion by ruddy ducks. *Oecologia* 76:27–36.

Touhara, K., & Vosshall, L. B. 2009. Sensing odorants and pheromones with chemosensory receptors. *Annual Review of Physiology* 71:307–332.

Tracy, B. F., & McNaughton, S. J. 1995. Elemental analysis of mineral lick soils from the Serengeti National Park, the Konza prairie, and Yellowstone National Park. *Ecography* 18:91–94.

Tregenza, T., & Wedell, N. 2000. Genetic compatibility, mate choice and patterns of parentage: Invited review. *Molecular Ecology* 9:1013–1027.

Treves, A., & Naughton-Treves, L. 1999. Risk and opportunity for humans coexisting with large carnivores. *Journal of Human Evolution* 36:275–282.

Triblehorn, J. D., Ghose, K., Bohn, K., Moss, C. F., & Yager, D. D. 2008. Free-flight encounters between praying mantids (*Parasphendale agrionia*) and bats (*Eptesicus fuscus*). *Journal of Experimental Biology* 211:555–562.

Trillmich, F., & Wolf, J. B. W. 2008. Parent–offspring and sibling conflict in Galápagos fur seals and sea lions. *Behavioral Ecology and Sociobiology* 62:363–375.

Trivers, R. L. 1971. The evolution of reciprocal altruism. *Quarterly Review of Biology* 46:35–57.

———. 1972. Parental investment and sexual selection. In *Sexual selection and the descent of man, 1871–1971*, ed. B. Campbell, 136–179. Chicago: Aldine.

———. 1974. Parent-offspring conflict. *American Zoologist* 14:249–264.

Trivers, R.L., & Hare, H. 1976. Haplodiploidy and the evolution of the social insect. *Science* 191:249–263.

Trumbo, S. T. 2002. Hormonal regulation of parental behavior in insects. In *Hormones, brain and behavior*, vol. 3, ed. D. W. Pfaff, A. P. Arnold, A. M. Etgen, S. E. Fahrbach, R. L. Moss, & R. R. Rubin, 115–139. New York: Academic Press.

———. 2012. Patterns of parental care in invertebrates, in Royle, Smiseth, & Kölliker, eds, *The evolution of parental care*, Oxford: Oxford University Press. 81–100.

Vadas, R. L., Jr. 1994. The anatomy of an ecological controversy: Honey-bee searching behavior. *Oikos* 69:158–166.

Vahed, K. 2007. All that glitters is not gold: Sensory bias, sexual conflict and nuptial feeding in insects and spiders. *Ethology* 113:105–127.

Vakirtzis, A. 2011. Mate choice copying and nonidependent mate choice: A critical review. *Annales Zoologici Fennici* 48:91–107.

Valcu, C.-M. & Kempenaers, B. 2015. Proteomics in behavioral ecology. *Behavioral Ecology* 26:1–15.

Vale, G.A. 1980. Flight as a factor in the host-seeking behavior of tsetse flies (Diptera: Glossinidae). *Bulletin of the Entomological Research* 70:299–307.

Valone, T. J. 1989. Group foraging, patch estimation, and public information. *Oikos* 56:357–363.

———. 2006. Are animals capable of Bayesian updating? An empirical review. *Oikos* 112:252–259.

———. 2007. From eavesdropping on performance to copying the behavior of others: A review of public information use. *Behavioral Ecology and Sociobiology* 62:1–14.

Valone, T. J., & Lima, S. L. 1987. Carrying food items to cover for consumption: The behavior of ten bird species feeding under the risk of predation. *Oecologia* 71:286–294.

van den Berghe, E. P., & Gross, M. R. 1989. Natural selection resulting from female breeding competition in a Pacific salmon (Coho: *Oncorhynchus kisutch*). *Evolution* 43:125–140.

van der Staay, F. J. 2006. Animal models of behavioral dysfunctions: Basic concepts and classification, and an evaluation strategy. *Brain Research Reviews* 52:131–159.

Van Dyck, H. 2012. Changing organisms in rapidly changing anthropogenic landscapes: The significance of the "Umwelt" concept and functional habitat for animal conservation. *Evolutionary Applications* 5:144–153.

van Langeveld, F., Ettema, J. A., Donners, M., Walisdevries, M. F., & Groenendijk, D. 2011. Effect of spectral composition of artificial light on the attraction of moths. *Biological Conservation* 144:2274–2281.

van Oudenhove, L., Billoir, E., Boulay, R, Bernstein, C., & Cerdá, X. 2011. Temperature limits trail following behaviour through pheromone decay in ants. *Naturwissenschaften* 98:1009–1017.

Varricchio, D. J., Moore, J. R., Erickson, G. M., Norell, M. A., Jackson, F. D., & Borkowski, J. J. 2009. Avian paternal care had dinosaur origin. *Science* 322:1826–1828.

Vehrencamp, S. C. 1983. Optimal degree of skew in cooperative societies. *American Zoologist* 23:327–335.

Verbeek, M. E. M., Drent, P. J., & Wiepkema, P. R. 1994. Consistent individual differences in early exploratory behaviour of male great tits. *Animal Behaviour* 48:1113–1121.

Verner, J. 1964. Evolution of polygamy in the long-billed marsh wren. *Evolution* 18:252–261.

Verner, J., & Wilson, M. F. 1966. The influence of habitats on mating systems of North American passerine birds. *Ecology* 47:143–147.

Verrell, P. A. 1986. Wrestling in the red-spotted newt (*Notophthalmus viridescens*): Resource value and contestant asymmetry determine contest duration and outcome. *Animal Behaviour* 34:398–402.

Vickery, W. L., Giraldeau, L.-A., Templeton, J. J., Kramer, D. L., & Chapman, C. A. 1991. Producers, scroungers, and group foraging. *American Naturalist* 137:847–863.

Vila-Gispert, A., Moreno-Amich, R., & García-Berthou, E. 2002. Gradients of life-history variation: An intercontinental comparison of fishes. *Reviews in Fish Biology and Fisheries* 12:417–427.

Vincent, A. C. J. 1995a. A role for daily greetings in maintaining seahorse pair bonds. *Animal Behaviour* 49:258–260.

———. 1995b. Trade in seahorses for traditional Chinese medicines, aquarium fishes, and curios. *TRAFFIC Bulletin* 15:125–128.

———. 2008. Conservation on coral reefs: The world as an onion. *American Fisheries Society Symposium* 49:1435–1467.

Vincent, A., Evans, K., & Marsden, A. D. 2005. Home range behaviour of the monogamous Australian seahorse, *Hippocampus whitei*. *Environmental Biology of Fishes* 72:1–12.

Visalberghi, E., Addessi, E., Truppa, V., Spagnoletti, N., Ottoni, E., Izar, P., & Fragaszy, D. 2009. Selection of effective stone tools by wild bearded capuchin monkeys. *Current Biology* 19:213–217.

Vitt, L. J., & Price, H. J. 1982. Ecological and evolutionary determinants of relative clutch mass in lizards. *Herpetologica* 38:237–255.

Vleck, C. M. 2001. Hormonal control of incubation behaviour. In *Avian incubation: Behaviour, environment and evolution*, ed. D. C. Deeming, 54–62. Oxford, UK: Oxford University Press.

von Frisch, K. 1956. *Bees: Their vision, chemical senses, and language*. Ithaca, NY: Cornell University Press.

———. 1967. *The dance language and orientation of bees*. Cambridge, MA: Harvard University Press.

———. 1971. *Bees: Their vision, chemical senses, and language*. 2nd ed. Ithaca, NY: Cornell University Press.

von Frisch, K., Wenner, A. M., & Johnson, D. L. 1967. Honeybees: Do they use direction and distance information provided by their dances? *Science* 158:1072–1077.

Von Uexküll, J. 1909. *Umwelt und Innenwelt der Tiere*. Berlin: J. Springer.

Vos, P., Hogers, R., Bleeker, M., Reijans, M., van de Lee, T., Hornes, M., Frijters, A., et al. 1995. AFLP: A new technique for DNA fingerprinting. *Nucleic Acids Research* 23:4407–4414.

Walcott, C. 1996. Pigeon homing: Observations, experiments and confusions. *Journal of Experimental Biology* 199:21–27.

———. 2005. Multimodal orientation cues in homing pigeons. *Integrative and Comparative Biology* 45:574–581.

Wallace, A. R. 1878. *Tropical nature, and other essays*. London: Macmillan.

Wang, J., Wurm, Y., Nipitwattanaphon, M., Riba-Grognuz, O., Huang, Y. C., Shoemaker, D., & Keller, L. 2013. A Y-like social chromosome causes alternative colony organization in fire ants. *Nature* 493:664–668.

Ward, A. J. W. & Hart, P. J. B. 2003. The effects of kin and familiarity on interactions between fish. *Fish and Fisheries* 4:348–358.

Ward, M. P., & Schlossberg, S. 2004. Conspecific attraction and the conservation of territorial songbirds. *Conservation Biology* 18:519–525.

Warner, R. R. 1988. Traditionality of mating-site preferences in a coral reef fish. *Nature* 335:719–721.

———. 1990. Male versus female influences on mating-site determination in a coral reef fish. *Animal Behaviour* 39:540–548.

Warren, W. C., Clayton, D. F., Ellegren, H., Arnold, A. P., Hillier, L. W., Künstner, A., Searle, S., et al. 2010. The genome of a songbird. *Nature* 464:757–762.

Washburn, M. F. 1908. *The animal mind*. New York: MacMillan.

Watson, J. B. 1913. Psychology as the behaviorist views it. *Psychological Review* 20:158–177.

———. 1924. *Behaviorism*. New York: People's Institute.

Watts, C. R., & Stokes, A. W. 1971. The social order of turkeys. *Scientific American* 224:112–118.

Weber, J. N., & Hoekstra, H. E. 2009. The evolution of burrowing behavior in deer mice (genus *Peromyscus*). *Animal Behaviour* 77:603–609.

Weber, J. N., Peterson, B. K., & Hoekstra, H. E. 2013. Discrete genetic modules are responsible for complex burrow evolution in *Peromyscus* mice. *Nature* 493:402–405.

Wehner, R. 1998. Navigation in context: Grand theories and basic mechanisms. *Journal of Avian Biology* 29:370–386.

———. 2001. Polarization vision—a uniform sensory capacity? *Journal of Experimental Biology* 204:2589–2596.

Weimerskirch, H., Martin, J., Clerquin, Y., Alexandre, P., & Jiraskova, S. 2001. Energy saving in flight formation. *Nature* 413:697–698.

Welke, K., & Schneider, J. M. 2009. Inbreeding avoidance through cryptic female choice in the cannibalistic orb-web spider *Argiope lobata*. *Behavioral Ecology* 20:1056–1062.

Wenner, A. M. 2002. The elusive honey bee dance "language" hypothesis. *Journal of Insect Behavior* 15:859–878.

Wenner, A. M., Well, P. H., & Johnson, D. L. 1969. Honey bee recruitment to food sources: Olfaction or language? *Science* 164:84–86.

Werdenich, D., & Huber, L. 2006. A case of quick problem solving in birds: String pulling in keas, *Nestor notabilis. Animal Behaviour* 71:855–863.

West, S. A., Griffin, A. S., & Gardner, A. 2007. Evolutionary explanations for cooperation. *Current Biology* 17:R661–R672.

West-Eberhard, M. J. 1984. Sexual selection, competitive communication and species-specific signals in insects. In *Insect communication*, ed. T. Lewis, 283–324. New York: Academic Press.

Westendorp, R. G. J., & Kirkwood, T. B. L. 1998. Human longevity at the cost of reproductive success. *Nature* 396:743–746.

Westneat, D. F., & Stewart, I. R. K. 2003. Extra-pair paternity in birds: Causes, correlates, and conflict. *Annual Review of Ecology and Systematics* 34:365–396.

Westoby, G. W. M., & Partridge, K. J. 1986. Human homing: Still no evidence despite geomagnetic controls. *Journal of Experimental Biology* 120:325–331.

Wheeler, B. C. 2009. Monkeys crying wolf? Tufted capuchin monkeys use anti-predator calls to usurp resources from conspecifics. *Proceedings of the Royal Society B* 276:3013–3018.

Whiten, A., Goodall, J., McGrew, W. C., Nishida, T., Reynolds, V., Sugiyama, Y., Tutin, C. E. G., Wrangham, R. W., & Boesch, C. 1999. Cultures in chimpanzees. *Nature* 399:682–685.

Whiten, A., & van Schaik, C. P. 2007. The evolution of animal "cultures" and social intelligence. *Philosophical Transactions of the Royal Society B* 362:603–620.

Widdig, A. 2007. Paternal kin discrimination: the evidence and likely mechanisms. *Biological Review* 82:319–334.

Wilkinson, G. S. 1984. Reciprocal food sharing in the vampire bat. *Nature* 308:181–184.

———. 1990. Food sharing in vampire bats. *Scientific American* 262:76–82.

Williams, G. C. 1966. *Adaptation and natural selection: A critique of some current evolutionary thought*. Princeton, NJ: Princeton University Press.

Williams, T. M., Selegue, J. E., Werner, T., Gompel, N., Kopp, A., & Carroll, S. B. 2008. The regulation and evolution of a genetic switch controlling sexually dimorphic traits in *Drosophila. Cell* 134:610–623.

Wilson, A. B., Ahnesjö, I., Vincent, A. C. J., & Meyer, A. 2003. The dynamics of male brooding, mating patterns, and sex roles in pipefishes and seahorses (family Syngnathidae). *Evolution* 57:1374–1386.

Wilson, D. S., & Wilson, E. O. 2007. Rethinking the theoretical foundation of sociobiology. *Quarterly Review of Biology* 82:327–348.

Wilsson, E., & Sundgren, P.-E. 1997. The use of a behavior test for the selection of dogs for service and breeding. I. Methods of testing and evaluating test results in the adult dog, demands on different kinds of service dogs, sex and breed differences. *Applied Animal Behavior Science* 53:279–295.

Wiltschko, R., & Wiltschko, W. 1981. The development of sun compass orientation in young homing pigeons. *Behavioral Ecology and Sociobiology* 9:135–141.

———. 1989. Pigeon homing: Olfactory orientation—a paradox. *Behavioral Ecology and Sociobiology* 24:163–173.

———. 2009. Avian navigation. *Auk* 126:717–743.

Wiltschko, W., Munro, U., Ford, H., & Wiltschko, R. 1998. Effect of a magnetic pulse on the orientation of silvereyes, *Zosterops l. lateralis*, during spring migration. *Journal of Experimental Biology* 201:3257–3261.

Wiltschko, W., & Wiltschko, R. 2005. Magnetic orientation and magnetoreception in birds and other animals. *Journal of Comparative Physiology A* 191:675–693.

Wingfield, J. C., Hegner, R. E., Dufty, A. M., Jr., & Ball, G. F. 1990. The "challenge hypothesis": Theoretical implications for patterns of testosterone secretion, mating systems, and breeding strategies. *American Naturalist* 136:829–846.

Wingfield, J. C., Jacobs, J. D., Tramontin, A. D., Perfito, N., Meddle, S., Maney, D. L., & Soma, K. 2000. Toward an ecological basis of hormone-behavior interactions in reproduction of birds. In *Reproduction in context*, ed. K. Wallen & J. Schneider, 85–128. Cambridge, MA: M.I.T. Press.

Wingfield, J. C., Lynn, S. E., & Soma, K. K. 2001. Avoiding the "costs" of testosterone: Ecological bases of hormone-behavior interactions. *Brain, Behavior and Evolution* 57:239–251.

Wisby, W. J., & Haesler, A. D. 1954. Effect of olfactory occlusion on migrating silver salmon (*O. kisutch*). *Journal of the Fisheries Research Board of Canada* 11:472–478.

Wisenden, B. D. 1999. Alloparental care in fishes. *Reviews in Fish Biology and Fisheries* 9:45–70.

Witzgall, P., Kirsch, P., & Cork, A. 2010. Sex pheromones and their impact on pest management. *Journal of Chemical Ecology* 36:80–100.

Wolf, M., van Doorn, G. S., Leimar, O., & Weissing, F. J. 2007. Life-history trade-offs favour the evolution of animal personalities. *Nature* 447:581–584.

Wolf, M., & Weissing, F. J. 2012. Animal personalities: Consequences for ecology and evolution. *Trends in Ecology and Evolution* 27:452–461.

Wolff, J. O., & Macdonald, D. W. 2004. Promiscuous females protect their offspring. *Trends in Ecology and Evolution* 19:127–134.

Wong, B. M., & McCarthy, M. 2009. Prudent male mate choice under perceived sperm competition risk in the eastern mosquito fish. *Behavioral Ecology* 20:278–282.

Wright, T. F., Dahlin, C. R., & Salinas-Melgoza, A. 2008. Stability and change in vocal dialects of the yellow-naped amazon. *Animal Behaviour* 76:1017–1027.

Wu, L-Q., & Dickman, J. D. 2012. Neural correlates of magnetic sense. *Science* 336:1054–1057.

Wynne, C. D. L. 2004. The perils of anthropomorphism. *Nature* 428:606.

———. 2007. What are animals? Why anthropomorphism is still not a scientific approach to behavior. *Comparative Cognition & Behavior Reviews* 2:125–135.

Wynne-Edwards, K. E., & Timonin, M. E. 2007. Paternal care in rodents: Weakening support for hormonal regulation of the transition to behavioral fatherhood in rodent animal models of biparental care. *Hormones and Behavior* 52:114–121.

Wynne-Edwards, V. C. 1962. *Animal dispersion in relation to social behavior.* Edinburgh: Oliver and Boyd.

Yack, J. E., & Dawson, J. W. 2008. Insect ears. In *The senses: A comprehensive reference*, vol. 3, ed. Hoy, R. R., Sheperd, G. M., Basbaum, A. I., Kaneko, A., & Westheimer, G., 35–54. New York, NY: Elsevier.

Yager, D. D. 2012. Predator detection and evasion by flying insects. *Current Opinion in Neurobiology* 22:201–207.

Yamaguchi, N., Dugdale, H. L., & MacDonald, D. W. 2006. Female receptivity, embryonic diapause, and superfetation in the European badger (*Meles meles*): Implications for the reproductive tactics of males and females. *Quarterly Review of Biology* 81:33–48.

Yamamoto, Y., Hino, H., & Ueda, H. 2010. Olfactory imprinting of amino acids in lacustrine sockeye salmon. *PLoS ONE* 5:e8633.

Yang, G., Pan, F., & Gan, W-B. 2009. Stably maintained dendritic spines are associated with lifelong memories. *Nature* 462:920–925.

Yanoviak, S. P. 2006. Canopy Arthropod Biology— https://canopyants.net/research/gliding-arthropods/ant-videos/ (accessed July 8, 2016).

Yanoviak, S. P., Dudley, R., & Kaspari, M. 2005. Directed aerial descent in canopy ants. *Nature* 433:624–626.

Yasué, M., Quinn, J. L., & Cresswell, W. 2003. Multiple effects of weather on the starvation and predation risk trade-off in choice of feeding location in redshanks. *Functional Ecology* 17:727–736.

Yerkes, R. M. 1913. The heredity of savageness and wildness in rats. *Journal of Animal Behavior* 3:286–296.

Yin, S., Fernandez, E. J., Pagan, S., Richardson, S. L., & Snyder, G. 2008. Efficacy of a remote-controlled, positive-reinforcement, dog-training system for modifying problem behaviors exhibited when people arrive at the door. *Applied Animal Behaviour Science* 113:123–138.

Yom-Tov, Y. 2001. An updated list and some comments on the occurrence of intraspecific nest parasitism in birds. *Ibis* 143:133–143.

Zahavi, A. 1975. Mate selection: A selection for a handicap. *Journal of Theoretical Biology* 53:205–214.

Zarrow, M. X., Gandelman, R., & Denenberg, V. H. 1971. Prolactin: Is it an essential hormone for maternal behavior in the mammal? *Hormones and Behavior* 2:343–354.

Zeifman, D. M. 2001. An ethological analysis of human infant crying: Answering Tinbergen's four questions. *Developmental Psychobiology* 39:265–285.

Zhao, G. Q., Zhang, Y., Hoon, M. A., Chandrashekar, J., Erlenbach, I., Ryba, N.J.P., & Zuker, C. S. 2003. The receptors for mammalian sweet and umami taste. *Cell* 115:255–266.

Zink, A. 2003. Quantifying the costs and benefits of parental care in female treehoppers. *Behavioral Ecology* 14:687–693.

Zuk, M., Thornhill, R., Ligon, D. J., & Johnson, K. 1990. Parasites and mate choice in red jungle fowl. *American Zoologist* 30:235–244.

Answers to Selected Questions

Answers to odd-numbered questions are available online in the Instructor's Support Package at www.oup-arc.com/nordell.

Chapter 1: The Science of Animal Behavior

2. (a) Behavior. (b) Behavior. (c) Behavior. (d) Behavior. (e) Behavior.
4. A hypothesis is an explanation that makes a prediction. An educated guess does not make a prediction.
6. This is a hypothesis based on anthropomorphism and therefore is very difficult to test. However, to begin to address this question, you could examine the levels of stress hormones in birds at bird feeders and birds not at bird feeders and compare the two. This test would indicate stress levels, but not happiness.

Chapter 2: Methods for Studying Animal Behavior

4. European wildcat and domestic cat, cheetah and cougar, jaguar and lion.

Chapter 3: Evolution and the Study of Animal Behavior

2. The conclusion is premature because we do not know whether aggressive behavior is heritable in this population. If aggression in this species is heritable, then the conclusion is valid. However, selection cannot affect the frequency of a trait in a population if the trait is not heritable.
4. Male success in territory acquisition is based on individuals' ability to compete for territories and mates. There are only a finite number of territories available, which limits the number of successful male breeders.

Chapter 4: Behavioral Genetics

2. Conduct a parent-offspring regression analysis of boldness behavior. For example, you could measure boldness in adult females and their offspring to see if there was a positive correlation between the two. If so, you would conclude that boldness is heritable.
4. In the study of social organization in fire ants, you could reasonably conclude that variation in the genotype at the *Gp-9* locus affects behavioral organization more strongly than do environmental factors.
6. Lower heritability for a behavioral trait means that more of the variation in the behavior is due to environmental than genetic variation. This suggests that behaviors are more affected by environmental variation than morphology, perhaps because behaviors develop in more variable environments.

Chapter 5: Sensory Systems and Behavior

2. Dogs cannot perceive long wavelengths associated with the colors red and orange because they do not have photoreceptor cells that detect these wavelengths of light.
4. Infrasounds attenuate more slowly than higher frequencies as they travel through the environment.
6. In each case, the researchers added a stimulus to the environment (substrate-borne vibrations or weak electric fields) and observed changes in behavior in response to the added stimulus.

Chapter 6: Communication

2. If only the largest, most dominant males can produce low-frequency sounds, then females can use male vocalizations to assess male quality (i.e., body size). If females prefer to mate with dominant males, then we can predict that females will exhibit a mating preference for males that produce the lowest-frequency vocalizations.
4. Allow half the test females to observe a fight between two males and the other half to observe nothing. Immediately after the male contest, allow the female to exhibit a mating preference for one of the two males. If females prefer the dominant male without seeing him win a fight, it is likely that he produces chemical pheromones that advertise his status. If females select the dominant male only after watching him win, then it appears that they are gaining information about relative fighting ability via eavesdropping. For an example, see Alquiloni, Buřič, & Gherardi (2008).
6. The squirrels eavesdrop on blue jay vocal signals to learn about the presence of potential cache robbers, which changes their behavior. Among fighting fish, a focal male changes its behavior because of the presence

of a female but does not change its behavior in response to the presence of a male bystander. Thus, in the first example, the eavesdropper's behavior changes, while in the latter example, the focal male changes its behavior only in the presence of a female bystander.

Chapter 7: Learning, Neuroethology, and Cognition

2. This mutant should have no ability to learn: it can't benefit from experiences because it can't remember them.

4. Local enhancement occurs when individuals focus their attention on others and use their behavior to learn about the location of a resource such as a food patch. Public information is the use of information derived from the behavior of others to learn about the quality of a resource like a food patch.

6. There were likely two changes: (1) an increased number of neurotransmitters were released when Max learned the new behavior; and (2) new dendritic spines formed when the behavior was learned, some of which still persist. Only the second change is associated with lifelong memory.

Chapter 8: Foraging Behavior

2. Experiment A: The patch with heavy sand has a higher energetic cost of foraging and is predicted to have a higher GUD. Experiment B: The jays will have higher GUDs in the cold, shady patch than in the warm, sunny patch, because the energetic cost of foraging is higher in the cold, shady patch. For a comparable study, see Kilpatrick (2003).

4. The patch-use model predicts that increases in travel time will increase patch-use time. You can predict that shrews will spend more than one minute and harvest more than nine items from patches when travel distance (time) is increased.

6. Both models assume that individuals maximize their fitness by maximizing energy intake rate.

Chapter 9: Antipredator Behavior

2. If a predator is not present and predation risk is low, then territory size will be large. If a predator is present and predation risk is high, then territory size will be small, because individuals will want to reduce the amount of time defending the territory and susceptible to predators. You could set up aquaria with predator cues (visual or chemical), or

with predators both present and absent, and measure the size of the territories defended by males.

4. The dilution effect states that the probability of being killed declines as group size increases. The group size effect is not applicable to this situation, because the antipredator benefit of the group size effect is that larger groups spot predators sooner than smaller groups (even though each individual scans less and feeds more), and here the predator has already been spotted. The selfish herd is also not applicable, because this antipredator behavior requires that individuals simply move toward the center of a group.

6. Scientific Process 9.1: Null hypothesis (1): There is no difference in food in the mudflat and salt marsh habitats. Null hypothesis (2): Ambient temperature does not affect the feeding decisions of redshanks. Scientific Process 9.2: Null hypothesis: The presence of a predator will not affect the mating behavior of water striders. Scientific Process 9.3: Null hypothesis: Ground squirrels do not increase blood flow to their tail when attacked by rattlesnakes.

8. Both the pursuit-deterrence and alarm signal hypotheses may explain this behavior. You could distinguish these explanations by determining whether squirrel chuck calls and tail-wagging behavior occur most often when kin are nearby.

Chapter 10: Dispersal and Migration

2. Natal dispersal and proximate mechanisms: The density of individuals, hunger levels, and the number of aggressive interactions could all be proximate mechanisms for determining the level of competition. Pheromones and visual recognition could be used to determine the relatedness of others as a proximate mechanism to initiate natal dispersal (i.e., inbreeding avoidance). Natal dispersal and ultimate mechanisms: Competition reduces fitness, and so individuals that disperse to areas of lower competition should have higher fitness. Inbreeding depression reduces fitness. Individuals that avoid breeding with close kin will have higher fitness than those that breed with relatives.

4. There must be a positive year-to-year correlation in breeding site quality. A good (or poor) site this year must have a high probability of also being a good (or poor) site the next year.

6. As the climate warms, overwinter survivorship of residents should increase. This change could result in higher fitness for residents compared to migrants and could lead to a reduction in the proportion of migrants in the population.

8. Work on hCRY2 suggests that magnetoreception is light dependent, yet in all studies of humans, subjects were blindfolded. Experiments with human subjects who are not blindfolded might yield different results.

Chapter 11: Habitat Selection, Territoriality, and Aggression

2. The predation risk pattern described suggests that parrots exhibit an Allee effect: fitness increases as group size increases. Therefore, you might predict that parrots will exhibit conspecific attraction.

4. Patch A: 20 ducks; Patch B: 10 ducks; Patch C: 6 ducks. To solve the problem, first convert each patch to the number of items per minute. This yields 10/min, 5/min, and 3/min, respectively, for a total of 18 food items offered each minute in the patches. There are 36 ducks, and all must obtain food at the same rate, which can be calculated to be 2/min.

6. For humans, a short line is a rich patch, while a long line is a poor patch. Humans will sort themselves between these lines according to the IFD model. Individuals must be free to move between lines and must be able to assess the quality (or speed) of each line. Students could view lines at local stores or their university bookstore and see what parameters are important for moving through lines quickly. For example, is it the number of items in a cart or the number of individuals in each line that is most important? Recent work indicates that it is best to choose the line with the fewest individuals, no matter how much they have in their carts, because the paying process requires so much more time than scanning objects.

8. Tinbergen's four questions were: What is the mechanism that causes the behavior? How does the behavior develop? What is the function of the behavior? How did the behavior evolve?

Oyegbile and Marler examined the mechanism that causes winners of fights to have a higher probability of winning future fights. In this proposed mechanism, testosterone level is higher in winners, and high testosterone levels enhance the likelihood of winning an aggressive encounter.

Verrell studied the duration of wrestling contests in newts. The mechanism (factor) associated with longer contests was the body mass of the female contested. Ultimately, larger females are associated with higher reproductive fitness for males, and so selection should favor higher energetic investment (longer fights) in contests over larger females.

Chapter 12: Mating Behavior

2. Because females can take up to four days to become sexually mature and males mature in one day, some females likely had less opportunities for mating than others, including males. This would bias the results toward males having higher reproductive success, since they were all sexually mature at the time of the experiments.

4. Observations demonstrated that males varied naturally in terms of exhibiting mate guarding. Those that had the highest rate of paternity for their nest also exhibited high levels of mate guarding but other factors may explain this correlation. The experiment manipulated the amount of time a male was away from his mate. Males that were experimentally removed suffered higher levels of cuckoldry, providing stronger support for the hypothesis that mate guarding minimizes cuckoldry.

6. The fitness of the territorial strategy appears to be higher than the fitness of the roving tactic because it results in more offspring produced. This holds true even in a population with similar numbers of territorial and roving males. It is therefore likely that the roving tactic is a conditional strategy.

8. Dugatkin (1992) placed a rival female next to the nonpreferred male and allowed them to interact through a glass partition to simulate mate choice by a rival. Place et al. (2010) allowed subjects to view dating interactions. Observation of mutual interest in the date was used to simulate mate choice by a rival.

Chapter 13: Mating Systems

2. In an extra-pair mating system, individuals typically form pair-bonded social associations, and social and genetic mating systems differ. In a promiscuous mating system, pair bonds are not formed.

4. Polygynous species have higher variance in male mating success than do monogamous

species. Common crossbills should exhibit lower variance in male mating success than Gunnison's sage grouse.

6. If individual females independently seek out tents, this behavior could be an example of resource defense polygyny, in which the resource needed by females is the tent structure. If females associate in harems that a male defends from rivals, this system would be described as female defense polygyny.

Chapter 14: Parental Care

2. The individual with the increased flight costs should reduce its level of care. This experiment has been conducted several times— see, for example, Tieleman et al. (2008).

4. When environmental variation is large, resources may be very low in some years. When resources are low, an entire synchronous clutch may fail because of insufficient food if all chicks are equal competitors. Asynchronous clutches establish a competitive hierarchy so that the first chick to hatch is a superior competitor for food. In years of low resources, there may be enough food to successfully raise one offspring, which results in higher parent fitness compared to that which would result from a synchronous clutch.

6. Males with small clutches face higher predation risk when defending a clutch. As a result, the benefits of defense may be smaller than the costs, and so the males will not provide parental care. This prediction could be tested by manipulating predation risk for male frogs to examine its effect on frogs defending egg clutches. Those exposed to higher predation risk would be predicted to exhibit lower levels of care.

8. Such an analysis eliminates all women who died during childbirth and focuses on women who have come to the end of their lifetime reproductive effort.

Chapter 15: Sociality

2. Groups find patches faster, so you might predict larger groups for the treatment with the hidden patches.

4. First, subordinate individuals that stay in a group can obtain other fitness benefits (e.g., reduced predation risk). Second, subordinate individuals may stay in a group because they may gain future benefits via social queuing.

6. Females are more susceptible than males to pathogens.

Answers to Scientific Process Box "Evaluate" Questions

[SP 1.1]

The figure should look like 1.7b with 'Number of worms' as the X axis label and 'Number of robins' as the Y axis label.

[SP 1.2]

Possible answers involve variation in vegetation or soil type in yards.

[SP 2.1]

One controlled for the presence of a novel object (compared to a novel predator) in the arena, while the other simply had no object in the arena.

[SP 3.1]

Yes, because exploratory behavior is a heritable trait.

[SP 3.2]

These fish spent more time and energy defending the larger territory, which reduced their energy available for growth.

[SP 4.1]

Full siblings were chosen so that individuals were very similar with respect to their genetic composition.

[SP 4.2]

The lines would not be parallel but instead would show many crossing lines.

[SP 5.1]

Conduct an experiment in which you inhibit different sensilla in different individuals and carry out the same trials. The experimental antlion larvae that do not respond to the needle vibrations provide information on which sensilla are used to detect vibrations made by prey.

[SP 6.1]

These females have higher reproductive success.

[SP 6.2]

Individuals with high opercular flare rates are assumed to have high fighting ability because they are in the best physiological condition.

[SP 7.1]

There must be some fitness cost associated with a large HF size, so selection has not favored the evolution of large HF size in those species that do not rely heavily on finding stored food.

[SP 7.2]

Exposure to a novel predator odor did not result in reduced feeding rate compared to exposure to a novel nonpredator odor.

[SP 8.1]

Visual cues seem most important, while olfactory cues appear to be the least important.

[SP 8.2]

You would need to compare the rate of change in search times over trials when trout hunt for conspicuous or cryptic prey.

[SP 8.3]

The GUDs were equal for each of the three feeder types on each day, as predicted by Brown's model, which assumes foragers maximize fitness by maximizing energy intake rate.

[SP 9.1]

The trade-off was between energy intake rate and predation risk. Individuals could experience low predation risk only by feeding in a poor habitat, leading to low energy intake rate. To obtain higher energy intake rates, they were exposed to higher predation risk. There was no habitat with both rich enough food to allow high energy intake rates and low predation risk.

[SP 9.2]

Both activity level and mating behavior should be even lower than the levels observed in this experiment.

[SP 9.3]

Increasing the temperature of their entire body could be too energetically costly. Furthermore, many snakes locate food by detecting infrared radiation, so increased body temperature might not act as an antipredator deterrent.

[SP 10.1]

One potential answer is that few available breeding sites were available on the pond.

[SP 10.2]

One potential answer is that in the absence of information about the sun's location, individuals rely on other sensory information to orient north, perhaps based on magnetoreception.

[SP 11.1]

The dominant competitor would occupy the habitat patch with the highest amount of resources (food delivery rate) and prevent others from occupying that patch so that fewer than predicted individuals would settle there and more than predicted would settle in the poorer habitat patch.

[SP 11.2]

One potential explanation includes reduced predation risk via the group size effect.

[SP 12.1]

Male reproductive success could be higher if more ornamented females produce more eggs.

[SP 12.2]

You could test young (naïve) and older (experienced) females. Copying would be predicted to be more prevalent for younger females.

[SP 13.1]

The hypothesis would not be supported if growth rates of individuals were high in small pools of water.

[SP 14.1]

One would conclude that parental care is not influenced by certainty of paternity.

[SP 14.2]

Posssible answers are reduced survivorship until the next year or reduced immune system function.

[SP 14.3]

Data on the feeding rates of these chicks as their parents brought food to the nestlings.

[SP 15.1]

One potential answer is that the chimps avoided the nasty person because of wariness toward a non-cooperative human and not because of their reputation.

Credits

Index